The
Pleasures
of
Counting

To judge in this [utilitarian] way demonstrates ... how small, narrow and indolent our minds are; it shows a disposition always to calculate the reward before the work, a cold heart and a lack of feeling for everything that is great and honours mankind. Unfortunately one cannot deny that such a mode of thinking is common in our age, and I am convinced that this is closely connected with the catastrophes which have befallen many countries in recent times; do not mistake me, I do not talk of the general lack of concern for science, but of the source from which all this has come, of the tendency to look out everywhere for one's advantage and to relate everything to one's physical well being, of indifference towards great ideas, of aversion to any effort which derives from pure enthusiasm.

Gauss

We agreed then on the good things we have in common. On the advantage of being able to test yourself, not depending on others in the test, reflecting yourself in your work. On the pleasure of seeing your creature grow, beam after beam, bolt after bolt, solid, necessary, symmetrical, suited to its purpose; and when it's finished, you look at it and you think that perhaps it will live longer than you, and perhaps it will be of use to someone you don't know, who doesn't know you. Maybe, as an old man you'll be able to come back and look at it, and it will seem beautiful, and it doesn't really matter so much that it will seem beautiful only to you, and you can say to yourself 'maybe another man wouldn't have brought it off'.

Primo Levi

To be placed on the title-page of my collected works: Here it will be perceived from innumerable examples what is the use of mathematics for judgement in the natural sciences, and how impossible it is to philosophise correctly without the guidance of Geometry, as the wise maxim of Plato has it.

Galileo

Mathematicians are people who devote their lives to what seems to me a wonderful kind of play.

Constance Reid

The Pleasures of Counting

T. W. KÖRNER

Trinity Hall
Cambridge

PUBLISHED BY THE PRESS SYNDICATE OF THE UNIVERSITY OF CAMBRIDGE
The Pitt Building, Trumpington Street, Cambridge, CB2 1RP, United Kingdom

CAMBRIDGE UNIVERSITY PRESS
The Edinburgh Building, Cambridge CB2 2RU, United Kingdom
40 West 20th Street, New York, NY 10011–4211, USA
10 Stamford Road, Oakleigh, Melbourne 3166, Australia

First published 1996

Printed in the United Kingdom at the University Press, Cambridge

Typeset in Times 11/14pt

A catalogue record for this book is available from the British Library

Library of Congress Cataloguing in Publication data available

ISBN 0 521 56087 X hardback
ISBN 0 521 56823 4 paperback

CONTENTS

Preface   viii

I The uses of abstraction

1 Unfeeling statistics   **3**
  1.1 Snow on cholera   3
  1.2 An altar of pedantry   14

2 Prelude to a battle   **21**
  2.1 The first great submarine war   21
  2.2 The coming of convoy   25
  2.3 The second submarine war   32

3 Blackett   **38**
  3.1 Blackett at Jutland   38
  3.2 Tizard and radar   44
  3.3 The shortest wavelength will win the war   50
  3.4 Blackett's circus   57

4 Aircraft versus submarine   **62**
  4.1 Twenty-five seconds   62
  4.2 Let's try the slide-rule for a change   73
  4.3 The area rule   79
  4.4 What can we learn?   87
  4.5 Some problems   93

II Meditations on measurement

5 Biology in a darkened room   **101**
  5.1 Galileo on falling bodies   101
  5.2 The long and the short and the tall   105

6 Physics in a darkened room   **116**
  6.1 The pyramid inch   116
  6.2 A different age   127

7   Subtle is the Lord   **137**
  7.1   Galileo and Einstein   137
  7.2   The Lorentz transformation   141
  7.3   What happened next?   149
  7.4   Does the earth rotate?   154

8   A Quaker mathematician   **159**
  8.1   Richardson   159
  8.2   Richardson's deferred approach to the limit   164
  8.3   Does the wind have a velocity?   176
  8.4   The four-thirds rule   186

9   Richardson on war   **194**
  9.1   Arms and insecurity   194
  9.2   Statistics of deadly quarrels   198
  9.3   Richardson on frontiers   208
  9.4   Why does a tree look like a tree?   215

III The pleasures of computation

10   Some classic algorithms   **231**
  10.1 These twice five figures   231
  10.2 The good old days   237
  10.3 Euclid's algorithm   242
  10.4 How to count rabbits   250

11   Some modern algorithms   **258**
  11.1 The railroad problem   258
  11.2 Braess's paradox   268
  11.3 Finding the largest   275
  11.4 How fast can we sort?   282
  11.5 A letter of Lord Chesterfield   292

12   Deeper matters   **298**
  12.1 How safe?   298
  12.2 The problems of infinity   305
  12.3 Turing's theorem   311

IV Enigma variations

13   Enigma   **319**
  13.1 Simple codes   319
  13.2 Simple Enigmas   331
  13.3 The plugboard   338

14 The Poles   **348**
14.1 The plugboard does not hide all finger-prints   348
14.2 Beautiful Polish females   353
14.3 Passing the torch   362

15 Bletchley   **368**
15.1 The Turing bombes   368
15.2 The bombes at work   377
15.3 SHARK   381

16 Echoes   **391**
16.1 Hard problems   391
16.2 Shannon's theorem   398

V The pleasures of thought

17 Time and chance   **413**
17.1 Why are we not all called Smith?   413
17.2 Growth and decay   422
17.3 Species and speculation   433
17.4 Of microorganisms and men   444

18 Two mathematics lessons   **452**
18.1 A Greek mathematics lesson   452
18.2 A modern mathematics lesson I   459
18.3 A modern mathematics lesson II   464
18.4 A modern mathematics lesson III   471
18.5 A modern mathematics lesson IV   477
18.6 Epilogue   481

19 Last thoughts   **488**
19.1 A mathematical career   488
19.2 The pleasures of counting   492

Appendix 1. Further reading   **494**
A1.1 Some interesting books   494
A1.2 Some hard but interesting books   501

Appendix 2. Some notations   **508**

Appendix 3. Sources   **511**

Bibliography   522
Index   529
Acknowledgements   534

PREFACE

This book is meant, first of all, for able school children of 14 and over and first year undergraduates who are interested in mathematics and would like to learn something of what it looks like at a higher level. There exist several books with a similar aim. I remember with particular pleasure from my own childhood the book *From Simple Numbers to the Calculus* by Colerus with its uncompromising opening:

> Mathematics is a trap. If you are once caught in this trap you hardly ever get out again to find your way back to the original state of mind in which you were before you began to investigate mathematics.

In Appendix A1.1 I list and discuss some of them. However, the aim is so worthwhile and the number of such books so limited that I feel no hesitation in adding one more.

In many American universities there are courses known universally, if not officially, as 'Maths for poets'. This book does not belong to that genre. It is intended, rather, as 'Maths for mathematicians' — for mathematicians who know very little mathematics as yet, but who, perhaps, will one day give lectures which the present author attends open-mouthed with admiration†.

I hope that this book will also be enjoyed by my fellow professionals and by those general readers who value mathematics without fearing it‡. Both groups will have to indulge in some judicious skipping (the professionals will not need to have Cantor's diagonal argument explained to them or to be told that Kolmogorov was a great mathematician; the general reader may tiptoe past the scarier algebra). My colleagues are perfectly capable of deciding by themselves whether or not to read this book, but two analogies may be of assistance to the general reader.

First she might like to consider why 'fly on the wall' documentaries showing life in medical school or on a warship are interesting to many who are not doctors or sailors. Listening to a mathematician talking to mathematicians about things that interest mathemati-

†When Wiles was a humble graduate student at Cambridge I had already reached the dizzying heights of a lectureship. Twenty years later I was in the back row when he announced his solution of the three centuries old Fermat problem.

‡The desire for large sales is shared by authors and publishers. David Tranah, to whom, like many other CUP mathematics authors, I owe an immense debt of gratitude, suggested that a different title would help. For once, I did not take his advice and the title *The Joy of x* remains available.

cians may well be more enlightening than listening to mathematicians speaking to non-mathematicians about things that they hope may be interesting to non-mathematicians. Alternatively she may consider the choice facing tourists in some exotic city. They may dine in fine restaurants where everything is scrupulously clean and the service is excellent but the cuisine has been modified to suit international tastes. Alternatively they can go to a local taverna where the waiters are rushed off their feet and, in any case, speak very little English, where glimpses of the kitchen fail to inspire confidence in its hygiene, and some of the dishes look very strange indeed. This book is the local taverna. The disadvantages are genuine but so is the cooking.

Even those for whom this book is intended should not necessarily expect to understand all of it. I have pitched the level of exposition at the level I would expect of beginning students of mathematics at Trinity Hall, and, if I were talking to them, I would not be surprised to have to give extra explanations on points not taught to this or that student at this or that school. Only an exceptionally well taught (perhaps even overtaught) 14 year old could expect to understand everything in this book (though I hope that any persevering 14 year old should end by understanding much of it). Professional mathematicians consider a mathematics book worthwhile if they understand *something* new after reading it and excellent if they understand a fair amount that is new to them; anything more would lead them to suspect that the material was too easy to be worthwhile.

If there is something that you do not understand you should (if you can) ask someone like your school teacher about it. If there is no one to ask, read on and perhaps understanding will come. If this does not work, try some other part of this book. Although the book is complete without the exercises, I hope that you will glance at them. Some, like Exercise 9.2.3, are simple commentaries on the text. Others, like Exercise 11.4.14 and Exercise 16.2.13, make use of mathematics met near the end of a school or near the beginning of a university course; they require substantially more knowledge than is assumed in the body of the book. I believe that such exercises have been clearly sign-posted. Appendix 2 may help if you are baffled by my notation.

Montaigne feared that 'some may assert that I have merely gathered here a big bunch of other men's flowers, having furnished nothing of my own but the string to hold them together'. This book is such a bunch of flowers and I hope that the reader may be sufficiently intrigued by some topic or quotation to explore the garden from which that particular flower came†. In any case I hope that she will see that the characteristic sound of mathematics is not that of a chorus speaking in solemn unison but the babble of individual voices.

Equally, since I have addressed the reader as I would my own students or colleagues, I have not sought to hide the fact that I hold opinions on many topics. In this and several other respects I have chosen to write a book which is *intended* for schoolchildren but not *suitable* for them.

I should like to thank Dr A. Altman, Mr A. O. Bender, Mr A. Cummins, Dr T. Gagen, Miss J. Gog, Mr T. Harris, Mrs E. Körner, Mrs W. Körner, Mr M. D. K. Lightfoot, Miss K. Maunder, Mr G. McCaughan, Dr J. R. Partington, Prof Sir B. Pippard, Miss C. Salmond, Dr G. Sankaran, Mr T. Wakeling and Dr P. Whittington for reading first drafts of this book and Mrs M. Storey for copy-editing the second draft. It would certainly have been a much worse book if I had not followed most of their advice and probably a much better one if I had followed all of it. My e-mail address is twk@pmms.cam.ac.uk and I shall try to keep a list of corrections accessible from my Web home page‡. The computational number theorist Bryan Birch, when asked which programming language he used, replied 'Graduate student'. I should like to thank Mr G. McCaughan for Tables 8.1 and 8.2 and for computer generating many of the figures. To these named persons I must add the designer§, technicians and everybody else at CUP who have laboured to convert a turbid stream of electronic impulses into an elegant book.

Many authors end their introductions with a dedication 'To my spouse for putting up with me during the writing of this book.' Since the last six words of such a dedication have always seemed to me unnecessarily limiting, I simply dedicate this book to my wife Wendy with love.

Trinity Hall, Cambridge
September 1995

T. W. Körner

---

†Hence 'the habit of fitting out the most trivial quotation with a reference as though it were applying for a job.' Most of the references, including this one, will be found in Appendix 3. Numbers in square brackets like [144] refer you to the list of books in the Bibliography.

‡http://www.pmms.cam.ac.uk/home/emu/twk/.my-home-page.html

§'I think you will like [my verses] – when you see them on a beautiful quarto page where a neat rivulet of text shall mumur thro' a meadow of margin ... they will be the most elegant thing of their kind.'

PART I

The uses of abstraction

CHAPTER 1

Unfeeling statistics

1.1 Snow on cholera

Mathematics is, at least in part, the science of abstraction. Mathematicians look at the rich complexity of the real world and replace it with a simple system which, at best, palely reflects one or two aspects of it. Roads become lines, towns become points, weather becomes a series of numbers (temperature, wind-speed, pressure, ...) and human beings become units. The object of the first part of this book is to show how useful such abstraction can be.

In 1818 Europe became aware of a terrifying epidemic raging in parts of India. The disease, previously unknown to European science, struck suddenly, manifesting itself in violent diarrhoea and vomiting followed by agonising muscular cramps. An early description tells how

> The eyes surrounded by a dark circle are completely sunk in the sockets, the skin is livid ... the surface [of the skin] is now generally covered with cold sweat, the nails are blue, the skin of the hands and feet are corrugated as if they had been long steeped in water.

Often the skin turned blue or black and sometimes the convulsions were so severe that the body was contracted into a ball which could not be straightened out after death. The disease was named 'Cholera Morbus' and killed perhaps half its victims.

Most probably cholera had always existed in India but the movement of armies and the increase in long-distance trade brought about by the expansion of the British and Russian empires now allowed it to spread. In Russia infected villages were surrounded by troops with orders to shoot anyone trying to leave. Spain made it a capital offence to leave an infected town. In spite of all efforts to contain the disease, it swept through Europe in the years 1830–2 killing one citizen in 20 in Russia, one in 30 in Poland and Austria and many in every European country before 'burning itself out'. In 1848 it returned and there were epidemic outbreaks in Britain in most of the years up to 1855.

What was the cholera, how was it spread, how could it be prevented and how cured? It was a disease of the poor which also killed the rich — but that was and is true of most epidemic diseases. Many people, particularly reformers, felt sure that it was associated with dirt, poor sanitation, bad water, bad air, bad diet and crowded living conditions. Beyond that there was no agreement. Was it contagious? If so, how could it evade all attempts at quarantine and why were doctors and clergymen who attended the sick so often spared? Was it caused by a poisonous miasma created by some process of fermentation in the presence of bad drains or stagnant water? That would help explain why it was a summer disease — except in Scotland where the outbreaks occurred in winter. Was the miasma, which some experts claimed to have seen, electrical or was it something to do with ozone?

In 1849 Dr John Snow published yet another theory. Snow had worked his way up from fairly humble beginnings to the top of the medical profession, becoming one of the founding fathers of anaesthesia. (In 1853 he was the doctor chosen to give chloroform to Queen Victoria in childbirth.) He was a shy, diffident man, wholly immersed in his work† and devoted to the relief of suffering. Snow, as an expert on respiration, rejected the miasmatic theory. If the disease were due to a 'miasma', surely the lungs would be affected first. Since cholera was primarily a disease of the alimentary canal, the cholera producing material must be swallowed 'and the increase of the ... cholera poison must take place in the interior of the stomach and the bowels'. Unless strict cleanliness was observed, the cholera poison, once excreted, would be transferred to hands and thence to food and drink ready to infect further victims.

†Though, according to his biographer, 'in the last few years of his life he so far threw off restraint as to visit the opera occasionally.'

> If the cholera had no other means of communication than those which we have been considering, it would be constrained to confine itself to the crowded dwellings of the poor, and would be continually liable to die out accidentally in a place, for want of opportunity to reach fresh victims; but there is often a way open for it to extend itself more widely, and to reach the well-to-do classes of the community; I allude to the mixture of the cholera evacuations with the water used for drinking and culinary purposes, either by permeating the ground, and getting into wells, or by running along channels and sewers into the rivers from which entire towns are supplied with water.

Snow showed that his theory was in accordance with most of what was known about the disease although he could only speculate about the nature of the cholera poison which 'having the property of reproducing its own kind, must necessarily have some sort of structure, most likely that of a cell'. However, in the last 200 years legions of clever doctors had produced ingenious theories about this or that disease each in accordance with most known facts but, in the end, contributing no permanent knowledge of cause, prevention or cure.

Snow was not just an ordinary clever doctor. In the words of his sonorous Victorian biographer:

> During subsequent years, but specially during the great epidemic outbreak of the disease in London in 1854, intent to follow out his grand idea, he went systematically to work. He laboured personally with untiring zeal. No one but those who knew him intimately can conceive how he laboured, at what cost and at what risk. Wherever cholera was visitant, there was he in the midst. For the time he laid aside as much as possible of the emoluments of practice; and when even, by early rising and late taking rest, he found that all that might be learned was not, from the physical labour implied, within the grasp of one man, he paid for qualified labour.

Part of his time he spent in the traditional way, listening to the arguments of his opponents and collecting information about particular cases but most of his effort went into collecting statistics — that is counting large numbers of cases†. Moreover, rather than just collecting statistics, as many of his contemporaries did, in the hope that something might emerge, he sought statistical evidence for or against his particular theory of cholera.

†The idea of applying statistics to practical problems was of course very much in the air, so much so that some modern historians feel that Snow was merely part of the 'Spirit of the Age'. However, Snow did the work and it seems unfair that the 'Spirit of the Age' should get the credit.

At the time, London was supplied by various private water companies which drew their water from different sources and supplied different districts. His first statistical analysis tabulated the number of deaths per thousand in the various districts supplied by each water company for the epidemics of 1832 and 1848. The results for the 1848 epidemic are shown in Figure 1.1.

At first sight the results seem conclusive evidence in favour of a waterborne source for cholera. All the districts which suffered worst were supplied wholly or partly by the Southwark and Vauxhall water company. However, the districts supplied by the Southwark and Vauxhall company have other features in common. William Farr, of the Registrar-General's office, who was responsible for the collection of the statistics on which the table was based had also analysed them and found a strong correlation between height above sea level and deaths from cholera. (Taking the epidemics of 1849 and 1853–4 together he found that the mortality in the lowest parts of London was 15 times the mortality in the highest.) For Farr the Southwark and Vauxhall company districts were distinguished by their low-lying nature rather than their water supplier.

The interpretation of the table is made still harder by the fact that some districts have more than one water supplier. Thus, for example, a commercial war broke out between the (predecessor of the) Southwark and Vauxhall and the Lambeth water companies with the result that many districts were supplied by pipes of both water companies. (Eventually peace was restored and the companies celebrated their reconciliation by raising their rates by 25%.) However, Snow realised that

Figure 1.1: Snow's table of cholera deaths and water supply for London, 1849.

District.	Population in the middle of 1849.	Deaths from Cholera.	Deaths by Cholera to 10,000 inhabits.	Annual value of House & Shop room to each person in £.	Water Supply.
Rotherhithe	17,208	352	205	4.238	Southwark and Vauxhall Water Works, Kent Water Works, and Tidal Ditches.
St. Olave, Southwark . .	19,278	349	181	4.559	Southwark and Vauxhall.
St. George, Southwark . .	50,900	836	164	3.518	Southwark and Vauxhall, Lambeth.
Bermondsey	45,500	734	161	3.077	Southwark and Vauxhall.
St. Saviour, Southwark . .	35,227	539	153	5.291	Southwark and Vauxhall.
Newington	63,074	907	144	3.788	Southwark and Vauxhall, Lambeth.
Lambeth	134,768	1618	120	4.389	Southwark and Vauxhall, Lambeth.
Wandsworth	48,446	484	100	4.839	{Pump-wells, Southwark and Vauxhall, river Wandle.
Camberwell	51,714	504	97	4.508	Southwark and Vauxhall, Lambeth.
West London	28,829	429	96	7.454	New River.
Bethnal Green	87,263	789	90	1.480	East London.
Shoreditch	104,122	789	76	3.103	New River, East London.
Greenwich	95,954	718	75	3.379	Kent.
Poplar	44,103	313	71	7.360	East London.
Westminster	64,109	437	68	4.189	Chelsea.
Whitechapel	78,590	506	64	3.388	East London.
St. Giles	54,062	285	53	5.635	New River.
Stepney	106,988	501	47	3.319	East London.
Chelsea	53,379	247	46	4.210	Chelsea.
East London	43,495	182	45	4.823	New River.
St. George's, East . .	47,334	199	42	4.753	East London.
London City . . .	55,816	207	38	17.676	New River.
St. Martin	24,557	91	37	11.844	New River.
Strand	44,254	156	35	7.374	New River.
Holborn	46,134	161	35	5.883	New River.
St. Luke	53,234	183	34	3.731	New River.
Kensington (except Padding-ton)	110,491	260	33	5.070	West Middlesex, Chelsea, Grand Junction.
Lewisham	32,299	96	30	4.824	Kent.
Belgrave	37,918	105	28	8.875	Chelsea.
Hackney	55,152	139	25	4.397	New River, East London.
Islington	87,761	187	22	5.494	New River.
St. Pancras	160,122	360	22	4.871	{New River, Hampstead, West Middlesex.
Clerkenwell	63,499	121	19	4.138	New River.
Marylebone	153,960	261	17	7.586	West Middlesex.
St. James, Westminster .	36,426	57	16	12.669	Grand Junction, New River.
Paddington	41,267	35	8	9.349	Grand Junction.
Hampstead	11,572	9	8	5.804	Hampstead, West Middlesex.
Hanover Square & May Fair .	33,196	26	8	16.754	Grand Junction.
London	2,280,282	14137	62	—	

this intermingling represented not a problem but an opportunity when, in 1852, the Lambeth company moved their water works up river clear of the London sewage whilst the Southwark and Vauxhall continued to draw its water downstream. Snow wrote that this

> ... admitted of the subject being sifted in such a way as to yield the most incontrovertible proof on one side or the other. In the sub-districts enumerated in [Figure 1.2] as being supplied by both Companies the mixing of the supply is of the most intimate kind. The pipes of each company go down all the streets, and into nearly all the courts and alleys. A few houses are supplied by one company and a few by the other, according to the decision of the owner or occupier at the time. ... Each company supplies both rich and poor, both large houses and small; there is no difference either in the condition or occupation of the persons receiving the water of the different companies ...
>
> The experiment, too was on the grandest scale. No fewer than three hundred thousand people of both sexes, of every rank and station, from gentlefolks down to the very poor, were divided into two groups without their choice, and, in most cases without their knowledge; one group being supplied with water containing the sewage of London, and, amongst it whatever might have come from cholera patients, the other group having water quite free from such impurity.

When cholera returned to London in 1854 Snow determined to visit every house in these districts where a cholera death occurred and record the name of the supplying company. As he remarks 'The enquiry was necessarily attended with a good deal of trouble' and in many cases it proved impossible to find anyone who knew the name of the water supplier. Fortunately a simple chemical test based on the high salt content of the Southwark and Vauxhall company's water ('part of that' he wrote 'which has passed through the kidneys ... of two millions and a quarter of the inhabitants of London') enabled him to deal with these cases as well. He communicated his initial findings to Farr 'who was much struck with the result' and arranged official assistance for the last three weeks of the investigation. Snow summarised his results in Figure 1.2. (A star * indicates that the apportionment of deaths among water sources within the sub-district involved a small amount of estimation.)

It is a maxim among practical statisticians that 'The data you need are not the data you have, the data you have are not the data you want and the data you want are not the data you need.' Although the total number of houses supplied by each company was known, this total was not broken down by district. The simplicity of Snow's 'grand experiment' was thus marred by the fact that he could not directly separate those districts supplied by both companies from those supplied by only one or the other. However, all the districts are contiguous and show similar death rates for the previous 1848 epidemic (when both companies drew from similar sources), and the effect shown in Table 1.1 is so marked that it can hardly be due to some subtle difference between the districts served†.

Just as Snow was completing his case by case enquiries into the districts south of the Thames detailed in his tables, there was a dreadful outbreak of cholera north of the Thames in Soho. 'Within two hundred

†When a district by district breakdown of the number of houses supplied by each company later became available, Snow reanalysed his data but the pattern was unaltered. He calculates the relative mortality of consumers of the two water suppliers as being in a ratio of about 1 to 6.

Figure 1.2: Snow's table of cholera deaths for districts with mixed water supply, 1854.

Sub-Districts.	Popula-tion in 1851.	Deaths from Cholera in the seven weeks ending 26th August.	Water Supply.				
			Southwark & Vauxhall.	Lambeth.	Pump-wells.	River Thames and ditches.	Unascertained.
*St.Saviour,Southwark	19,709	125	115	—	—	10	—
*St. Olave, Southwark	8,015	53	43	—	—	5	5
*St. John, Horsleydown	11,360	51	48	—	—	3	—
*St.James,Bermondsey	18,899	123	102	—	—	21	—
*St. Mary Magdalen .	13,934	87	83	—	—	4	—
*Leather Market	15,295	81	81	—	—	—	—
*Rotherhithe	17,805	103	68	—	—	35	—
*Battersea .	10,560	54	42	—	4	8	—
Wandsworth	9,611	11	1	—	2	8	—
Putney .	5,280	1	—	—	1	—	—
*Camberwell	17,742	96	96	—	—	—	—
*Peckham .	19,444	59	59	—	—	—	—
Christchurch,Southwk.	16,022	25	11	13	—	—	1
Kent Road	18,126	57	52	5	—	—	—
Borough Road .	15,862	71	61	7	—	—	3
London Road .	17,836	29	21	8	—	—	—
Trinity, Newington .	20,922	58	52	6	—	—	—
St. Peter, Walworth .	29,861	90	84	4	—	—	2
St. Mary, Newington.	14,033	21	19	1	1	—	—
Waterloo Road (1st) .	14,088	10	9	1	—	—	—
Waterloo Road (2nd)	18,348	36	25	8	1	2	—
Lambeth Church (1st)	18,409	18	6	9	—	1	2
Lambeth Church(2nd)	26,748	53	34	13	1	—	5
Kennington (1st)	24,261	71	63	5	3	—	—
Kennington (2nd)	18,848	38	34	3	1	—	—
Brixton .	14,610	9	5	2	—	—	2
*Clapham .	16,290	24	19	—	5	—	—
St.George,Camberwell	15,849	42	30	9	2	—	1
Norwood .	3,977	8	—	2	1	5	—
Streatham	9,023	6	—	1	5	—	—
Dulwich .	1,632	—	—	—	—	—	—
Sydenham .	4,501	4	—	1	2	—	1
	486,936	1514	1263	98	29	102	22

and fifty yards [roughly 250 metres] of the spot where Cambridge Street
meets Broad Street there were upwards of five hundred fatal attacks
of cholera in five days. The mortality in this limited area probably
equal[led] any that was ever caused in this country even by the plague;
and was much more sudden, as the greater number of cases terminated
in a few hours.' Snow's biographer writes:

> While then the vestrymen [parish authorities] were in solemn deliberation
> they were called to consider a new suggestion. A stranger had asked, in
> modest speech, for a brief hearing. Dr Snow, the stranger in question,
> was admitted and in few words explained his view .... He had fixed
> his attention on the Broad Street pump as the source and centre of the
> calamity. He advised the removal of the pump handle as the grand
> prescription. The vestry was incredulous but had the good sense to carry
> out the advice. The pump handle was removed and the [epidemic] was
> stayed.

In his account of the outbreak Snow presents his evidence for believing
that the Broad Street pump was the cholera source in the form of a
map (see Figures 1.3 and 1.4).

Although the Soho area had piped water, the supplying companies
only turned on their mains for two hours a day and not at all on
Sundays. Many of the inhabitants thus drew some or all of their
supplies from one of the water pumps marked on Snow's map. The
recorded deaths from cholera during the epidemic are represented by a
black bar indicating the location of the house in which the fatal attack
started.

I began this chapter by talking about abstraction. The twentieth
century reader brought up in a culture of statistics, graphs and maps
may not grasp the degree of abstraction represented by Snow's map.
She should pause and try to imagine the agony and sorrow associated
with each black bar and the fear and terror of a time when the dead
'were removed wholesale in dead-carts for want of sufficient hearses to
convey them'. Snow's map is a representation from which all that is

Table 1.1. *Cholera deaths by water company, 1854.*

Company	Number of houses	Deaths from cholera	Deaths per 10000 houses
Southwark and Vauxhall company	40046	1263	315
Lambeth Company	26107	98	37
Rest of London	250243	1422	59

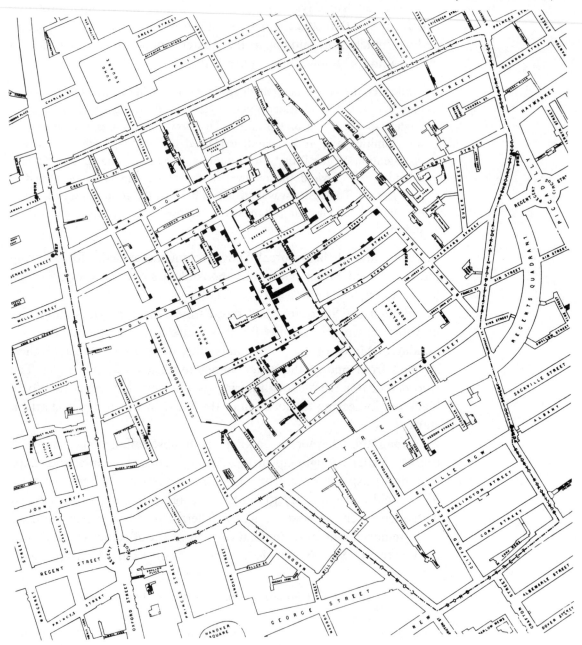

Figure 1.4: Central portion
of Snow's map.

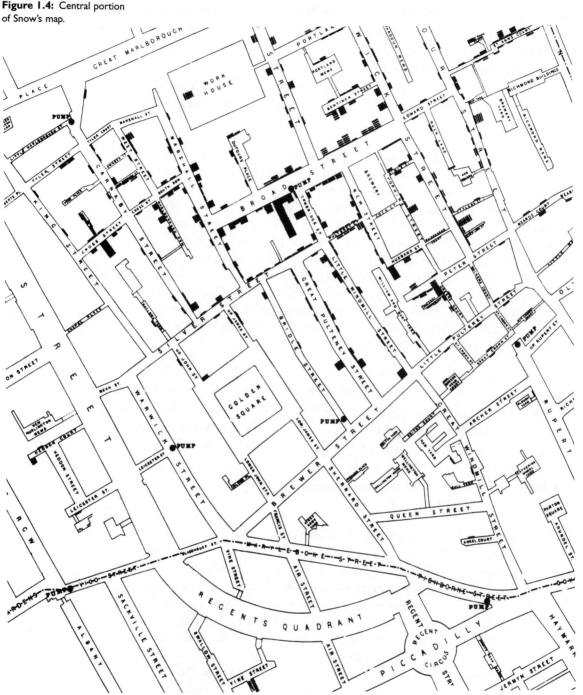

humanly important and all that is individual has been drained. It is, as it were, a shadow of a shadow and yet, by reducing all this human suffering to little black bars on a sheet of paper, Snow saved thousands upon thousands from similar deaths.

The remarkable thing about the Soho outbreak was its restricted geography. The map almost speaks for itself but Snow comments on it as follows:

> It requires to be stated that the water of the pump in Marlborough Street, at the end of Carnaby Street, was so impure that many people avoided using it. And I found that the persons who died near this pump in the beginning of September, had water from the Broad Street pump. With regards to the pump in Rupert Street, it will be noticed that some streets which are near to it on the map, are in fact a good way removed, on account of the circuitous road to it. These circumstances being taken into account, it will be observed that the deaths either very much diminished, or ceased altogether, at every point where it becomes decidedly nearer to send to another pump than the one in Broad Street. It may also be noticed that the deaths are most numerous near to the pump where the water could be more readily obtained.

Snow gives other significant pieces of evidence. The Broad Street Brewery had its own well and allowed each workman a certain amount of malt liquor. None of its 70 or so employees fell seriously ill. A factory on the same street kept two tubs of water permanently filled from the Broad Street pump for its workpeople to drink from. Eighteen out of its 200 employees died of cholera. Snow also recounts the striking story of Mrs Eley, the widow of the founder of the factory, who had retired to Hampstead West End and

> ... had not been in the neighbourhood of Broad Street for many months. A cart went from Broad Street to West End every day, and it was the custom to take out a large bottle of the water from the pump at Broad Street, as she preferred it. The water was taken on Thursday, 31st August, and she drank of it in the evening, and also on Friday. She was seized with cholera on the evening of the latter day, and died on Saturday .... A niece, who was on a visit to this lady, also drank of the water; she returned to her residence, in a high and healthy part of Islington, was attacked with cholera, and died also. There was no cholera at the time either at West End or in the neighbourhood where the niece died.

(However, a female servant who also drank the water did not contract the disease.)

Because we know that germs exist and cause disease we are happy to accept Snow's evidence at face value. Cholera can spread through the water system and in the case of the Soho outbreak the source must have been the Broad Street pump. His contemporaries were less happy to accept a theory which postulated an invisible life form. A further problem with Snow's theory in so far as it concerned the Broad Street pump was that he was unable to explain how it became contaminated. The pump-well was opened and carefully inspected without revealing any 'hole or crevice by which impurity might enter' and chemical tests failed to reveal any particular contamination. It is thus not surprising that the government inspectors appointed to report on the outbreak entirely rejected Snow's theories.

> The extraordinary eruption of cholera in the Soho district which was carefully examined ... does not appear to afford any exception to generalisations respecting local states of uncleanliness, overcrowding, and imperfect ventilation. The suddenness of the outbreak, the immediate climax and short duration, all point to some atmospheric or other widely diffused agent still to be discovered, and forbid the assumption, in this instance, of any communication of the disease from person to person either by infection or contamination of water with the excretions of the sick.

Although the vestrymen joined the inspectors in rejecting Snow's condemnation of the pump, they seem to have felt that more investigation was called for and they set up their own committee of enquiry. In the face of governmental obstruction and local doubts (would not an enquiry damage slowly returning confidence and trade?), the committee pressed ahead with a door to door survey of the entire district. The curate of a local church, the Reverend Henry Whitehead, volunteered for the massive task of surveying Broad Street itself and managed to interview well over half its original inhabitants. (A good account of his work is given in [97].) Initially opposed to Snow's theories, he was forced to change his mind as the evidence accumulated and he found that of the 137 persons who drank water from the pump 80 developed cholera whereas of the 297 who did not drink the water only 20 did. Finally he discovered the key to the mystery. Just before the outbreak a baby girl had died of what could have been cholera in a house with a privy only a few feet away from the Broad Street pump. Digging revealed that crude and wrongly constructed drainage provided a nearly direct route for the contamination of the pump water. Moreover, the day that the pump handle was removed another occupant (the baby's father) contracted cholera and, in all probability, it had only been Snow's timely intervention which had prevented a second outbreak.

Snow died, hard at work on a book on anaesthesia, four years later. His book on cholera which cost him £200 to prepare and publish sold only 56 copies but 'Dr Snow's Theory of Cholera' gradually won acceptance among many of those concerned with the disease. Among those convinced by Snow was William Farr. When an explosive epidemic of cholera hit East London in 1866 he was 'thus prepared ... to scrutinise the water supply' and, despite denials by water company officials, traced the source of the epidemic to open ponds, tainted by sewage from a nearby river, which were being used as emergency reserves. This practice was stopped and the epidemic ceased.

The New York epidemic in the same year was also handled according to Snow's ideas. It was not necessary to accept, or even to understand, the theoretical underpinnings of Snow's views to follow his practical recommendations. 'Boiling drinking water or disinfecting clothing and bedding were measures that any alert physician or board of health could carry out. At least there was no harm in trying.' Although the city had greatly increased in size and had certainly not become any healthier in general, the number of deaths was one tenth that in the previous epidemic of 1849.

The movement for clean water and proper sewerage, which was fuelled by many other sources, reached a successful conclusion in Europe before the end of the century. In 1883 Koch isolated the bacillus for choleras, providing the causative agent which Snow had been unable to give.

The last major European outbreak of cholera occurred in Hamburg in 1892. The merchant oligarchy which ruled the city had repeatedly postponed expensive changes in its water supply system and the inhabitants drank untreated water from the Elbe. The neighbouring town of Altona had a water filtration plant. A street divided the two cities. On one side the cholera raged unchecked; the other side was spared almost completely. John Snow had won his argument.

1.2 An altar of pedantry

The *Wall Street Journal* is a newspaper which prides itself on its hard-headedness†. On 2 June 1987 it ran an article entitled 'Human Sacrifice'.

> Last Friday an advisory panel of the Food and Drug Administration decided to sacrifice thousands of American lives on an altar of pedantry.
>
> Under the klieg lights of a packed hearing room at the FDA, an advisory panel picked by the Agency's Center for Drugs and Biologics declined to recommend approval of tPA, a drug that dissolves blood clots after heart attacks. In a 1985 multicenter study conducted by the US National Heart Lung and Blood Institute, tPA was so conclusively effective at this that the trial was stopped. This decision to withhold it

†In my experience people who claim to be hard-headed tell us more about their hearts than their heads.

from patients should be properly viewed as throwing US medical research into a major crisis.

Heart disease dwarfs all other causes of death in the industrialised world, with some 500 000 Americans killed annually; by comparison some 20 000 have died of AIDS. More than a thousand lives are being destroyed by heart attacks every day. In turning down treatment with tPA, the committee didn't dispute that tPA breaks up the blood clots impeding blood flow to the heart. But the committee asked that Gentech, which makes the genetically engineered drug, collect some more mortality data. Its submission didn't include enough statistics to prove to the panel that dissolving blood clots actually helps people with heart attacks.

Yet on Friday, the panel also approved a new procedure for streptokinase, the less effective clot dissolver — or thrombolytic agent — currently in use. Streptokinase previously had been approved in an expensive, specialised procedure called intracoronary infusion. An Italian study, involving 11 712 randomised heart patients at 176 coronary-care units in 1984–1985, concluded that administering streptokinase intravenously reduced deaths in that group by 18%. So the advisory panel decided to approve intravenous use of streptokinase, but not approve the superior thrombolytic tPA. This is absurd.

Indeed, the panel's suggestion that it is necessary to establish the efficacy of thrombolysis stunned specialists in heart disease. Asked about the committee's justification for its decision, Dr Eugene Braunwald, chairman of Harvard Medical School's department of medicine, told us: 'The real question is, do you accept the proposition that the proximate cause of a heart attack is a blood clot in the coronary? The evidence that it is is overwhelming, *overwhelming*. It is sound basic medical knowledge, it is in every text-book of medicine. It has been firmly established in the past decade beyond any reasonable question. If you accept the fact that this drug [tPA] is twice as effective as streptokinase in opening closed vessels and has a good safety profile then I find it baffling how that drug was not recommended for approval.'

Patients will die who would otherwise live longer. Medical research has allowed statistics to become the supreme judge of its inventions. The FDA, in particular its bureau of drugs under Robert Temple, has driven that system to its absurd extreme. The system now serves itself first and people later. Data supersede the dying.

The advisory panel's suggestion that tPA's sponsor conduct further mortality studies poses grave ethical questions. On the basis of what medicine already knows about tPA, what US doctor will give a randomised heart-attack victim a placebo or even streptokinase? We'll put it bluntly: Are American doctors going to let people die to satisfy the Bureau of Drugs' chi-squared studies?

Friday's tPA decision should finally alert policy makers in Washington and the medical research community that the theories and prac-

tices now controlling drug approval in this country are significantly
flawed and gone grievously wrong in the FDA bureaucracy. As an
interim measure FDA commissioner Frank Young with Gentech's as-
sent could approve tPA under the agency's new experimental drug rules.
Better still, Dr Young should take the matter in hand, repudiate the
panel's finding and force an immediate reconsideration. Moreover, it
is about time Dr Young received the clear, public support of Health
and Human Services Secretary Dr Otis Bowen in his efforts to fix the
FDA.

If on the other hand Drs Young and Bowen insist that the actions
of bureaucrats are beyond challenge, then perhaps each of them should
volunteer to personally administer the first randomised mortality trials
of heart-attack victims receiving the tPA clot buster or nothing. Alternatively, coronary-care units receiving heart-attack victims might use a
telephone hot line to ask Doctor Temple to randomise the trial himself
by flipping a coin for each patient. The gods of pedantry are demanding
more sacrifice.

As most readers of this article would have known, there were other
issues at stake as well. The fortunes of the pioneering biotechnology
company Gentech depended on tPA. By withholding approval of the
new drug the bureaucrats of the FDA could gravely damage the growth
of the new 'genetic engineering' technology. Fortunately for Gentech,
approval was not long withheld and within a couple of years half the
company's income came from tPA.

The causes of the newspaper's anger go back a long way. One place
to start is with Koch's claim to have discovered bacilli which caused
cholera. Koch's claim was also not accepted immediately. One of his
opponents drank a beaker full of Koch's bacilli to prove the falsity of
the theory and then, in a 'graphic demonstration of the power of a
German professor over his assistants' had the experiment repeated by
his assistant. Both became ill, the professor mildly so, the assistant
more seriously, but both survived. Since we now know that Koch was
right, every author who recounts the story feels obliged to explain the
professor's survival. Unfortunately each gives a different explanation.

However, the professor's experience must have been widely, if un-
knowingly shared (for example by the widow Eley's servant). Snow had
to admit that most of the customers of the Southwark and Vauxhall
company did not go down with cholera and that not everybody who
drank from the Broad Street pump died as a result. In the weakest part
of his paper he suggests that the 'cholera poison' is particulate (like,
say, tapeworm eggs) and so not everybody who swallowed contaminated
water would swallow the 'poison'.

Nowadays medical scientists are happy to accept that biological systems do not exhibit the same direct association of cause and effect as that shown in the isolated and simplified demonstrations of the physicist. Only the cholera bacillus causes cholera but whether someone exposed to the cholera bacillus gets cholera depends on many factors. (For example, since the bacillus requires an alkaline environment it is usually killed by the stomach acids before it reaches the favourable environment of the intestines.) Some people are allergic to eggs, most are not. A reasonable dose of aspirin reduces a headache, but may kill one unfortunate individual in a million. A particular drug cures one case but fails to cure an apparently similar case.

Because of this variability the only way to be sure that one treatment is better than another is by trying both treatments out on a large number of cases, some patients getting the first treatment and some the second. Since it is well established that the mere fact of being treated makes people better (what, after all, are doctors but faith healers armed with penicillin?), trials may be held in which one group receives pills containing the drug under test and the other group apparently identical pills (the placebos referred to above) which do not contain the drug. Only if the drug does better than the placebo is it considered effective. There is another subtle problem associated with such trials. Suppose that an established procedure *A* is being tested against a new procedure *B*. Doctors may tend to use *A* in cases where they believe it will work and reserve *B* for the otherwise hopeless cases where 'there is nothing to lose'. The new treatment may therefore appear less effective simply because it is mainly given to the iller patients. To avoid this, treatments must be assigned at random, essentially by 'flipping a coin for each patient'. The commentator of the *Wall Street Journal* is not the only person to find the apparatus of 'placebos' and 'randomisation' repellent — similar criticisms attend each trial of potential AIDS drugs.

The ISIS† (International Study of Infarct Survival) trials provide instructive examples of modern medical trials. Most heart attacks are caused by blood clots blocking the supply of blood to the heart muscles. It is natural to try the effect of administering 'clot dissolving' drugs but such drugs may cause bleeding and there is, at least in theory, the possibility that a bleed into the brain may cause a serious stroke (cerebral haemorrhage) and it is by no means clear that the risks of side effects will outweigh the benefits. A series of trials on the 'clot buster' streptokinase produced conflicting results and the ISIS-2 trial was intended to settle the matter by using much larger samples. The study also considered the effect of aspirin which was known to reduce

†The Isis is a river that flows through Oxford just as the Cam flows through Cambridge.

3   Although the reduction in deaths is only about five patients per
       100, heart attacks are so common that the suggested change in
       treatment would save thousands of lives a year.

Each of these arguments has weaknesses but most people will find at
least one of them convincing.

Another argument against these trials might be that they only told
doctors what they already knew. The Chairman of Harvard Medical
School's Department of Medicine quoted above would hardly have
been surprised by the success of a 'clot buster', even an old-fashioned
one like streptokinase, in reducing deaths. The effectiveness of aspirin
was, however, a welcome surprise as may be seen from the fact that
the proportion of British cardiologists routinely giving aspirin to heart-
attack patients rose from 10% in 1987 (the year before the report
appeared) to 90% in 1989.

The ISIS-2 trials were followed by two new large scale trials (ISIS-3
and GISSI-2) of the various 'clot busters' available. The combined
results for streptokinase (SK) and tPA are shown in Figure 1.6 (all
patients were given aspirin). There is no evidence that one treatment
is better than another in preventing stroke deaths. There is, however,
one significant difference in the two treatments which, for some reason
or another, the *Wall Street Journal* failed to mention. In 1989 tPA cost
$2250 a dose and streptokinase cost $80 a dose (and you could get 50
aspirin tablets for a dollar). At the time the new report appeared tPA
was the most widely used clot buster in the USA and it was calculated
that the use of tPA rather than streptokinase was costing over $100
million a year.

When we are ignorant of fundamental mechanisms we must fall back
on statistics. The Chairman of Harvard Medical School's Department
of Medicine may have believed that 'sound basic medical knowledge'
called for ever more powerful clot busters — the figures prove otherwise.
(Incidentally, though tPA was no more effective than streptokinase in
its intended purpose it did, as theory suggested it might, provoke more
strokes. About two more patients in 1000 suffered disabling strokes
under the tPA regime†.) We have no very clear idea why aspirin
is so effective, but our figures show that it is. Our dependence on
statistics is galling. We can never be entirely sure, even with the best
conducted survey, that some subtle failure in design may not vitiate
our conclusions. We can never be comfortable when men play at gods,
giving one treatment to one patient and some other treatment (or no
treatment at all) to another simply on the throw of a die. But we have
no real choice.

†A more dramatic example
is given by the CAST
(Cardiac Arrythmia
Suppression Trial).
Arrythmias (irregularities
of the heart-beat) are
associated with the sudden
death of heart-attack
patients. The widely
accepted preventive
treatment was to give
patients the antiarrythmic
drugs encaide and
flecainide, and it is
estimated that about
500 000 new patients per
year received the drugs in
the US alone. The
treatment was so firmly
established that many
doctors argued that
placebo-controlled trials
would be unethical. When
the trial actually took
place, it was found that,
just as expected, the drugs
did suppress arrythmias
but sometimes did so in the
most direct and permanent
way. Those given the drugs
had three times the death
rate of those given the
placebo.

CHAPTER 2

Prelude to a battle

2.1 The first great submarine war

On 10 April 1917 Admiral Jellicoe, First Sea Lord, commander of a British Navy, which had held undisputed command of the seas for three generations, handed over a memorandum to the naval representative of his new American ally. The memorandum showed the British and neutral shipping losses of the last months: 536 000 tons in February, 603 000 tons in March and a predicted 900 000 tons in April. The American admiral recalls what followed.

> It is expressing it mildly to say that I was surprised by this disclosure. I was fairly astounded; for I had never imagined anything so terrible. I expressed my consternation to Admiral Jellicoe.
>
> 'Yes,' he said quietly as though he was discussing the weather and not the future of the British Empire. 'It is impossible for us to go on with the war if losses like this continue ... '
>
> 'It looks as though the Germans are winning the war,' I remarked.
>
> 'They will win, unless we can stop these losses — and stop them soon,' the Admiral replied.
>
> 'Is there no solution for the problem?' I asked.
>
> 'Absolutely none that we can see now.'

(The quotation above is taken from A. J. Marder's magisterial history *From the Dreadnought to Scapa Flow* on which the whole of this chapter relies.)

The German submarines which produced this disaster were not what we would now call true submarines but were, essentially, submersible torpedo boats. Underwater they used engines powered by electric batteries which were capable of giving a range of about 50 miles [80 kilometres] at a speed of about 4 knots [7.5 kilometres per hour] or permitting a dash of about 15 miles [24 kilometres] at 8 knots [15 kilometres per hour]. The batteries had to be recharged on the surface using separate diesel engines. These diesels also gave the submarines a surface range of, typically, 7500 miles [12 000 kilometres] at 7 knots [13 kilometres per hour] and a maximum surface speed of 16 knots [30 kilometres per hour].

Only very slow merchant ships travelled at less than 8 knots and destroyers had speeds of more than 30 knots. Once surfaced, the submarine was vulnerable even to the light gun of an armed merchant ship. The submarine was therefore effective only as a weapon of ambush, sinking its prey without warning and without allowing the chance of surrender. Since it had room only for its own crew and since the submarine was so vulnerable when surfaced, no attempt could be made to rescue the victims. The submarine, if it was to fulfill its potential as a weapon, had to break, in the most shocking manner, with the customs of the sea and the laws of war.

The effectiveness of the submarine as a commerce destroyer, provided it was allowed to sink without warning, only became clear after the outbreak of war but, from then on, the German High Command pressed for unrestricted submarine warfare against Britain. A short period of such warfare gave promising military results but included the sinking of the passenger liner *Lusitania* with the loss of 124 American lives. This incident outraged United States opinion which had before been strongly against involvement in any European conflict. Partly because of this and partly because Germany lacked enough submarines to make her submarine blockade fully effective, the experiment was terminated.

By the beginning of 1917 the German Navy had many more submarines, the German economy was deteriorating under the pressures of war and the British naval blockade, and the German Army could not promise a decisive victory on land. The chief of the German Naval Staff wrote:

> A decision must be reached in the war before the Autumn of 1917, if it is not to end in the exhaustion of all parties and consequently disastrously for us. Of our enemies, Italy and France are economically so hard hit that they are only upheld by England's energy and activity. If we can break England's back, the war will at once be decided in our favour. Now England's mainstay is her shipping, which brings to the British Isles the necessary supplies of food and materials for war industries, and ensures their solvency abroad ...
>
> I do not hesitate to assert that, as matters now stand, we can force England to make peace in five months by means of an unrestricted U-boat campaign. But this holds good only for a really unrestricted campaign ....
>
> A further condition is that the declaration and commencement of the unrestricted U-boat war should be simultaneous, so that there is no time

for negotiations, especially between England and the neutrals. Only on these conditions will the enemy and the neutrals be inspired with ... terror ....

Unrestricted U-boat warfare ..., begun early enough, [will] ensure peace before the next harvest, that is before 1 August; we have no alternative. In spite of the danger of a break with America, unrestricted submarine war, begun soon, is the right means to bring the war to a victorious end for us. Moreover it is the only way.

His advice was taken and, on 1 February 1917, German submarines once again began sinking without warning. On 6 April, the United States provoked by this resumption (and the discovery of secret attempts by Germany to draw Mexico into a war against her) declared war on Germany. If Britain could hold out long enough, the fresh American resources might settle the European conflict, but on 18 April the British Secretary of State for War wrote to his chief general: 'The situation is very bad and we have lost command of the sea.'

Since the professionals had lost control of the situation, they were bombarded with suggestions from outsiders ranging from training cormorants to land on enemy periscopes (this is said to have failed because the cormorants were too good; having been trained with British submarines they refused to use German periscopes), through filling the North Sea with barrels of Eno's Fruit Salts (the bubbles were to force the submarines to surface), to a revival of the obsolete eighteenth century practice of convoy.

In the convoy system merchant ships were only allowed to sail in large groups under the escort of warships. At first sight this seems an obvious measure but there were several cogent objections.

1   Convoy was a defensive rather than an offensive measure. The job of the navy was to patrol the sea lanes to keep them clear of hostile forces.

2   The delay involved in assembling a large number of merchant ships would reduce the total tonnage transported each month.

3   The arrival of a large convoy at a port would overtax the loading and unloading facilities. (This was a very important point. Marsden suggests that the failure of the fragmented British port and railway system to adapt to wartime conditions was almost as damaging to the economy as the German submarine offensive.)

4   A convoy can only travel as fast as its slowest ship. Not only does this again reduce the total tonnage transported each month, but it also means that the faster ships cannot use their speed as protection against attack.

5   Both merchant and navy captains believed that a large number
     of merchant and naval escort ships would be unable to keep in
     the tight formations demanded by convoy. When a group of ten
     masters of merchant ships was consulted, they replied that two or
     three ships was the maximum that 'might be able' to sail together
     and keep station.

6   A convoy might well prove an easier rather than a harder target
     for a submarine. Admiralty instructions at the beginning of 1917
     made the point explicitly.

     The system of ... convoy is not recommended in any area where
     submarine attack is a possibility. It is evident that the larger the
     number of ships forming the convoy, the greater is the chance of a
     submarine being enabled to attack successfully and the greater the
     difficulty of the escort in preventing such an attack. ... A submarine
     [can] remain at a distance and fire her torpedo into the middle of
     a convoy — with every chance of success.

7   Finally, even if convoy was considered a desirable measure, where
     were the naval escorts to come from? Admiralty statistics (supplied
     by the customs authorities) showed that over 5000 ships entered or
     left British ports each week. To convoy numbers of that magnitude
     was an impossible task.

The men who commanded the British Navy were able and energetic.
They and their immediate predecessors had ruthlessly modernised a
dozy and old-fashioned fleet which rested on laurels half a century
old. They realised that modern warfare required mastery of technology
and industrial methods, and ensured that the British fleet acquired that
mastery. Faced with the new threat of the submarine they sought an
answer in new technology — the depth charge (a compound of 'applied
chemistry, synthetic earthquake and sudden death'), the hydrophone,
hunter-killer submarines, air patrols, deep mines (the British mine
proved ineffective, so an exact copy of a German type was used) and
other devices. Trade routes were patrolled with everything that could be
spared — but whilst the Germans lost about three submarines a month
they built seven a month to replace them, and British and neutral losses
went on rising.

In retrospect we can identify the missing weapon in the Admiralty's
armoury. In May 1917, after, but as a result of, the crisis described here,
Sir Eric Geddes, a railway engineer by training, and manager of the
North Eastern Railway before the war, was appointed to the Board of
Admiralty (and became the first landsman since the seventeenth century

to be appointed admiral). Among his first acts was to set up a Statistical Department under one of his North Eastern Railway colleagues.

One admiral wrote

> We have been upside down here ever since the North Eastern Railway took over the management, but manage to worry along somehow doing our jobs. Geddes is mad about statistics and has forty people always making graphs and issuing balance sheets full of percentages, etc. Unfortunately worrying about what happened last month does not help the present or the future and wastes a great deal of time. It may be well enough in a Life Assurance business or a railway. We do not get anything done quicker now than this time last year and most things a good deal slower but there is more made of them on paper.

Marder comments that

> [this] inability to see that foresight must be based on sound knowledge of the past is very illuminating.

Elsewhere he states that

> [the Admiralty] had not at any time during the war made any serious study of the problem of trade protection. The Admiralty had accepted it as a principal responsibility of the Navy *without determining the precise extent of that responsibility.* 'It is as if an insurance company agreed to insure a man's life without troubling to find out his age, occupation or state of health.'

The Admiralty had fully grasped the role of scientific thought in building a modern fleet, but failed to see that similar thought was needed to employ it most effectively.

Exercise 2.1.1 *Think carefully about the arguments against convoy given above. Which seem to you valid, which invalid and which partially valid? Give your reasons. Are there any other arguments for or against which ought to have been considered?*

2.2 The coming of convoy

What were the arguments of those who favoured convoy? Convoy was a traditional form of protection, used most recently during the Napoleonic wars when, indeed, the Admiralty was given power to enforce convoy by law. But the experience of the days of sail could not, by itself, carry conviction in the age of steam. The proponents of convoy could point out that the main battle fleet was always protected by a screen of destroyers — but, again, the circumstances were very different from those involved in the day to day escort of merchantmen. Finally, and most convincingly, they could point out that troop-ships were always convoyed and that none had been lost.

From early in 1917 convoy advocates could point to one further piece of evidence. British coal was vital for French industry which needed to import at least $1\frac{1}{2}$ million tons a month supplied by 800 colliers. By December 1916 the disruption caused by submarine warfare had reached such levels that French factories were having to shut down. Under French pressure a system of convoys was instituted which restored the regularity of supplies. Contrary to expectations the convoys enjoyed an extraordinary immunity to submarine attack. In April nearly 2600 ships sailed in the coal convoys and five were lost to submarines. Similar results followed the earlier but less remarked introduction of convoy on the Dutch route (the 'Beef Trip').

Of course, the experience of short routes close to naval bases might not apply to long ocean routes. However, the Scandinavian routes which brought iron ore from Narvik had been suffering terrible losses (one ship in four failed to survive the round trip). Once more the experimental introduction of convoy produced staggering gains. In the first month the loss rate fell more than a hundredfold to 0.24% and port congestion decreased. Initially there was a daily sailing of six merchant ships under heavy escort but, with experience, this was modified to a five day cycle with convoys of between 20 and 50 ships without an increase in escort strength and without incurring heavier losses.

In the previous section I outlined several objections to convoy. Let us see how experience answered them.

1   *Convoy is a defensive rather than an offensive measure.* It is very hard to find submarines in the immensity of the ocean and even if they are sighted by a patrol they have a very good chance of getting away. However, a submarine which does not sink merchant ships is useless. If the merchant shipping sails in convoy, then the submarine must come to the convoy and, having come, must brave the escorts to carry out its purpose. Thus convoy is an offensive measure forcing the submarine to fight rather than flee. It is no accident that many of the great battles of the era of sail and several of the surface naval battles of World War II were fought round convoys.

2   *Delays involved in assembling convoys.* In practice it turned out that independent sailings were subject to such disruption by reports of U-boats that convoys actually speeded up departure.

3   *Delays involved in unloading convoys.* It turned out that it was

the *unpredictability* of arrival of independently routed ships which disrupted the (relatively) smooth running of the ports and the railway systems which served them. The high probability that convoys would arrive and leave on schedule was a remedy for, and not a cause of, port congestion.

4 *A convoy can only travel as fast as its slowest ship.* The statistics of the two World Wars show that a ship is safer in a convoy even if the convoy's speed is substantially lower than that achievable by the ship on its own. A faster ship sailing alone will, of course, deliver more tonnage per month, but only until it is sunk.

Exercise 2.2.1 *(i) A convoy sails at the speed of its slowest ship and has to be assembled. For these and other reasons, independently sailing ships complete their voyages more quickly than those in convoy. Suppose that merchant ships sailing independently take 75% of the time to complete their voyage but lose 14% of their number to submarines on each voyage, whilst convoyed ships lose 5% per voyage. (These figures are not out of line with those experienced on the Atlantic run during the first years of the Second World War. The First World War figures are much more favourable to convoy.) We start with a given fleet of merchant ships and must decide whether to use convoy for all of them or let them all sail independently. We can produce ships quickly enough to replace all those lost in convoy. Show that in the time it takes to make three convoy voyages, an independently sailing fleet will have made more voyages than a convoyed one but the position will be reversed for the time it takes to make six. What will happen over a long time?*

(ii) The situation calls out for the use of differential equations. Suppose that we can produce merchant ships at a rate μ (ships per unit time) and that we lose merchant ships at a rate λ (ships per ship afloat per unit time). Explain briefly why, with this model, the size $x(t)$ of our fleet is governed by the differential equation

$$\dot{x} = -\lambda x + \mu,$$

and deduce that

$$x(t) = \frac{\mu}{\lambda} + \left(X - \frac{\mu}{\lambda}\right)e^{-\lambda t},$$

where X is the size of the fleet when $t = 0$. What happens to the size of our fleet when t is large?
Suppose that the ships of our fleet can make κ voyages per unit time. Show that if $T \geq 0$ our fleet will make

$$\frac{\mu\kappa}{\lambda}T + \frac{\kappa}{\lambda}\left(X - \frac{\mu}{\lambda}\right)(1 - e^{-\lambda T})$$

voyages in the time between $t = 0$ and $t = T$. If we decide on convoy we take $\kappa = \kappa_C$ and $\lambda = \lambda_C$ whilst if we decide on independent sailings then $\kappa = \kappa_I$ and $\lambda = \lambda_I$. Show that if

$$\frac{\kappa_C}{\lambda_C} > \frac{\kappa_I}{\lambda_I}$$

then convoy is better in the long run but that if

$$\kappa_I > \kappa_C$$

then independent sailings are best in the short run.

(iii) Explain the conclusions of (ii) to a bright naval officer who is allergic to mathematics†.

(iv) In both (ii) and (iii) we made implicit simplifications of the real life situation to make calculations possible. Identify as many as you can and comment on how much you think they matter.

5   *Problems in keeping formation.* To quote Marder:

> The fears adumbrated by the masters and naval officers proved to be quite groundless for (as the evidence of the many successful troop convoys since August 1914 should have indicated), when straggling all over the place would make their ship a gift to a U-boat, the masters quickly learned the art of sailing in tightly packed formation.

6   *Convoy would provide an easier target for submarines.* This 'common sense' view fails to take into account three factors.

(*a*)   The ocean is so large and a convoy occupies so little space in it that a convoy is almost as hard to find as a single ship. If a single ship is visible to a submarine at 10 kilometres and a convoy at 12 kilometres the chance of the submarine finding the convoy is only 20% larger than that of finding the single ship. (In addition a convoy could be more easily diverted by radio away from known or suspected U-boat locations.)

(*b*)   Attacking an escorted convoy is more dangerous and more difficult than attacking a single ship. Even if a U-boat manages to get into a position to attack it will normally only manage to sink one or two ships.

(*c*)   In the worst case when the U-boat manages to make repeated attacks it carries only a limited number of torpedoes and can sink only a limited number of ships.

Dönitz, who was to lead the German U-boat offensive in the Second World War, and served as a U-boat captain in the First World War, recalls the effect of the introduction of convoy.

†In January 1901, *The Times* carried editorials on the new 'Submarine Boat' being experimented with by the US and French Navies and on the excessive mathematical knowledge required of British naval officers. The leader writer thought that they should be allowed to substitute 'Latin and the scholarly study of a modern Romance language'. Although none of the admirals who took part in the ensuing correspondence thought that Latin should enter the naval syllabus, several agreed that there should be less mathematics. Vice Admiral Sir Cyprian Bridge, a former director of naval intelligence' wrote that 'There may have been British naval officers whose mental facilities have been strengthened by a course of mathematical study, but personally I have never met one, and they are probably as rare as the great auk, or interpreters in Chinook. ... Will it be believed, say, fifty years hence that young sea officers were kept for months out of sight of the sea to study a subject which no one contends will be the slightest use to more than an infinitesimal minority of them?'

The oceans at once became bare and empty; for long periods at a time the U-boats, operating individually, would see nothing at all; and then suddenly, up would loom a huge concourse of ships, thirty or fifty or more of them, surrounded by an escort of warships of all types†. The solitary U-boat which most probably had sighted the convoy purely by chance, would then attack, thrusting again and again and persisting if the commander had strong nerves, for perhaps several days and nights, until physical exhaustion of both commander and crew called a halt. The lone U-boat might well sink one or two of the ships or even several, but that was but a poor percentage of the whole. The convoy would steam on. In most cases no other German U-boat would catch sight of it, and it would reach Britain bringing a rich cargo of foodstuffs and raw materials safely to port.

†Dönitz substantially exaggerates the size of the convoys (the largest First World War convoy had 48 ships and the average Atlantic convoy in that war had 15) and the strength of their escorts but the exaggeration is itself significant of how a convoy appeared to the submariner.

Variations on the phrase 'The oceans at once became bare and empty' occur repeatedly in the memoirs of the time. This is partly because memoir writers read one another but mainly because it expressed the truth. Instead of a steady stream of individual ships submarines now had to deal with a much smaller number of convoys which were almost as hard to find and substantially harder to attack.

Exercise 2.2.2 *Consider a large square of side a containing a small circle of radius r.*

(i) If a point is chosen at random in the square what is the probability that it lies inside the circle? If r is increased by 20% show that the probability that the point lies within the circle increases by about 40%.

(ii) If a line is chosen at random parallel to one of the sides of the square what is the probability that it crosses the circle? If r is increased by 20% show that the probability that the line crosses the circle increases by 20%.

(iii) Do you think (i) or (ii) is more relevant to 6(a) above?

So far, so good. But, even if convoy was desirable, how could the Admiralty solve the final problem?

7 *The number of ships sailing each week overwhelmed the Navy's ability to provide escorts.* Remember that the Admiralty's published statistics showed that over 5000 ships entered or left British ports each week. Commander Henderson was head of the organisation for the French coal convoys and therefore had good contacts with the Ministry of Shipping from whom he was able to obtain figures giving the exact number of arrivals and departures of ocean-going

merchant ships each year. He discovered that the Admiralty's figures showed the number of vessels of all nationalities of over 100
tons which had entered or left port during a week. 'If a dredger
sailed from Yarmouth to Felixstowe, it was counted once. If it
sailed back again the next day it was counted again. If it then
went to Harwich it was counted a third time. If it made the reverse
journey inside the same week, it would be counted six times.' If
the list was restricted to true ocean-going steamers of over 1000
tons the 5000 ships a week dwindled to the manageable number of
300 ships a week (or a little over 20 arrivals and departures a day).
The new figures simultaneously showed that convoy was possible
and cast a grim light on the existing situation. The news that
German submarines were now sinking over 10 ships a day, which
looked tolerable against a figure of 5000 arrivals and departures
a week, looked very different when measured against a figure of
20 arrivals and departures a day. (In fact the British were losing
10% of their total ocean-going shipping each month.)

Propaganda has been defined as 'that branch of the art of lying which consists in very nearly deceiving your friends without
quite deceiving your enemies'. The Admiralty figures had been
deliberately compiled to reassure neutral shipping and the world
in general. 'It was, in fact,' as one officer involved told Marder,
'similar to our casualty lists and bulletins and like these told no
untruths, but not necessarily the whole truth.' But such misleading figures take on a life of their own in a heavily centralised
bureaucracy like the Admiralty where:

> Congestion at the top and inertia at the bottom prevented inves-
> tigation of alternative courses of action. The ship was, as it were,
> running onto the rocks because the Captain and Navigator were
> too busy to lay off the new course. Nor was there any systematic
> attempt to analyse or evaluate the result of current operations.
> Valuable information, such as the decisive success of convoy on the
> [Dutch] route, was frequently ignored among the mass of material
> marked to the decisional authorities. It is one thing to read a re-
> port and quite another to select and mentally register its important
> points. The continual perusal of dockets, telegrams and reports,
> dealing with a wide variety of subjects acts like a drug. It dulls the
> perceptive faculties and paralyses power of criticism and selection
> behind a deceptive façade of hard work.

Although the Admiralty authorities must have known that their figures hid the true numbers of ships involved, they made no attempt

to find those true figures and apparently assumed that the figure of 5000 arrivals and departures simply exaggerated some other very large number†.

Henderson transmitted these figures to his Chief of Department but, at the same time, and contrary to all discipline, contacted the Prime Minister Lloyd George directly to urge the adoption of convoy. What happened next is clouded by inaccurate recollection and overdramatic memoirs which attest, once again, that victory has many fathers whilst defeat is an orphan. However, increasing desperation, Lloyd George's energy and Henderson's figures impelled the Admiralty into action. The entry of the United States with its promise of more ships and, in particular, more destroyers provided a convenient rationale for the drastic change of policy.

Mistakes were made (initially only inward bound ships were convoyed, protecting cargoes but only half protecting ships) but the introduction of convoy proceeded remarkably smoothly. The decision to convoy was taken at the beginning of May and by November Geddes, speaking 'like a company managing-director at a shareholder's meeting,' was able to tell the House of Commons that 'In September 90 per cent of the total vessels sailing in all Atlantic trade was convoyed, and since the convoy system was started — and it has been criticised in some quarters — the total percentage of loss per convoyed vessel through the danger zone is 0.5%, or 1 in 200.' Although the rate of replacement for merchant ships was only to match the rate of sinking in the closing months of the war, the crisis was over.

> It was a war-winning victory, which received very little publicity, at the time or afterwards. Those who took part in it were hardly aware that they were winning. Those who directed it often doubted their own success.

Between February 1917 and the end of the war, 83 958 ships were convoyed in the Atlantic and home waters of which 260 were sunk. (Interestingly only five ships in all were sunk in convoys which had both an air and surface escort.) By comparison, from November 1917 when the statistics were first gathered until the end of the war, 48 861 ships sailed independently of which 1 497 were sunk. Although much more effort was put into 'Hunting Forces' and 'Standing Patrols' than into convoy escorts, 'defensive' convoy escorts sank the same number of submarines as their 'offensive' counterparts.

Between the wars none of the three major navies, those of Britain, the USA and Japan, devoted much thought to the lessons of the submarine war. We shall discuss some of the reasons in the next section, but one contributory factor was the feeling that 'proper' naval warfare should

consist of the clash of great battle-fleets. On the British side there may
also have been an unwillingness to recognise how close they had been
to disaster and how ineffective many of the Navy's favourite nostrums
had been.

> Truth-loving Persians do not dwell upon
> The trivial skirmish fought near Marathon.
> As for the Greek theatrical tradition
> Which represents that summer's expedition
> Not as a mere reconnaissance in force
> By three brigades of foot and one of horse
> (Their left flank covered by some obsolete
> Light craft detached from the main Persian fleet)
> But as a grandiose, ill starred attempt
> To conquer Greece — they treat it with contempt;
> And only incidentally refute
> Major Greek claims, by stressing what repute
> The Persian monarch and the Persian nation
> Won by this salutary demonstration:
> Despite a strong defence and adverse weather
> All arms combined magnificently together.

Towards the end of the First World War Dönitz was captured when
the submarine he commanded was sunk in the Mediterranean. He was
a prisoner in Gibralter when news of the armistice came through. As
he and a fellow German officer watched the wild celebrations 'with
infinitely bitter hearts' a British Captain joined them.

> Dönitz waved his arm to encompass all the ships in the roads, British,
> American, French, Japanese, and asked if he could take any joy from a
> victory which could only be attained with the whole world for allies.
> 'Yes,' the Captain replied after a pause, 'it's very curious.'

2.3 The second submarine war

Naval officers in previous centuries had never troubled to find out why
convoys were so successful. They just knew they were, and used them. But
in 1917 a Commander Rollo Appleyard ... sat down in the Admiralty
to calculate how and why convoys worked. Using ships' logs, eye-witness
accounts, diagrams of U-boat attacks, diagrams of convoy formations,
columns, and escort dispositions and with an accurate knowledge of
the capabilities of merchant ships, escorts, U-boats and their torpedoes,
Appleyard made an analysis, the first in naval history, of the principles
of convoy attack and defence.

Appleyard's work was as good a piece of 'operational research' as anything produced in the Second World War and contained many clear recommendations but as far as many naval officers were concerned its tables, algebraic formulae and diagrams formed 'An account of the Trojan War written by Euclid'. A further obstacle was provided by the fact that the books which set forth his conclusions

> formed part of a Technical History series which were officially classified as Confidential Books. 'CBS' books were kept locked up in a safe except when actually in use. ... Court-Martials were held, careers were broken when CBS went missing.
>
> Thus ... by treating these documents as priceless, which indeed they were, the Navy deprived them of their intended audience, making it almost impossible for anybody at all to read them and quite impossible for them to reach the readership they deserved. Finally, in 1939 they were declared obsolete and ordered to be destroyed†.

†Under these circumstances the reader will not be surprised to learn that I have relied on the account of Appleyard's work given in Chapter 9 of [258] and have not seen the original.

Another 'lost work' was a publication entitled *Notes on Aids to Submarine Hunting* which summarised the First World War experience of the Airship Department. When the Royal Air Force was set up as a separate arm in 1918 it immediately began to battle with the two older services for control over everything that flew and access to any money that was going. For the RAF there was no history before 1918 and the naval histories of the First World War dealt with nothing above the mast head.

‡Perfect pitch is fairly common but Paget could identify the component frequences in a mixture such as a spoken vowel. He used this gift in constructing models of the human vocal apparatus which produced the consonants and vowels of human speech.

The reason why the Admiralty thought Appleyard's conclusions obsolete can be traced back to late 1915 when watchers on the Firth of Forth would have seen Sir Richard Paget, Secretary to the submarine and electrical section of the Admiralty Board of Invention, being hung over the side of a boat with his head under water whilst Sir Ernest Rutherford, Nobel prize winner and the greatest experimental physicist of the age, held his legs. Paget had an extraordinarily good ear‡ and the experiment sought to find out if the noises made by a submarine had some unique 'signature' of frequencies. The results were disappointing but Rutherford in Britain and Langevin in France developed what is now known as SONAR (SOund NAvigation and Ranging, by analogy with RAdio Detection And Ranging) but, until the Second World War, was known as ASDIC (from the Anti-Submarine Detection Committee)§. ASDIC came just too late for the First World War but was developed throughout the inter-war years and was widely believed to be the complete answer to the submarine.

§Although Rutherford put enormous energy into this work he did on one occasion decline to attend a committee meeting, writing that 'I have been engaged in experiments which suggest that the atom can be artificially disintegrated. If it is true it is of far greater importance than a war.'

To understand the links between the submarine battles of the First and Second World Wars it is important to realise that Germany was forbidden by the peace treaty to maintain a large navy and to operate

any submarines whatsoever. Although some clandestine evasion of the
treaty took place, it was not until after Hitler's rise to power in 1933 that
the construction of a new 'Great Power' navy and submarine fleet was
begun. On the victor's side, the lack of obvious enemies and the cost
of the previous war led to substantial disarmament and low military
expenditure throughout the 1920s†. This meant that the U-boats of the
Second World War were not very different from those of the first. It
also meant that Germany entered the war with far fewer submarines
and Great Britain with far fewer convoy escorts than they needed for
the battles ahead‡.

It is always possible to criticise military planners in retrospect. Either
'they failed to learn the lessons of the previous conflict' or 'they prepared
to fight the last war rather than the next'. We know that the key naval
battles for Britain and Germany would be those for the Atlantic convoys
— the Admiralties of the two nations did not even know for sure who
they were arming against. Until very late in the day, Hitler hoped
to avoid war with Britain, or at least to delay it for several years.
German naval planning centred on a world class surface fleet to be
ready in 1947. British naval planning had to deal with two major likely
opponents, Germany and Japan, on opposite sides of the world. Neither
side considered the possibility that Germany would control the Atlantic
sea coast of Europe from the North Cape to the Pyrenees. Both sides
exaggerated the effectiveness of ASDIC.

A minority within the German Navy still believed in the submarine
as a war winning weapon. Led by Dönitz, they had developed new
tactical doctrines which they believed would counter both ASDIC and
convoy. First they noted that ASDIC was a method for detecting
submerged submarines. This might well rule out the use of submarines
as 'intelligent mines' waiting outside enemy ports, but the experience
of the First World War had shown that surfaced submarines could use
their superior speed to carry out very effective night attacks on convoys.
Single U-boats would find it hard to evade convoy defences, and even
if they were successful could only sink one or two ships; but groups,
hunting as packs, could overwhelm the defence and destroy a whole
convoy. Here is Dönitz again.

> A U-boat attacking a convoy on the surface and under cover of darkness
> ... stood very good prospects of success. The greater the number of
> U-boats that could be brought simultaneously into the attack, the more
> favourable would become the opportunities offered to each individual
> attacker. In the darkness of the night sudden violent explosions and
> sinking ships cause such confusion that the escorting destroyers find

†Military historians tend to
imply that this low military
expenditure was
irresponsible, but it is
difficult to see what benefit
Britain would have derived
from an arms race with the
United States or what
difference a few more
battleships and several
extra squadrons of obsolete
planes would have made
ten years later. The rusting
hulks of the Soviet Navy
demonstrate that high
military expenditure may
decrease rather than
increase security.

‡Dönitz refers to the lack
of German submarines as
'one of the most tragic
situations in naval history',
but this is a matter of
opinion.

their liberty of action impeded and are themselves compelled by the accumulation of events to split up.

However, before a convoy could be attacked it had to be found. The obvious solution was to use long patrol lines of individual submarines. Once a convoy was spotted it could be shadowed whilst other submarines were directed towards it by wireless.

The supporters of surface ships were quick to point out a major weakness of the scheme. It was a saying in the German Navy that 'All radio traffic is high treason.' During the build-up and the attack itself, a large number of radio signals had to pass between shore and submarine and, much worse, between submarines and shore. Even outside this critical period the Dönitz system required occasional situation reports to be radioed from submarine to land. If the signals could be decoded, the enemy would be at a major advantage, but, even if they remained unbroken, they would give away the submarine's position. To this Dönitz could reply that new machine-generated coding systems meant that his submarine force could change its codes easily and regularly. Even if the enemy cracked the code for one period they would be plunged back into darkness with the change of code. During the previous war, the British had broken the German naval codes but the old-fashioned methods (in particular the failure to change the system used often enough) of the British Navy made it likely that this time the position would be reversed. As to radio signals giving away submarine positions, experiments with shore based direction finding indicated errors of the order of hundreds rather than tens of kilometres making radio direction finding tactically useless. There remained the final problem that the radio traffic involved in setting up an attack would show that the convoy was under threat and allow its defences to be reinforced. This objection did not impress Dönitz and those who thought like him, since ships could not travel fast enough to reinforce in time and, if aircraft did turn out to be a problem, attacks could be made outside the range of their bases.

The German submariners would have a further advantage in the coming battle. Dönitz combined absolute loyalty to the Nazi State with great tactical ability and a remarkable ability to inspire his crews. But again there was a possible weak point. During the war an officer was assigned to Dönitz's staff while recovering from wounds.

His position, newly created, was 'Opponent's Representative'; his task, to examine the intelligence the Allies might have, and from this deduce their next moves. Taking his work seriously, he was not 'super-respectful' ... He made three presentations to Dönitz and [his second in command]

Godt, which apparently so disturbed them that his position was immediately abolished, never to be resurrected, and he was posted back to
sea.

The first two years of war vindicated Dönitz's views almost completely. At the beginning of the war, Britain imported 55 million tons
of goods a year including all her oil, most of her raw materials and
half her food. Her merchant fleet had a carrying capacity of about 17
million tons. By March 1941, her imports were running at about 30
million tons a year and her citizens were rationed to 2 ounces of tea
and one egg per fortnight, 4 million tons of shipping had been lost and
2.5 million tons of shipping was under or awaiting repair. If Britain
had needed to replace the shipping lost from her own resources, her
situation would already have been hopeless, but she had acquired three
million tons which fled from occupied Europe and placed her future
hopes in the goodwill and untried ship building capacity of the United
States (which had produced a grand total of 28 ocean-going ships in
1939). In the meantime the U-boat fleet would continue to grow. To
Churchill, contemplating the depressing prospect,

Figure 2.1: The price of
petrol has been increased by
one penny. "Official."

[the] only thing that ever really frightened me during the war was the
U-boat peril. ... It did not take the form of flaring battles and glittering
achievements. It manifested itself through statistics, diagrams and curves
unknown to the nation, incomprehensible to the public.

How much would the U-boat warfare reduce our imports and shipping? Would it ever reach the point where our life would be destroyed?
Here was no field for gestures or sensations; only the slow, cold drawing
of lines on charts, which showed potential strangulation. Compared with
this there was no value in brave armies ready to leap upon the invader,
or in a good plan for desert warfare. The high and faithful spirit of
the people counted for nought in this bleak domain. Either the food,
supplies, and arms from the New World and from the British Empire
arrived across the oceans, or they failed.

Dönitz's task was to sink ships faster than the rest of the world could
build them until there were no longer enough ships left afloat on the
oceans to supply Britain. From 1941 until the middle of May in 1943,
he believed that he would succeed.

CHAPTER 3

Blackett

3.1 Blackett at Jutland

In 1919 the British Admiralty decided that young officers whose education had been interrupted by the war should be sent to Cambridge for a six month course of general studies. Among them was Patrick Blackett who had seen action at the age of 17 in a naval battle off the Falklands. When the two greatest battle fleets the world had ever seen clashed off Jutland, he was a 19 year old gunnery officer on HMS *Barham*, flagship of the British Fifth Battle Squadron. The battle involved 110 000 men, of whom about 9000 were killed.

He remembered passing

> the spot where the [battle-cruiser] *Queen Mary* had disappeared. The patch of oily water, where a dozen survivors of the crew of 1200 were clinging to pieces of wreckage, as I saw it through the periscope of the front turret of the *Barham*, gave me a strong awareness of the danger of assuming superiority over the enemy in military technique. ...

> In the first decade of this century, belief in the technical superiority of the British Navy was almost an article of national faith. This faith was shaken at Jutland, with the loss of three British battle-cruisers by explosions caused by enemy gun-fire. No major German ship blew up — in fact, none were sunk during the action, though one was so badly damaged that it was later sunk by the crew. What was wrong with the British battle-cruisers? The answer is simple. They had not been designed structurally to survive hits by enemy projectiles of the same type as they themselves were designed to fire. Their defensive strength had been unduly sacrificed to offensive power. Luckily, this defect was confined to the battle-cruisers ... .

> The new German Navy, without tradition or experience, had proved itself superior in gunnery and in ship construction. It was only the marked superiority in numbers of British ships that sent the German High Sea Fleet scuttling back to harbour, and so brought strategic victory.

Later he served on the destroyer *Sturgeon* which within 12 hours of its transfer to a new base incurred grave displeasure by moving at full speed across the Admiral's bows without permission and sinking a U-boat in full view of the assembled fleet.

From a strategic viewpoint Jutland had been a British success. 'The German fleet,' as an American paper put it, 'had assaulted its jailer and then returned to its cell.' None the less the Navy remained unhappy with its performance. Blackett shared the Navy's belief that the faulty design of the British warships and in particular the battle-cruisers lay behind the failure to inflict a crushing defeat on the German fleet. However, Marder, in his great work on the British Navy in World War I, concludes that the design of the ships was sound and the problem lay elsewhere.

The explosion of three British battle-cruisers was almost certainly due to flash or flame from the gun turrets passing down the ammunition hoists and reaching the magazine. In theory, the various precautions built into the turret complex should have prevented this. However, the apparent safety of the new cordite explosives and the desire to achieve high rates of fire had led to a general bypassing of safety precautions both at a ship level where, for example, magazine doors were left open, and in the design of the charges themselves. The German Navy had learned of the dangers when one of its ships nearly blew up in an earlier battle. But why had so few German ships been sunk?

For the British Navy in 1914 its big guns were everything. The gunnery branch was the surest road to promotion. British ships were more lightly armoured than their German counterparts in order that they might carry heavier guns.

> Guns, big guns, long ranges (battle practice was held at 14 000 yards [13 000 metres] in 1913 and the battle cruisers fired at 16 000 yards [14 500 metres] in the spring of 1914), good shooting — these were the articles of faith for the naval officer, and to the proper development of this method of attack all systems of tactics were directed.

As ranges increase the path of a shell changes. Air resistance has more time to operate and the horizontal component of the shell's velocity at the end of its flight becomes steadily smaller compared with the vertical component. Thus the shell hits the target's side armour at a very oblique angle. When this problem was raised, it was argued that the existing armour-piercing shell which worked when the shell struck armour plate at right angles would also work well for oblique impact. Requests for realistic trials were turned down on the grounds of expense. The more expensive lesson of Jutland showed that the shells were in fact 'wretchedly inefficient' at oblique impact. Even after Jutland there was a half-year delay in remedying the situation. The commander of the main British fleet left a letter only to be opened if he were killed in action. 'If I am killed', he told his staff captain, 'so also will you be, probably. This letter, if the Fleet ever fails to do all that is expected of it in action, will place the blame where it should lie, and not on the shoulders of the officers and men of the Fleet when I am unable to defend them.'

The failure of shell design to keep pace with changes in range was the kind of serious but understandable mistake which often accompanies rapid technological change. The system of 'proving' (i.e. testing) the quality of shells admits of no excuse. Sir Francis Pridham, who was ultimately responsible for these tests from 1941 to 1945, explained how they worked.

> The system of Proof was more or less dictated by the shell makers and was such that in the opinion of the Ordnance Board, the only assumption that could make sense of the procedure was that all shell were good shell and that failures were few and due to the machinations of malignant spirits.
>
> The shell were manufactured in 'Lots', of four hundred, each Lot being subdivided into Sub-lots of one hundred. When a Lot was brought forward by the shell maker for Proof, two shells were picked out at random from Sub-lot No. 1, to be tested at an armour plate of specified thickness, at a specified striking velocity and at a specified angle of impact. If the first shell fired succeeded in penetrating the plate whole, the full Lot of [399] shell passed into service. If this first shell failed the second [from the same sub-lot] was fired. If this was successful the full Lot of [now] 398 were accepted. If the second shell also failed, the Sub-lot was sentenced 'Reproof' and the shell maker was given the option of withdrawing the whole lot from Proof, or of allowing the remaining 300 [three sub-lots] to proceed to Proof. Needless to say they generally chose the latter. Proof then commenced with the next Sub-lot of 100, acceptance being governed by the same procedure as that for Sub-lot No. 1.

Exercise 3.1.1 *(i) Show that if 50% of the shells are duds there is a probability of 3/4 that the full lot of 399 or 398 shells will be accepted.*
(ii) Show that if 84% of the shells are duds the probability of at least 298 shells being accepted is greater than 1/2.

He continues:

> The Ordnance Board's Professor of Statistics, having been given the results of Proof firings of the heavy shell then in the Fleet, calculated that from 30% to 70% were probably dud shell, but the data from these Proof firings was insufficient to enable him to give a nearer approximation.

Marder states that the system just described continued in operation until 1944.

Exercise 3.1.2 *The problem of testing shells is a non-trivial one. To begin with, we have to decide what quality of shell is technically possible. It is not an easy matter to build a shell that can be fired 10 kilometres, penetrate a certain thickness of armour plating and then explode. We must expect the process to fail from time to time. Suppose that shells are produced in batches of 100 and we believe that the proportion of duds can be reduced to 10% (all the figures in the exercise are fictitious) but we are prepared to accept anything less than a 20% failure rate.*

(i) Suppose we test n out of the batch and if any fail we reject the batch. Show that in order to reduce the probability of accepting a batch with a 20% failure rate to about 1/10 we must take n = 10.

(ii) Show that if we take n = 10 we will reject over 65% of the batches in which the proportion of duds is 10%.

Thus, taking n = 10, even if all our batches are of the best quality, we shall use 10% of each batch in testing and reject 65% of all batches. Show that 31.5% of our output of shells will actually reach the battleships.

We cannot get round this problem by pure thought. Our best hope is to study the production of the shells themselves. If, as seems likely, the quality of shells does not vary randomly between batches but that, for example, certain factories consistently produce good shells, then we can modify our procedures to take account of this. Batches from factories which have done well in past tests might have fewer test firings but results from several batches would be aggregated to provide a check that quality is being maintained.

What moral can we draw from all this? The simplest moral for the young to draw is that their grandfathers were fools. But, as Marder remarks,

> [t]hose who used the many instances of naval unpreparedness and the horrendous catalogue of mistakes and shortcomings with which to belabour the politicians and the Naval High Command during and after the war were faced at the outset with the embarrassing fact that Britain had actually won the war and the victory of the Navy had been extraordinarily complete.

He goes on to praise the officers of that Navy for 'their technical competence and their energy, courage, determination, perseverance and unwavering confidence'. At one end of the Naval School at which Blackett studied was emblazoned the proud boast: 'There is nothing the Navy cannot do.'

How could such a great navy have inferior shells? The immediate
cause was the gap between the scientists and engineers who produced
the weapons and the navy which was to use them. The Navy understood
its need for the most modern equipment, the biggest guns and the fastest
ships†. But the use of the new equipment, guns and ships was the sole
province of the professional naval officer. A 1917 minute conveys the
nature of the problem.

†'Man invents, monkeys
imitate' thundered Fisher,
the great moderniser of the
British Navy.

> In the course of the discussion which ensued Sir E. Rutherford, Mr
> Threlfall, Professor Bragg and the Secretary referred to the feeling which
> was general among the scientific members of the committee that their
> utility to the navy in connection with the anti-submarine question was
> largely impaired by lack of information as to the actual service conditions
> and as to the nature of the methods already employed for the detection
> and destruction of enemy submarine depots. [The scientists have] been
> informed for example that it was not possible to put them in touch
> with the anti-submarine depots. It was pointed out that in scientific
> research it was found to be essential that the researcher should have
> the widest knowledge and personal experience of the difficulties to be
> solved. Commodore Hall dissented and expressed the view that the only
> information necessary to be given was that the enemy submarines were
> in the sea, and that means were required to detect their presence.

Organisations which depend on the instant implementation of decisions, like hospitals and ships, require well defined hierarchies and
unquestioning obedience to certain orders. They are liable to carry
both hierarchy and obedience to unnecessary and sometimes dangerous
extremes. Marder gives the example of a gunnery officer who in 1911

> reported to his Captain that a changing gunnery technique rendered
> certain gunnery practices obsolete and wasteful of time and ammunition.
> He therefore suggested the introduction of more realistic firings, including
> divisional practices in unfavourable conditions of wind and weather such
> as would probably occur in battle. [The Commander of the Home
> Fleet] expressed general agreement with these proposals but personally
> informed the officer that the First Sea Lord ... favoured the retention of
> these obsolete firings and if the paper were forwarded its implied criticism
> of Admiralty policy might prejudice his promotion to commander. He
> advised him to withdraw the paper.

Later Blackett was to play a leading role in the Battle of the Atlantic,
a struggle whose scale, whether measured in length of days, size of
battlefield, in total forces involved, in loss of life or in sheer destruction,
dwarfed Jutland. But even then, Blackett was not merely concerned
with the immediate problem of winning the war. His naval experiences

†'He could not understand
why it was necessary to
walk several yards from
the men before shouting
commands.' Blackett
needed no artificial aids to
make his personality felt.

had shifted him strongly to the left†. For him the problems of excessive hierarchy, mismanagement of technical resources and resistance to change which he found in the Navy reflected the problems of society at large. To find and solve the shell problem would have required the Navy to have thought in a certain way, and to think in that way would have called too much into question. To find a solution to Britain's slow technological decline and to the gross inequalities in its society would also require new ways of thought and the calling of many things into question.

Blackett's first night in Cambridge was a memorable one for him. 'He had never before, he liked to recall, heard intellectual conversation — Osborne and Dartmouth (his naval school and college) and destroyers at sea had not provided that. The first night's talk persuaded Blackett (and this was characteristic of him) that other people had had much worse wars than he had.' (The quotation, and the biographical details of this chapter and the succeeding ones, are taken from the obituary notice by Sir Bernard Lovell.) He visited the Cavendish Laboratory to see what a scientific laboratory was like and within three weeks resigned from the Navy to become an undergraduate studying first mathematics and then physics. In 1921 he became a research student of Rutherford.

> Even without the far-sighted action of the Admiralty, I would surely have found a place in some university or some technologically orientated firm. But to be deposited, so to speak, on the shores of Cambridge just as the Cavendish Laboratory was rising under Rutherford's inspired direction to great heights of eminence was luck indeed for me. So I owe much to the Admiralty.

In his 1948 Nobel Prize lecture Blackett explained how Rutherford assigned him a research problem.

> In 1919 Sir Ernest Rutherford made one of his (very numerous) epoch-making discoveries. He found that the nuclei of certain light elements, of which nitrogen was a conspicuous example, could be disintegrated by the impact of fast alpha particles from radioactive sources, and in the process very fast protons were emitted. What actually happened during the collision between the alpha particle and the nitrogen nucleus could not, however, be determined by the scintillation method then in use. What was more natural than for Rutherford to look to the Wilson cloud method to reveal the finer details of this newly discovered process?

The first student given this task built a cloud chamber but had to return home to Japan unexpectedly with the work hardly started. Blackett inherited the apparatus and the project. He estimated on the basis of scintillation results that a million alpha particles fired into nitrogen should yield about 20 of the desired events but

> ... provided by Rutherford with so fine a problem, by C. T. R. Wilson with so powerful a method, and by nature with a liking for mechanical gadgets, I fell with a will to the task of photographing some half million alpha-ray tracks.

Blackett automated the cloud chamber so as to get 270 photographs an hour and took about 23 000 photographs. Each photograph contained on average about 18 alpha-ray tracks. Among the 23 000 × 18 tracks examined there were 8 events of the desired type.

Ten years later, working with Occhialini, he adapted the cloud chamber to work with cosmic rays. Since performing a cloud chamber expansion on the off-chance that a cosmic ray is passing through the apparatus is very inefficient, they arranged detectors above and below the chamber. The experiment was triggered only when the detectors fired simultaneously, thus making the cosmic rays 'take their own photograph'. In this way they observed the creation of electron–positron pairs. Further work showed that the gamma rays (light) could give rise to electron–positron pairs and that, symmetrically, the collision of an electron and a positron could give rise to gamma rays — a vivid illustration of Einstein's theory that 'material' particles and 'immaterial' light radiation are just different forms of energy†. However, the growing European crisis now called Blackett away from pure physics, first to organise help for academic refugees and then to prepare for a war that he believed was inevitable.

3.2 Tizard and radar

In 1934 it was becoming clear that Nazi Germany must be considered the 'ultimate potential enemy' and that Britain was essentially defenceless against the growing German Air Force. No point in Britain was more than 70 miles from the sea, there was no way of intercepting bombers by night, and the only way of intercepting bombers by day was by keeping fighters in the air continuously — a system that was bound to fail sooner rather than later and, when it failed, lead to the destruction of the entire fighter force on the ground. An internal memorandum of the Air Ministry put the matter bluntly.

> It was clear that the Air Staff had given conscientious thought and effort to the design of fighter aircraft, to methods of using them without early warning, and to balloon defences. It was also clear, however, that little or no effort had been made to call on science to find a way out. ... unless

†The prediction of the positron by Dirac before its discovery by Anderson and its confirmation by Blackett and Occhialini was a triumph of the theorists. Rutherford grumbled 'It seems to me that in some way it is regrettable that we had a theory of the positive electron before the beginning of the experiments. Blackett did everything possible not to be influenced by the theory, but the way of anticipating results must inevitably be influenced to some extent by the theory. I would have liked it better if the theory had arrived after the experimental facts had been established.'

science evolved some new method of aiding air defence, we were likely to lose the next war.

In true British fashion, it was decided to set up a small committee to consider the matter. The defence of Thermopylae was to be organised by a committee of four. All were scientists: Wimperis, who represented the Air Ministry, the chairman Tizard who had worked with aircraft during the First World War at a time when scientists were their own test pilots and had since worked in both academia and the scientific civil service, Hill, a Nobel prizewinner for physiology who, like Tizard, had First World War experience of defence research, and 'Professor Blackett F.R.S. who was a Naval Officer before and during the War and has since proved himself by his work at Cambridge as one of the best of the younger scientific leaders of the day.' Members were not paid and the committee was to be purely advisory. Tizard's biographer writes

> Throughout the whole of its existence ... it lacked executive power, so that when Tizard wished new devices to be tested, planes flown, staff to be utilised, or Air Force machinery set into motion these matters had to be organised by persuasion, on the 'old boy net', through the goodwill of individual Commanders — and sometimes in the face of determined opposition from those resolutely preparing for the last war. The Committee had no offices and much of the confidential interviewing that the work demanded was carried out by Tizard privately at his flat .... The 'secretariat' for some while consisted solely of Rowe [Wimperis's assistant] who for much of the Committee's life was also the secretary of numerous other bodies. In the early days even a typist was rarely available, although on occasion kindly Commanders would offer to get memoranda and reports handled.

At its first meeting the committee heard a proposal from Watson-Watt of the National Physical Laboratory.

> One idea which appeared to have a special fascination for the Air Ministry was the concept of a death ray which would either claw an aircraft out of the sky or burn up the occupants at the turn of the switch. ... [The proponents of such proposals] usually referred to black boxes which generated weird mixtures of ultraviolet rays, X-rays and radio waves. The schemes had two things in common: one was the claim that sheep or pigeons had dropped dead on being subjected to the mysterious rays; the other was that vast sums of money would be required before the inventors could proceed with their death-dealing devices.

Watson-Watt had been consulted as to the practicality of such a device and asked his assistant Wilkins to calculate the power required. Wilkins worked out the radio power required to raise the temperature of a man (represented for the purposes of calculation by 75 kilograms of water in a body of cross-sectional area of 1 square metre), standing 600 metres away from the transmitter, by 2 degrees centigrade in 10 minutes. The answer of many thousands of kilowatts ruled out any sort of death ray, but Wilkins realised that it might be possible to locate aircraft by detecting the radio waves that they reflected. Even this would be hard since the reflected signal would have only 10^{-19} of the power of the transmitted signal. However, Watson-Watt was convinced that it could be done and put the case to the Tizard committee.

Faced with a choice between a proposal which might work and other possibilities which almost certainly would not, the committee threw its weight behind Watson-Watt. Within weeks the Air Ministry had provided £10 000 (the cost of three fighters) for experiments and by the end of the year the Treasury had secretly agreed to provide £10 000 000 to build a chain of radar stations† to specifications which existed only on paper. (The first stations were handed over in mid-1937 and, for various reasons, one of which we shall discuss later, initially gave such disappointing results that the abandonment of the project was considered. Fortunately the problems were rapidly resolved.)

The committee was not helped by the addition of a loose cannon to their number. Blackett recalled the episode as follows.

> Although the Air Defence Committee started up in January 1935 with only Tizard, Hill, Wimperis and myself as members, in July of the same year the Secretary of State for Air, under pressure from Mr Winston Churchill, enlarged it by the addition of Professor F. A. Lindemann (afterwards Lord Cherwell). It was not long before the meetings became long and controversial; the main points of dispute concerned the priorities for research and development which should be assigned to the various projects which were being fathered by the Committee. For example, Lindemann wanted higher priority for the detection of aircraft by infra-red radiation and for the dropping of parachute-carrying bombs in front of enemy night bombers and lower priority for radar, than the other members thought proper. On one occasion Lindemann became so fierce with Tizard that the secretaries had to be sent out of the committee room so as to keep the struggle as private as possible. In August 1936, soon after this meeting, A. V. Hill and I decided that the Committee could not function satisfactorily under such conditions; so we resigned. A few weeks later Lord Swinton re-appointed the original committee without Lindemann, but with Professor E. V. Appleton, and a few months later with Professor T. R. Merton as additional members‡.

The material successes of the Air Defence Committee are startling enough. In January of 1935 Wilkins and Watson-Watt had suggested

†The document describing how the radars should be sited was sent to the Air Ministry for approval. It came back untouched except for one condition – that the choice of site '... should not gravely interfere with the grouse shooting.'

‡I have chosen the most dramatic version of events. Lindemann's defenders agree that he behaved extremely badly on the committee but claim that he did so at other times and over other issues.

on the basis of a back of the envelope calculation that radio waves could be used to detect aircraft. By September of 1939 the whole of the east coast of Great Britain was under continuous radar watch. Early warning radars which could detect aircraft at a range of 100 miles were backed by shorter-range sets capable of tracking individual aircraft. The whole system had been built in absolute secrecy. But Tizard and his colleagues had built more than a radar chain, they had constructed a new system of air defence.

Up to now fighters had operated on their own initiative, finding or failing to find their targets by eye. The new scheme called for them to be directed by radio onto targets located by radar. In 1936 Tizard convinced the authorities that trials should be carried out to see how the, as yet non-existent, radar system should be used. Participants were told that by some unexplained means they would get a fifteen minute warning of the approach of enemy aircraft and from then on would receive approximate details of height, speed and course. (The 'attacking force' transmitted constant radio signals which were tracked by a direction-finding station which telephoned the results through.)

Exercise 3.2.1 *(i) Consider a simplified two dimensional problem in which the bomber is travelling with constant speed u along the x axis in such a way that its position is given by $(ut, 0)$ at time t. At time $t = 0$ a fighter is at (x_0, y_0). The fighter has constant speed v and travels in such a way that its position is $(x_0 + vt \cos \theta, y_0 + vt \sin \theta)$. Under which conditions on u, v, x_0 and y_0 can the fighter intercept the bomber and how should θ be chosen to achieve the interception?*

(ii) At time 0 a bomber is at (x_1, y_1) travelling with speed u in a direction at angle ϕ to the x axis and a fighter of speed v is at (x_0, y_0). Can the fighter intercept the bomber and how should it fly to achieve the interception?

(iii) Restate part (ii) in a three dimensional context and get as far as you can with it. (If you use vectorial methods remember that, in the end, the result must be translated into usable form.)

†The Americans call such exercises 'Strapped Turkey Tests', the idea being that a turkey is strapped to a table and a blunderbuss fired at it from a distance of one metre. This shows 'that the concept is feasible'.

Within a few weeks radio directed fighters were intercepting successfully almost every time. However, this record was achieved against bombers flying in a straight line† and the moment the targets were allowed to change course and altitude the success rate was greatly reduced. Evasive flying meant that the complicated and time consuming calculations involved in producing interception courses had to be repeated over and over again and that the information on which these calculations were based was probably out of date by the time it was used. Tizard produced a simple solution to this problem based on the fact that fighters travel faster than bombers. If the bomber is at A travelling along the line AD and the fighter is at B, then the controller should direct the fighter along the line BC intersecting AD at a point C such that AB is the base of the isosceles triangle ACB (i.e. the two angles $\angle BAC$ and $\angle CBA$ are equal, see Figure 3.1).

Exercise 3.2.2 *(i) Explain why, if neither bomber nor fighter now alter course, the fighter will arrive at the 'rendezvous' C before the bomber.*

(ii) By drawing accurate diagrams, investigate what will happen if the bomber continues on a straight line course but the controller alters the fighter's course at regular intervals in accordance with Tizard's rule.

(iii) Explain why, even if the bomber alters course from time to time, a controller altering the fighter's course at regular intervals in accordance with Tizard's rule will still achieve interception. (You will need to assume that the controller alters the fighter's course at substantially shorter intervals than the bomber changes course. However, a bomber taking continuous violent evasive action against an unseen opponent will not make much progress towards the target.)

Later in the trials a harassed controller discovered that the 'Tizzy Angle' could be judged by eye and a system of interception was established which was to last until the 1960s together with its jargon of 'scramble', 'angels' and 'vectors' (course to steer).

In mid-1937 three radar stations had been brought into operation only to reveal a new problem.

An aircraft over the North Sea was commonly observed by two of the stations and often by all three of them. If each station told the same story of an aircraft position all was well but … this rarely happened; small

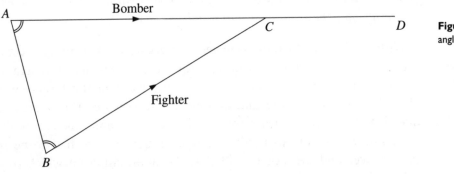

Figure 3.1: The Tizard angle.

errors of calibration, minor aerial troubles and errors made by observers combined to give indications which often deserved the RAF epithet of 'a dog's dinner'.

In fact this problem occurs whenever we make repeated observations. Because of inevitable errors these observations are *bound to be contradictory* and we have to merge these contradictory observations into a consistent picture. A 'filter room' was set up to reconcile data from the radar stations and pass on its best estimate to the controllers.

Obviously the idea of radar could be applied in other ways. Tizard later recalled

> ... going home one day after a visit to the experimental station ... and thinking that however important RDF, as it was then called, was to the Air Force, it must be just as important, if not more so, to the Navy. Full of this, I sought an early interview with a naval expert. I told him of the work, and drew upon my imagination a little. I was even bold enough to say that in a few years RDF would make accurate blind firing possible from ship to ship. He listened to me patiently then he said: 'May I ask if you have ever seen a warship?' I said 'Yes, many times.' 'Well' he said, 'if you had observed them closely enough you would know that there was no room on them for any more aerials.' 'Well' I answered, 'my advice is that you take some of them off.'

However, it was the British reluctance to fight a night battle which had allowed the German fleet to escape at Jutland and, slowly at first, and then with increasing enthusiasm the Admiralty added their support to a weapon which would allow night engagements. (Their foresight was rewarded in early 1941 when primitive radar allowed the ambush and destruction of three Italian cruisers off Cape Matapan.)

The first chain of radar stations were distinguished by their 100 metre steel masts. Ship mounted radars would have to be a little more compact. But as Bowen relates,

> Tizard was ... the one who in 1936 first suggested that an attempt be made to make a radar set small enough to go into an aircraft. As usual, his reasoning was crystal clear, and a beautiful example of how he always anticipated future events. He argued as follows. The success of early warning radar meant that the German daylight attack would be beaten back .... They would then turn to night bombing. Under these conditions, ground control could put the fighters within three or four miles of the bomber, but the fighters could not see them until they were within 500 or 1000 feet [150 or 300 metres].

How could a radar system be made small enough to be put in an aircraft?

The key lies in the wavelength used. Just as a violin string can be persuaded to produce notes whose frequencies are integral multiples of its fundamental frequency (harmonics) but cannot produce notes of a lower frequency, so a radio transmitter cannot produce high power radio-waves of a wavelength much greater than its size. In the same way an effective receiving system for radio waves must have size comparable with or greater than the wavelength of the radio waves involved. More generally the 'typical length' associated with a radar system is the wavelength of the radio-wave used.

However, it becomes harder and harder to produce sufficiently powerful transmitters as the wavelength is reduced. Admiralty and industrial scientists managed to produce a system which worked on 1.5 metres rather than the 13 metres wavelength of the first chain. This ten-fold miniaturisation meant that radar could be mounted in ships and (with great difficulty†) in aircraft. Further, the transmission systems were now small enough to be rotated and used as searchlights rather than having to illuminate the whole sky. The transmitting masts were replaced by the parabolic reflectors of modern radar.

†When the Swordfish torpedo-bombers were equipped with radar they flew in pairs, one to carry the radar and the other the torpedo.

3.3 The shortest wavelength will win the war

There is a further reason for wishing to reduce the wavelength. Consider the functions shown in Figure 3.2. The first function $f_1(x) = \sin 2\pi x/\lambda$ represents an endless train of waves of wavelength λ. The second function f_2 is a 'pulse' of length many times λ and it is not hard to assign it a 'typical' wavelength λ. However, as we reduce the pulse length (as for the functions f_3 and f_4) it becomes harder and harder to see what might be meant by the pulse having 'typical' wavelength λ. We conclude that

(1) if we can say that a pulse has 'typical' wavelength λ it must have length substantially larger than λ.

Once again we see the 'typical length' associated with a radar system is the wavelength of the radio-wave used. In particular, because we cannot distinguish very well between parts of a single pulse,

(2) measurements of distance made using a radar system which operates on wavelength λ are liable to errors of about λ or greater.

Our inability to measure distances beyond a certain degree of accuracy limits our ability to measure bearings (angles).

Exercise 3.3.1 *Suppose that we observe target C from two points A and B a distance a apart and we wish to compute the bearing $\theta = \angle ABC$ from the known ranges $r_1 = BC$ and $r_2 = AC$. We suppose that r_1 and*

r_2 are very much bigger than a (*for example, that a is about 1 metre and r_1 and r_2 are larger than a kilometre*).

(*i*) *Suppose first that $r_1 > r_2$. Take X on the line CB so that $CX = r_2$. Figure 3.3 which is not drawn to scale may help. Explain why $\angle AXC = \angle CAX$. Explain also why $\angle BCA$ is very small and deduce that $\angle AXC$ is very close to a right angle. Conclude that $\angle AXB$ is very close to a right angle and use trigonometry to show that, to a very close approximation,*

$$\cos\theta = \frac{r_1 - r_2}{a}.$$

(*ii*) *Show that the same conclusion holds if $r_2 \geq r_1$.*

(*iii*) *Deduce that if we make an error ϵ in measuring $r_1 - r_2$ then the resulting error in calculating $\cos\theta$ is ϵ/a. Thus the error in calculating $\cos\theta$ is inversely proportional to a, the length of the baseline AB. In*

Figure 3.2: What is the wavelength?

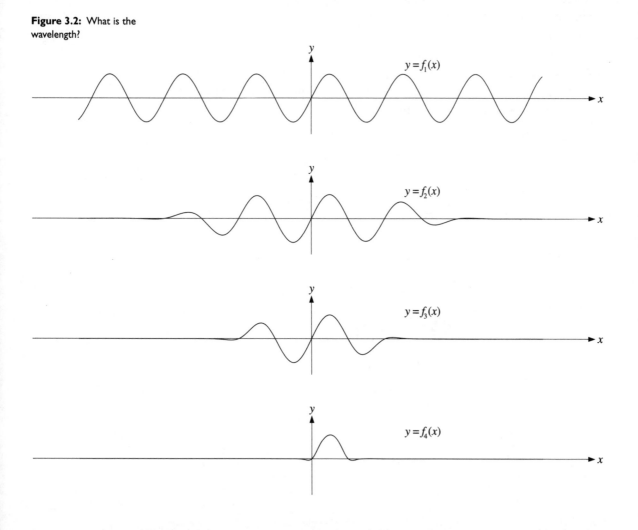

other words, the angular discrimination possible for a given size of error
is directly proportional to the length of the baseline.

(iv) (This needs a little calculus.) Suppose $3\pi/4 \geq \theta \geq \pi/4$. Show
that the error in calculating θ is also inversely proportional to a.

For radar, the 'length of baseline' corresponds to 'size of receiver' and, if
we keep the wavelength constant, the angular discrimination achievable
is proportional to the size of the receiver. Thus

$$\text{angular discrimination} \propto \frac{\text{size of receiver}}{\text{wavelength}}.$$

This relation turns up in the design of every optical and radio device
from radio telescopes to electron microscopes. As 'Heisenberg's inequality' it takes centre stage in quantum mechanics and as 'Nyquist's
bound' it governs how fast information can be transmitted by a given
communication system. In the context of radar, it means that if we
reduce the wavelength by a factor of ten while keeping the size of our
apparatus constant, we can improve our angular discrimination by a
factor of ten.

The first radar mounted in British ships worked on a wavelength
of 7.5 metres and gave ranges to within an accuracy of 50 metres at
13 000 metres, but bearing accuracy only to within 10° (though some
clever technology introduced a little later could have reduced the error

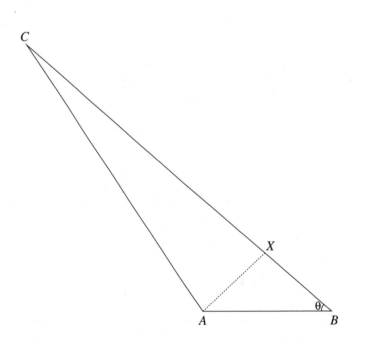

Figure 3.3: Bearing from range.

to within $1°$). This meant that radar ranging was already more accurate than was possible with optical range finders.

Exercise 3.3.2 *Suppose that we observe target C from two points A and B a distance a apart and we wish to compute the range $r = AC$ from the known bearings $\theta_1 = \angle ABC$ and $\theta_2 = \angle CAZ$ where Z lies on the line AB produced. We suppose that r is very much bigger than a (for example, that a is about 20 metres and r is larger than 5 kilometres) and that $3\pi/4 \geq \theta_1 \geq \pi/4$.*

(i) Suppose first that $\pi/2 \geq \theta_1 \geq \pi/4$. Take X on the line CB so that $CX = r$. Figure 3.4 which is not drawn to scale may help. Explain why $\angle BCX = \theta_1 - \theta_2$ and show that

$$AX = 2r \sin \frac{\theta_1 - \theta_2}{2}.$$

Explain why $\angle CXA$ is very close to a right angle. Explain also why $\angle BCA$ is very small and deduce that $\angle BAX$ is very close to $\pi/2 - \theta_1$. Conclude that $\angle AXB$ is very close to a right angle and use trigonometry to show that, to a very close approximation, $a = AX \cos \theta_1$. Deduce that, to a very close approximation,

$$a = 2r \sin \frac{\theta_1 - \theta_2}{2} \sin \theta_1$$

and give a formula for r in terms of a, θ_1 and θ_2. It is well known that, when ψ is small, $\sin \psi = \psi$ to a very close approximation. Use this fact

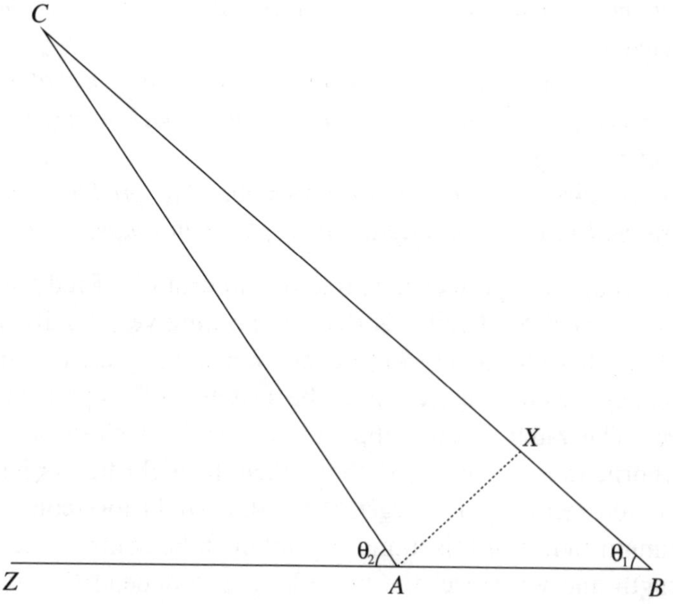

Figure 3.4: Range from bearing.

to obtain

$$r = \frac{a \sin \theta_1}{\theta_1 - \theta_2}$$

to a very close approximation.

(ii) *Show that the same conclusion holds if* $3\pi/4 \geq \theta_1 \geq \pi/2$.

(iii) *(This needs a little calculus.) Let us write* $\phi = \theta_1 - \theta_2$ *so that*

$$r = \frac{a \sin \theta_1}{\phi}.$$

Show that if we make an error of $\delta\phi$ *in measuring* ϕ *but measure a and* θ_1 *exactly, then the corresponding error* δr *in measuring r is given approximately by*

$$\delta r = -\frac{a \sin \theta_1}{\phi^2}\delta\phi.$$

Are errors in measurement of θ_1 *important in comparison with errors in measuring* ϕ?

Once again the error is inversely proportional to a, the length of the baseline, so we want a baseline as long as possible. Optical range finders on battleships were in effect two telescopes mounted as far apart as possible. The distance to the nearest stars is measured by observing them at six month intervals and so using a diameter of the earth's orbit as a baseline.

(iv) *A major problem with this method of range finding (which was, of course, the only one available before radar) is that, as the formula for* δr *given in (iii) shows, the error is inversely proportional to the square of* ϕ *and* ϕ *itself is going to be very small. Show, using the first displayed formula in (iii), that (if* θ *is fixed)* ϕ *is inversely proportional to r. Deduce that the error* δr *in measuring the range r increases as the square of the range r.*

Explain why, in principle, we would expect errors in radar ranging (where we use the time for the signal to bounce back to us) to be independent of the range.

In modern surveying we use laser techniques to produce light pulses which are used in the same way as radio pulses in radar.

There was another pointer towards the importance of reducing wavelength. The principle of radar is that of shouting very loudly and then listening very hard for a very weak echo. Whilst the pulse is transmitted the extremely sensitive receiver must be switched off to prevent it being damaged. The radar is thus blind to echoes from close objects. For early airborne radars, this meant that, when the night-fighter had closed to within 400 metres of its target, the radar would lose contact. Since the minimum radius of this 'blind region' must be comparable with the pulse length and we expect the pulse length to depend linearly on the wavelength, this gives another strong reason for wishing to reduce the wavelength. (Later British and German experience showed that night-fighters could operate very effectively with a blind region of radius 200 metres and this reason lost most of its force.)

Trials before the war had shown that airborne radar could detect ships at sea but a surfaced submarine remained a very small target to be distinguished from the 'clutter' of waves surrounding it so that the shorter the wavelength available the easier this would be. Thus the requirements of the night-fighter, the convoy escort and the anti-submarine aircraft were identical. In British radar circles it was a commonplace that 'The side which has power on the shortest wavelength will win the war.' Encouraged by Tizard, work began at Birmingham University which led in 1940 to the development by Boot and Randall of the cavity magnetron, which was to be the heart of a ten centimetre wavelength radar. 'Including the two halfpennies used to seal the end plates, the prototype cavity magnetron cost an estimated £200: had it cost £2 million it would still have been a bargain.' (When, in late 1940, Tizard persuaded the British Government to share (almost) all its scientific secrets with the still neutral United States, there was much doubt on the American side whether the secrets would be worth the trouble of revealing. However, when the contents of Tizard's black box, which contained, among other things, a cavity magnetron, were displayed, it was seen to be what Roosevelt's scientific adviser, Vannevar Bush, called 'the most valuable cargo ever brought to our shores'.)

All this was well in the future when, at the end of 1939, the Germans scuttled their pocket battleship *Admiral Graf Spee* outside Montevideo harbour rather than send it out to meet what they believed to be a superior British force. The British Naval Attaché forwarded reports of a 'range-finder on top of the superstructure, which was continuously in use at sea and kept revolving all the time by a motor'. Careful study of pre-war photographs showed that, sometime in 1938, a new cylindrical structure had been built on top of the existing range-finder house. Unfortunately the wreck was guarded by the Uruguayan Navy and closer inspection seemed impossible. At this point Señor Vega-Helgura, the representative of a Montevideo engineering firm, offered a substantial sum for the wreck as salvage, and his friend the German ambassador closed the deal. As it happened another friend of Señor Vega-Helgura was in Montevideo on a visit. Since he represented Messrs Thos. Ward Ltd, ship-breakers of Sheffield, what could be more natural than that he should inspect the wreck and send home 'samples' of the scrap on offer? His reports and the 'samples' carried a most unwelcome message. Not only did the Germans have radar, but their radar was in several ways more advanced than the British. (For more details, together with a note from the Treasury objecting to the expenditure of £14 000 on the purchase of a burnt-out battleship, consult *Radar At Sea* by D. Howse.)

The idea of radio echo location had occurred to many people at many times but only at the end of the 1920s was the technology in place to make it practical. During the 1930s Germany, the United States, the Netherlands, France, the Soviet Union, Italy and Japan all took the first steps towards radar, with Britain a relative late-comer to the race†. Under the impulsion of the Tizard committee the British had caught up with Germany by the outbreak of war and both countries were substantially ahead of other competitors.

It is generally agreed that the quality of German radar was superior to that of British radar. German radar had been built by engineers, British radar had been lashed together by physicists. In part this represented a deliberate choice of a forced march towards new technology (Watson-Watt's motto of 'Second best tomorrow' rather than 'Best, but next week'), but in part it reflected Britain's industrial weakness and in particular the lack of sufficient suitable engineers‡. The strength of the British system lay not in the radar itself, but in its integration into the air defence system. (Thus the British refused to believe that the Germans had radar because German fighters took so long to scramble, whilst the desultory German search for a possible British radar was restricted to a system using the single figure metre wavelengths which a radar 'ought to use' rather than the primitive 25 metre wavelengths that the first chain actually used.)

The methods Tizard had used to discover how radar could be used came to be called 'operational research'. Those who used the phrase found it hard to define what exactly it meant and to explain what was new about it. Certainly it involved the application of science not merely to the invention of weapons, but to the choice of tactics in their use. However, it also required the kind of collaboration between scientists and military typified by the radar 'Sunday Soviets' in which senior scientists, Staff Officers, junior research workers and serving officers 'straight from the heat of battle' met informally and where anyone could say anything to anyone. Whatever 'operational research' meant precisely, it was something that could be copied, and the idea of operational research spread through the British and then the American armed services. Nothing comparable occurred in Germany. Possibly a regime which celebrated the triumph of the will rather than the intellect was unfavourable to such developments but, in any case, Hitler's Third Reich produced no figures comparable to Tizard or Blackett.

†Some embarrassment was caused by the discovery of a secret patent submitted by Butement and Pollard in 1931 giving the details of a workable radar system. They had been allowed to do experiments at Woolwich — provided they were not in working hours. When they offered to give a demonstration they were told that: 'There was no War Office requirement.'

‡R. V. Jones gives an account of the attempt during the 1930s by the self-made millionaire Lord Nuffield to found an Oxford College devoted to engineering. 'According to what I heard at the time this prospect alarmed the strong humanist element in Oxford, headed by the Vice-Chancellor, Lord Lindsay, who sought to palliate the engineering onslaught by persuading Nuffield to broaden his objective. There would be less opposition to the foundation of a new college, he said, if Nuffield could disguise his intentions by replacing specific mention of engineering by some more subtle wording. Engineering was a science, but it made a more direct impact on society, and so it might be fairly described as "social science". Therefore, if Lord Nuffield would specify social science as the primary interest, there would be much less opposition to its creation. It was only after the college had been founded and staffed not with engineers but social scientists that Lord Nuffield realised that he had been outwitted.'

In January 1943, the Germans shot down a British bomber carrying a new radar. Examination showed it to operate at an incredible ten centimetre wavelength. Göring commented bleakly, 'I expected them to be advanced, but frankly I never thought they would get so far ahead. I did hope we could at least be in the same race.' The German military had not asked for such a radar and, since the proper role for German science was to supply what the military asked for, it had not produced such a radar†.

†Apparently a start was made in 1942 on a research programme aimed at producing a one centimetre German radar. However, it was argued that at such short wavelengths radio waves would not scatter in all directions from a target but be reflected (like visible light on a polished surface) away from the receiver ('mirrored away'). There was a violent row which resulted in all centimetric research being expressly forbidden. Since the memories of boffins are fallible in the same way and for the same reasons as those of generals the exact details of this episode will probably never be known.

†R. V. Jones gives a rather chilling account of an occasion when Tizard asked him to bear a conciliatory message to Lindemann whose 'only response was to give a mild snort and say, "Now that I am in a position of power, a lot of my old friends have come sniffing around."''

3.4 Blackett's circus

When Churchill became prime minister he brought Lindemann with him as scientific adviser. Whilst it was important that Churchill should have advice from a competent scientist whom he knew and trusted, it was unfortunate that Lindemann never forgave those whom he saw as enemies†. Whilst the Battle of Britain was being fought and won using the defence system it had created, the Tizard committee was quietly dissolved.

For a few months Blackett became scientific adviser to the general in charge of anti-aircraft guns.

> My immediate assignment was to assist the Service staff to make the best use of the gun-laying radar sets ... which were then being delivered to the AA batteries round London. ... The small group of young scientists whom I hurriedly collected to work with me and which included physiologists, an astronomer and a mathematician as well as physicists soon found themselves studying, both at HQ and on the gun sites, a variety of problems connected with the operational use of radar sets, guns and predictors.

The 'hurriedly collected' group and its successors became known as Blackett's circus.

> Immense scientific and technical brilliance had gone into the rapid design and manufacture of the [gun-laying radar] sets; likewise at a more leisurely pace into the construction of the guns and predictors. Understandably, but unfortunately, partly through shortage of scientific and technical personnel but also partly through a certain lack of imaginative insight into operational realities, hardly any detailed attention had been paid to how actually to use the [radar] data to direct the guns until the Battle of Britain was in progress. Thus the first months of the AA battle against the night bomber were fought with highly developed radar sets and guns, but with the crudest and most improvised links between them.

When firing at aircraft we must aim at where the aircraft will be when the shell explodes, not where it is now — that is we must predict its future course from its past†. An anti-aircraft gun could hurl a shell some 8000 metres but the shell would take 25 seconds to cover the distance during which an aircraft could move three kilometres. The predictors in use had been constructed to use accurate direct visual observation and not the crude data available from the early metre wavelength radar. The problem may be illustrated by the following simple model. Suppose we are interested in a function $f(t)$ but we can only observe $f(t) + e(t)$ where $e(t)$ is the 'error in observation'. Suppose further we may only observe $f(t) + e(t)$ for $0 \le t \le 1$ and we wish to predict $f(2)$. Let us write $T(f + e)$ for our *estimate* of $f(2)$ from our observations. Under many circumstances

$$T(f + e) = T(f) + T(e).$$

†During attacks on the *Bismarck*, Swordfish torpedo-bombers met a terrifying wall of fire from the battleship's anti-aircraft guns but survived. I have been told that the predictors on the *Bismarck's* guns had been adjusted for attack by modern fast monoplanes and thus fired just ahead of these old-fashioned slow biplanes. It is a plausible story but I have found no written source.

Thus our estimate of $f(2)$ given observations of $f + e$ is the sum of the estimate $T(f)$ of $f(2)$ given error-free observations of f and the estimate $T(e)$ of $e(2)$ given error-free observations of e. But errors in observation are unpredictable so $T(e)$ is unrelated to $e(2)$. As we make T 'better' in an attempt to get $T(f)$ closer to $f(2)$, so $T(e)$ will tend to behave more and more wildly as it attempts to fit the essentially random e. To sum up, sophisticated prediction methods applied to accurate data may give useful accurate predictions, simple prediction methods applied to rough data may give useful rough predictions but sophisticated prediction methods applied to rough data will almost certainly produce rubbish.

Exercise 3.4.1   *(i) Suppose x_1, x_2, ... , x_n are distinct real numbers. Show that if*

$$g_j(t) = \prod_{i \ne j} \frac{t - x_i}{x_j - x_i}$$

(i.e. $g_j(t)$ is the product of the $n - 1$ terms $(t - x_i)/(x_j - x_i)$ with $i \ne j$ and $1 \le i \le n - 1$), then g_j is a polynomial of degree $n - 1$ and

$$g_j(x_j) = 1$$
$$g_j(x_i) = 0 \ \ if \ i \ne j.$$

 (ii) If y_1, y_2, ... , y_n are real numbers and we set

$$P(t) = \sum_{j=1}^{n} y_j g_j(t),$$

show that P is a polynomial of degree at most n − 1 with

$$P(x_i) = y_i$$

for $1 \leq i \leq n$.

(iii) [This part is not central to the argument.] Let P be as in (ii). Suppose that Q is another polynomial of degree at most n − 1 with

$$Q(x_i) = y_i$$

for $1 \leq i \leq n$. *Show that P − Q is a polynomial of degree of at most n − 1 which takes the value 0 at n points and deduce that P = Q. [Combined with (ii) this tells us that there is a unique polynomial P of degree n − 1 with* $P(x_i) = y_i$ *for* $1 \leq i \leq n$.]

(iv) Consider a sequence of numbers $\ldots, a_m, a_{m+1}, a_{m+2} \ldots$ *such as the number of washing machines produced by a country in the mth month, or the value of the world 100 metre sprint record at the end of year m. We know* a_r *for* $r \leq m$ *and would like to guess* a_{m+1}. *One natural way forward (though as we shall see, not necessarily a very good one) is to consider the polynomial P of degree n − 1 which has*

$$P(i) = a_{m-n+i}$$

for $1 \leq i \leq n$ *and guess that* a_{m+1} *will be close to*

$$T_n(a_{m-n+1}, a_{m-n+2}, \ldots, a_m) = P(n).$$

Show that

$$T_0(a_m) = a_m,$$

so our T_0 *prediction is that* a_{m+1} *will be close to* a_m *and 'tomorrow will be like today'. The 'tomorrow will be like today' method of forecasting is very powerful and it took many years before scientific weather forecasting could beat it. When we look at forecasting methods we should, of course, look at the proportion of successful predictions but we should also compare that proportion with the proportion of successful predictions based on the rule 'tomorrow will be like today'. Show that*

$$T_1(a_{m-1}, a_m) = 2a_m - a_{m-1},$$
$$T_2(a_{m-2}, a_{m-1}, a_m) = 3a_m - 3a_{m-1} + a_{m-2},$$

and find T_3, T_4 *and* T_5.

(v) If you know the binomial theorem you should now be able to guess the general formula for T_n *and prove it.*

(vi) [This part is not central to the argument.] Let

$$S_n(a_{m-n}, a_{m-n+1}, \ldots, a_m, a_{m+1}) = a_{m+1} - T_n(a_{m-n}, a_{m-n+1}, \ldots, a_m).$$

Show that

$$S_n(a_{m-n}, a_{m-n+1}, \ldots, a_m, a_{m+1})$$
$$= S_{n-1}(a_{m-n+1} - a_{m-n}, a_{m-n+2} - a_{m-n+1}, \ldots, a_{m+1} - a_m).$$

Show that if $a_r = \sum_{k=0}^n b_k r^k$ *for all integers r and some fixed* b_0, b_1, ..., b_n, *then* $a_r - a_{r-1} = \sum_{k=0}^{n-1} c_k r^k$ *for all integers r and some fixed* c_0, c_1, ..., c_{n-1} *(which need not be found explicitly). Deduce that if* $a_r = \sum_{k=0}^n b_k r^k$ *for all integers r and some fixed* b_0, b_1, ..., b_n, *then*

$$S_n(a_{m-n}, a_{m-n+1}, \ldots, a_m, a_{m+1}) = 0$$

and the nth prediction method works perfectly, i.e.

$$T_n(a_{m-n}, a_{m-n+1}, \ldots, a_m) = a_{m+1}$$

for all m.

   By using part (iii), or otherwise, show conversely that if the nth prediction method works perfectly for some sequence a_r, *then there exist* b_0, b_1, ..., b_n *such that*

$$a_r = \sum_{k=0}^n b_k r^k$$

for all integer r.

   (vi) Apply the prediction methods T_0, T_1, T_2, *and so on to some appropriate sequences. If you are too lazy to go and find some for yourself, here are two taken from the nearest books I have to hand. The first sequence consists of the values of* $\tan 0.20$, $\tan 0.21$, $\tan 0.22$ *and so on correct to five figure accuracy: 0.20271, 0.21314, 0.22362, 0.23414, 0.24472, 0.25534, 0.26602, 0.27676, 0.28755, 0.29841, 0.30934, 0.32033. The second sequence is the total budget expenditure in billions of dollars of New York State for the years 1966–7, 1967–8 and so on: 4.0, 4.6, 5.5, 6.2, 6.7, 7.4, 7.8, 8.5, 9.7, 10.7, 10.8. However, you will find it more interesting and more convincing to choose your own sequences. You will find, I hope, that, as you increase n, the predictions first improve and then, after a certain* n_0 *depending on the series, deteriorate. You will also find, I hope, that the critical value* n_0 *is larger for the kind of 'regular' series you find in mathematical tables and smaller for 'real data' you find in physics or economics texts.*

   (vii) Let $e_0 = 1$ *and* $e_r = 0$ *otherwise (we might suppose that* e_r *represents the constant zero sequence with an error of 1 when r = 0). Examine the effect of applying the prediction methods* T_0, T_1, T_2, *and so on to the sequence* e_r. *If* a_r *is a sequence, compare the effect of applying the prediction methods* T_0, T_1, T_2, *and so on to the sequences* a_r *and* $a_r + e_r$. *Use this to explain the phenomena described in the last two sentences of part (vi).*

We may sum up by quoting a Chinese proverb. 'Prophecy is very diffi-cult, particularly when it concerns the future.'

The first step that Blackett's circus took to palliate the problem

was to help work out in a week or two the best method of plotting the [radar] data and of predicting the future enemy position for the use of the guns on the basis only of pencil and paper, range and fuse tables. The second task was to assist in the design of simple forms of plotting machines which could be manufactured in a few weeks. The third stage was to find means of bringing the existing predictors into use in conjunction with radar sets. This was found to be possible if, by intensive training of the predictor crews, the inaccurate radar data could be *smoothed* manually. A special school was set up by AA Command to work out the methods of doing this and to give the necessary training. The fourth stage was to attempt to modify the predictors to make them handle the rough [radar] data more effectively. This proved possible with the Sperry predictor leading to what was known as the *amputated* Sperry which played a useful though limited role as an alternative to the use of plotting methods.

†However, it was not until 1944 with the use of the more accurate ten centimetre radar sets and a new electronic predictor developed by the Americans that the problem was completely solved just in time to master the menace of the V1 cruise missile.

Blackett was able to report that whilst at the start of the blitz 20 000 AA shells were fired for each bomber shot down, the number of 'rounds per bird' had fallen to 4000 by the following summer†. We conclude this chapter with an example of a general problem in a particular form.

On looking into the rounds per bird achieved by the different regional defences, it was noticed with surprise that the coastal batteries appeared to be shooting twice as well as those inland; their rounds per bird were only about one half. All sorts of ... hypotheses were considered as explanations of this strange result. Were the coastal guns better sited, or did the radar work better over sea? Perhaps the enemy aircraft flew lower and straighter than over the land. Then suddenly the true explanation flashed to mind. Over the land a battery, say, in the Birmingham area, which claimed an enemy aircraft destroyed, would have this claim checked by the search for and identification of the crashed machine. If no machine was found, the claim was disallowed. But, in general, no such check could be made of the claims of the coastal batteries, since an aircraft coming down at sea would not be traceable. The explanation of the apparent better shooting of the coastal batteries turned out to be due, therefore, to an overestimate by the coastal batteries (as by almost all other batteries) of their claims of enemy aircraft destroyed, by a factor of about two.

CHAPTER 4

Aircraft versus submarine

4.1 Twenty-five seconds

During the Second World War the aircraft of Coastal Command were responsible for the aerial war against the submarine. That war began badly when one of its aircraft attacked a surfaced Royal Navy submarine HMS *Snapper*. The attack was extremely accurate and an 'anti-submarine bomb' hit *Snapper* square on the base of the conning tower. The submarine suffered no damage beyond the loss of four electric light bulbs. Fortunately a more effective weapon was developed in the air-dropped depth charge.

A typical encounter between an aircraft and a submarine was a race with its timing determined by the ability of a U-boat to submerge very rapidly. (In training exercises submarines submerged fully in 25 seconds from the sounding of the warning gong.) The moment the aircraft spotted the submarine, it headed straight towards it and dropped depth charges over the point where the submarine was last seen. The 'stick' of depth charges fell in a more or less straight line along the line of flight of the aircraft. The depth charges then sank to predetermined depth and exploded. If the U-boat was within a certain distance of the centre of the explosion (the lethal radius; this was about six metres) it would almost certainly be sunk. Within a somewhat larger distance the U-boat might be damaged but, if none of the depth charges exploded close enough, the attack had failed completely. Figure 4.1 shows in schematic form a bird's eye view (i.e. in plan) of such an attack.

It required about 170 man hours by maintenance staff and other personnel to produce one hour's operational flying, and, on average, perhaps 200 hours of flying to result in an attack. The few minutes of an attack thus represented the result of at least 34 000 man hours work. In 1941 about 2–3% of attacks were successful — so about one and a half million man hours were required to sink one submarine. What could be done?

In his 1941 exposition of the ideas of operational research, Blackett wrote:

'New weapons for old' is apt to become a very popular cry. The success of some new devices has led to a new form of escapism which runs somewhat thus: 'Our present equipment doesn't work very well; training is bad, supply is poor, spare parts non-existent. Let's have an entirely new gadget!' Then comes the vision of a new gadget, springing like Aphrodite from the Ministry of Aircraft Production, in full production, complete with spares and attended by a chorus of trained crews.

He concludes that

... relatively too much scientific effort has been expended hitherto in the production of new devices and too little in the proper use of what we have got.

In that year Blackett was given the opportunity to vindicate his views when Coastal Command invited him to set up an operational research group to study their problems. (When he left the Anti-Aircraft Command the Commanding General lamented 'They have stolen my

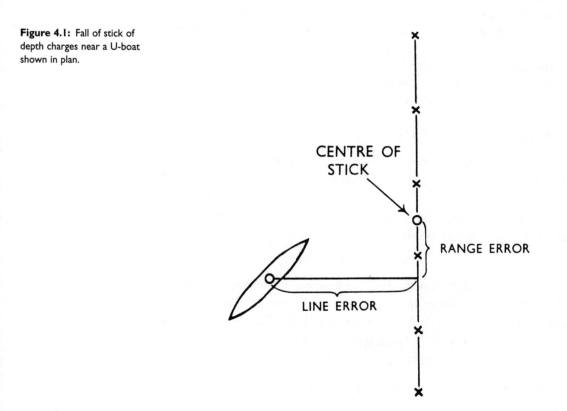

Figure 4.1: Fall of stick of depth charges near a U-boat shown in plan.

magician.') At the end of the war his successor C. H. Waddington (whom we shall meet later as an eminent biologist interested in morphogenesis and as the father of Dusa Macduff) wrote an account of the work done.

> At that time [1946], it got as far towards publication ... as corrected galley proofs. But this turned out to be a ghost edition. Changes in the political climate, the fall in temperature towards the Cold War, led the Security Authorities to withdraw permission for publication.

The book *OR In World War 2* eventually appeared in 1973. This chapter is closely based on Waddington's account† and I hope that some readers will be sufficiently intrigued to try to obtain the original book. They will find the effort involved amply repaid.

†The figures in this chapter are redrawn from the same source.

What could be done to improve matters without new gadgets, or even with new gadgets? The reader should make a list of all the possibilities that occur to her.

One night in April 1941 Blackett was in the anti-submarine operations room.

> On a large wall map were displayed the guessed positions of U-boats in the Atlantic. From the recorded number of hours flown by Coastal Command aircraft over the relevant area, I calculated in a few lines of arithmetic on the back of an envelope the number of U-boats which should have been sighted by aircraft. The number came out about four times the actual sightings. This discrepancy could be explained either by assuming the U-boats cruised submerged or by assuming that they cruised on the surface and in about four cases out of five saw the aircraft and dived before being seen by aircraft. Since U-boat prisoners asserted that U-boats seldom submerged except when aircraft were sighted, the second explanation was probably correct.

(Recall that submerged U-boats travelled very slowly and so U-boats had to make as much as possible of their voyage on the surface.)

Exercise 4.1.1 *Blackett's back of an envelope calculation was probably based on the* approximate *formula*

$$N = 2svHD$$

where N is the number of submarines that would be sighted if they did not dive, s is the maximum distance at which an aircraft should spot a submarine, v is aircraft speed, H number of hours flown and D is the U-boat density (number of U-boats per unit area of ocean). Justify the formula and discuss the approximations made.

Further information was obtained by analysing all the reported sightings up to May 1941.

It appeared that in nearly 40% of the cases the U-boat was already diving when first seen, which meant that it had seen the aircraft first. In a further 20%, the U-boat had already submerged (leaving only its periscope visible) when seen.

Thus 60% of those submarines actually seen had spotted the aircraft before themselves being spotted. In an attempt to estimate how many U-boats went completely undetected a histogram was plotted of the estimated time between the aircraft spotting the submarine and the submarine spotting the aircraft. The result is given in Figure 4.2. The figure shows a range of values from a maximum of 5 minutes (the aircraft spots the submarine at maximum range and the submarine fails to observe the aircraft before being attacked) and a minimum value of −6 minutes (the submarine spots the aircraft at maximum range of its visibility). The smooth bell-shaped curve seems to represent the known data quite well and suggests that about 2/3 of the submarines (corresponding to the shaded area under the curve) escaped detection†.

†A related problem occurs in peace time. Suppose that we wish to build flood defences capable of dealing with the highest sea levels we may expect in the next 200 years. If we have records of the sea levels over the last 50 years, how can we predict the highest sea level in the next 200 years?

How then could one raise the chance of the aircraft sighting the submarine first ... ? All the obvious courses of action were considered and recommended where necessary — better lookout drill for the air crews, better binoculars, etc. Then the best direction of aircraft course in relation to the sun was considered. If the aircraft flew down sun, the U-boat crew might have more difficulty in seeing it. Discussing these questions one day in Coastal Command, a Wing Commander said casually: 'What colour are Coastal aircraft?' Of course I knew they were mainly black as they were mainly night bombers such as Whitleys. But before the question was asked me, I had missed the significance of the fact. Night bombers are painted black so as to reflect as little light as possible from enemy searchlights. When there is no artificial illumination by search-

Figure 4.2: Difference in spotting times, aircraft versus U-boat.

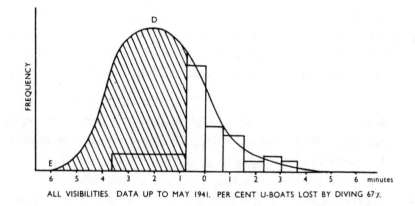

ALL VISIBILITIES. DATA UP TO MAY 1941. PER CENT U-BOATS LOST BY DIVING 67%.

lights an aircraft of any colour flying at moderate or low height, both by day and by night, is normally seen at a great distance as a dark object against a lighter sky.

Under normal conditions for the North Atlantic an aircraft will be seen against an overcast cloudy sky with no direct sunlight. Its underside is illuminated by light reflected from the sea, which is about 1/20th the intensity of the skylight. Thus even if the underside reflects all the light falling on it, it would still appear dark against the background.

This reasoning suggested that aircraft of Coastal Command should be repainted white. Model and full-scale tests were made of the average sighting distances of white and black aircraft which suggested that white aircraft should be able to approach about 20% nearer with the same chance of being spotted. Combined with the kind of estimates illustrated in Figure 4.2 this suggested an increase of about 40% in the number of U-boats sighted before they themselves dived†. In fact, by mid 1943, the same methods of estimation as we described earlier showed that

> The number of escaping U-boats had fallen from some 66% to a figure that rose from 10% at low visibilities to 35% at high visibilities. This very satisfactory result was probably due to improvements in aircraft look-out and to the effectiveness of [the repainting], though a deterioration in the standard of U-boat watch may also have played a part.

The reader may feel that I have not played fair in the presentation of the problem. I did not tell her which colour the aircraft were painted and my emphasis on not using new gadgets might be thought to exclude a new coat of paint for existing machines. However, in the course of this chapter, she has been given all the information required to propose one of the most successful operational research recommendations of the war. Perhaps she might like to close the book at this point and think.

One of the best ways of analysing an organisational problem is to ask not 'What can we improve?' but 'What can we change?'. When we find something we can change we can *then* ask what effect changing it will have. Let us return to the description of the encounter of aircraft and submarine in the second paragraph of this chapter. We list some possibilities.

1 *Improve ability of aircraft to spot submarine or reduce the ability of the submarine to spot the aircraft.* We have already discussed this.
2 *Use faster aircraft.* This is a 'gadget' solution.
3 *Improve accuracy of bombing.* We shall discuss this later.
4 *Change the depth at which the depth charges are set to explode.*

†If we wish to be wise after the event, we could remark that gulls and other sea-birds which, presumably, have the same problems as anti-submarine aircraft, also have white plumage. Since even a white painted aircraft appeared dark against the daytime Atlantic sky it was suggested that even better results would be obtained by using artificial lighting. With the correct illumination the aircraft would be effectively invisible. 'The Americans ... with their genius for nomenclature, christened the project "Yahoody the Impossible"' but nothing came of it.

Other possibilities may occur to the reader but suggestion 4 is particularly interesting since, like the colour of our aircraft, the depth setting
of the depth charges is under our control in a way that the speed
of our aircraft or the accuracy of our bombing is not. The Japanese
changed the setting of their depth charges when Congressman Jackson
May, member of the Military Affairs Committee of the US House of
Representatives, returning from a tour of the Pacific war zone, gave
a press conference at which he said that American submarines were
surviving well because the Japanese were setting their depth charges to
go off at too shallow a depth.

How should we set our depth charges? Roughly speaking the choice
resolves itself into shallow (7.5 metres, effective against surfaced or
submerging submarines), medium or deep (30 metres). It is not possible
to change the setting of the depth charges during the attack, so the
decision must be made before the aircraft takes off. Since an aircraft
can only carry a few depth charges and the depth charge will destroy
a submarine only if it explodes close to it (i.e. is positioned correctly
both in depth and plan), a 'compromise' in which we set different
depth charges to explode at different depths would be a 'compromise'
between the best solution (whatever that might be) and the worst.
We must therefore choose unequivocally between shallow, medium and
deep. Which does the reader choose and why? Remember that a
U-boat dives as soon as it sees an attacking aircraft and tries to go as
deep and as far away from where it was as possible.

There is an old joke about a policeman who sees a drunk on his
hands and knees under a street lamp. 'What are you doing?' asks the
policeman. 'I'm looking for my keys.' 'Where did you lose them?' 'Over
there.' 'Then why are you looking over here?' 'Because it is light here.'

The correct choice of depth setting was found by E. J. Williams (in
peace time an expert on the quantum theory of atomic collisions) and
gives an unexpected twist to the joke. We quote Blackett again.

> On the assumption that a U-boat would, on the average, sight the
> attacking aircraft some two minutes before the attack, and that in this
> time it could dive to about 100 feet [30 metres] depth, the Coastal
> Command and Admiralty orders were that depth charges were to be set
> to explode at 100 feet depth.
>
> Williams spotted a fallacy in the argument leading to the 100-foot
> depth setting. It might be true that *on the average* a U-boat might sight
> the aircraft a long way off and so manage to get to 100 feet before the
> attack. However, just in those cases the U-boat had disappeared out of
> sight of the aircraft for so long that the air crew could not know where
> to drop the depth charges, so that the effective accuracy in plan of the

attack was very low. Williams drew attention to the few cases when the U-boat failed to see the aircraft in time and so was on the surface when attacked. In these cases the bombing accuracy in plan was high, as the U-boat was visible at the time of attack. However, Williams pointed out that in just these cases the explosion of the depth charges at 100 feet would fail to damage seriously the U-boat as the radius of lethal damage of the depth charges was only about 20 feet [6 metres]. Thus the existing method of attack failed to sink deep U-boats owing to low bombing accuracy and failed to sink shallow U-boats due to the depth setting.

In other words, the drunk is right to look under the street light. Only a few submarines will be on the surface when attacked but since these are the only ones we have any chance of hitting we should concentrate on these by setting the depth charges at shallow. As far as can be judged, the change in setting multiplied the number of submarines sunk per hundred attacks by a factor of four.

> Captured German U-boat crews thought that we had introduced a new and much more powerful explosive. Actually we had only turned a depth-setting adjuster from the 100-foot to the 25-foot mark.

A simple change of tactic had converted the role of aircraft from submarine scarecrow to submarine killer.

We now turn to the problem of bombing accuracy (noting that the new restriction to surfaced or submerging submarines simplifies matters considerably). At first sight there was no problem since aircraft crews taking part in trials demonstrated impressive accuracy. However, the results in actual attacks were disappointing. This could mean that accuracy deteriorated substantially under operational conditions or that there was something wrong with the depth charges†. To resolve the question it was decided to mount cameras in the planes to take pictures at definite intervals during the attack. It is, of course, much easier to make such a decision than to implement it. The reader is invited to reflect on the problems involved. She should also ask herself how she would set about interpreting the data produced.

One way of using the data is to superimpose the results of all the attacks on one diagram. (We ignore the geometric and other problems involved. A discussion will be found in Section 7.3 of Waddington's book.) The result of doing this with the first 16 attacks to be analysed is shown in Figure 4.3. The dots give the centre points of the sticks of depth charges and the arrows the direction of attack. The radius of the outer ring is 300 feet (roughly 90 metres). We see at once that the centre points are rather scattered (the errors turned out to be

†At various times during the two World Wars the German, British and American navies used torpedoes of defective design. The Admiralties of the three nations were united in blaming failure of torpedo attacks on the submariners rather than the torpedoes.

about three times greater than the air-crews expected) but, apart from exhortation, is there anything that can be done about it?

Something rather interesting appears if instead of just plotting the individual centres we also plot their mean. Formally, if the individual centres have Cartesian coordinates (x_1, y_1), (x_2, y_2), ... (x_n, y_n), their mean point, or centre of gravity, is $(\bar{x}, \bar{y})$, with

$$\bar{x} = \frac{x_1 + x_2 + \ldots + x_n}{n}, \qquad \bar{y} = \frac{y_1 + y_2 + \ldots + y_n}{n}.$$

The result is shown in Figure 4.4. The hatched circle indicates the mean point and distances are marked in feet. From this diagram it was clear that

> [t]he distribution of the mid-points of the sticks was not centred on the conning tower, but on a point some 60 yards [55 metres] ahead. At the time the tactical instructions, and the training in low level bombing, both laid considerable stress on the need to aim-off ahead so as to allow for the forward travel of the U-boat during the fall of the depth charges. It

Figure 4.3: Sixteen attacks on submarines.

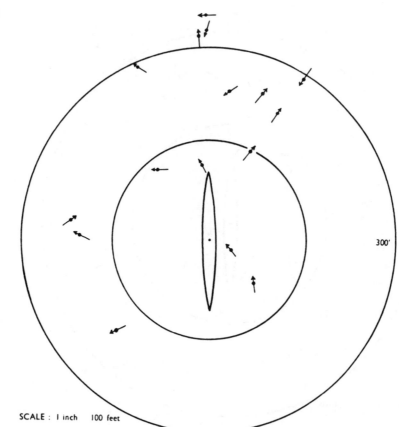

SCALE : 1 inch   100 feet

300'

was clear that this aim-off was being overdone. It is notoriously difficult to estimate distances at sea by eye when travelling at the speed of an aircraft (a difficulty which also accounts for the optimistic estimates by pilots of their bombing accuracy). It could be shown that in this case the cure was worse than the disease; a greater percentage of hits would be obtained if pilots forgot all about the forward travel of the U-boat, and aimed baldly at the conning tower. The suggestion was adopted and in later analyses the systematic error forward had disappeared.

The result can be seen in Figure 4.5. The removal of this systematic error produced (as far as can be judged) a 50% increase in kills.

By the beginning of 1945 the percentage of U-boat kills per attack had risen to over 40%. Some of this improvement was due to improved technology and some, presumably, to deterioration in the standard of U-boat crews, but much must be attributed to scientific effort expended

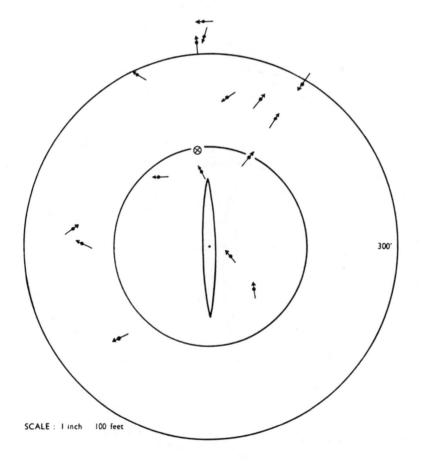

Figure 4.4: Mean centre point for sixteen attacks on submarines.

SCALE : 1 inch   100 feet

300'

not on the production of new devices but on finding the proper use of existing ones — in two words operational research.

We conclude this chapter by discussing a diagram along the lines of the two previous figures. Figure 4.6 is a composite of the positions of U-boats sighted by aircraft escorting convoys between July and December of 1942. (The inner circle has radius about 18.5 kilometres and the convoy direcion is indicated by the arrowhead.) Once again the reader is invited to pause and see what useful information she can extract from this diagram.

[We] had always hoped to be able to discover the tactics of the pack by plotting the relative positions of sightings. Evidence could be found for the presence of shadower U-boats lying about 8 to 16 miles [13 to 26 kilometres] ahead and within 45° of the convoy course. This seemed sensible enough, but a great many of the sightings were made over the

Figure 4.5: Mean centre point for attacks after aim-off countermanded.

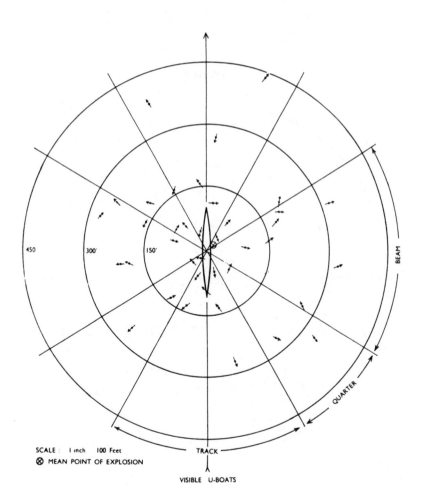

whole area behind the convoy, out to about 30 miles [50 kilometres] or more; there seemed no reason why U-boats should get into such a position, and it was not until we realised that the packs we discovered had already had their positions disturbed by partial or complete breaking of their contact with the convoy that we could understand why this tactically unfavourable position had such a high U-boat density.

In August 1942, Dönitz told a Swedish journalist that 'the aeroplane can no more eliminate the U-boat than the crow can fight a mole'. But to evade an aircraft the submarine had to dive. Once under water it was effectively deaf and blind, it could not communicate with the rest of the pack or track the convoy, and its speed fell well below that of its prey. Air escort prevented U-boats from shadowing and getting into attack position and unless U-boats attacked convoys they were useless. So long as the submarine had to spend much of its time surfaced to recharge batteries and so long as the speed of a submerged submarine

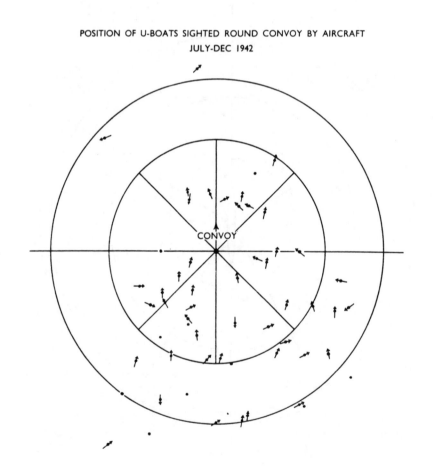

POSITION OF U-BOATS SIGHTED ROUND CONVOY BY AIRCRAFT
JULY-DEC 1942

Figure 4.6: Position of U-boat sightings in relation to convoy.

fell (except for short dashes) below that of even the slowest convoy, aircraft could now checkmate the submarine.

The success of the Allies' war against the submarine was not a foregone conclusion. The Japanese lost their war. They failed to introduce merchant convoys until the end of 1943, failed to keep convoy radio silence and failed to develop effective anti-submarine tactics. They lost seven-eighths of their merchant shipping.

On the 30 April 1945, U-2511 became the first Type XXI U-boat to embark on an operational cruise. It could maintain its maximum underwater speed of 18 knots [33 kilometres per hour] for one and a half hours, an underwater speed of 12–14 knots [about 26 kilometres per hour] (substantially faster than most convoys) for 10 hours and an underwater speed of 6 knots [11 kilometres per hour] for up to 48 hours. The use of a schnorkel removed any need for surfacing to recharge its batteries and allowed it to travel indefinitely at 12 knots [22 kilometres per hour] with only a small schnorkel head above the sea surface. Except in calm seas the task of searching for such a target among the waves by eye or by radar 'could be likened to a golfer looking for a ball in a limitless field of daisies'. On 2 May, U-2511 was detected by a British patrol group but easily evaded it in a high speed underwater dash. At midnight on 5 May, Dönitz, now Hitler's successor ('by merit raised to that bad eminence'), ordered the German surrender and the message reached the submarine as it prepared to attack a British cruiser at point blank range. The captain carried out a mock attack and returned to base undetected. The war had ended just as the advantage was about to move from the aircraft back to the submarine.

4.2 Let's try the slide-rule for a change

After nine months with Coastal Command Blackett was asked to set up an operational research group at the Admiralty. The war against the U-boat remained his main concern but now his task was to consider it from a general standpoint. What, for example, was the best division of limited shipbuilding resources between merchant ships to carry cargoes and escort vessels to protect them?

To answer this question we must know how many extra merchant ships in a convoy would be saved by each extra escort ship assigned to each convoy. Presumably the average number $f(n)$ of ships lost from a convoy is a function of n, the number of escort vessels, but how can we calculate f? Blackett's answer is that we cannot, but that we do not need to. If f is well behaved we expect f to be locally approximately linear with

$$f(x + h) \approx f(x) + Ah,$$

for some constant A, when h is reasonably small. (If you know a little calculus you will write $A = f'(x)$. We return to this point in Section 8.2.) The single number $-A$ (observe that we expect $A < 0$) tells us the number of merchant ships saved by the addition of an escort vessel. If we plot the average number of losses against the number of escorts we may hope that they will lie *very roughly* on a line of slope A. (It is not hard to point out the problems raised by this line of approach, but can the reader suggest a better method? 'If I could, I would ... readily offer a candle to Saint Michael and another to his dragon.') Analyses of this type indicated that a convoy with nine escort vessels suffered on average 25% less losses than one with six. Thus, with the pattern of convoys then in force, each additional escort put into service could be expected to save between two and three merchant vessels a year. However, the unambiguous conclusion that it paid to build more escorts at the expense of less merchant ships had to be balanced against

> the practical difficulty of changing the shipyards rapidly over from the building of merchant vessels to the building of escort vessels. As so often occurs with the predictions of economic theory, the theoretically optimum production programme cannot be realised quickly in practice.

Since convoys must travel at the speed of the slowest ship the Admiralty organised fast and slow convoys. Blackett's group found that fast (nine knot) convoys lost only about 3/5 the number of ships as slow (seven knot) convoys *provided that the convoy had air cover*. A convoy with air cover for eight hours per day lost, on average, only 2/5 the number of ships of a convoy without air cover. The reason for the startling gain associated with a small increase in convoy speed is revealed in Figure 4.6. A U-boat could only keep up with a fast convoy by travelling on the surface most of the time. Even an unsuccessful attack by aircraft forced it to submerge and lose contact with the convoy†.

Speed alone was no protection. In November 1940 Lindemann, Churchill's personal scientific adviser, persuaded the Cabinet that ships capable of speeds greater than 12 knots should not sail in convoy‡. The results of the next six months showed that ships capable of 15 knots suffered the same loss rates when sailing independently as did ordinary convoys. Anything slower did much worse. On the key North Atlantic run 13.8% of ships sailing independently were lost compared with 5.5% of those sailing in convoy. Only the giant liners whose speed

†Most people would consider that there are few more unmilitary objects than the blimp (non-rigid airship) and most officers in the US Navy would have agreed with this assessment. However, the company that manufactured them was influential and the Navy had to adopt some 30 of them for coastal convoys. As might be expected, no submarine was ever sunk by a blimp, but, equally significantly, not a single ship out of the 89 000 they escorted was lost to direct submarine attack.

‡'His recommendation ... horrified us' wrote the convoy planning officer at the time. ' ... I used to pray at night that he would be run over by a bus.' Lindemann also changed the method of classifying shipping losses so that ships initially sailing in convoy but later detached were *included* as convoy losses – thus overcounting convoy losses and undercounting independent losses.

of 30 knots meant that they could outrun a surfaced submarine were safer outside convoys than inside. The decision was reversed.

Exercise 4.2.1 *The figure of 5% losses per convoy appears small and reflects the fact that many convoys evaded attack. However, each cargo carried to Britain required two convoy voyages, the eastward one carrying the cargo and a westward empty one. The passage of 10 cargoes thus required 20 convoys.*

(i) If sunk ships are replaced, how many will have to be found (as a proportion of the original ships involved) in 20 convoys?

(ii) If a sailor signs on for 20 convoy voyages what is the probability that he will be torpedoed on one of these voyages?

Since every available ship had to be used, there was no way to increase convoy speed but there were ways to increase air cover. At the time shore based aircraft could provide cover except over 'The Gap', a patch of ocean between Iceland and Newfoundland about two or three days' sailing wide. It was here, in what his men called 'The Devil's Gorge' and the Allied seaman 'Torpedo Junction', that Dönitz concentrated his attack.

The British, like the Germans, were strip mining their economy to fight the war. Even so, resources were finite and every aircraft assigned to Coastal Command was one less for Bomber Command. Harris, the head of Bomber Command, was convinced that Germany could be rapidly defeated by bombing alone and was supported in this view by Lindemann. To these men the Battle of the Atlantic was a defensive distraction from the great offensive bomber battle they believed would shortly win the war.

Harris wrote that Coastal Command was 'merely an obstacle to victory'.

> While it takes approximately 7000 hours of flying to destroy one submarine at sea, that was approximately the amount of flying required to destroy one third of Cologne. ... The purely defensive use of air power is grossly wasteful. The naval employment of aircraft consists of picking at the fringes of enemy power, of waiting for opportunities that may never occur, and indeed probably never will, of looking for needles in a haystack.

He was supported by Lindemann.

> It is difficult to compare quantitatively the damage done to any of forty-odd big German cities of a 1000 ton raid with the advantages of sinking one U-boat out of 400 and saving three or four ships out of 5500. But

it will surely be held in Russia as well as here that the bomber offensive must have more immediate effect on the course of the war in 1943.

Blackett replied that it was perfectly possible to make quantitative comparisons. From the convoy statistics

[a] long-range aircraft, such as a Liberator I, operating from Iceland and escorting convoys in the middle of the Atlantic, *saved* at least half a dozen merchant ships in its service lifetime of some thirty flying sorties†. If used for bombing Berlin, the same aircraft in its service life would drop less than 100 tons of bombs and kill not more than a couple of dozen enemy men, women and children and destroy a number of houses. No one would dispute that the saving of six merchant vessels and their crews and cargoes was of incomparably more value to the Allied war effort than the killing of some two dozen enemy civilians, the destruction of a number of houses, and certain very small effect on production.

†Blackett's figure of 30 sorties should remind us that during the course of the war Coastal Command lost 1777 aircraft and 5866 pilots and air-crew.

The reader will notice that Blackett's aircraft is not wandering round the ocean on the off-chance of spotting a submarine, as Harris suggests, but staying close to the convoy where the submarines must come, nor even sinking U-boats as Lindemann describes but simply rendering them useless for their only purpose.

Ultimately the decision had to be Churchill's and he was bombarded with arguments from all sides.

After some intense argument, Harris burst out with: 'Are we fighting this war with weapons or slide-rules?' To this, after a shorter than usual pause and puff of his cigar, Churchill said: 'That's a good idea; let's try the slide-rule for a change.'

Unfortunately it was Lindemann's slide-rule, Harris's mastery of obstructive tactics and Churchill's own preference for offence over defence which prevailed for a year during which nearly 8 million tons of shipping (1664 ships) were lost, of which over 6 million tons were sunk by submarines, and only 7 million tons of shipping built.

It appeared‡ that the only aircraft with sufficient range to cover the air gap would be the VLR (very-long-range) version of the American Liberator (B24). Inter-service rivalry was as strong in the US as it was in Britain and Admiral King in charge of the American naval effort was preoccupied with the Pacific war and inclined to let the British pull their own chestnuts out of the fire. Eventually the British Admiralty had to appeal directly to the American President. Roosevelt himself intervened and, by special presidential edict, a substantial number of VLR Liberators were assigned to Coastal Command. At the same time a couple of 'escort carriers' (merchant ships converted into aircraft

‡After the war the British Lancaster bomber was developed into the Shackleton, an extremely successful long-range anti-submarine aircraft. Sometimes described as 'ten thousand rivets flying in close formation' it remained in service in an anti-submarine role until the 1970s and, as an early warning aircraft, until 1990.

carriers) became available for convoy duties. We shall describe the effect of the closing of the 'Air Gap' in Section 13.2.

A detective novel is an obviously artificial construct in which all statements are clues, red herrings or mere decoration. But all narrative involves selection and imposes an artificial order. The reader knows, with a certainty denied to the participants, that the Allies will win the Battle of the Atlantic. Even if she knew nothing of Blackett before, she knows by now that Blackett is the hero of the narrative and will, usually, be proved right. The chaos of reality cannot be recreated†. What I have done, however, is to bombard her with facts in an attempt to produce at least some of the obscurity against which real decision making occurs. I now ask that she should stop, close the book and think whether there was anything else that Blackett's circus should have considered. We shall return to this question in the next section but before going on to that section we close with an application of mathematics to warfare which dates back to 1916.

†A remarkable attempt is made in the *The Great Devonian Controversy* by Martin Rudwick. Who would have thought that a 450 page book on geology in the 1830s could grip like a thriller?

Exercise 4.2.2 *(Lanchester's theorem; this exercise requires elementary calculus.) Lanchester was a brilliant and unorthodox engineer who, among many other things, wrote a paper in 1895 in which he laid down the fundamental conditions for sustained flight. A revised version was considered unsuitable for the Royal Society and so submitted to the Physical Society which rejected it in 1897. Amplified to form two volumes published in 1907 and 1908 it formed the text-book of aircraft designers for two decades. (The Wright brothers' first flight took place in 1903.) In 1916, he published a book‡ Aircraft In Warfare [136] in which he propounded 'Lanchester's N-square Rule'. He contrasts the old conditions under which 'it was not possible by any strategic plan or tactical manoeuvre to bring other than approximately equal numbers of men into the actual fighting line: one man would ordinarily find himself opposed to one man' with modern conditions when 'the concentration of superior numbers gives an immediate superiority in the active combatant ranks, and the numerically inferior force finds itself under far heavier fire, man for man, than it is able to return'.*

‡Based on articles written just before the outbreak of war and containing discussions of anti-submarine tactics, air-launched torpedoes and strategic bombing.

(i) Consider a battle between two armies armed with rifles, the Blue army with b(t) men at time t and the Red army with r(t) men at time t. In a given short time the number of effective hits by the Red army will be proportional to the number of men in that army. Consequently the losses of the Blue army will be proportional to the number of men in the Red army.

$$b'(t) = -cr(t),$$

where c is a strictly positive constant. Similarly

$$r'(t) = -kb(t),$$

where k is another strictly positive constant. Show that

$$\frac{d}{dt}(kb(t)^2 - cr(t)^2) = 0$$

and deduce that

$$kb(t)^2 - cr(t)^2 = kb(0)^2 - cr(0)^2.$$

Conclude that the Blue army will win a battle of annihilation if

$$kb(0)^2 - cr(0)^2 > 0.$$

How many men will it have left? (Throughout the question we assume that $b(0), r(0) > 0$.)

(ii) For simplicity suppose that $c = k$. Suppose that Red has 7000 men divided into two armies of 4000 and 3000 men. What will happen if Blue's army of 5000 men attacks first one Red army and then, after that battle is decided, attacks the second Red army? What happen if Red manages to unite his two armies before Blue attacks? Lanchester illustrates his point with reference to the opposing British and German Navies. The pre-war plans of the German Navy had called for it to nibble away at British strength in a series of small battles and the British had countered by gathering its major units into a single Grand Fleet. He summarises his conclusion by saying that the fighting strength of a force varies as the square of its numerical strength.

(iii) (This is not central to the argument.) Returning to part (i) show that

$$r''(t) = ckr(t)$$

and that

$$r'(0) = -kb(0).$$

Hence find $r(t)$ and show that, if $kb(0)^2 - cr(0)^2 > 0$, there is a $t_0 > 0$ such that $r(t_0) = 0$. What does t_0 represent?

(iv) However, much of modern warfare consists in firing in the general direction of the enemy in the hope of hitting something. Explain why, under these circumstances a more plausible set of equations is

$$b'(t) = -cr(t)b(t), \quad r'(t) = -kr(t)b(t),$$

with $c, k > 0$. Show that now

$$kb(t) - cr(t) = kb(0) - cr(0),$$

so that, under these circumstances, it is more reasonable to say that the fighting strength of a force varies directly with its numerical strength.

(v) (This is not central to the argument.) For simplicity suppose that, in part (iv), $k = b = 1$ and $b(0) - r(0) = 1$. Show that $b(t) = 1 - r(t)$ and deduce that

$$r'(t) = -(1 - r(t))r(t).$$

Hence show that

$$\frac{r(t)}{1 - r(t)} = \frac{r(0)}{1 - r(0)}e^{-t}$$

and that $r(t) > 0$ for all $t \geq 0$ but $r(t) \to 0$ for all $t \geq 0$.

Show that the same conclusions hold for general (strictly positive) values of k, c, $b(0)$ and $r(0)$ provided $kb(0) - cr(0) > 0$. What happens if $kb(0) - cr(0) < 0$ and if $kb(0) - cr(0) = 0$?

Lanchester's N-square rule was much praised in military circles, possibly because it appeared to give scientific backing for standard strategic rules. Blackett comments, however, that 'it is generally very difficult to decide whether, in any particular case, such a law applies or not.' (For example, is a battle between escorts and U-boats closer to (i) or (iv)?)

> *Such a priori investigations are of very little use for handling a complicated event like, say a mass U-boat battle against a convoy. They are of some use sometimes in handling selected parts of such battles, for instance, to calculate the chance of a U-boat penetrating an escort screen of a given number of escort vessels under given weather conditions. [They] are nearly always useful and necessary to study the performance of the actual weapons, for example to calculate the chance of killing a U-boat with, say, a 14 pattern depth charge attack, under certain assumed conditions of plan and depth errors.*

The two chapters of Lanchester's book dealing with the N-square rule are reprinted in Volume IV of Newman's The World of Mathematics.

4.3 The area rule

As we saw in the last section, Blackett's team investigated how convoy losses depended on the number of escorts provided, the speed of the convoy and the provision or not of air cover. If we think carefully about how we might analyse the information available, it becomes clear that there is a further important variable involved — the size of the convoy. Moreover, it is a variable under our control, and the key to operational research is the identification of such variables.

The reader should pause to think what general considerations should govern the size of a convoy. Would she use a small number of large convoys or a large number of small convoys or some other system? If she wanted further information what would it be and how would she use it?

The obvious place to start was with the existing rules.

Under the exigencies of a very critical situation, the organisation of convoys and their escort vessels was inevitably, to a considerable extent, a matter of chance. But certain broad principles to govern their organisation had been laid down by the Admiralty. Generally speaking, large convoys were thought to be relatively dangerous and small convoys relatively safe. A convoy of 40 merchant vessels was considered to be about the best size and convoys of more than 60 ships were prohibited. As regards the required number of escort vessels for a convoy of given size, a rough and ready guide was provided by the long-standing $3 + N/10$ rule. This laid down a minimum of 3 escort vessels for a very small convoy, and one additional escort vessel for every 10 ships in the convoy. Thus a convoy of 20 ships ($N = 20$) would have 5 escorts and a convoy of 60 ships have 9 escorts. The implication of this rule whose origin was never traced†, was that this number of escort vessels would make convoys of different size equally safe, that is the same average percentage losses would be expected.

However the Admiralty $3 + N/10$ rule could be shown to be not consistent with the rule that small convoys were safer than large. For consider the alternative of running (*a*) three convoys of 20 ships, each with five escorts given by the rule and (*b*) one convoy of 60 ships with all the available 15 escorts. Clearly the large convoy, according to the rule, would be much safer. For the rule for a 60-convoy gave only nine escorts as necessary for equal safety with the small convoys, whereas 15 would be available by pooling the three separate escort groups of five each.

When the actual records of ships lost in convoys of different size were looked into, it was found surprisingly that in the previous two years large convoys had suffered much smaller relative losses than small convoys. The figures were startling. Dividing convoys into those smaller and larger than 40 ships it was found that the smaller convoys, with an average size of 32 ships had suffered an average loss of 2.5 per cent, whereas the large convoys with an average size of 54 ships, had suffered only a loss of 1.1 per cent. Thus large convoys appeared to be in fact over twice as safe as small convoys.

Though the statistics seemed quite reliable, the scientists [in the operational research group] felt it necessary to make as sure as was humanly possible that large convoys were in fact safer than smaller ones before attempting to convince the Admiralty that their long-founded preference for small convoys was mistaken. After all, statistics can be in error — particularly through chance fluctuations of the relatively small numbers involved in such calculations. Perhaps the lower losses of the large con-

†This remark of Blackett's brings to mind the doubtless apocryphal story of an operational research investigation into the firing of field artillery. The scientists found that, in addition to the crew of five who actually manned the gun there was a sixth soldier who stood at attention throughout the firing but who appeared to have no other function. Past and present instruction manuals gave no clues and questioning serving and retired officers proved equally unhelpful until a Boer war veteran was consulted. 'The sixth soldier holds the horses.'

voys in the previous two years had been due to chance. We felt that if we could find a rational explanation of why large convoys should be safer than small ones, it would strengthen the case for a change of policy.

Can the reader provide the 'rational explanation' required?

Blackett continues:-

... an intensive study of all available facts about the U-boat campaign against the convoys was undertaken. Of great use were the accounts of prisoners of war from sunken U-boats of the detailed tactics pursued by the U-boats in their 'wolf-pack' attacks on the convoys. After several weeks of intensive research analysis, and discussion, the following facts emerged. The chance that a given merchant ship would be sunk in any voyage depended on three factors: (*a*) the chance that the convoy in which it sailed would be sighted; (*b*) the chance that having sighted the convoy, a U-boat would penetrate the screen of escort vessels round it; and (*c*) the chance that when a U-boat had penetrated the screen the merchant ship would be sunk. It was found (*a*) that the chance of a convoy being sighted was nearly the same for large and small convoys; (*b*) that the chance that a U-boat would penetrate the screen depended only on the linear density of escorts, that is, on the number of escort vessels for each mile of perimeter to be defended; and (*c*) that when a U-boat did penetrate the screen the number of merchant ships sunk was the same for both large and small convoys — simply because there were always more than enough targets†.

†The normal U-boat carried 14 torpedoes, and so, assuming that a submarine expended three torpedoes to sink one ship, it had a maximum 'bag' of about five ships. In both World Wars, a U-boat which succeeded in attacking ships in convoy sank on average one ship in the course of the attack.

‡Recall the formula 'circumference = $2\pi \times$ radius' if necessary.

§Recall the formula 'area = $\pi \times$ radius^2' if necessary.

Consider a convoy with circular perimeter patrolled by escorts. If we double the radius of the circle then since the perimeter depends *linearly* on the radius‡ we will require twice as many escorts to keep the number of escort vessels for each mile of perimeter the same. On the other hand the area of a circle varies as the *square* of the radius§ so that the number of vessels convoyed is quadrupled. Using the facts stated by Blackett at the end of the last quotation we see that (*a*) the probability of being sighted is essentially the same for both convoys, (*b*) if the convoy is spotted the chance that a submarine will penetrate the defensive screen is the same for both convoys and (*c*) the number of ships sunk by any submarine which penetrates the defensive screen is the same for both convoys. Thus both convoys will lose (on average) the same number of ships and so the larger convoy will have (on average) one quarter the percentage losses of the small. Further, the number of escorts per merchant ship used for the larger convoy is half that of the smaller. In other words, if we quadruple the size of each convoy we will suffer one quarter of our previous losses and require only one half of our previous escort force!

Exercise 4.3.1 *Show that for convoys with circular perimeter with a fixed number of escort vessels per mile of perimeter*

$$\text{number of escorts required per merchant ship} \approx AN^{-1/2},$$

and

$$\text{expected percentage losses} \approx BN^{-1}$$

where N is the number of ships in the convoy and A and B are constants. What can you say for convoys sailing in different formations (e.g. squares)?

There was now statistical evidence and theoretical backing for the proposition that big convoys were safer than small ones. But, as Blackett says,

> before we advised the Navy to make this major change, we had to decide whether we really believed in our analysis. I personally convinced myself that I did, by the conviction that if I were to send my children across the Atlantic during the height of the U-boat attacks I would send them by a big rather than by a small convoy.

The reader will have noticed that points (*a*) and (*c*) were those we gave in Section 2.2 as the reasons why convoy worked. The 'area rule' had already been given in Appleyard's forgotten survey of the lessons of the First World War. Had it been remembered in 1940, it would have greatly reduced the strain on the very limited number of escorts available. Blackett calculates that, had the policy of large convoys been adopted in the spring of 1942 rather than the spring of 1943, the actual loss of five million tons of merchant shipping would have been reduced to four million (a saving equivalent to about two hundred ships) and considered the failure of his group to investigate the effects of varying convoy size earlier to have been its most serious mistake.

As it was, the introduction of larger convoys of up to 160 merchant ships came after the Battle of the Atlantic had been won and so the theory was not tested directly. The reduction in the number of escorts required for Atlantic convoy duties helped swell the number of anti-submarine vessels in support of the Normandy invasion.

Exercise 4.3.2 *We have only considered the convoy as a method for saving ships. However, it is also a method for destroying submarines. Although the most important ratio to both sides was*

$$\frac{\text{monthly merchant shipping tonnage lost}}{\text{monthly merchant shipping tonnage built}},$$

Dönitz measured the tactical success by the exchange ratio

$$\frac{\text{number of U-boats lost in attacking a convoy}}{\text{number of ships sunk by the attack}}.$$

Increasing the number of escorts per convoy increased the chance that an attacking U-boat would be sunk but also made the job of a surviving U-boat harder. Waters has written a fascinating study of convoy statistics since the thirteenth century. He claims that (in the absence of air cover) the effect of escorts can be summed up in two laws

$$(1) \qquad K = C_1 U E,$$

$$(2) \qquad L = C_2 \frac{U}{E},$$

where L is the number of merchant ships lost, U the number of U-boats attacking, E the number of escorts, K the number of U-boats destroyed and C_1 and C_2 are constants. The equations were found to hold for 1941–2 with $C_1 \approx 1/100$ and $C_2 \approx 1/5$. (Law (1) seems very plausible, provided $C_1 E$ is not too big, but I am more sceptical of the wide applicability of (2).) Show that, if the two laws hold,

$$\text{exchange ratio} = C_3 E^2,$$

where C_3 is a constant. Show that, for the 1941–2 constants, the exchange ratio rises above 1 if five or more escorts are employed. If we use large convoys then we can employ more escorts per convoy and obtain better exchange ratios for any convoy that comes under attack.

Before being too critical of the Admiralty's preference for small convoys it must be remembered that the majority of naval experts in pre-war Britain and Germany expected the chief threat to convoy to come from surface raiders. A submarine can only fire its limited number of torpedoes but a pocket battleship can annihilate an entire convoy. When the *Admiral Scheer* attacked convoy HX84 only the heroic sacrifice of the armed merchant vessel *Jervis Bay* and failing light kept losses down to six ships.

Exercise 4.3.3 *(i) Suppose that a battleship will annihilate any convoy it finds and that the chance of the battleship finding a convoy is independent of its size. Show that the average percentage loss is now independent of convoy size.*

(ii) Consider how Blackett's points (a), (b) and (c) would be modified for the age of sail. (It may be helpful to note that when the danger consisted of individual privateers, convoys of over 200 merchant ships might only have one escort, but when powerful opposition was expected, the main battle fleet might be used.)

By the time Blackett's team began its investigations, the Battle of the Atlantic had resolved itself into a long succession of battles between convoys and submarines for which statistics could be gathered and tactics investigated. It should also be remembered that in a long war public opinion becomes reconciled to occasional disasters and tactics can be chosen to give 'best averages' rather than 'avoiding worst outcomes'†. However, we can echo Blackett on the importance of subjecting

> as many as possible of the rules and dogmas of a fighting Service or Command to critical but sympathetic analysis. In nine cases out of ten, the rules or dogmas were found to be soundly based; in the tenth, sometimes, changed circumstances had made the rules out of date.

Exercise 4.3.4 *Consider a company whose main business is that of operating N machines which are liable to rare but expensive accidents each of which costs the company £k. Clearly, the larger the company the better able it is to absorb the cost of accidents, but if it does not take out insurance and there are $N/10$ accidents or more in a year the company will go out of business. Insurance is available but will cost £2Nk/1000 a year. The probability of any particular machine having an accident in any particular year is 3/2000. The company wishes the probability that it is forced out of business by accidents to be less than 1/100 in any particular year but otherwise wishes its average costs to be as low as possible. Show that a small company with $N = 10$ will take out insurance but that a large company with $N = 100$ will not. (There are not many industries of this type, but the kind of considerations presented here apply to airlines.)*

The struggle to secure some very-long-range aircraft for convoy defence was part of a larger argument over the bomber offensive against German cities. Blackett was temperamentally opposed to 'a major military campaign against the enemy's civilian population rather than against his armed forces. During my youth in the Navy in World War I such a campaign would have been inconceivable.' He and Tizard believed that supporters of this campaign had grossly exaggerated its probable effectiveness. They analysed Lindemann's estimates of the destruction of German housing in the next 18 months. Tizard concluded that the estimate was five times too high, Blackett that it was six times too high. Lindemann replied blandly that his calculations had not been made for statistical analysis, but 'partly to save the Prime Minister the trouble of making arithmetical calculations' and was 'intended to show that we can really do a lot of damage by bombing built-up areas with the sort of air force which should be available.'

†When the USA entered the war the US Navy made the expensive mistake of delaying convoy for coastal traffic. When one British officer commented on this he was told 'that the US Navy did not enjoy the respect of the American public as was the case with the Royal Navy and the British. If an American ship was sunk even in a weakly escorted convoy there would be an outcry for someone in the Navy office to be hung.'

The story goes that at that time in the Air Ministry it was said of anyone who added two and two together to make four, 'He is not to be trusted; he has been talking to Tizard and Blackett.' Less agreeable stories circulated: that anyone who made such calculations must be defeatist.

In the end I find it hard to believe that the scientific arguments had much influence one way or the other. Half-memories of books like H. G. Wells's *The War In the Air*, immediate memories of the bombing of London, the fear that Stalin might make a separate peace if he felt that Russia was bearing the full brunt of the fighting, the desire to undertake important operations independently of the Americans, fear of the casualties involved in a possibly unsuccessful landing on the Continent, inter-service rivalries and the feeling that if enough high explosive was dropped on Germany something must give† — each of these by itself carried more emotive power than 'mere figures' and together they proved irresistible.

†Discussions of German naval strategy, and in particular the two submarine wars often refer to a 'fatal belief in simple, preferably "ruthless" plans on which the enemy would allow himself to be impaled.'

> When our brother Fire was having his dog's day
> Jumping the London streets with millions of tin cans
> Clanking at his tail, we heard some shadows say
> 'Give the dog a bone' — and so we gave him ours;
> Night after night we watched him slaver and crunch away
> The beams of human life, the tops of topless towers.
>
> Which gluttony of his for us was Lenten fare
> Who mother-naked, suckled with sparks were chill
> Though cotted in a grille of sizzling air
> Striped like a convict — black, yellow and red;
> Thus were we weaned to knowledge of the Will
> That wills the natural world but wills us dead.
>
> O delicate walker, babbler, dialectician Fire,
> O enemy and image of ourselves,
> Did we not on those mornings after the All Clear,
> When you were looting shops in elemental joy
> And singing as you swarmed up city block and spire,
> Echo your thought in ours? 'Destroy! Destroy!'

However, Blackett did not feel that the outcome was inevitable.

I confess to a haunting sense of personal failure, and I am sure that Tizard felt the same way. If only we had been more persuasive and had forced people to believe our simple arithmetic, if we had fought officialdom more cleverly and lobbied ministers more vigorously, might we not have changed the decision?

It is now clear that Blackett and Tizard were right about the effects of the bombing offensive but there remain arguments for its strategic

necessity. The reader who wishes to pursue the matter further could start
with the books of Max Hastings (*Bomber Command*) who concludes that
'the cost of the bomber offensive in life, treasure and moral superiority
over the enemy tragically outstripped the results that it achieved' and
of John Terraine (*The Right Of The Line*) who argues that there was
no other option and concludes with Kitchener's words that 'We cannot
make war as we ought; we can only make it as we can'.

After the war Blackett used his position as government adviser to
argue against Britain seeking to build its own atomic weapons. His
obituarist remarks that:

> In retrospect, it seems remarkable that Blackett was unique among his
> distinguished contemporaries in realising that Britain could not remain
> in this technological race, and furthermore had the immense courage to
> press his views at the highest level of government.

The interested reader will find the full text of his dissent in Volume 1 of
Gowing's *Independence and Deterrence*. He was not quite alone; Tizard
objected to the high priority given to atomic weapons:

> We persist in regarding ourselves as a Great Power, capable of everything
> and only temporarily handicapped by economic difficulties. We are not
> a Great Power and never will be again. We are a great nation, but if we
> continue to behave like a Great Power we shall soon cease to be a great
> nation. Let us take warning from the fate of the Great Powers of the
> past and not burst ourselves with pride (see Aesop's fable of the frog).

To the end of his life Blackett argued that atomic weapons were in-
appropriate instruments of policy for a second-class power like Britain.
It must be stated that, even now, the majority, and possibly even an
overwhelming majority, of his countrymen disagree with him — though
they might be hard pressed to point to any specific benefit which the
possession of atomic weapons has conferred on their country†.

Blackett's very publicly expressed doubts about the nuclear policy of
the United States caused that country to exclude him from entry for
some years. On one occasion he was arrested when a plane he was
on had to make an unscheduled refuelling stop in the United States.
Presumably this did not greatly worry the man who, according to the
citation for his American Medal for Merit, had

> played a decisive role in the scientific activities of his own country ....
> His brilliant analysis and interpretation of operational data was a major
> factor leading to the Allied success in the anti-submarine campaign. The
> value of his work to the United States was outstanding, and his advice
> and counsel to us were invaluable contributions to the war effort.

† A man snaps his fingers every few moments. When asked why he does it he replies 'I do it – snap – to keep the – snap – elephants away.' 'But there are no elephants within a thousand miles of here.' 'Yes – snap – you see how – snap – effective it is.'

In due course his views ceased to be seen as dangerously heretical, and were accepted as merely heretical, leaving the way clear for him to become a respected elder statesman of the British scientific community.

As Professor of Physics at Manchester he encouraged the development of radio astronomy and the building of some of the earliest electronic computers (both, incidentally, examples of beating swords into ploughshares, radio astronomy springing from radar, and electronic computers from the descendants of the code breaking machines described in Chapter 15). His personal research changed to the field of geomagnetism where he designed and built extremely sensitive magnetometers. Using these he could measure weak magnetic fields in old sedimentary rocks. The direction of these fields would (subject to various caveats) show the direction of the Earth's magnetic field at the time when the rocks were laid down. In this way he obtained evidence for the, then highly controversial, theory of continental drift†.

It would be tedious to list the advisory committees, chairmanships and public lectures in which he attempted sometimes successfully and sometimes unsuccessfully to harness scientific thought and technological advance in the service of his country and mankind in general. A golden thread links the gunnery officer at Jutland, the pupil of Rutherford, the wartime boffin and Baron Blackett of Chelsea, holder of the Order of Merit‡ and President of the Royal Society.

> I cannot praise a fugitive and cloistered virtue, unexercised and unbreathed, that never sallies out and sees her adversary, but slinks out of the race, where that immortal garland is to be run for not without dust and heat.

To which Blackett might have added Milton's injunction to remember that

> Ease and leisure is given thee for thy retired thoughts out of other men's sweat.

4.4 What can we learn?

Here is Blackett recalling a discussion at Coastal Command in the spring of 1941.

> The long-range German aircraft, the Focke-Wolf 200, were taking heavy toll of our shipping west of Ireland. To meet this there were available only a few Beaufighters. The problem was how best to use these fighters. A strong case was argued by the operations staff for the following procedure. Suppose the FW200s were known to be operating mainly in

†The theoreticians denied the possibility of continental drift on the grounds that no conceivable mechanism for it existed. This time it was the experimentalists who had the last word.

‡As its name implies, this is distinguished among the various British Orders by being given for merit.

an area west of Ireland 200 by 200 miles. Suppose further that a single Beaufighter could 'sweep' a lane 20 miles wide, that is, that an enemy aircraft could be expected to be sighted at a distance of 10 miles. It was argued that the best tactics would be to wait until all the available ten Beaufighters were serviceable, and then to fly them equally spaced over the area, so that the ten swept lanes each 20 miles wide would cover the whole area of supposed operations of the FW200s. In this way it was hoped the area would be 'swept clean', that is, any aircraft operating that day would certainly be sighted.

The disadvantage of only flying when all ten fighters were serviceable was clearly that on most days no fighters would be out at all; moreover perhaps the day the fighters flew the enemy would not. The alternative was, of course, every day to fly any fighter when serviceable, even when only one, so as to have a chance, even though a small one, of sighting an enemy aircraft every day. The controversy between the exponents of the two tactics resolved itself into how to compare the theoretical certainty of sighting any enemy aircraft flying on the few days when all ten fighters were flown with the sum of the small chances of sighting the enemy on all the days when any aircraft were flown. The view of the Operations Research Section was that the two tactics gave about the same overall chances of sighting the enemy, assuming that the enemy flew every day and that the same amount of flying was done, but that in practice for many reasons there was a very strong case for flying every day with all available aircraft however few.

After 'many lively discussions' the Operations Research Section's view prevailed.

A few days later, I was met with the pleased tribute from the Controller of Operations: 'I say, Blackett, I am so glad you explained to me all about probability. As soon as the war is over I am going straight to Monte Carlo and then I really will win.'

Exercise 4.4.1 *Explain clearly why the argument given above is consistent with the argument set out in the next quotation. In hunting for U-boats the fundamental measure of effective flying is 'miles flown' and not 'hours flown'.*

Officers who had been tempted into speaking of hours flown, and were then called to order, sometimes tried to justify themselves by the argument that, since the U-boats were assumed to surface at random, they were just as likely to appear at any point of the ocean surface; and that therefore one could theoretically hunt them just as well by waiting over one spot in a blimp as by flying around; so it was, they argued, the hours and not the distance which is important. This would, of course, be true enough if all one were interested in were to see a U-boat break surface; it is not true if one wants to catch them during the period in which they remain on the surface.

Like Blackett's Controller of Operations we are interested in how many of the lessons of the U-boat war can be applied in peace. The investigations into best maintenance intervals certainly can.

The first study undertaken in the middle of 1943 tackled perhaps the most fundamental point of all; do inspections succeed in reducing the liability of the aircraft to break down? The flying period after an [inspection] was divided into ten-hourly intervals, and the number of repairs or failures per 100 flying hours determined for the first ten-hour period, the second [ten-hour] period and so on. ... [The figures show] that the rate of failure or repair is highest just after an inspection, and thereafter falls, becoming constant after about 40–50 flying hours.

†However, important detail has been omitted and anyone wishing to make serious use of the data must consult Waddington's book, page 64.

Figure 4.7 gives an idea of the kind of magnitudes involved†.

Figure 4.7: Rate of incidence of repairs against time of last inspection.

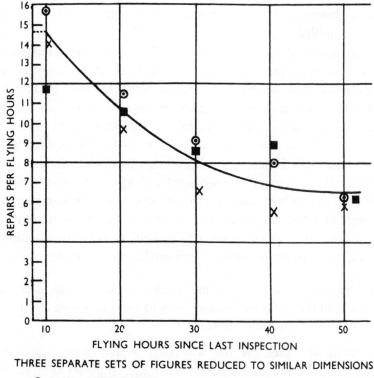

Two conclusions immediately follow from this most unexpected result. Firstly, ... inspection tends to *increase* breakdowns and this can only be because it is doing positive harm by disturbing a reasonably satisfactory state of affairs. Secondly, there is no sign that the rate of breakdown is beginning to increase again after 40–50 flying hours, when the aircraft is coming due for its next [inspection].

The second of these points, although perhaps the less striking, is probably the more important in practice. Without a fuller study of the nature of the failures and repairs needed just after inspection, it might be argued that they were matters of minor petty unserviceability, and were a small price to pay for the rectification of items on which the safety of the aircraft depended, and which would have tended to fail if the inspection had not been done. If, however, these fundamental items really exist, and are becoming due for inspection at the scheduled intervals, one would expect signs of an increasing failure rate as the time of inspection is approached. The lack of such signs is an indication that the inspections are being done too close together.

In other words, 'If it ain't broke don't fix it.' It was by no means easy to extend the intervals between inspections (who wishes to give an order for less maintenance?) but eventually the length of the maintenance cycle was doubled.

The scheduling of inspections turns out to be a complicated problem. Some components are 'inspectable' because examination can reveal warning signs (thus we can measure the depth of tread on tyres) and others (like fuses) are 'uninspectable' because we have no means of testing whether they are likely to fail in the near future. If an item is uninspectable we can either wait for it to fail or replace it on a regular schedule. It is natural to think in terms of the 'lifetime' of each component (though even here we must distinguish between items like engine parts, for which the appropriate measure is presumably flying time, items like undercarriage parts, for which the appropriate measure is presumably the number of takeoffs and items liable to corrosion, for which the appropriate measure is presumably time elapsed). Thus

[an] attempt was made to set a life to certain parts, but, for nearly all sorts of fault found, the part was not very likely to go wrong and just as likely to do so at one moment as the next.

Our intuition, based on the way that simple mechanical devices and complicated biological systems wear out, suggests that as things get older they grow more liable to fail. However, many electronic devices have the property that their propensity to failure remains the same however long they have been in use. At first sight this seems an attractive property, but the following example shows the kind of problem it throws up.

Consider a factory with 100 lighting units which must be kept on at all times. We can choose to use ordinary bulbs which have negligible probability of failure for 1000 hours but which tend to burn out shortly thereafter or neons which have an average future lifetime of 5000 hours *however long they have been used*. Each bulb and each neon costs £5. It costs £100 to get a team out to change a lighting unit but once the team is out it can change as many units as we wish at a further cost of £5 a unit. If we use ordinary bulbs, we can arrange for a team to come in every 1000 hours and replace all the bulbs at a cost of £500 for bulbs and $£100 + £500 = £600$ for labour. The total cost is £1100 and since the process must be repeated every 1000 hours the average cost per hour of replacing bulbs will be £1.1. On the other hand, if we use neons, then there is no sense in replacing working neons since the new neons are just as prone to fail as the old. We must thus wait until one breaks down and replace it. Since we have 100 neons and each has an average lifetime of 5000 hours this will occur, on average, once every 50 hours and will cost us £105 for labour and £5 for the bulb. We will thus have to spend, on average, £110 every 50 hours and so the average cost per hour of replacing neons will be is £2.2. Although neons last on average five times as long as ordinary bulbs, their average replacement cost is twice as much.

Exercise 4.4.2 *In this exercise you should use the fictional data of the previous paragraph.*

(i) Show that, if we have only 8 lighting units, then it is cheaper to use neons.

(ii) Suppose we have 100 working neons. By using the fact that the average future lifetime of a working neon is 5000 hours whatever has happened so far, show that the average time until 10 of the neons will have failed is

$$\frac{5000}{100} + \frac{5000}{99} + \frac{5000}{98} + \ldots + \frac{5000}{92} + \frac{5000}{91} > 500$$

hours.

(iii) Show that, if the factory can operate provided at least 90 of the lighting units are working, then it is cheaper to use neons, sending a team in to replace failed neons whenever only 90 of them are working.

(iv) Suppose, as before, that we must keep all the lighting units on but that we equip each unit with two neons in such a way that the lighting unit gives a satisfactory light until both *neons have failed. Show that it*

is almost as cheap to use this system (sending in a team to replace both neons when a unit fails) as to use bulbs. Show that, if we use three neons per unit, then it is cheaper to use neons than to use bulbs.

As I have noted earlier, Coastal Command was engaged in a constant struggle with Bomber Command to obtain more aircraft. It was suggested that Coastal Command could make do without more aircraft by the simple expedient of getting more effort (that is flying hours) out of its existing ones. In a wartime context the argument turned out to be fallacious.

> Suppose the effort per aircraft were doubled, [without increasing the rate at which aircraft were supplied to Coastal Command], then the accident rate would be doubled, the wastage rate due to enemy action would be presumably doubled and the rate at which aircraft wore out with use would be doubled. If the replacement rate remained the same, therefore, the front line strength would rapidly decline. For a short period the total effort would be doubled and with it the total wastage rate; as the front line strength declined, however, effort and wastage would fall, until finally equilibrium would again be reached with the replacement rate. ... the point of equilibrium would be at half the original front line strength, and this strength would produce the same total strength as before. Thus an improvement in the efficiency with which the aircraft were used would lead to a temporary increase in flying effort and a permanent economy in overhead expenses (since the same effort would come from half the original force) but no extra flying could be carried out over a long period unless the replacement of wastage was accelerated so as to meet the greater losses that must be expected. If Coastal Command were to fulfill an increased offensive role against U-boats, it would have to get a larger supply of aircraft.

Exercise 4.4.3 *Why does the argument just given not apply to a commercial airline fleet in peacetime?*

Blackett's circus contained two future Nobel Prizewinners (Blackett and Kendrew), five Fellows of the Royal Society and several future professors. In view of this, the reader may ask whether the arguments I have shown here are typical. Surely such a high-powered group used high-powered mathematical arguments? I think it would be fair to say that, although they also conducted some quite complicated investigations, their greatest successes were arrived at using the kind of simple mathematical tools exhibited here†. One reason must be that, just as it is unprofitable to use precision tools in stone quarrying, so it is unprofitable to use delicate mathematical techniques on imprecise data. Another is that, in the rapidly changing circumstances of the

†I have underplayed the statistical and probabilistic side of the work but it, too, depended on the use and understanding of relatively simple ideas.

submarine war, a reasonable proposal today is much more use than the best proposal tomorrow. In operational research, as in much human endeavour, the first 10% of the effort produces 90% of the result. In peacetime, under stable conditions, it may be very much worth while, or even essential, to use sophisticated tools to produce the full 100%, but not in war.

If the results were produced with such simple tools, why was such a high-powered group needed to produce them? I think the answer lies in the difference between 'knowledge' and 'competence'. To an 8 year old, multiplication may appear a difficult operation. To a 16 year old, algebraic manipulation may seem hard but multiplication is trivial. At 18, calculus is hard but algebraic manipulation presents few problems. After a year or so of university mathematics, multi-dimensional calculus is hard but the one-dimensional calculus of school is reassuringly familiar; and so on. We lack the confidence and competence to make use of our highest levels of knowledge but the possession of the highest levels gives us the confidence required to make use of the lower levels. When we cannot speak a foreign language or can only speak it badly, even buying a railway ticket becomes a tiring and emotional experience. When we can speak a language well (and the reader certainly speaks her own language well), we can concentrate on our thoughts and not worry about the language in which we express them. For Blackett's team, their mathematics was a well known language which left them free to concentrate on the essentials of their problems.

There is a constant demand for students with degrees in the 'numerate' subjects like mathematics, physics and engineering. Some employers require specialist knowledge, but the great majority are not interested in the highest levels of mathematical knowledge attained by their future employees, whether in Galois theory or in particle physics. What they want is the competence and confidence at lower levels (those of a university first year, say) indicated by knowledge of the higher levels.

4.5 Some problems

If you go into any mathematics or engineering library today you will find a large section on operational research. In it are books on maintenance cycles, on the theory of networks (see Section 11.1 and Section 11.2), on spare part inventories and so on. Some of the applications, for example work on organisation of hospital services, would please Blackett; others, for example work on the organisation of advertising campaigns, would certainly leave him cold. Overall he might well be disappointed at how little of the work displayed contributes directly 'to the great task of increasing the efficiency of our social system and the well-being of our population'.

The examples in this book show how much can be done with mathematics. Why can we not do more? If the reader looks closely, she will see that the examples we have chosen so far — the conquest of cholera, the war against the submarine — are ones where there is a single agreed goal. Cholera kills the poor, but it also kills the rich; it is in no one's interest to have an epidemic of cholera. In the same way, survival in a modern war becomes the overriding single interest for all. However, it is rare for societies to have simple agreed goals.

At the simplest level, it is unlikely that any change will benefit all members of a society. If we replace several small hospitals by a large one, then we may get greater efficiency and better treatment, but people without cars will find it much harder to reach the hospital. Mathematics does not enable us to make moral choices and overenthusiastic use of mathematics can obscure the moral choices we must make.

More interestingly, societies often wish to pursue several objectives, for example, liberty, equality and efficiency at the same time. There will always be a ready audience for those who proclaim that one of these goals dominates all others, or who preach the even more comforting doctrine that all of society's goals are perfectly compatible. In spite of the libraries of books written along these lines, I suspect that most societies, and certainly all those societies in which I would wish to live, pursue objectives which are, logically speaking, incompatible.

How are we to apply the 'scientific method' to a society whose goals are contradictory? Are we to declare our objectives to be 50% liberty, 20% equality and 30% efficiency — and, if we do, what would we mean by this?

It is sometimes held that the pursuit of contrary goals is a sign of the irrationality of society as compared with the rationality of individuals (often exemplified by the speaker or his audience) but observation shows that individuals also pursue incompatible goals†. It is generally believed that careful choice of diet and constant exercise can extend our expected lifetime. The more pleasures we deny ourselves the longer we live to regret their loss. How should we choose between a short happy life and a longer less happy one‡? Do we multiply our length of life in days by our happiness quotient?

Even when we restrict our attention to very simple things, human wishes are remarkably resistant to mathematics. Suppose that you were kidnapped, flown long distances in cramped conditions, driven along snowy roads, hustled into a machine that took you to the top of a mountain, then had slippery planks strapped to your feet and were

† I have an economist acquaintance, who, for the honour of his profession, tries always to act as an economic man but even he tires of this occasionally.

‡ Both for ourselves and for other people. See, for example, the papers in *Costs, Risks and Benefits of Surgery* edited by Bunker, Barnes and Mosteller.

pushed onto a long, steep, snow covered slope. Your view of the risks involved might well be more pessimistic than if you chose to go on a skiing holiday. It is well known that people are prepared to accept much higher levels of risk in things that they do voluntarily (rock climbing, smoking or driving cars) than in situations where they have no choice (nitrate pollution of water, risks concerned with atomic power). We can hardly call this irrational, but it certainly runs contrary to the mathematician's simple minded equation

$$\text{value of safety improvement} = \frac{\text{number of lives saved}}{\text{cost of safety improvement}}.$$

Inoculation against a deadly infectious disease may itself carry a certain risk of death. So long as the disease is prevalent, parents will be anxious to have their children inoculated but when (perhaps as the result of the inoculation programme itself) the disease becomes less prevalent parents become less willing to have their children inoculated. According to the obvious 'rational' mathematical model, parents will want their children inoculated if the probability of death from the disease for an uninoculated child is greater than the probability of death from the inoculation. In fact, many parents will refuse to have their children inoculated unless the probability of death from the disease for an uninoculated child is very much greater than the probability of death from the inoculation. Presumably they are applying the rule that it is worse to cause harm by action than by inaction, and I would not care to argue that such a rule is irrational.

Good teachers lead their pupils into the dark and then out of it. The rest of us can at least obey the first part of this instruction. It is not my purpose here to suggest solutions but only to propose problems. My final problem concerns the silent partner in every decision. When people speak of our duty to posterity we may be tempted to echo Addison's old College Fellow.

> 'We are always doing', says he 'something for Posterity, but I would fain see Posterity do something for us.'

However, we are constantly urged to refrain from destroying rain forests and poisoning the atmosphere 'for the sake of posterity', so it is surely worth asking how we should act towards posterity.

Before asking how we should act, it is worth asking how we do act. A witty economist when asked 'How much is posterity worth?' replied 'Three per cent per annum.' By 'Three per cent per annum' he meant the real interest (allowing for inflation) that investors then expected on their investment. If instead of consuming the price of 100 loaves of bread now, we invest the same sum, we will receive the price of 3 loaves of bread every year forever. Alternatively, if we reinvest the interest every year, then after 100 years, through the miracle of compound interest, we will have the price of about 1922 loaves. A more shocking way of presenting the same result is that one death now is equivalent to 19 deaths in 100 years time. This way of putting it makes it clear that the 'Three per cent per annum' is the bribe the future pays us not to consume. Most of us regret the fact that the dodo is extinct and, no doubt, for a sufficient sum of money the sailors who drove it to extinction could have been persuaded to spare it. But the same sum of money invested over 350 years at a real rate of return of 3% would have multiplied in value by a factor of about 31 000. If £10 000 (at present values) would then have sufficed to save the dodo, the choice taken by our ancestors on our behalf thus lay, apparently, between the dodo and £310 000 000. If we could take the decision on our own behalf, we might well, however regretfully, sacrifice the dodo.

> 'I weep for you' the Walrus said,
> 'I deeply sympathise.'
> With sobs and tears he sorted out
> Those of the largest size,
> Holding his pocket-handkerchief
> Before his streaming eyes.
>
> 'O Oysters,' said the Carpenter,
> 'You've had a pleasant run!'
> Shall we be trotting home again?'
> But answer came there none —
> And this was scarcely odd, because
> They'd eaten every one.

Exercise 4.5.1 *Check the various statements about compound interest made above.*

It would thus appear that the lower the real rate of interest the greater the say the future has in our decisions†. Is there an 'ethical' rate of interest which gives the future the 'correct' say in our decisions? Perhaps we should plan using an ethical rate of 0% so that one loaf set aside now is equivalent to one loaf in 100 years and one life saved now is equivalent to one saved in 100 years? Let us return to the £310 000 000 we apparently obtained by investing £10 000 over 350 years. Where did we invest it? In the last 350 years, banks have folded, houses have gone up in flames, stock market bubbles have burst, industries have been created and destroyed, great States have gone bankrupt and Empires

†The less it costs to bribe people the more people you can bribe.

vanished from the map. Our money has probably vanished unnoticed in some dimly remembered financial catastrophe. Part of the 3% per annum was a bribe to sacrifice present consumption for future benefits but part was a payment for assuming unknown risks.

When we build a canal, a railway or a road, we build for the future, but the future is unpredictable†. The men who built the canals thought they were building forever. Within less than a century the railways had driven most canals out of business and a century later the car and the aeroplane dominated the transport scene. Our actions are like shouts echoing through an empty building, becoming more and more distorted and finally dying away into silence. Seen in this way the 'ethical rate of interest' represents the uncertain effects of our actions. An 'ethical rate of interest' of 2% per annum balances 100 certain loaves of bread today against an uncertain 724 loaves of bread in 100 years time.

† A large and rich Cambridge college, which has already survived 500 years and has every intention of surviving another 500, recently interviewed four sets of investment advisors. Each group was asked what they considered 'long-term' to mean. Three said five years and the fourth said three days.

Exercise 4.5.2 *Rework the discussion of the dodo with a 2% per annum interest.*

The reader will hardly need to be told that the figure of 2% 'ethical rate of interest' was plucked out of the air. The figure of 3% 'acceptable real rate of return' is more soundly based in that most people who actually deal with long-term investments believe that good management will produce a return of between $2\frac{1}{2}$% and 4%. But just as we do not know the future so we do not know how much we do not know about the future. Most economists believe that the past is a good guide to the future. They point to past predictions that Britain would run out of oak trees (for building ships) or that the world would run out of coal or oil or mercury. In every case alternatives or new sources have been found. They believe that the 3% per annum rule has served posterity well in the past and will serve it equally well in the future.

When lemmings migrate, they make their way in a straight line across country swimming across any body of water they meet. They cross many rivers and lakes successfully but, when they reach the sea, the rule which has served them well in the past fails them and they drown.

PART II

Meditations on measurement

CHAPTER 5

Biology in a darkened room

5.1 Galileo on falling bodies

†It would be nice if there were a direct link with the modern publishers Elsevier. Unfortunately the Elzivir family firm lasted from 1581 to 1712, and the present holders of the name began publishing in 1880.

‡For example, Galileo was forbidden to visit nearby Florence to consult doctors about his approaching blindness. He had to obtain special permission to leave his house even to attend church services in Holy Week and had to promise not to talk to anybody.

In 1638 the Dutch publishers Elzivir† published a book by Galileo entitled *Dialogues Concerning Two New Sciences*. Since the Catholic Church had put Galileo under permanent house arrest and forbidden the publication of any book written by him, the work is introduced by a preface in which Galileo expresses surprise that a manuscript intended for a few private friends should have found its way into the hands of the printers. In spite of the difficult circumstances of its composition‡, the book sparkles with good humour. It takes the form of a dialogue between three friends: Salviati, who puts the point of view of Galileo's new physics, Simplicio, who puts the old point of view and Sagredo, who represents the intelligent layman. Here they discuss Aristotle's view that things fall at a speed proportional to their weight.

SALVIATI ... I greatly doubt that Aristotle ever tested by experiment whether it be true that two stones, one weighing ten times as much as the other, if allowed to fall at the same instant, from a height of, say, 100 cubits, would so differ in speed that, when the heavier had reached the ground, the other would not have fallen more than 10 cubits.

SIMPLICIO His language would seem to indicate that he had tried the experiment, because he says: *We see the heavier*; now the word *see* shows that he had made the experiment.

SAGREDO But I, Simplicio, who have made the test can assure you that a cannon ball weighing one or two hundred pounds or even more, will not reach the ground by as much as a span ahead of a musket ball weighing only half a pound, provided both are dropped from a height of 200 cubits.

SALVIATI But even without further experiment, it is possible to prove clearly, by means of a short and conclusive argument that a heavier

101

body does not move more rapidly than a lighter one provided both bodies are of the same material, and in short such as those mentioned by Aristotle. But tell me, Simplicio, whether you admit that each falling body acquires a definite speed fixed by nature, a velocity which cannot be increased or diminished except by the use of force [*violenza*]† or resistance.

SIMPLICIO There can be no doubt but that one and the same body moving in a single medium has a fixed velocity which is determined by nature and which cannot be increased except by the addition of momentum [*impetio*] or diminished except by some resistance which retards it.

SALVIATI If then we take two bodies whose natural speeds are different, it is clear that on uniting the two, the more rapid one will be partly retarded by the slower, and the slower will be somewhat hastened by the swifter. Do you not agree with me in this opinion.

SIMPLICIO You are unquestionably right.

SALVIATI But if this is true, and if a large stone moves with a speed of, say, eight while a smaller moves with a speed of four, then when they are united, the system will move with a speed less than eight; but the two stones when tied together make a stone larger than that which before moved with a speed of eight. Hence the heavier body moves with less speed than the lighter; an effect which is contrary to your supposition. Thus you see how, from your assumption that the heavier body moves more rapidly than the lighter one, I infer that the heavier body moves more slowly.

SIMPLICIO I am all at sea because it appears to me that the smaller stone when added to the larger increases its weight and by adding weight I do not see how it can fail to increase its speed or, at least, not to diminish it.

SALVIATI Here again you are in error, Simplicio, because it is not true that the smaller stone adds weight to the larger.

SIMPLICIO This is indeed, quite beyond my comprehension.

SALVIATI It will not be beyond you when I have once shown you the mistake under which you are labouring. Note that it is necessary to distinguish between heavy bodies in motion and the same bodies at rest. A large stone placed in a balance not only acquires additional weight by having another stone placed upon it but even by the addition of a handful of hemp its weight is augmented six to ten ounces according to the quantity of hemp. But if you tie the hemp to the stone and allow them to fall freely from some height, do you believe that the hemp will press down upon the stone and thus accelerate its motion or do you think the motion will be retarded by

†Here the translator is struggling with the problem that words like 'force' which in Galileo's time only had a non-technical meaning have now, as a result of the success of the theories of Galileo and Newton, acquired an extra, technical, meaning.

a partial upward pressure? A man always feels the pressure upon his shoulders when he prevents the motion of a load resting on him; but if he descends just as rapidly as the load will fall how can it gravitate or press upon him? Do you not see that this would be the same as trying to strike a man with a lance when he is running away from you with a speed which is equal to, or even greater than, that with which you are following him? You must therefore conclude that during free and natural fall, the small stone does not press upon the larger and consequently does not increase its weight as it does when at rest.

SIMPLICIO But what if we should place the larger stone upon the smaller?

SALVIATI Its weight would be increased if the larger stone moved more rapidly; but we have already concluded that when the small stone moves more slowly it retards to some extent the speed of the larger, so that the combination of the two, which is a heavier body than the larger of the two stones, would move less rapidly, a conclusion which is contrary to your hypothesis. We infer therefore that large and small bodies move with the same speed provided that they are of the same specific gravity.

SIMPLICIO Your discussion is really admirable; yet I do not find it easy to believe that a bird-shot falls as rapidly as a grind-stone.

SALVIATI Why not say a grain of sand as rapidly as a grindstone? But Simplicio, I trust you will not follow the example of many others who divert the discussion from its main intent and fasten upon some statement of mine which lacks a hair's-breadth of the truth and under this hair, hide the fault of another which is as big as a ship's cable. Aristotle says that, 'an iron ball of one hundred pounds falling from a height of one hundred cubits reaches the ground before a one-pound ball has fallen a single cubit'. I say that they arrive at the same time. You find on making the experiment, that the larger outstrips the smaller by two finger-breadths that is, when the larger has reached the ground, the other is short of it by two finger-breadths: now you would not hide behind these two fingers the ninety-nine cubits of Aristotle, nor would you mention my very small error and at the same time pass over in silence his very large one. ...

Galileo thus attacks Aristotle's view that heavy bodies fall faster than light ones in two ways. Through the mouth of Sagredo he reports an experiment which shows that 'a cannon ball weighing one or two hundred pounds or even more, will not reach the ground by as much as a span ahead of a musket ball weighing only half a pound, provided both are dropped from a height of 200 cubits'†. 'But', he adds through the mouth of Salviati, 'even without further experiment, it is possible to prove clearly, by means of a short and conclusive argument, that a heavier body does not move more rapidly than a lighter one provided both bodies are of the same material.' Paraphrasing his argument, let us consider two identical lumps of material falling side by side. Whether they are linked to form a single lump or are completely disconnected we would expect the rate of fall to be the same. Since the single lump has twice the mass (to use modern terminology) of each of the two separate lumps we see that the rate of fall is the same for a lump of mass $2m$ as for a lump of mass m. Similar arguments show that all lumps of the same material fall in the same way irrespective of their mass‡.

Galileo's 'thought experiment' does two things. First it establishes that Aristotle's view must be wrong. Second, it suggests what hypothesis we ought to make — that all bodies, no matter what their composition, fall in the same way. Of course, it does not prove that the hypothesis is true. As Simplicio says, it is not true 'that a bird-shot falls as rapidly as a grindstone' and we can only expect our hypothesis to correspond to the exact truth for bodies falling in a vacuum. Even there Galileo carefully restricts his argument to bodies of the same material. However, if we believe with Galileo that nature is described by simple mathematical laws, the extension of our hypothesis *as a hypothesis* to bodies of different materials is the correct way forward. Since we also agree with Galileo that experiment and not authority is the final judge of theory§, we wish to test the hypothesis by experiment. Galileo's hypothesis has been the subject of a series of subtle and difficult experiments by the Hungarian physicist Eötvös and others, and, so far, has passed with flying colours.

Galileo goes on to discuss what we would now call the effects of air resistance. His main method is to consider fall through the more resistive medium of water though he also uses observations on the effect of cannon fire. His conclusions include a clear statement of the meaning of what we now call terminal velocity. He reports observations and experiments with balls rolling down inclined planes which lead him to conclude that bodies falling in vacuum experience constant acceleration and concludes in a mathematical tour de force by demonstrating that the path of a projectile is a parabola. Although, as Galileo pointed out, the study of motion was a very old one 'concerning which the books written by philosophers are neither few nor small', all this was new and, as he claimed, he had 'opened up to this vast and most excellent science,

†It used to be fashionable to doubt whether such experiments were, in fact, performed, but historical research reveals so many instances that anyone passing under a tall building must have seemed in constant danger from falling weights dropped by enquiring philosophers.

‡Compare Blackett's demonstration, in Section 4.3, that the the $3 + N/10$ rule was inconsistent with the view that large convoys should be avoided.

§In a letter written late in life, Galileo claimed to be a true follower of Aristotle (a claim which rather startled his correspondent) because of the great care he took to use correct reasoning and because ' ... among the safe ways to pursue truth is the putting of experience before any reasoning, we being sure that any fallacy will be contained in the latter, at least covertly, it not being possible that a sensible experience is contrary to truth. And this also is a precept much esteemed by Aristotle and placed [by him] far in front of the value and force of the authority of everybody in the world. ... not only should we not yield to the authority of another, but we should deny authority to ourselves whenever we find that sense shows to the contrary.'

of which my work is only the beginning, ways and means by which other minds more acute than mine will explore its remote corners' — though I doubt that there have been many minds more acute than that of Galileo.

5.2 The long and the short and the tall

The first of Galileo's 'Two New Sciences' was dynamics; the second deals with the strength of materials. Here Galileo again used the kind of thought experiment which he applied to falling bodies. Consider, for example, a vertical pillar of square cross-section with side a (in appropriate units) supporting the greatest weight W (in appropriate units) it can bear before collapsing. We can place four such pillars side by side to form a new vertical pillar of square cross-section with side $2a$ and just supporting four loads each of weight W, or, equivalently a single load of weight $4W$. The old pillar had cross-sectional area a^2 and just supported a load of weight W; the new pillar has cross-sectional area $4a^2$ and can just support a load of weight $4W$. Thus, for a pillar of square cross-section, quadrupling the cross-sectional area quadruples the load it can bear. This strongly suggests, and further argument confirms, that the strength of a pillar of fixed length is proportional to its cross-sectional area. If we ignore the weight of the pillar we would expect the maximum load it could support not to depend on its height, but if we take its own weight into account we would expect the maximum load to decrease with height.

Suppose now that we have a pillar of height h and cross-sectional area A which can support a weight W. If we construct a pillar of the same shape made out of the same material but with doubled dimensions (so that it has height $2h$ and cross-sectional area $4A$) the argument shows that the maximum weight it can support will be at most $4W$. Galileo's analysis is more subtle and deals with more difficult problems† but we have already seen enough to follow his conclusion.

†In one case incorrectly, but pioneers are judged by what they get right, not what they get wrong.

> From what has already been demonstrated, you can plainly see the impossibility of increasing the size of structures to vast dimensions either in art or in nature; likewise the impossibility of building ships, palaces or temples of enormous size in such a way that their oars, yards, beams, iron bolts and, in short, all their other parts will hold together; nor can nature produce trees of extraordinary size because the branches would break down under their own weight; so also it would be impossible to build up the bony structures of men, horses or other animals so as to hold together and and perform their normal functions if these animals were to be increased enormously in height; for this increase in height can only be

accomplished by employing a material which is harder or stronger than usual, or by enlarging the size of the bones, thus changing their shape until the form and appearance of the animal suggests a monstrosity.

Consider, for example, a temple with a roof supported by columns. If we multiply all the linear dimensions by some large λ then the volume and so the weight of each part of the temple will be multiplied by λ^3 but the cross-section of the supporting columns will only be multiplied by λ^2 so their strength will in turn be multiplied by a factor of at most λ^2. If we make λ too big the columns must fail†. In the same way a giant constructed in the same way as a man but with all his linear dimensions multiplied by some large λ would have λ^3 times the weight of a man but the bones in his legs would be at most λ^2 times as strong as those of a man.

†It is said that the first question to be asked when visiting a medieval cathedral is 'When did the central tower collapse?'

In his classic essay *On Being The Right Size* Haldane recalls

> ... Giant Pope and Giant Pagan in the illustrated *Pilgrim's Progress* of my childhood. These monsters were not only ten times as high as Christian, but ten times as wide and ten times as thick, so that their total weight was a thousand times his, or about eighty to ninety tons. Unfortunately the cross-sections of their bones were only a hundred times those of Christian, so that every square inch of giant bone had to support ten times the weight borne by a square inch of human bone. As the human thigh-bone breaks under about ten times the human weight, Pope and Pagan would have broken their thighs every time they took a step. This was doubtless why they were sitting down in the picture I remember. But it lessens one's respect for Christian and Jack the Giant Killer.

Similar simple computations shed light on the working of the heart. The heart expels blood using the contraction of the surrounding muscle. Since muscle can only contract to well over half its resting length (and works best when asked for less substantial contractions), it would appear, at first sight, that the heart cannot be very efficient. However, since the volume of the heart varies as the *cube* of its linear dimensions, a contraction to 0.8 of resting length of the surrounding muscle would produce a reduction to $0.8^3 \approx 0.5$ of the volume and pump out about half the contents of the heart.

Exercise 5.2.1 *(i) Show that with the same model, a contraction to 0.6 of resting length of the surrounding muscle would pump out about three quarters the contents of the heart.*

(ii) More realistically, suppose that, when the heart is full, the volume of the surrounding muscle is twice that of the blood it contains. Show

that to pump out half the contents of the heart now requires only a 6%
contraction of the muscle surrounding and to expel all the blood requires
a contraction of about 13%†.

†This example is taken
from the charming book
Vital Circuits by Vogel
which shows how much
light a little physics and a
little mathematics can shed
on the workings of our
circulation.

Although it lies outside the main concerns of this chapter, I cannot
resist including an application of number theory to biology. Some
bamboos live for 80 years or more without flowering, then flower, set
seed, covering the ground beneath thickly with seeds, and die. Other,
more common bamboos flower and seed more frequently but do so
in synchrony, all seeding at the same time. At first sight this seems a
curious and wasteful procedure since many more seeds are produced
than could possibly produce new bamboos. The key to the riddle lies
not in the bamboos themselves, but in the numerous insects and birds

‡Though readers of Robert
Sheckley's witty *Dimension
of Miracles* will not need to
be told this.

which can consume the seeds — predators, as the biologists call them‡.
The bamboos are providing so much seed in such a short time that the
predators cannot eat all of it and some must survive.

Promising as this strategy of 'predator satiation' appears, it will not
work if the bamboos seed every year because the seed eaters will
then adjust their own breeding seasons so that their young can take
advantage of an annual feast. Few species of bamboo flower more often
than once in 15 or 20 years. There is a species of cicada in the Northern
United States which follows a similar pattern, living underground as
'nymphs' for 17 years and then emerging above ground in millions and,
in the space of a few weeks, completing their life cycle by becoming
adult, mating, laying eggs and dying. A related species in the South
does the same but with a cycle time of 13 years.

> Why do we have 13 and 17 year cicadas, but no cycles of 12, 14,
> 15, 16 or 18? Thirteen and 17 share a common property. They are
> large enough to exceed the life cycle of any predator, but they are also
> prime numbers (divisible by no integer smaller than themselves). Many
> potential predators have 2 to 5 year life cycles. Such cycles are not
> set by the availability of periodical cicadas (for they peak too often in
> years of non-emergence), but cicadas might be eagerly harvested when
> the cycles coincide. Consider a predator with a cycle of five years:
> if cicadas emerged every 15 years, each bloom would be hit by the
> predator. By cycling at a large prime number, cicadas minimise the
> number of coincidences (every 5×17 or 85 years, in this case). Thirteen-
> and 17-year cycles cannot be tracked by any smaller number.

We now return to the main theme of this chapter. When a sphere of
mass m falls, two forces act on it, the constant force mg due to gravity
and a resisting force F proportional to the square of its velocity v and
to its area A, i.e. $F = CAv^2$ for some constant C. If these two forces

balance, there is no acceleration and the sphere drops with constant 'terminal velocity' V. The balancing condition tells us that

$$mg = CAV^2$$

and so

$$V = \frac{m^{1/2}g^{1/2}}{C^{1/2}A^{1/2}}.$$

If the sphere has radius r and density ρ then

$$V = Kr^{1/2}\rho^{1/2}g^{1/2}$$

for some constant K. Thus the sphere will have terminal velocity proportional to the square root of its radius. More generally we expect the terminal velocity of similarly shaped objects to vary in proportion to the square root of their length (so that if the linear dimensions are reduced by a factor of 100 the maximum speed they reach is reduced by a factor of 10). Haldane put the matter with typical vividness.

> Gravity, a mere nuisance to Christian, was a terror to Pope, Pagan and Despair. To the mouse and any smaller animal it presents practically no dangers. You can drop a mouse down a thousand-yard mine shaft; and, on arriving at the bottom, it gets a slight shock and walks away. A rat is killed, a man is broken, a horse splashes. For the resistance presented to movement by the air is proportional to the surface of a moving object. Divide an animal's length, breadth and height each by ten; its weight is reduced to a thousandth, but its surface only to a hundredth. So the resistance to falling in the case of a small animal is relatively ten times greater than the driving force.
>
> An insect, therefore, is not afraid of gravity: it can fall without danger, and it can cling to the ceiling with remarkably little trouble. It can go in for elegant and fantastic forms of support like that of the daddy-long-legs. But there is a force which is as formidable to an insect as gravitation to a mammal. This is surface tension. A man coming out a bath carries with him a film of water about one-fiftieth of an inch [half a millimetre] in thickness. This weighs roughly a pound [half a kilogram]. A wet mouse has to carry about its own weight in water. A wet fly has to lift many times its own weight and, as everyone knows, a fly once wetted by water or any other liquid is in a very serious position indeed. An insect going for a drink is in as great danger as a man leaning over a precipice in search of food. If it once falls into the grip of the surface tension of the water — that is to say, gets wet — it is likely to remain so until it drowns. A few insects, such as water-beetles, contrive to be unwettable; the majority keep well away from their drink by means of a long proboscis.

Haldane's examples show the biological importance of the fact that surface increases like the square L^2 of the typical length L whilst volume

†Provided that length, breadth and height increase together. Snakes and worms avoid some of the problems of increasing size considered in this chapter.

increases more rapidly as the cube L^3 of that length†. This fact is also very important for warm blooded animals which are designed to keep at constant temperature. We would expect heat loss to be proportional to the surface area so the rate of heat production for the animal should be proportional to L^2. Since the volume (and so mass M) of an animal is proportional to L^3 this means that the rate of heat production for an animal should be proportional to $M^{2/3}$. Notice that this means that the rate of heat production per unit mass varies as $M^{-1/3}$ — so a gram of mouse will be 20 or so times as metabolically active as a gram of elephant.

We thus expect a relation of the form

$$P = cM^{\alpha}$$

where P is the metabolic rate, M the mass, c is constant and $\alpha = 2/3$. The natural way to investigate relations like this is to take logarithms, obtaining

$$\log P = \log c + \alpha \log M.$$

If we plot $\log P$ against $\log M$ we should see a straight line of slope α.

Biologists 'combining the ingenuity of Daedalus with the patience of Job' have measured the metabolic rate of a large number of warm blooded animals with the results shown in Figure 5.1 (copied from Schmidt-Nielsen's book *Scaling* which I have used extensively in this section). The 'mouse-to-elephant curve' as it is called is extraordinarily regular but does not quite fit our prediction since the 'best fitting' line shown has slope 0.74 (which we may as well round up to 3/4 for the purposes of discussion) rather than 2/3.

Several ingenious theories have been put forward to account for this discrepancy. For example, it was suggested that the metabolic rate is directly proportional to surface area but that, since animals come in different shapes (an elephant is not a scaled-up mouse), surface area grows more slowly than $M^{2/3}$. Some idea of the difficulty of verifying biological theories of this type is given in the following quotation.

Many methods have been used for measuring surface areas. An animal can be skinned and the area of the skin determined, but how does one know how much to stretch the skin? The animal can be divided into a number of cylinders and cones, and the area of each determined separately. The animal can be covered with paper and the area of paper afterward determined, either by planimetry or by weighing the paper. The surface area of a cow has been determined by covering the animal with an ink roller, counting the revolutions of the roller ... Various authors disagree even when only one animal is involved. [For example, values for the surface area of a rat of given weight may differ by a factor of two.] ... It is uncertain just what is meant by surface area. Does the 'true' surface area of an animal include the skin area between the legs that is not exposed to the outside? Does it include the ears, and if so, both sides?

I would argue that rather than seeking a single reason for nature's apparent preference for a 3/4 rather than a 2/3 power law, we should pause to think whether we have allowed our mathematical enthusiasm to carry us a little too far. Here are my reasons.

Closeness of exponents   An elephant weighs about 10^5 times the weight of a mouse. The ratio $R_{3/4}$ of the metabolic rates for the two predicted by the 3/4 rule is given by

$$R_{3/4} = \left(\frac{\text{mass of elephant}}{\text{mass of mouse}} \right)^{3/4}$$

and ratio $R_{2/3}$ of the metabolic rates for the two predicted by the 2/3 rule is given by

$$R_{2/3} = \left(\frac{\text{mass of elephant}}{\text{mass of mouse}} \right)^{2/3}.$$

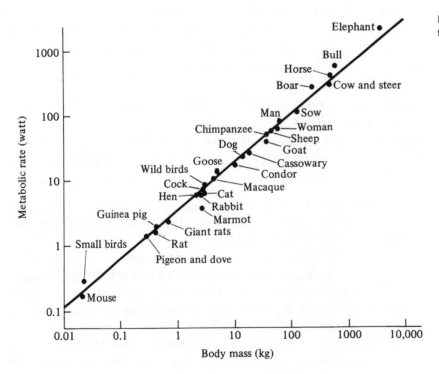

Figure 5.1: Metabolic rates for mammals and birds.

Thus

$$\frac{R_{3/4}}{R_{2/3}} = \left(\frac{\text{mass of elephant}}{\text{mass of mouse}}\right)^{3/4-2/3}$$

$$= \left(\frac{\text{mass of elephant}}{\text{mass of mouse}}\right)^{1/12}$$

$$\leq (10^5)^{1/12} = 10^{5/12} \approx 2.6$$

and the metabolic rate of the elephant predicted by looking at the mouse and applying the 3/4 rule is only about 5/2 times as large as that predicted by applying the 2/3 rule. It is hard to believe that such a small difference is important in this context.

Non-uniqueness of causes   In order to produce heat, animals must burn oxygen. If the elephant had the same metabolic rate per unit mass as a mouse, it would have an overall metabolic rate about 20 times as large as its actual value. Not only would it have to get rid of 20 times as much heat per unit time but it would have to consume 20 times as much oxygen per unit time. I suspect that most engineers would prefer the task of redesigning the elephant's cooling system to cope with a 20-fold increase in heat output to that of redesigning its breathing apparatus to cope with a 20-fold increase in oxygen consumption. Since the rate of oxygen absorption is also proportional to the surface area used, simple scaling arguments based on oxygen consumption also give a 2/3 power rule. Should we say then, that the mouse-to-elephant curve depends on heat loss or that it depends on oxygen consumption? Surely this is a false dichotomy. All we can say is that any animal whose metabolic rate departed greatly from the 2/3 rule would have problems *both* with heat balance and oxygen supply. Nor should we expect that we have identified all the problems which would face such an animal.

Non-uniqueness of purpose   If you examine a screwdriver you will find that the shaft is far thicker than it needs to be to avoid twisting when screwing in a screw or even when unscrewing a rusty screw. This seems very wasteful until you need it to lever off the top of a can of paint†. (Laithwaite contrasts this with the tea stirrer found in some fast food outlets ' ... a plastic strip weakened by a slot in it so that it can never be a screwdriver, or a chisel, is too thick to be a tooth pick, not sharp enough to be a scriber, cannot pick locks, peel apples or open packets of sliced bacon. It can only stir drinks.')

On the whole, however, our machines, from jumbo jets to traffic lights and from television sets to soft-drink dispensers, are designed to do just

†The soldier's habit of using sub-machine-guns as bottle-openers lead to the incorporation of a purpose-built bottle-opener in at least one type.

one thing and to do it under fairly benign conditions. Animals must do many things and do them under a wide range of conditions. The passage from Haldane shows that human bones are ten times as strong as they need to be for standing still. However, we need to walk and, in the past, human beings needed to run and jump. This puts our bones under strong bending stresses and brings them much closer to their safety limits. (The reader will note that we showed that the strength under *compression* of bones is proportional to their cross-sectional area. Galileo showed that the same was true for *bending* though there are still other ways in which failure can occur and for which the area rule fails.) Because bone strength increases with cross-sectional area whilst weight increases with volume, we expect the proportion of skeleton in the body of an animal to increase with size and indeed a mouse is about 4.5% skeleton, a cat is about 7% skeleton and an elephant about 12%. However, this increase in proportion is nowhere near enough to maintain the relative strength of the skeleton — the elephant is a fragile creature compared to a mouse, at least in so far as its bone structure is concerned. On the other hand the elephant has a much less strenuous life than a mouse and so puts much less strain on its bones. A mouse which saunters across an open space runs the risk of ending up as someone's dinner — the elephant can take as long as it wants. The mouse's metabolic rate per unit mass is 20 times that of the elephant so it burns up and must consume 20 times as much food per unit mass each day (a mouse must consume about a quarter of its body weight in food each day). Small mammals live lives of frenetic activity.

Variation in design   The simple scaling arguments we have used ignore the possibilities of changes in design. For example, animals may have fur or use fat as insulation. (Large tropical animals, like the elephant, are more concerned with getting rid of heat than conserving it and have no fur.) Notice that just as larger convoys required a proportionally smaller effort to defend, so large animals can insulate themselves more easily than small ones. The mass of a centimetre thickness of insulation is proportional to the surface area insulated and thus follows the familiar 2/3 power rule (i.e. the weight of insulation is proportional to $M^{2/3}$ where M is the total body weight). There are no small mammals in arctic regions.

Having said all this, it is still clear that scaling arguments do provide substantial insight into the way animals live. We end our discussion with three examples from animal locomotion — diving, climbing and jumping. (The reader will not need to be warned that the calculations involved are of the back of an envelope type and statements of the form '*a* equals *b*' should be read '*a* equals *b* approximately' or even 'with a bit of luck, *a* and *b* will be of the same order of magnitude'.)

Diving Consider a diving mammal which must carry down with it all the oxygen it needs. We would expect the amount of oxygen it can carry to be proportional to its volume and so to its mass M. It then consumes oxygen at a rate proportional to its metabolic rate M^α so the time it can stay submerged will be proportional to

$$\frac{M}{M^\alpha} = M^{1-\alpha}.$$

The 3/4 rule suggests that $\alpha = 3/4$ and so the time it can stay submerged is proportional to the *fourth* root $M^{1/4}$ of its mass and this may be one reason why whales are so large. (This example is taken from Maynard Smith's *Mathematical Ideas in Biology*. Since he is writing for biologists the author keeps the mathematics simple and so well within the capabilities of most readers of this book.)

Climbing If we raise a body of mass M through a height of h we increase its potential energy by Mgh. If an animal of mass M climbs with steady velocity v then it will need to expend energy at a rate Mgv. If its metabolic rate is cM^α then

proportion of total energy produced used in climbing

$$= \frac{\text{rate of energy expenditure on climbing}}{\text{metabolic rate}}$$

$$= \frac{Mgv}{cM^\alpha} = KvM^{1-\alpha},$$

for some constant K. Experiments cited in [211] suggest that climbing vertically at 2 kilometres per hour requires a hardly noticeable increase in oxygen consumption of 23% for a mouse but a very substantial 189% for a chimpanzee. The ease with which a squirrel scampers up tree trunks is genuine; for an animal of such small size it makes little difference whether it runs up or down.

Jumping The flea can jump to a height more than 100 times its own body length whilst, even in my athletic prime, I doubt that I could clear a high jump of half my body length. What do the scaling arguments of this chapter have to say about this? The energy E for a jump comes from shortening the muscles in the animal's body and the energy so produced is proportional to the volume of muscle available. The volume of muscle available is in turn proportional to the volume of the animal's body and so to its mass. Thus

$$E = KM$$

for some constant K. Since the increase in potential energy for a jump of height h is Mgh we have

$$KM = E = Mgh$$

and so

$$h = K^{-1}g.$$

In other words the height an animal can jump is independent of its size. In fact the maximum heights jumped by fleas, men, kangaroos, jerboas and grasshoppers lie within a factor of 3 of each other†.

Exercise 5.2.2 *(i) Use scaling arguments to discuss the fact that children seem to find falling over less painful than adults.*

(ii) How can pole vaulters jump three times their own height?

We have found plausible (though not necessarily correct) reasons for various numerical facts in biology. No discussion would be complete without mentioning one set of numerical coincidences for which no really convincing explanation has been given. Small mammals have high metabolic rates: they breathe fast to obtain the oxygen they must burn, their hearts beat rapidly to pump the oxygen bearing blood to their muscles, they live fast and die soon. Large mammals take things slowly and die late. Nielsen puts the matter clearly (remember that we are not talking about exact equalities but about orders of magnitude).

> A 30 gram mouse that breathes at a rate of 150 times per minute will breathe about 200 million times during its 3-year life; a 5-ton elephant that breathes at the rate of 6 times per minute will take approximately the same number of breaths during its 40-year lifespan. The heart of the mouse ticking away at 600 beats per minute, will give the mouse some 800 million heartbeats during its lifetime. The elephant, with its heart beating 30 times per minute, is awarded the same number of heartbeats during its life.

The number of heartbeats per lifetime is roughly the same (about 10^9) for most mammals‡. Why?

The poet Petrarch ridicules the ideas of natural history held by one of his contemporaries§ and goes on to say that

> ... even if they were true, they would not contribute anything whatsoever to the blessed life. What is the use — I beseech you — of knowing the nature of quadrupeds, fowls, fishes and serpents and not knowing or

†However, the flea's performance is still extraordinary. Because the flea is so small the distance over which it can accelerate before takeoff is tiny (less than 1 millimetre) so the takeoff time is very short (less than 1 millisecond) and the acceleration correspondingly large (over 200g, i.e. 200 times the acceleration due to gravity). Since muscle contraction cannot proceed this fast, the flea stores energy by compressing a piece of elastic material and releases it by tripping a release mechanism.

‡Humans are a bit on the high side, with more than double this figure.

§He had made the mistake of calling Petrarch 'certainly a good man but a scholar of poor merit.'

even neglecting man's nature, the purpose for which we are born, and whence and whereto we travel?

There is some truth in this, but a journey is not defined by its destination and good travellers seek to understand the country through which they pass.

CHAPTER 6

Physics in a darkened room

6.1 The pyramid inch

We now take up again some of ideas raised in the previous chapter which the reader may wish to reread. In Section 5.1 we saw how Galileo showed the implausibility of Aristotle's suggestion that the speed with which a body falls is proportional to its weight. Galileo's argument is an example of 'physics in a darkened room' where we attempt to find, by pure thought alone, restrictions which must apply to any possible physical law — a dangerous but fascinating pastime. One example is given by 'dimensional analysis'.

Consider the formula for the area A of a rectangle with sides of length a and b:

$$A = ab. \tag{6.1}$$

If we measure a and b in centimetres we obtain a value for A (measured in square centimetres) which is $10\,000 = 100^2$ times the value for A (measured in square metres) which we obtain if we measure a and b in metres. We say that A has the dimensions of length squared and write

$$[A] = L^2. \tag{6.2}$$

Formally this means that if we use two units of length (call them the new unit and the old) such that the old unit is K times the new, then measuring A in the new units produces an answer K^2 times that of the old. As the reader knows the result applies more generally to the area of all reasonable shapes. In the same way volume has the dimension of length cubed.

Now suppose we wish to find the volume V of a cone of height h standing on an elliptical base with semi-axes of length a and b. If someone suggests that

$$V = \pi h a^2 b^2$$

116

we know immediately that the suggestion cannot work since, if we use the new and old units of the previous paragraph

$$V_{\text{new}} = K^3 V_{\text{old}}$$

but

$$h_{\text{new}} a_{\text{new}}^2 b_{\text{new}}^2 = K^5 \pi h_{\text{old}} a_{\text{old}}^2 b_{\text{old}}^2.$$

Thus if the formula gives the correct answer in one set of units it cannot give the correct answer in the other. The same argument may be put more briefly by observing that

$$[V] = L^3, \quad [ha^2b^2] = [h][a]^2[b]^2 = LL^2L^2 = L^5.$$

The two sides of the putative equation 'have different dimensions' and we say that the formula is 'dimensionally incorrect'.

Exercise 6.1.1 *Show that the formula (which is, in fact, correct)*

$$V = \frac{\pi}{3} hab$$

is dimensionally correct.

In 1864 The Astronomer Royal of Scotland, Charles Piazzi Smyth, published a book *Our Inheritance in the Great Pyramid* which Martin Gardner calls a classic of its kind. The first casing stone for the great pyramid had just been unearthed.

> The stone measured slightly more than twenty-five inches, and Smyth concluded that the length was none other than the sacred cubit. If we adopt a new inch — Smyth calls it the 'Pyramid inch' which is exactly one twenty-fifth of the width of the casing stone, then we obtain the smallest divine unit of measurement used in the monument's construction. It is exactly one ten-millionth of the earth's polar radius. Somehow, it had been passed on through the generations, but in the process altered slightly, making the British inch a trifle short of the sacred unit. Many years later a number of other casing stones were dug up. They had entirely different widths. By that time, however, the Pyramid inch had become so firmly established in the literature of Pyramidology that devotees merely shrugged and admitted that the first casing stone just 'happened' to be a cubit wide.

A society was formed in the United States 'to work for the revision of measuring units to conform to sacred pyramid standards and to combat the "atheistic metrical system" of France. President James A. Garfield was a supporter of the Society, though he declined to serve as president.' Gardner quotes part of one of its rallying songs:

> Then down with every 'metric' scheme,
> Taught by the foreign school,
> We'll worship still our Fathers' God!
> And keep our Fathers' 'rule'!
> A perfect inch, a perfect pint,
> The Anglo's honest pound,
> Shall hold their place upon the earth,
> Till time's last trump shall sound!

If I were a pyramidologist of this kind, then I would not be surprised if formulae of physics which were correct when expressed in sacred inches ceased to be valid in the metric system. I would expect the laws of nature to take their correct form when expressed in the correct, God-given, English units and I would expect meaningless (or at least ugly) results if someone tried to write them in the atheistic, French metric units. Since I do not believe that nature has a preferred set of units, I believe that the laws of nature do not depend on a choice of units and I express this belief by demanding that the formulae of physics should be dimensionally correct.

In Newtonian mechanics and related systems, all quantities are expressed in terms of mass, length and time. For example, density may be expressed in kilograms per cubic metre (or grams per cubic centimetre). We say that density ρ has dimensions ML^{-3} and write

$$[\rho] = ML^{-3}.$$

The mass m of a ball of radius r and density ρ is given by

$$m = \frac{4\pi}{3}r^3\rho, \tag{6.3}$$

and we see that, as we require, the formula is dimensionally correct with

$$\left[\frac{4\pi}{3}r^3\rho\right] = [r^3][\rho] = [r]^3[\rho] = L^3ML^{-3} = M = [m].$$

In the same way velocity has dimensions LT^{-1} (so that we measure velocity in metres per second or kilometres per hour) and acceleration has dimensions LT^{-2}. Newton's second law says that the acceleration a of particle of mass m under a force F is given by

$$F = ma.$$

Since we demand that the formulae of physics should be dimensionally correct we must have

$$[F] = [m][a] = MLT^{-2}.$$

†One of Ramsey's sons was the remarkable mathematician and philosopher Frank Ramsey. The other was Archbishop of Canterbury.

Exercise 6.1.2 *In Ramsey's text-book* Dynamics† *we are asked to prove that the minimum velocity v required to project a particle from a height h to fall at horizontal distance a from the point of projection is given by*

$$v = (g((a^2 + h^2)^{1/2} - h))^{1/2}. \tag{6.4}$$

Check that the given answer is dimensionally correct. (Remember that g is the acceleration due to gravity.)

Experience soon shows that, given any dimensionally correct formula, we can manipulate it into a form in which all the quantities which appear are dimensionless (i.e. do not vary in size if we change the units). For example the formula

$$s = ut + \tfrac{1}{2}at^2$$

which gives the distance travelled in a time t by a particle with initial velocity u and steady acceleration a may be rewritten as

$$\pi_1 + \tfrac{1}{2}\pi_2 - 1 = 0$$

where

$$\pi_1 = \frac{ut}{s}, \ \pi_2 = \frac{at^2}{s}$$

are dimensionless. (For example $[\pi_1] = [u][t][s]^{-1} = LT^{-1}TL^{-1} = L^0T^0$; the reader should check that π_2 is dimensionless.) The reader may convince herself of this at her leisure‡ but for the moment she should accept this as a fact and see what consequences follow.

‡It may help to recall the *bon mot* I heard from a Russian physicist: 'Proofs in physics follow the standards of British justice and hold the accused innocent until proved guilty. Proofs in mathematics follow the standards of Stalinist justice and hold the accused guilty until proved innocent.'

Since the formulae of physics are dimensionally correct, it follows that they can all be rewritten in terms of dimensionless quantities. Thus each formula of physics can be rewritten in the form

$$f(\tau_1, \tau_2, \ldots, \tau_k) = 0$$

where τ_1, τ_2, ..., τ_k are all dimensionless quantities. Equation (6.1) becomes

$$f(\tau_1) = 0$$

with $f(\tau_1) = \tau_1 - 1$ and $\tau_1 = A/ab$. Equation (6.3) becomes

$$f(\tau_1) = 0,$$

with $f(\tau_1) = \tau_1 - 4\pi/3$ and $\tau_1 = m/(r^3\rho)$ and Equation (6.4) becomes

$$f(\tau_1, \tau_2) = 0,$$

with $f(\tau_1, \tau_2) = \sqrt{(\sqrt{(\tau_1^2 + \tau_2^2)} - \tau_2)} - 1$ and $\tau_1 = ga/v$, $\tau_2 = gh/v$.

So far what we have done may seem rather less exciting than pyramidology (or even watching the grass grow), but we can now draw some interesting conclusions.

The simple pendulum   Consider a pendulum consisting of a heavy weight (of mass m, say) suspended by a long light string or rod (of length l, say). If we set it gently swinging, we know that it will swing back and forth in a time t. (We call t the period of the pendulum. A pendulum clock is a machine for counting the swings of a pendulum.) Clearly m, l, t and g (the acceleration due to gravity) are related by some formula, but which? By the discussion of the previous paragraph we can rewrite it so as to involve only dimensionless variables. What dimensionless variables can we form from m, l, t and g? Since

$$[m] = M, \ [l] = L, \ [t] = T, \ \text{and} \ [g] = LT^{-2}$$

it is not hard to see that the only dimensionless variable we can form from these is

$$\tau = g l^{-1} t^{-2}.$$

(Of course, 2τ, τ^{-1} and so on are also dimensionless but they are not really different variables.) Thus the equation of the pendulum has the form

$$f(\tau) = 0.$$

We know that (unless f has some very special form) the equation $f(x) = 0$ fixes x (as a root of f). Thus we deduce that

$$\tau = A$$

for some constant A and so

$$g l^{-1} t^{-2} = A$$

whence

$$t^2 = A \frac{g}{l}.$$

A must be positive so we may take $C = A^{1/2}$ and obtain

$$t = C \sqrt{\frac{g}{l}}.$$

Thus the period of a simple pendulum does not depend on the mass of the bob and varies inversely with the square root of its length. Quadrupling the length of a pendulum doubles the period of swing.

A hovering helicopter   Since the kinetic energy E of a body of mass m and velocity v is given by the formula $E = \frac{1}{2} m v^2$ we have

$$[E] = [m][v]^2 = M(LT^{-1})^2 = ML^2 T^{-2}.$$

The principle of conservation of energy tells us that all forms of energy are interconvertible and thus must all have dimension $ML^2 T^{-2}$. Since the power of a motor is the usable energy produced per unit time, power must have dimensions $(ML^2T^{-2})T^{-1} = ML^2T^{-3}$.

Suppose we wish to find out how much power P is required to allow a helicopter of mass m to hover in still air. Presumably P depends on the length l of the rotor blades, the weight $W = mg$ of the helicopter and the density ρ of the air. We have

$$[P] = ML^2T^{-3}, \ [l] = L, \ [W] = MLT^{-2}, \ \rho = ML^{-3}$$

and a little investigation soon convinces us that, essentially, the only dimensionless variable we can form is

$$P^2W^{-3}\rho l^2.$$

The same argument as for the simple pendulum now gives

$$P^2W^{-3}\rho l^2 = A$$

and

$$P = CW^{3/2}l^{-1}\rho^{-1/2} = Cm^{3/2}g^{3/2}l^{-1}\rho^{-1/2},$$

for some constants A and C. We thus see why helicopters have long blades.

Exercise 6.1.3 *(Waves in deep water) The wavelength λ of a series of waves in deep water is the distance between successive crests. As a first approximation it seems reasonable to assume that the velocity v of the waves depends only on the wavelength λ, the acceleration due to gravity g and the water density ρ. Show that the velocity is proportional to the square root of the wavelength.*

If you stir a cup of coffee and leave it, the motion of liquid dies down quite rapidly because of frictional forces. It turns out that the correct measure of this frictional tendency is the *coefficient of viscosity* η of the liquid. All fluids (that is liquids or gases) have such a coefficient of viscosity. Air has a low coefficient of viscosity and treacle a very high one†. The only fact we shall use about η is its dimensions

$$[\eta] = ML^{-1}T^{-1}.$$

†According to materials scientists, glass is a liquid with an extremely high coefficient of viscosity. Very old panes of glass are thicker at their lower ends because of the slow downward flow of the glass.

Flow in a circular pipe   Consider a long cylindrical pipe of length l with circular cross-section of radius a. Suppose that a liquid of coefficient of viscosity η flows steadily along the pipe. We are interested in the rate Q (measured in volume per unit time) of flow of the liquid as a function of the pressure drop P per unit length. Since pressure is given by force per unit area

$$[P] = [\text{force}][\text{area}]^{-1}L^{-1} = (MLT^{-2})(L^2)^{-1}L^{-1} = ML^{-2}T^{-2}.$$

Further

$$[Q] = L^3 T^{-1}, \quad [a] = L, \quad [\eta] = ML^{-1}T^{-1}$$

and a little investigation soon convinces us that, essentially, the only dimensionless variable we can form is

$$P\eta^{-1}Q^{-1}a^4,$$

so by our usual argument

$$Q = \frac{Ca^4 P}{\eta},$$

for some constant C. Thus, for a given pressure gradient P, the volume flow Q increases with the fourth power of the radius. Doubling the diameter of an oil pipeline enables it to carry 16 times as much.

We have in our bodies two rather large and important pipes — the left and right coronary arteries which supply blood to the heart muscles. When fatty material is deposited on their walls, they narrow and the formula above shows that their carrying capacity drops as the fourth power of the radius. There is a substantial factor of safety built into the system but the fourth power law means that once the danger level is reached things can get worse very rapidly.

Wind tunnels The equations governing the behaviour of fluids are notoriously difficult to solve. It is said that the pure mathematicians who study them keep a bucket of water by the side of their blackboards as a reminder that nature has already found some solutions. One way of getting round this problem is the use of scale models.

Suppose we wish to find the lift F (a force) on the wing of a low-speed aircraft. We assume that viscosity and pressure changes may be neglected and that F depends only on a a representative length (say the wing span), ρ the air density and v the velocity of the air flow past the wing. The reader should apply dimensional analysis to show that

$$F = A\rho a^2 v^2,$$

for some unknown constant A. By doing an experiment on a scaled model in a wind tunnel we can determine A and so find out the lift on a full scale wing.

The bifilar pendulum The bifilar pendulum consists of a horizontal bar of length b and mass m suspended from fixed supports by two equal vertical strings to each end of the bar and allowed to make small vibrations. At school we were set as an experimental task to find how the period t varied with b and m. With the confidence of youth and the arrogance of a pure mathematician, I first performed a dimensional analysis of the type above and then spent a painful hour trying to get the pendulum to behave as predicted.

However, as the reader may already have spotted, there is another length involved — the length a of the strings. The quantities involved have dimensions given by

$$[a] = L, \ [b] = L, \ [g] = LT^{-2}, \ [m] = M, \ [t] = T.$$

We can form *two* dimensionless quantities

$$\tau_1 = \frac{a}{b}, \ \tau_2 = \frac{gt^2}{a}$$

neither of which depends on the other. There are many other dimensionless quantities we can form (for example, $\tau_3 = gt^2/b$) but a little experiment will convince the reader that they are all functions of τ_1 and τ_2 (for example $\tau_3 = \tau_1\tau_2$). We say that τ_1 and τ_2 form a complete set of dimensionless quantities for the problem. (Note that τ_1 and τ_3 also form such a set. There are many different complete sets for the same problem.) Since we believe that all our formulae can be written in dimensionless form we obtain

$$f(\tau_1, \tau_2) = 0.$$

At first sight this looks useless, but this is not so. If τ_1 is fixed then τ_2 is determined as a root of $f(\tau_1, x) = 0$. Brushing aside any problems due to the possibility of several roots, we conclude that τ_2 is function of τ_1, i.e.

$$\tau_2 = F(\tau_1).$$

It follows that

$$\frac{gt^2}{a} = H(ab^{-1})$$

for some appropriate function H, and so

$$t = \sqrt{\frac{a}{g}} G(ab^{-1})$$

for some appropriate function G. Thus, if I know the value of t for some value of a and b, I know it for all values of a and b which have the same ratio ab^{-1}. This is of no help in trying to find the behaviour of the bifilar pendulum without doing any experiments at all but, once we bite the bullet, it does greatly reduce the number of experiments needed.

Exercise 6.1.4 *A traditional mechanics question considers a ball of mass m thrown vertically upwards with a velocity u. You are told that when the ball has speed u it experiences a resistive force ku^α in the opposite* *direction to its motion, and asked to find the time t it will take to return to the point of projection. Find the dimensions of k and show that*

$$t = \frac{u}{g} F\left(\frac{ku^\alpha}{mg}\right)$$

for some suitable function F.

Supersonic wind tunnel   Suppose that we are interested in the drag D (a force) on a body of characteristic length a moving with high velocity v in a gas which has density ρ and which exerts a pressure P at rest. We would expect the effect of viscosity to be negligible and thus hope that the named quantities are the main ones involved. The reader should check that the two dimensionless quantities

$$\tau_1 = \frac{v\rho^{1/2}}{p^{1/2}}, \quad \tau_2 = \frac{D}{\rho v^2 a^2}$$

form a complete system, and so, arguing as for the bifilar pendulum,

$$\tau_2 = G(\tau_1),$$

and

$$D = \rho v^2 a^2 F\left(\frac{v\rho^{1/2}}{p^{1/2}}\right)$$

for some function F.

   The quantity $(p/\rho)^{1/2}$ has the dimensions of velocity. In fact it is known that the speed of sound c in the undisturbed gas is given by

$$c = \left(\frac{\gamma p}{\rho}\right)^{1/2}$$

for some constant γ (depending on the gas) and so

$$D = \rho v^2 a^2 G\left(\frac{v}{c}\right)$$

for some function G. (The dimensionless quantity v/c is called the Mach number and will thus be familiar to devotees of films about test pilots.) The reader will see that it is easy to scale up the result of this sort of wind tunnel experiment.

Ship motion   In building ships we are interested in the power P required to keep a ship of length l moving at constant velocity v. Presumably this will depend on the density ρ and viscosity η of the water. Since some of the work done consists in making waves, we must also consider the acceleration g due to gravity. Now

$$[P] = ML^2 T^{-3}, \ [l] = L, \ [v] = LT^{-1},$$
$$[\rho] = ML^{-3}, \ [\eta] = ML^{-1}T^{-1}, [g] = LT^{-2}$$

and a little investigation shows that we now require *three* dimensionless variables to form a complete set. One possible choice (among many) is

$$\tau_1 = \frac{v}{(lg)^{1/2}}, \ \tau_2 = \frac{vl\rho}{\eta}, \ \tau_3 = \frac{P}{\rho l^2 v^3}.$$

The same kind of arguments as we used in the case of the bifilar pendulum now show the existence of a function f such that

$$f(\tau_1, \tau_2, \tau_3) = 0,$$

and a function F such that

$$\tau_3 = F(\tau_1, \tau_2).$$

We write

$$\mathrm{Fr} = \tau_1 = \frac{v}{(lg)^{1/2}}, \ \mathrm{Re} = \tau_2 = \frac{vl\rho}{\eta}$$

and obtain

$$P = \rho l^2 v^3 F(\mathrm{Fr}, \mathrm{Re}).$$

Fr is called the Froude number and is associated with wave resistance. Re is called the Reynolds number and is associated with viscous resistance†.

In practice we cannot change η as we would wish, so for the purposes of discussion let us assume that we cannot change η at all. Under these circumstances Re/Fr is proportional to $l^{3/2}$ and it is impossible to scale down to a model while keeping both the Froude and Reynolds number the same. Unfortunately it turns out that both wave and viscous resistance are important and so modellers have to resort to various tricks of the trade (for example, by trying to find the wave resistance and viscous resistance separately and using experience to estimate their joint effect). We note that unless they are near the water surface, fish and submarines will not generate waves and their behaviour will be governed by the Reynolds number‡.

†Because of the way that it is defined a *small* Reynolds number is associated with a *high* viscosity (treacle) and large Reynolds number with low viscosity.

‡Nuclear powered submarines are short and fat since they only surface to enter and leave harbour. On the surface they are exceedingly inefficient with most of their power going into making waves. The First and Second World War submarines spent most of their time on the surface and were long and thin to reduce wave resistance. Modern conventional submarines must spend time travelling near the surface whilst they recharge their batteries using a schnorkel and, although wave resistance at schnorkel depth is considerably less than at the surface, it is still important. Their shape is thus a compromise.

Exercise 6.1.5 *Consider a ship carrying cargo a distance d at a velocity v. It costs α dollars per unit time to run the ship (because of crew wages and so on) and β dollars per unit of energy. The power required to travel at velocity v is $P(v)$ so that the cost per unit time of propelling the ship at velocity v is $\beta P(v)$ dollars per unit of time. Find the cost $\kappa(v)$ of carrying the cargo over the required distance.*

If $P(v) = A + Bv + Cv^2$ over the range of v considered [A, B, C > 0] show that

$$\kappa(v) = d\left[\left(\frac{(\alpha + \beta A)^{1/2}}{v^{1/2}} - (\beta C)^{1/2}v^{1/2}\right)^2 + \beta B + 2(\beta C(\alpha + \beta A))^{1/2}\right].$$

Deduce that the cheapest way of carrying the cargo on the required route is to choose $v = v^*$ where

$$v^* = \left(\frac{\alpha + \beta A}{\beta C}\right)^{1/2}.$$

What happens to v^* as α increases?

For some time tankers which were to be manned by United States crews were built with more powerful engines than those sailing with British crews. The United States crews had higher wages and consequently higher v^*.

The increasing power of the computer may eventually make the job of the modellers obsolete together with their water channels, water flumes, water tunnels, wind tunnels, shock tunnels, ship towing tanks, high-speed railways, annular water channels, whirling arms, tethered dynamic models, steering basins, ditching tanks, dropped bodies, fired and rocket driven models, spinning tunnels, surge tanks and tidal model basins, but for many years their mixture of science, art and a dash of black magic were essential to the progress of engineering.

There are various points that I have brushed aside in this discussion. The more mathematical ones like the question of how to find a complete set of dimensionless variables or how we know that each complete set will have the same number of variables are resolved by the methods of abstract linear algebra as taught in the first or second year of a university mathematics course†.

A more important problem is the choice of variables. How do we know that the key variables for a hovering helicopter are the power, the length of the rotor blades, the weight of the helicopter and the density of air? Clearly the colour of the pilot's eyes is unimportant, but what about air pressure and viscosity? In real life, dimensional analysis begins with a careful study of the equations involved and depends on long experience.

Consider the example of the simple pendulum with which we began. I took as key variables the length l of the pendulum the mass m of the attached weight and the acceleration g due to gravity. I ignored the amplitude (length) of the swing because 'as everyone knows' the period t is independent of the amplitude (for small swings). However, the independence is not an obvious fact and was the first great discovery of Galileo (made at the age of 19). Before closing my eyes to meditate on the pendulum I took a surreptitious peek at the real world‡.

In 1953, Einstein composed a preface to Galileo's *Dialogue Concerning the Two Chief World Systems*. The modern master of physics in a darkened room wrote:

†The complex of ideas is called the Buckingham π theorem. Like many such results it exists in many different forms. The classical exposition is in Birkhoff's *Hydrodynamics* but Chapter 1 of Logan's *Applied Mathematics* may be more accessible.

‡We can get a little further using dimensional analysis as I shall show in Exercise 8.2.5.

There is no empirical method without speculative concepts and systems; and there is no speculative thinking whose concepts do not reveal, on closer investigation, the empirical material from which they stem.

6.2 A different age

How long would you expect a paper reporting a crucial experiment in physics to be and how would you expect it to be written? Here in its entirety is a paper entitled *Interference Fringes With Feeble Light* written by G. I. Taylor in 1909 (to be found in his collected works).

The phenomenon of ionisation by light and by Röntgen rays has led to a theory according to which energy is distributed unevenly over a wave-front. There are regions of maximum energy widely separated by large undisturbed areas. When the intensity of light is reduced these regions become more widely separated, but the amount of energy in any one of them does not change; that is they are indivisible units.

So far, all the evidence brought forward in support of the theory has been of an indirect nature; for all ordinary optical phenomena are average effects, and are therefore incapable of differentiating between the usual electromagnetic theory and the modification of it that we are considering. Sir J. J. Thomson, however, suggested that if the intensity of light in a diffraction pattern were so greatly reduced that only a few of these indivisible units of energy should occur on a Huygens zone at once, the ordinary phenomena of diffraction would be modified. Photographs were taken of the shadow of a needle, the source of light being a narrow slit placed in front of a gas flame. The intensity of light was reduced by means of smoked glass screens.

Before making any exposures it was necessary to find out what proportion of the light was cut off by these screens. A plate was exposed to direct gas light for a certain time. The gas flame was then shielded by the various screens that were to be used, and other plates of the same kind were exposed till they came out as black as the first plate on being completely developed. The times of exposure necessary to produce this result were taken as inversely proportional to the intensities. Experiments made to test the truth of this assumption showed it to be true if the light was not too feeble.

Five diffraction photographs were then taken, the first with direct light and the others with the various screens inserted between the gas flame and the slit. The time of exposure for the first photograph was obtained by trial, a certain standard of blackness being attained by the plate when fully developed. The remaining times of the exposure were taken from the first in the inverse ratio of the intensities. The longest time was 2000 hours or about 3 months. In no case was there any diminution in the sharpness of the pattern although the plates did not all reach the standard blackness of the first photograph.

In order to get some idea of the energy of the light falling on the plates in these experiments, a plate of the same kind was exposed at a distance of two metres from a standard candle till complete development brought it up to the standard of blackness. Ten seconds sufficed for this. A simple calculation will show that the amount of energy falling on the plate during the longest exposure was the same as that due to a standard candle burning at a distance slightly exceeding a mile. Taking the value given by Drude for the energy in the visible part of the spectrum of a standard candle, the amount of energy falling on 1 cm^2 of the plate is 5×10^{-6} ergs/sec and the amount of energy per cm^3 of this radiation is 1.6×10^{-16} ergs.

According to Sir J. J. Thomson, this value sets an upper limit on the amount of energy contained in one of the indivisible units above.

In 1905, Einstein, then 'Technical Expert, Third Class' at the Bern Patent Office, wrote papers announcing his theory of special relativity, a proof via Brownian motion of the physical existence of molecules and an explanation of the photo-electric effect in terms of a quantum theory of light. The scientific community was readily convinced of the truth of the first two theories but remained sceptical of the third for 20 years.

Einstein proposed that light was made up of individual particles or quanta which we now call photons. This seemed totally contrary, not simply to the immensely successful Maxwell theory which treated light as the wavelike propagation of an electromagnetic disturbance, but to a 100 years of observation of the wave character of light. Particles travel in straight lines but waves spread out and the cancellation of troughs with crests and the reinforcement of crests by crests and troughs by troughs creates the 'interference patterns' which characterise wave motion. If we shine a thin beam of light at a needle we see, not the sharp shadow that we expect a stream of particles to produce, but the interference patterns typical of wave propagation.

One way of reconciling Einstein's proposal with this observation was the suggestion that 'optical phenomena are average effects' and that large numbers of jostling photons somehow produce the observed wavelike phenomena. (After all, sound and water waves are produced by large numbers of particles.) When G. I. Taylor, fresh from Cambridge undergraduate study of mathematics and physics, sought a research project J. J. Thomson suggested that he investigate whether, when the intensity of the light beam is so reduced that the putative photons are widely separated, the interference patterns vanish. (After all, sound will not travel through a near vacuum, and a few molecules of water will not form a wave.) Taylor set up the experiment at home with equipment costing under a pound and showed, as the paper above reports, that incredibly weak light beams produce the same interference patterns as strong ones. In later life he claimed that the reason he chose the project was that it left him free to go on a month's sailing cruise while the experiment was running.

The experiment seems to rule out the existence of photons conclusively, but the truth is much stranger. Experiments in the 1920s showed what had to be interpreted as collisions between electrons and photons and we are now convinced that light is made up of photons. In G. I. Taylor's experiment, the photons would have an energy of about 3×10^{-12} ergs and so, looking at the last but one sentence of Taylor's paper, we see that his dimmest beam of light contained roughly one photon per 10 000 cubic centimetres. Somehow, individual photons retain wavelike characteristics†. Thus what J. J. Thompson and G. I. Taylor presumably considered as a journeyman experiment confirming, once again, a classical theory and squashing wild speculations which had somehow got into circulation is now considered a central experiment in the new theory which replaced it.

After writing his first paper in the manner just described, Taylor returned to his main interests which lay not in the 'high physics' of the search for fundamental laws but in the equally hard task of using them to understand the world around us. 'While still at school I came across Lamb's *Hydrodynamics* in my uncle['s] ... library and though I could not understand it I was fascinated by its subject and hoped I would be able one day to use it in understanding the mechanics of sailing boats, a subject in which I was already much interested from the practical point of view.'‡ Initially, he worked on meteorology (including a six month expedition to study icebergs following the *Titanic* disaster). At the beginning of the First World War he offered his services to the army to set up a weather forecasting unit in the field. The officer to whom he made this proposal

> did not seem to doubt that I could tell what the weather was going to be — as he might very well have done — but thought the knowledge would be of no value in the field. 'Soldiers don't go into battle under umbrellas, they go whether it is raining or not.'

Instead, he joined the Royal Aircraft Factory at Farnborough. Here, among other things, he was involved in a project (insisted on by higher authority) to develop a dart to be dropped from an aeroplane on enemy troops below. After a trial drop of a bundle of darts, he and a colleague

†If you wish to learn how modern physics deals with the problem, you could start with Feynman's little book *QED, The Strange Theory of Light and Matter*.

‡The quotation comes from Batchelor's obituary of G. I. Taylor which I have used extensively in this chapter.

went over the field and pushed a square of paper over every dart we could find sticking out of the ground. When we had gone over the field in this way and were looking at the distribution, a cavalry officer came up and asked us what we were doing. When we explained that the darts had been dropped from an airplane, he looked at them and, seeing a dart piercing every sheet remarked: "If I had not seen it with my own eyes I would never have believed it possible to make such good shooting from the air."

He adds that the darts were never used

but not apparently for the reason we had originally raised as an objection — that they would be inefficient: we were told they were regarded as inhuman weapons and could not be used by gentlemen.

Like many of the scientists at Farnborough including Tizard and Lindemann, he decided that he could do a better job if he learned to fly. He remembered his training plane with some affection. Its maximum level speed was 60 miles per hour (about 95 kilometres per hour) and it was said to stall at 37 miles per hour.

Though the engine frequently gave out in the air, its low landing speed made it possible to land safely in many grass fields. Also, if it did crash, the staging supporting the elevator in front could act as a shock absorber. These machines were held together by piano wires, and so many were needed that the mechanics used to say that when they erected the machine they released a canary between the wings — if it got out a wire was missing.

(This sort of structure seems old fashioned to us but was in fact a rational one to use at low speeds. It is only at high speeds that the streamlined monoplane is superior. The First World War and the new aeroplane made the fortunes of several piano wire manufacturers.)

My instructor was a flight sergeant who had been damaged in a crash caused by one of his pupils. He decided this must not happen again, so if any of them exerted appreciable force on the dual control column he reported they were heavy handed and would never make good pilots. As very few machines were then available for training and many people wanted to join the air service, the threat that one would be turned out was to be avoided at all costs and many of this man's pupils took over control of their machines for the first time on their first solo flight. The resulting erratic manoeuvres were sometimes spectacular. I came through much better than most because of the months I had spent at Farnborough; at least I knew what the controls were supposed to do.

Taylor's work at Farnborough gave him a lifelong interest in the strength of materials and fracture problems. At the end of the war he returned to Cambridge (where he formed part of the 'Trinity talking eightsome' — a group of Sunday golfers led by Rutherford) and a life of uninterrupted research.

The four volumes of his collected papers are full of titles like *Stability of a viscous fluid contained between two rotating cylinders* (which contains the classic demonstration of the extraordinary 'order in chaos' which accompanies the breakdown of streamline flow) and *The mechanism of plastic deformation in crystals* which one would expect to find published in the *Proceedings of The Royal Society* and similar journals. It is more surprising to find an article reprinted from the *The Yachting Monthly and Motor Boating Magazine* of April 1934. The article is entitled *The holding power of anchors* and announces the first advance in anchor design for 2000 years. The original article is too long to be reproduced here, but Taylor gave an account of his invention to the Cambridge undergraduate mathematics magazine *Eureka*.

> Archimedes was not only a mathematician who expressed his thoughts by means of figures written on sand but, as your name 'Eureka' reminds us, [one who] solved essentially mathematical problems without using figures or symbols. ...
>
> In 1923 I bought the 48 foot [14.5 metres] yacht 'Frolic' which weighed 20 tons [20 tonnes]and drew 8 ft 3 ins [2.5 metres] of water. Her big anchor weighed 120 lbs [55 kilograms]. While winding up the anchor the sails were not capable of controlling the boat until the anchor was nearly up to the surface, and when anchored in 10 fathoms [18 metres] or more close inshore with an onshore wind, the effort involved in winding up the anchor to get under control before drifting onshore was too much for me. This and some problems connected with seaplanes provided the incentive to think about the design of lighter anchors.
>
> The earliest anchors were simple stones so that the ratio holding power/weight (H/W) was less than the coefficient of friction measured in air, usually less than one. The Greeks realised that a much bigger H/W could be attained by using a hook which would dig into the ground and they, or their contemporaries, invented the stock, that is a long bar at right angles to the plane of the hook which prevents it turning out of the ground once it is in. Since the stock would hold the fluke (that is the bent up part of the hook) pointing upwards if it fell that way and so prevent it from acting, it was necessary to add a second hook on the opposite side of the shank, thus making the anchor symmetrical about two planes through the shanks. This second hook is necessarily at such an angle to the shank that it prevents the first from dragging the shank downwards. For this reason, the high values of H/W which

could perhaps be attainable by a single hook cannot be had with a single anchor.

My problem was therefore to think of a way in which a single hook without a stock could be made to dig into the seabed whichever way it fell, and be stable when pulled horizontally below the surface. The solution I came to is shown in the sketch [Figure 6.1]. The shank A is hinged to the fluke B by a pin C whose axis is shown as the broken line CE. The blades D and J are nearly portions of circular cylinders with a common generator FG. The sketch shows the anchor seen from above and lying as it falls with A, J and G on the ground. When the chain pulls, the point G begins to dig in because it is aiming obliquely downwards, and the lateral pressure turns the blade further downwards because the centre of lateral pressure is ahead of the line CE. As the blade buries itself the centre of lateral pressure moves backwards and when it passes the line CE the direction of rotation of the blades about the pin C reverses and after dragging a short distance the anchor assumes a position where the plane of symmetry is vertical. In this position the blades can pull the shank into the ground. Also the anchor is stable when pulled with horizontal shank and blades under the ground, for if it rolled slightly so that the blade J was lower than the blade D, J would be in ground which was deeper and therefore more difficult to move than that round D. Thus the blades would rotate about the pin in such a way that the point G turned downwards and the anchor would return to the symmetrical position. By experimenting with a model on a sandy beach I found that the anchor could be towed in a circle keeping under the surface. When I dug up the blades while performing this experiment, I found the blades banked over just like an aeroplane when it makes a turn, but remained symmetrical when pulled in a straight line.

Figure 6.1: The CQR anchor.

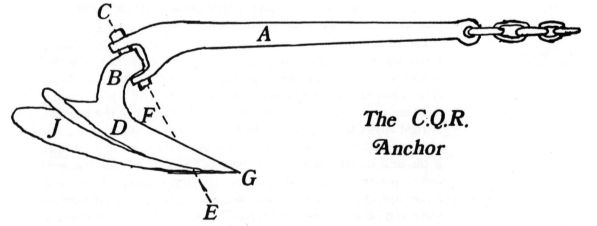

The C.Q.R.
Anchor

The maximum value of H/W varied with the nature of the seabed, in some grounds H/W was over 100 which is four or five times as great as that attainable with the traditional stocked anchor and 20 times that of the stockless anchor which all big steamships carry.

After inventing the anchor I, together with friends ... set up a small company to make them for our sailing friends. ... We called the company 'The Security Patent Anchor Co.' and would have liked to put the word 'secure' on the anchor, but it is not allowable to register a common word in that way, so we compromised and called the product 'CQR' For a long time afterwards people asked me what the Q stood for.

During the Second World War the CQR anchor was adopted by the navy for its torpedo boats and used to anchor the floating 'Mulberry' harbour used in the Normandy landings. Taylor himself worked on a large number of military problems to do with explosives, from those involved in constructing air-raid shelters through to depth charges and 'shaped charges'. In 1941 he was asked about the nature of the blast involving the release of a very large amount of energy in 'an infinitely concentrated form' (i.e. an atomic bomb).

His discussion of this problem drew heavily on dimensional considerations and we shall look at a simpler version using only dimensional arguments†. We assume that the energy E of the explosion is produced instantaneously and that the shock wave spreads out forming a sphere of radius r at time t. During the very early period that we are interested in, the pressure behind the shock wave will be hundreds of thousands times greater than in the undisturbed state, so the only other variable which is likely to be of importance is the initial air density ρ. The dimensions of the four identified quantities are given by

†Taylor's detailed computations were important in the decision to pursue the initial idea. Later he advised on the implosion mechanism for the bomb.

$$[E] = ML^2T^{-2}, \ [r] = L, \ [t] = T, \ [\rho] = ML^{-3}$$

and the only (independent) dimensionless quantity we can construct from them is (as the reader should check for herself)

$$E^{-1}\rho t^{-2}r^5.$$

Thus, by the arguments of the previous section,

$$\frac{\rho r^5}{Et^2} = C, \text{ a constant,}$$

so that

$$r = C\left(\frac{Et^2}{\rho}\right)^{1/5}$$

and we have shown, by dimensional arguments alone, that the radius of the wave front increases with the two-fifths power of time.

In 1950, when pictures of the first atomic test were declassified, Taylor published this work and compared his theoretical predictions with the photographic records which show that 'The ball of fire did ... expand very closely in accordance with the theoretical prediction made more than four years before the explosion took place.'† (See the two papers entitled *The formation of a blast wave by a very intense explosion* in his collected works.)

There are two ways of making progress in mathematics and physics. One is to do something better, the other is to do something for the first time. G. I. Taylor concentrated on doing things for the first time. He liked to think of himself as an amateur, an outsider like his grandfather George Boole, someone who showed that 'the pleasure and interest of being a scientist need not be confined to those gifted people who have the ability to pursue the highly specialised studies which are necessary for those who would reach the main frontiers of scientific advance.' As a final instance of Taylor's ability to find new and significant problems, let me cite his work on the swimming of microscopic organisms which he started when well past sixty.

When talking about the size of mammals in Section 5.2, I used the mouse and the elephant as examples of extremes, but mice and elephants

†Embarrassingly closely. Although the photographs had been declassified, the value of E was still a highly guarded secret. Taylor was able to obtain a value for C from other work, and his estimate for E was very close to the true one.

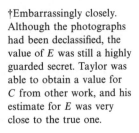

Figure 6.2: Escher's answer to the question: 'Why do few machines have legs and no animals have wheels?' (© 1995 M. C. Escher/ Cordon Art–Baarn–Holland.)

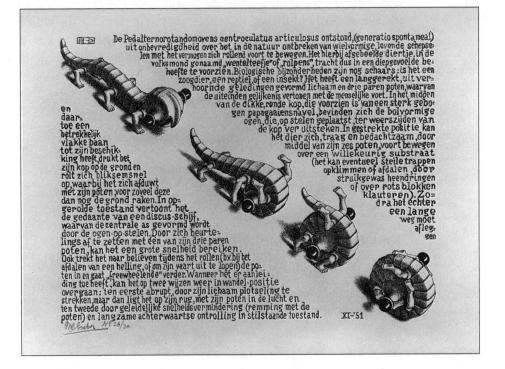

are both giants in the scale of living things. Since we belong among the giants, we are prejudiced in favour of size. Nowhere is this prejudice more misleading than when we use a microscope to look at tiny swimming creatures. We see a motion that 'looks wrong', that, to put it kindly (since we are kindly giants), lacks the purposeful elegance of a salmon. We are not surprised since, to us, small means crude and primitive.

The truth, as Taylor perceived, is rather different. As we noted in the previous chapter, the behaviour of swimming things is governed by the dimensionless Reynolds number

$$\text{Re} = \frac{vl\rho}{\eta}$$

where l is a typical length for the creature, v is its velocity and η and ρ are the viscosity and density of the ambient fluid. The smaller Re is, the more important are the effects of viscosity (friction). We can swim in water but it is proverbially difficult to move through treacle. For fish in water Re is usually of the order of many thousands, for a tadpole it is of the order of 10^2 and for bodies of the size of spermatozoa it is of the order of 10^{-3} or less. The question is thus not 'why do spermatozoa swim so oddly?' but 'how do they move at all?'. Taylor attacked the problem in two papers: *Analysis of the swimming of microscopic organisms* and *The action of waving cylindrical tails in propelling microscopic organisms.*

In his introduction to the second paper he explains clearly why microscopic creatures swim differently.

The self-propulsion of aeroplanes, ships and large fishes depends entirely on the inertia of the surrounding fluid. A propelling unit generates backward momentum which is exactly balanced by the forward momentum associated with fluid resistance. When microscopic organisms swim in water the forces due to viscosity are so much greater than those due to inertia that the latter can be neglected. Self-propulsion of a flexible body is achieved by so distorting its surface that the body must move

Figure 6.3: Working model of a swimming spermatozoon.

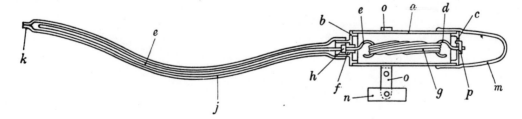

forward in order that the total force on it due to the viscous stress in the surrounding fluid may be zero.

To show that such motion is possible, he first describes a hypothetical, toroidal (doughnut shaped) creature which could move in such a manner by rotating about its circular axis like a smoke ring. (Unfortunately, Taylor's toroidal creature never came to Escher's† attention.) He then turns his attention to a simplified version of a spermatozoon.

In order to test his calculations, he constructed a working model (shown in Figure 6.3 mainly because I could not resist including it; the protuberance *n* is a counterweight to prevent the model rotating, for explanations of the rest of the figure you must go to Taylor's papers) which swam in a tank of glycerine ('the techniques for keeping the glycerine ... away from the hands and clothes are learned in the course of the work'). He made a film *Low Reynolds Number Flows* (with Educational Services Inc. in 1967) including shots of the model in action, which we were shown when I was an undergraduate. (The details of the film have long faded from my mind but the enthusiasm of the exposition remains fixed in my memory.)

G. I. Taylor pursued science for pleasure. Unlike Richardson (whom we shall meet in the next part of this book) and Blackett he did not worry about the uses science might be put to, but, like them, he was uninterested in personal power or gain. His biographer has left a charming picture of Taylor attending conferences as a grand old man when ' ... the organisers of [the] meeting expected this great authority to give a lecture on a topical subject and to make pronouncements on current research; instead there would be a refreshingly modest and enthusiastic (and sometimes incoherent) description of some neat little experiment on jets or waves on a sheet of liquid or peeling of a strip of adhesive', an individual amusing himself and gently mocking the brave new world of research teams, annual grant applications, computers, citation counting and boosterism.

†Escher was a Dutch artist whose work has a strong attraction for mathematicians. Although he had no formal mathematical training he drew some of his inspirations from browsing through the diagrams in mathematical and scientific books. Escher's mathematically most interesting works are his 'Periodic Drawings' based on a profound understanding of symmetry. (A collection of these was published under the auspices of the International Union of Crystallography in a book *Symmetry Aspects of M. C. Escher's Periodic Drawings* (author C. H. Macgillavry) which will well repay any work required to find it.) Conway, one of the great mathematical experts on symmetry, has ' ... a book of Escher's pictures on my piano. I try to ration myself to an Escher picture a day. Often I can't resist cheating and turning the page early, but I always insist on at least going out of the room first before I can turn the next page.'

CHAPTER 7

Subtle is the Lord

7.1 Galileo and Einstein

Galileo's troubles with the Catholic Church which led to his house imprisonment were caused by a book advocating (under the thinnest disguise) the Copernican view that it is the earth's rotation which causes the apparent motion of the sun in the sky and the stars in the heavens. One of the chief arguments against the earth's rotation is that we see no direct effects. To this Galileo replies through his mouthpiece Salviati as follows.

SALVIATI ... Shut yourself up with some friend in the main cabin below deck on some large ship, and have with you there some flies, butterflies and small flying animals. Have a large bowl of water with some fish in it; hang up a bottle that empties drop by drop into a wide vessel beneath it. With the ship standing still, observe carefully how the little animals fly with equal speed to all sides of the cabin. The fish swim indifferently in all directions; the drops fall into the vessel beneath; and, in throwing something to your friend, you need throw it no more strongly in one direction than another, the distances being equal; jumping with your feet together, you pass equal spaces in every direction. When you have observed all these things carefully (though there is no doubt that when the ship is standing still everything must happen in this way), have the ship proceed with any speed you like, so long as the motion is uniform and not fluctuating this way and that. You will discover not the least change in all the effects named, nor could you tell from any of them whether the ship was moving or standing still. In jumping, you will pass on the floor the same spaces as before, nor will you make larger jumps toward the stern than towards the prow even though the ship is moving quite rapidly, despite the fact that during the time that you are in the air the floor under you will be going in the direction opposite to your jump. In

throwing something to your companion, you will need no more force
to get it to him whether he is in the direction of the bow or the stern,
with yourself situated opposite. The droplets will fall as before into
the vessel beneath without dropping toward the stern, although while
the drops are in the air the ship runs many spans. The fish in their
water will swim towards the front of their bowl with no more effort
than towards the back, and will go with equal ease to bait placed
anywhere around the edge of the bowl. Finally the butterflies and
flies will continue their flights indifferently toward every side, nor will
it ever happen that they are concentrated toward the stern as if tired
out from keeping up with the course of the ship from which they
will have been separated during long intervals by keeping themselves
in the air. And if smoke is made by burning some incense, it will
be seen going up in the form of a little cloud, remaining still and
moving no more to one side than to the other. The cause of all these
correspondences of effect is the fact that the ship's motion is common
to all things contained in it and to the air also. That is why I said
you should be below decks; for if this took place above in open air,
which would not follow the course of the ship, more or less noticeable
differences would be seen in some of the effects noted. No doubt the
smoke would fall as much behind as the air itself. The flies likewise,
and the butterflies, held back by the air, would be unable to follow
the ship's motion if they were separated from it by a perceptible
distance. But keeping themselves near it, they will follow it without
effort or hindrance; for the ship, being an unbroken structure, carries
with it a part of the nearby air. For a similar reason we sometimes,
when riding horseback, see persistent flies and horseflies following
our horses flying now to one part of their bodies and now to another.
But the difference would be small as regards the falling drops, and as
to the jumping and throwing it would be quite imperceptible.

SAGREDO Although it did not occur to me to put these observations to
the test when I was voyaging, I am sure they would take place in the
way you describe. In confirmation of this I remember having often
found myself in my cabin wondering whether the ship was moving
or standing still; and sometimes at a whim I have supposed it going
one way when its motion was the opposite.

Nowadays we are carried in aeroplanes at many hundreds of kilometres an hour. In the smoking section we see little clouds, remaining still
and moving no more to one side than to the other. We walk along the
aisle in the direction of the plane's motion and back with equal ease.
When we pour drinks, the liquid falls vertically and does not stream out

behind us. Like Sagredo we do not actually have a flock of butterflies or a bowl of goldfish at hand but we are sure that what Galileo says would take place in the way he describes. Only when the plane hits turbulence and no longer moves with a steady, uniform velocity do we have any sensation of movement.

More than two centuries later, Galileo was echoed in a characteristic passage by the great physicist Maxwell.

> Our whole progress up to this point may be described as a gradual development of the doctrine of relativity of all physical phenomena. ...
>
> There are no landmarks in space: one portion of space is exactly like every other portion, so that we cannot tell where we are. We are, as it were, on an unruffled sea, without stars, compass, soundings, wind, or tide, and we cannot tell what direction we are going. We have no log which we can cast out to take a dead reckoning by; we may compute our rate of motion with respect to the neighbouring bodies, but we do not know how these bodies may be moving in space.

However, Maxwell's greatest achievement, his electromagnetic theory of light, ran contrary to the doctrine of relativity just described. Maxwell's theory requires the kind of mathematics developed in the first two years of a standard university mathematics course, so what follows is purely descriptive†. It is known that a changing magnetic field produces an electric field (this parallels the way that a generator works) and that a changing electric field produces a magnetic field. Maxwell's equations showed that it was possible for an electromagnetic disturbance to propagate through a vacuum in which a collapsing (and so changing) electric field gave rise to a magnetic field which collapsed in turn giving rise to an electric field whose collapse generated a magnetic field and so on. In a vacuum this disturbance moves with the speed of light and so Maxwell identified it with light. Experiment, in particular the generation of radio waves by Hertz, confirmed Maxwell's theory. For our purposes it is important to note that not only does Maxwell's theory tell us that such electromagnetic disturbances move with the speed of light but that *they can only move with the speed of light*.

As a 16 year old school boy, Einstein puzzled over this. ' ... the question came to me: If one runs after a light wave with [a velocity equal to the] light velocity, then one would encounter a time-independent wavefield. However something like this does not seem to exist.' If someone travelled at the speed of light along the path of a light beam, they would see the electric and magnetic fields alternately collapsing and growing *but the disturbance would appear not to move*. The standard

†And profoundly inadequate. As Hertz wrote 'To the question, "What is Maxwell's theory?" I know of no shorter or more definite answer than the following: Maxwell's theory is Maxwell's system of equations.' (See also, at greater length, the opening two paragraphs of Section 20.3 in Volume II of *The Feynman Lectures on Physics*.)

answer to this paradox was that it arose by the illegitimate extension of the principle of relativity from dynamics (which deals with the flight of butterflies and the fall of water drops) to electromagnetism. (The passage I quoted from Maxwell is entitled 'Relativity of dynamical knowledge'.) According to this answer, there is a particular stationary system in which Maxwell's equations hold exactly (and in which light travels at equal speed in all directions). Someone travelling with uniform non-zero velocity with respect to this system will find that Maxwell's equations do not hold exactly and that light does not travel at equal speed in all directions.

If we suppose the dockside to be the favoured system where Maxwell's equations hold exactly, Galileo's voyager will be able to tell that he is moving by subtle modifications of Maxwell's equations of which the easiest to detect may be changes in the velocity of light in different directions. Living as we do on a planet moving round a sun which is itself moving through space, we are in the same position as Galileo's voyager and, in principle, we can perform the same experiments. However, the speed of light is very large compared with the earth's velocity and Maxwell doubted whether such a delicate experiment would be possible. The challenge was taken up by the great American experimental physicist Michelson who designed his 'interferometer' for this purpose† but, although his experimental set-up was sensitive enough to detect the expected effects, none were observed. Ten years after his experiment and five years before Einstein's work, Michelson wrote:

> The experiment is to me historically interesting, because it was for the solution of this problem that the interferometer was devised. I think it will be admitted that the problem, by leading to the invention of the interferometer, more than compensated for the fact that this particular experiment gave a negative result.

One obvious solution might be to declare that the laws of electromagnetism behave like the laws of mechanics and are the same for everyone whether on the quayside or in the ship. Unfortunately, it turns out that Newton's laws of mechanics and Maxwell's laws of electromagnetism cannot both hold unaltered on the quayside and in the ship. One indication of what goes wrong appears when we recall our sarcastic dismissal of the 'pyramid inch' in Section 6.1. We do not believe that somewhere in the universe there is a 'pyramid inch' marked in some way to show that it and no other length is the fundamental unit of length in the universe. But, if the laws of electromagnetism hold everywhere, whether on the quayside or in the ship, then the speed of light is the same everywhere and gives us not a 'pyramid length' but a 'pyramid

†Many years later Einstein asked Michelson why he had spent so much time in trying to measure the speed of light to ever greater precision. Michelson replied in German 'Weil es mir Spaß macht.' *Because it is fun.*

speed'. There is thus a preferred unit of velocities and all velocities should be given as fractions of the speed of light. With this scheme the velocity u becomes dimensionless and so $LT^{-1} = [u]$ is dimensionless. In other words length and time have the same dimension.

We do not have to go this far to obtain a contradiction. The addition law for velocities in mechanics tells us that if an object travels northward with velocity u relative to a ship travelling northward with velocity v then the object travels northward with velocity $u + v$ relative to the fixed quayside. If the object is a beam of light (so that $u = c$) we have an immediate contradiction. Einstein has left us an account of his struggle with this problem. (Unfortunately, this reminiscence has been translated from German into Japanese and thence into English, but it brings us as close to the moment of discovery as we could reasonably hope. I have taken the liberty of straightening out some of the syntax so those interested in the original should consult the reference given in Appendix 3.)

This invariance of the speed of light was, however, in conflict with the well known rule for the addition of velocities in mechanics.

I had great difficulty in resolving the question of why the two [demands] were in conflict with each other. I wasted almost a year in fruitless considerations ... .

†Michele Besso, a friend from his student days and a colleague at the patent office.

Unexpectedly a friend of mine in Bern† then helped me. It was a very beautiful day when I visited him and began to talk with him as follows. 'I have recently had a question which was difficult for me to understand. So I have come here today to battle it out.' After a lot of discussions with him, I suddenly understood the matter. Next day I visited him again and said to him without greeting: 'Thank you. I've completely solved the problem.' My solution dealt really with the very concept of time, that is that time is not absolutely defined, but there is an inseparable connection between time and the signal velocity. With this conception, the foregoing extraordinary difficulty could be thoroughly solved. Five weeks after my recognition of this, the present theory of special relativity was completed.

In the next chapter we will look at part of Einstein's theory.

7.2 The Lorentz transformation

Let us return to our observer on the quayside and our observer on the ship. Suppose that the observer on the quayside has a coordinate system S consisting of orthogonal x, y, z coordinates‡ relative to axes fixed in the quayside *together with a time coordinate* t. He speaks of an event taking place at the point (x, y, z) at time t. The observer in the ship has a coordinate system S' consisting of orthogonal x', y', z' coordinates

‡That is to say, the familiar three dimensional Cartesian coordinates in which the x, y and z axes are at right angles.

relative to axes fixed in the ship *together with a time coordinate t'*. He speaks of an event taking place at the point (x', y', z') at time t'. If a certain event (say the flapping of a butterfly's wing or the explosion of a firework) occurs at the point (x, y, z) at time t in the coordinate system S of the observer on the quayside and at the point (x', y', z') at time t' in the coordinate system S of the observer on the ship we write

$$(x, y, z, t) \longleftrightarrow (x', y', z', t')$$

If we choose the coordinate systems S and S' at random we simply complicate the algebra without increasing the generality of our discussion. For simplicity we thus suppose that the ship has velocity u directed along the x axis and that the x and x' axes lie along the same line. We further suppose that the y and y' axes are parallel and that the z and z' axes are parallel. Finally we suppose that

$$(0, 0, 0, 0) \longleftrightarrow (0, 0, 0, 0)$$

(i.e. that the origins of the two coordinate systems coincide at zero time for both observers). The conditions we have imposed ensure that if

$$(x, y, z, t) \longleftrightarrow (x', y', z', t')$$

it follows that:

   (i) if $y = 0$ then $y' = 0$,
   (ii) if $z = 0$ then $z' = 0$,
   (iii) if $x = ut$, $y = 0$, $z = 0$ then $x' = y' = z' = 0$.

Condition (iii) follows from the fact that the ship has velocity u along the x axis.

In classical mechanics the conditions of the previous paragraph imply:

If $(x, y, z, t) \longleftrightarrow (x', y', z', t')$ then
$$x' = x - ut$$
$$y' = y$$
$$z' = z$$
$$t' = t.$$

Einstein suggested, in effect, that we should use a different rule for translating between the system S and the system S'.

If $(x, y, z, t) \longleftrightarrow (x', y', z', t')$ then
$$x' = px + qt$$
$$y' = y$$
$$z' = z$$
$$t' = sx + rt.$$

Is it possible to choose p, q, r and s so that the speed of light is the same for both observers? Remember that we have already demanded that, if

$$(x, y, z, t) \longleftrightarrow (x', y', z', t')$$

and $x = ut$, $y = 0$, $z = 0$, then $x' = 0$. Thus

$$0 = put + qt$$

and so

$$q = -pu. \tag{7.1}$$

Now consider a flash of light emitted in the positive x direction at the origin of S at time $t = 0$. It will travel outwards along the positive x axis at velocity c so its equation of motion for the observer on the quayside is

$$(x, y, z, t) = (ct', 0, 0, t').$$

However, $(0, 0, 0, 0) \longleftrightarrow (0, 0, 0, 0)$ and the speed of light is the same for the observer on the ship as it is for the observer on the quayside. Thus its equation of motion for the observer on the ship is

$$(x, y, z, t) = (ct', 0, 0, t').$$

Hence

$$(ct, 0, 0, t) \longleftrightarrow (ct', 0, 0, t'),$$

so that, using the suggested transformation,

$$ct' = pct + qt$$
$$t' = sct + rt$$

so that

$$c(sct + rt) = pct + qt,$$

and

$$sc^2 + rc = pc + q. \tag{7.2}$$

On the other hand if we consider a flash of light emitted in the *negative* x direction at the origin of S at time $t = 0$. we obtain

$$(-ct, 0, 0, t) \longleftrightarrow (-ct', 0, 0, t'),$$

so that

$$-ct' = -pct + qt$$
$$t' = -sct + rt,$$

whence

$$-c(-sct + rt) = -pct + qt$$

and

$$sc^2 - rc = -pc + q. \qquad (7.3)$$

Adding and subtracting the two equations (7.2) and (7.3) we see that

$$(sc^2 + rc) + (sc^2 - rc) = (pc + q) + (-pc + q)$$
$$(sc^2 + rc) - (sc^2 - rc) = (pc + q) - (-pc + q),$$

so that

$$2sc^2 = 2q$$
$$2rc = 2pc$$

and so

$$q = sc^2 \text{ and } r = p.$$

If we now use Equation (7.1) we obtain

$$r = p, \ q = -pu \text{ and } s = -\frac{pu}{c^2},$$

so our suggested translation rule now becomes:

If $(x, y, z, t) \longleftrightarrow (x', y', z', t')$ then
$$x' = p(x - ut)$$
$$y' = y$$
$$z' = z$$
$$t' = p\left(t - \frac{ux}{c^2}\right).$$

We started with four undetermined coefficients p, q, r and s and we now have only one. How can we find p? Remember that Einstein follows Galileo in saying that there is no reason to prefer the coordinate system of the quayside to that of the ship. Since the quayside is travelling with velocity $-u$ along the x' axis, exactly the same arguments with S and S' interchanged show that:

If $(x, y, z, t) \longleftrightarrow (x', y', z', t')$ then
$$x = p'(x' + ut')$$
$$y = y'$$
$$z = z'$$
$$t = p'\left(t' + \frac{ux'}{c^2}\right),$$

where p' is an, as yet, undetermined coefficient. It follows that, if

$$(x, y, z, t) \longleftrightarrow (x', y', z', t'),$$

then

$$x = p'(x' + ut') = p'p\left(x - ut - u\left(t - \frac{ux}{c^2}\right)\right) = p'p\left(x - \frac{u^2 x}{c^2}\right)$$

and so

$$p'p = \left(1 - \frac{u^2}{c^2}\right)^{-1}.$$

The symmetry between S and S' suggests that we should take $p = p'$ and so

$$p = \gamma_u$$

where

$$\gamma_u = \left(1 - \frac{u^2}{c^2}\right)^{-1/2}.$$

We have now arrived at the transformation law:

If $(x, y, z, t) \longleftrightarrow (x', y', z', t')$ then

$$x' = \gamma_u(x - ut)$$
$$y' = y$$
$$z' = z$$
$$t' = \gamma_u\left(t - \frac{ux}{c^2}\right).$$

These equations look unsymmetric because we have not had the courage of our convictions. If the speed of light is a fundamental unit we should use it to measure velocity. For the rest of this chapter we shall measure speed as a proportion of the speed of light so that $c = 1$. (Notice that if we keep the second as our unit of time then the distance between two points A and B will now be measured by the number of seconds it takes light to travel from A to B.)

Exercise 7.2.1 *The speed of light is approximately* 3×10^{10} *centimetres per second. Find the speed of a racing car as a proportion of the speed of light.*

In our new units the transformation law becomes:

$$\text{If } (x, y, z, t) \longleftrightarrow (x', y', z', t') \text{ then} \tag{7.4}$$
$$x' = \gamma_u(x - ut) \tag{7.5}$$
$$y' = y \tag{7.6}$$
$$z' = z \tag{7.7}$$
$$t' = \gamma_u(t - ux), \tag{7.8}$$

with $\gamma_u = (1 - u^2)^{-1/2}$.

We now know how to link coordinates for the two observers. What about velocities? Consider an object travelling at constant velocity in the quayside coordinates. For algebraic simplicity we shall suppose that it passes through the origin at time $t = 0$. The S coordinates of the object are thus

$$x = Ut$$
$$y = Vt$$
$$z = Wt,$$

where U, V, W are the 'components of velocity' and so $(U^2+V^2+W^2)^{1/2}$ is the speed of the object measured by the quayside observer. Applying the transformation laws given in Equations (7.5)–(7.8) we see that, for the observer on the ship

$$x' = \gamma_u(Ut - ut) = \gamma_u(U - u)t$$
$$y' = Vt$$
$$z' = Wt$$
$$t' = \gamma_u(t - uUt) = \gamma_u(1 - uU)t.$$

Thus $t = t'/(\gamma_u(1 - uU))$ and substitution gives

$$x' = \frac{U - u}{1 - uU}t'$$
$$y' = \frac{V}{\gamma_u(1 - uU)}t'$$
$$z' = \frac{W}{\gamma_u(1 - uU)}t'.$$

Thus, for the observer on board the ship, the particle also moves in a straight line with 'components of velocity'

$$U' = \frac{U - u}{1 - uU}, \quad V' = \frac{V}{\gamma_u(1 - uU)}, \quad W' = \frac{W}{\gamma_u(1 - uU)}$$

and speed

$$(U'^2 + V'^2 + W'^2)^{1/2} = \frac{((U - u)^2 + (1 - u^2)(V^2 + W^2))^{1/2}}{(1 - uU)}.$$

Exercise 7.2.2 *If $U = -u$, U has the size of the velocity of a racing car and U' is defined as in the paragraph above, estimate the value of*

$$1 - \frac{U - u}{U'}.$$

From the pre-Einsteinian point of view this looks rather peculiar (we expect $U' = U - u$, $V' = V$ and $W' = W$), but from our new Einsteinian standpoint the new expressions for U', V' and W' have a particular virtue. Suppose the object observed is a beam of light. Then $U^2 + V^2 + W^2 = 1$ and our transformation law gives

$$
\begin{aligned}
U'^2 + V'^2 + W'^2 &= \frac{(U - u)^2 + (1 - u^2)(V^2 + W^2)}{(1 - uU)^2} \\
&= \frac{(U - u)^2 + (1 - u^2)(1 - U^2)}{(1 - uU)^2} \\
&= \frac{(U^2 - 2uU + u^2) + (1 - u^2 - U^2 + u^2U^2)}{(1 - uU)^2} \\
&= \frac{u^2U^2 - 2uU + 1}{(1 - uU)^2} = 1,
\end{aligned}
$$

so the speed of light is the same for the two observers independent of the direction in which it travels.

Exercise 7.2.3 *Consider, as before, an object travelling at constant velocity in the quayside coordinates but omit the condition that it pass the origin at time $t = 0$. The S coordinates of the object are thus*

$$
x = x_0 + Ut
$$
$$
y = y_0 + Vt
$$
$$
z = z_0 + Wt.
$$

Show that, for the observer on board the ship,

$$
x' = x_0' + U't'
$$
$$
y' = y_0' + V't'
$$
$$
z' = z_0' + W't',
$$

where, as before,

$$
U' = \frac{U - u}{1 - uU}, \quad V' = \frac{V}{\gamma_u(1 - uU)}, \quad W' = \frac{W}{\gamma_u(1 - uU)}
$$

and where

$$
x_0' = \gamma_u x_0 + uU x_0, \quad y_0' = y_0 + uV x_0, \quad z_0' = z_0 + uW x_0.
$$

Let us return to our observers on the ship and on the quay. Suppose that they agree, when $(x, y, z, t) = (0, 0, 0, 0)$ for the quayside observer and $(x', y', z', t') = (0, 0, 0, 0)$ for the observer on the ship, to both let off fireworks after one unit of time has elapsed. Suppose further that the observer on the ship stays at the origin of the shipborne coordinate system S'. Then, by the Equations (7.5) and (7.8), the observer on the ship will have his time t' given by

$$
t' = \gamma_u(t - ux) = \gamma_u(t - u(ut)) = \gamma_u(1 - u^2)t = (1 - u^2)^{1/2}t
$$

and the observer at the quayside will (after allowing for the time the flash takes to reach him) say that the firework was let off on the ship after a time $\gamma_u = (1 - u^2)^{-1/2}$. This delay is called 'time dilation'. Since there is no preferred system in Einstein's universe, exactly the same argument shows that the observer on the ship will (after allowing for the time the flash takes to reach him) say that the firework was let off on the quayside after a time $\gamma_u = (1 - u^2)^{-1/2}$.

When cosmic rays hit the earth's atmosphere, large quantities of unstable particles called muons are created. They have a very short half-life and, if classical physics were correct, hardly any should be observed at sea level. However, they travel at such high speeds that time dilation has an important effect and the firework display of disintegration is delayed long enough for them to be observed at sea level.

Exercise 7.2.4 *Suppose that the velocity of a muon (relative to an observer on the earth) is 0.99 in light speed units. Show that its expected survival time (as measured by the observer) is increased by a factor of 7.1 compared with that of a stationary muon.*

The transformations (7.5)–(7.8) only marked the beginning of Einstein's task. He verified in detail that Maxwell's equations keep their form under these transformations (so that they appear the same whether on the quay or on the ship) and then, in a second paper, showed how Newton's equations of mechanics must be modified to keep their form under these new transformations. This paper contains the famous formula

$$E = mc^2.$$

It may seem sacrilegious to meddle with the most famous equation of the twentieth century, but if we stick to our principles and use units in which $c = 1$ this becomes

$$E = m,$$

that is

$$\text{Energy} = \text{mass.}$$

A brick is just energy in a rather concentrated form and a beam of light a rather diffuse form of mass. Nowadays the mass of elementary particles is given in electron-volts (a measure of energy)†.

Maja Einstein wrote, with the partiality of a sister, how, after the publication of his first relativity paper in the *Annalen der Physik*

> The young scholar imagined that his publication in the renowned and much-read journal would draw immediate attention. He expected sharp opposition and the severest criticism. But he was very disappointed. His

†We also measure distance in time units since a metre is now officially defined as the distance light travels in 1/299 792 458 of a second.

publication was followed by icy silence. The next few issues of the journal did not mention his paper at all. ... Some time after the appearance of the paper, Albert Einstein received a letter from Berlin. It was sent by the well known Professor Planck, who asked for clarification of some points which had been obscure to him. After the long wait this was the first sign that his paper had been at all read. The joy of the young scientist was especially great because the recognition of his activities came from one of the greatest physicists of that time.

7.3 What happened next?

In fact, the time was ripe for Einstein's theory. (The great Dutch physicist Lorentz had already written down the system of Equations (7.5)–(7.8) which are usually referred to as the Lorentz transformation, but was unable to grant equality of treatment to the two observers. 'The chief cause of my failure [to anticipate Einstein],' he wrote later 'was my clinging to the idea that only the variable t can be considered the true time and that my local time t' must be regarded as no more than an auxiliary mathematical quantity.') Among the early converts was the brilliant mathematician Minkowski, one of Einstein's professors†, who developed the modern geometrical version of special relativity and began a major public lecture on it with the words

> The views on space and time which I wish to lay before you have sprung from the soil of experimental physics, and therein lies their strength. They are radical. Henceforth, space by itself and time by itself are doomed to fade away into mere shadows, and only a kind of union of the two will preserve an independent reality.

†He is reported to have said 'Ach, der Einstein, der schwänzte immer die Vorlesungen – dem hätte ich das gar nicht zugetraut.' *Oh, that Einstein, always cutting lectures – I really would not have believed him capable of it.* (The anecdote is taken from Constance Reid's *Hilbert* where no source is given, but every teacher will vouch for its psychological plausibility.)

By 1910, special relativity was part of accepted physics among the rising generation (the same could not be said of Einstein's photon theory of light to which we referred in Section 6.2 — that theory was not to be generally accepted for another 15 years).

The theory of special relativity deals with frames of reference which are in uniform *non-accelerated* motion relative to one another (so called 'inertial frames'). What can we say about accelerated motion? Let us remember the passage of Galileo which I quoted in Section 5.1.

> Note that it is necessary to distinguish between heavy bodies in motion and the same bodies at rest. A large stone placed in a balance not only acquires additional weight by having another stone placed upon it but even by the addition of a handful of hemp its weight is augmented six to ten ounces according to the quantity of hemp. But if you tie the hemp to the stone and allow them to fall freely from some height, do you believe that the hemp will press down upon the stone and thus accelerate its

motion or do you think the motion will be retarded by a partial upward pressure? A man always feels the pressure upon his shoulders when he prevents the motion of a load resting on him; but if he descends just as rapidly as the load will fall how can it gravitate or press upon him? Do you not see that this would be the same as trying to strike a man with a lance when he is running away from you with a speed which is equal to, or even greater than that with which you are following him? You must therefore conclude that during free and natural fall, the small stone does not press upon the larger and consequently does not increase its weight as it does when at rest.

Now let Einstein speak.

When in 1907, I was working on a comprehensive paper on the special theory of relativity ... I had also to attempt to modify the Newtonian theory of gravitation in such a way that its laws would fit in the theory. Attempts in this direction did show that this could be done, but did not satisfy me because they were based on physically unfounded hypotheses.

Then there occurred to me the happiest thought of my life, in the following form. The gravitational field has only a relative existence in a way similar to the electric field generated by magnetic induction. *Because for an observer falling freely from the roof of a house there exists* — at least in his immediate surroundings — *no gravitational field*. Indeed, if the observer drops some bodies then these remain relative to him in a state of rest or uniform motion, independent of their particular chemical or physical nature (in this consideration the air resistance is of course ignored). The observer therefore has the right to interpret his state as 'at rest'.

Because of this idea, the uncommonly peculiar experimental law that in the gravitational field all bodies fall with the same acceleration attained at once a deep physical meaning. Namely that if there were to exist just one single object that falls in the gravitational field in a way different to the others, then with its help the observer could realise that he is in a gravitational field and is falling in it. If such an object does not exist however — [and experiment] has shown to great accuracy [that it does not,] then the observer lacks any means of perceiving himself as falling in a gravitational field. Rather he has the right to consider his state as one of rest and his environment as field-free relative to gravitation.

The experimentally known matter independence of the acceleration of fall [i.e. the fact that all bodies, whatever their composition fall with the same acceleration] is therefore a powerful argument for the fact that the relativity postulate has to be extended to coordinate systems which, relative to each other, are in non-uniform motion.

Thus Einstein decided that the laws of physics should take their simplest form when expressed relative to coordinate systems in free fall (i.e. moving like freely falling bodies). However, it took him nearly ten years to transform this insight into a satisfactory theory. Once again, we may quote his own words.

> I had excellent teachers (for example, Hurwitz, Minkowski), so that I really could have gotten a sound mathematical education. However, I worked most of the time in the physical laboratory, fascinated by the direct contact with experience. The balance of the time I used in the main in order to study at home the works of Kirchhoff, Helmholtz, Hertz, etc. The fact that I neglected mathematics to a certain extent had for cause not merely my stronger interest in the natural sciences than in mathematics, but also in the following strange experience. I saw that mathematics was split up into numerous specialities, each of which could easily absorb the short lifetime granted to us. Consequently I saw myself in the position of Buridan's ass which was unable to decide upon any specific bundle of hay†. This was obviously due to the fact that my intuition was not strong enough in the field of mathematics to differentiate clearly the fundamentally important, that which is really basic, from the rest of more or less dispensable erudition. Beyond this, however, my interest in the knowledge of nature was also unqualifiedly stronger: and it was not clear to me as a student that the approach to a more profound knowledge of the basic principles of physics is tied up with the most intricate mathematical methods. This dawned on me only gradually after years of independent scientific work.

† Buridan's ass is a figure of medieval philosophy. Placed between two equally attractive bundles of hay, the poor creature was unable to decide which to start with and consequently starved to death.

Both halves of Einstein's statement must be taken with a pinch of salt. Any student who can 'study at home the works of Kirchhoff, Helmholtz, Hertz, etc.' must have substantial mathematical knowledge and technique and the 'most intricate mathematical methods' which he used later are all drawn from one particular branch of mathematics. However the mathematical techniques of his early papers are those which would be known to any good mathematical physicist of the time and his remark that Minkowski's geometrical interpretation of relativity was 'überflüssige Gelehrsamkeit' ('superfluous learnedness') reflected the usual physicist's prejudices.

The decision to view nature by using freely falling coordinate systems created problems which simple mathematical techniques could not resolve. We can choose a freely falling coordinate system at the North Pole and we can choose a freely falling coordinate system at the South Pole *but we cannot choose a single coordinate system which is freely falling at both poles.* More generally, each little patch of space has its own freely falling coordinate system which fails to coincide with any other. We are in the position of a geographer with a room filled to the roof with little scraps of paper each containing a tiny portion of map.

The mathematical study of such a roomful of map fragments dates back to Gauss who used them to study surfaces. In 1854, when the young Riemann had to give a probationary lecture, he submitted three possible titles for it. Custom dictated that his first title should be chosen but Gauss chose the third: *On The Hypotheses That Underlie Geometry.* Riemann rose to the challenge and produced a lecture that is still read today. Extended mathematical metaphors are tiresome for experts and incomprehensible for novices, but, very roughly, Riemann showed how the roomful of maps could be treated as an object (surface or space) in its own right without reference to anything else. Riemann's ideas were developed by Beltrami, Christoffel and, particularly by the Italian Ricci and his pupil Levi-Civita into a system which permitted computation.

For most mathematicians in 1901 this 'absolute differential calculus' must have been one of those theories which are treated respectfully in public (clearly it was very hard and had a respectable pedigree) and sceptically in private. Very few people worked in the field and the complex notation hid the underlying ideas from the casual enquirer. (Even today, a first course in differential geometry contains a great deal of semolina and very little jam.) But, as Einstein relates:

> If all [accelerated] systems are equivalent, then Euclidean geometry can-
> not hold in all of them. To throw out geometry and keep [physical] laws
> is equivalent to describing thoughts without words. We must search for
> words before we can express thoughts. What must we search for at this
> point? This problem remained insoluble to me until 1912 when I suddenly
> realised that Gauss's theory of surfaces holds the key for unlocking this
> mystery. I realised that Gauss's surface coordinates had a profound sig-
> nificance. However I did not know at the time that Riemann had studied
> the foundations of geometry in an even more profound way. I suddenly
> remembered that Gauss's theory was contained in the geometry course
> given by Geiser when I was a student .... I realised the foundations of
> geometry have physical significance. My dear friend the mathematician
> Grossman was there when I returned from Prague to Zürich. From him
> I learned for the first time about Ricci and later about Riemann.

Grossman's first reaction was that 'this is a terrible mess with which physicists should not be involved' but he went on to collaborate with Einstein in applying tensor analysis to Einstein's problems. (The 'ab-solute differential calculus' is now called 'tensor analysis' since that is what Einstein called it.) Many difficulties remained but the general theory of relativity as it finally emerged in 1915 was a tensor theory.

Not every difficult student becomes an Einstein and not every orphan theory of pure mathematics turns out to be the language of nature. When such a miracle occurs, we should rejoice.

I have shown how Galileo and Einstein looked at falling bodies. Although the discussion does not demand it, I cannot resist quoting from a memoir of a friend of Newton's old age which was discovered and published in 1936. Here he recounts part of a visit 'to Sir Isaac at his lodgings ... [when I] dined with him and spent the whole day with him.' (I have made some minor changes but the spelling, punctuation and grammar remain a little strange to our eyes.)

> After dinner, the weather being warm, we went into the garden and drank tea, under the shade of some apple trees, only he and myself. Amidst other discourse, he told me, he was just in the same situation, as when formerly, the notion of gravitation came into his mind. It was occasion'd by the fall of an apple, as he sat in a contemplative mood. Why should that apple always descend perpendicularly to the ground, thought he to himself. Why should it not go sideways or upwards, but constantly to the earths centre?
>
> Assuredly, the reason is that the earth draws it. There must be a drawing power in matter: and the sum of the drawing power in the matter of the earth must be in the earths centre, not in any side of the earth. Therefore does this apple fall perpendicularly or towards the centre. If matter thus draws matter, it must be in proportion of its quantity. Therefore the apple draws the earth as the earth draws the apple†.

†Some historians had doubted the truth of Newton's apple. This lack of faith has never found any echo in Cambridge where the visitor will be shown two descendants of the original tree – unfortunately of two different varieties.

Just as the happy idea of Einstein was merely the germ of general relativity, so the realisation that 'the apple draws the earth as the earth draws the apple' was merely the germ of Newton's theory of *universal* gravitation, but it was the germ.

Every day, everyone sees something fall. Three men, Galileo, Newton and Einstein looked at what everyone sees every day and saw the plan of the universe. Will there some day be a fourth?

> To see a World in a Grain of Sand
> And a Heaven in a Wild Flower,
> Hold Infinity in the palm of your hand
> And Eternity in an hour.

Further reading

Abraham Pais has written a marvellous life of Einstein: *Subtle is the Lord ...* which makes it clear why Einstein was the greatest physicist of the twentieth century. Two simple texts which place relativity in its general physical context are Einstein and Infeld *The Evolution of Physics* and Born *Einstein's Theory of Relativity*. If the reader consults Einstein's original paper in the collection *The Principle of Relativity* [53] she may be surprised by how much she can understand. If you wish to learn more about the dual wave-particle nature of light, then I repeat my earlier recommendation of *QED, The Strange Theory of Light and Matter* which is written in Feynman's inimitable and lucid style. If you are interested in the general question of how far one can do physics in a darkened room, then the same author's *The Character of Physical Law* is full of insights.

7.4 Does the earth rotate?

The new theories and discoveries of Copernicus, Galileo and others found a pessimistic echo in Donne who wrote:

> And new philosophy puts all in doubt,
> The element of fire is quite put out;
> The sun is lost, and th'earth, and no man's wit
> Can well direct him where to look for it.
> And freely men confess that this world's spent,
> When in the planets, and the firmament
> They seek so many new; they see that this
> Is crumbled out again to his atomies.
> 'Tis all in pieces, all coherence gone;
> All just supply, and all relation:
> Prince, subject, father, son are things forgot,
> For every man alone thinks he has got
> To be a phoenix, and that then can be
> None of that kind of which he is, but he.
> This is the world's condition now ...

[From *An Anatomy of the World*. There can only be one phoenix alive at a time.]

Einstein's theories of relativity aroused similar, though less well expressed, sentiments. There is, perhaps, something vaguely absurd in one man, who has never understood Newton's mechanics or Euclid's geometry, violently defending them against the 'usurper Einstein', or another, who has never given the nature of time a moment's reflection, suddenly lamenting the loss of 'absolute time', whatever he now believes that to have been; and there is something a little sinister in someone who claims that the loss of faith in traditional institutions which followed the First World War was connected to a new theory of physics. However, these feelings were genuine and summed up in the phrase 'Einstein claims that everything is relative'.

If everything is relative, then perhaps Galileo's judges were right, and we can just as well say that the sun revolves round the earth as that the earth revolves round the sun. The reader who finds such a proposition plausible should at once head for the nearest large fun-fair. With luck she will find what is, in effect, a giant centrifuge. After paying an entrance fee, she will be ushered into a large circular room and told to stand with her back to the wall. As the room starts to spin, she will feel herself pushed against the wall as if by a giant hand and then suddenly see the room rotate by 90° so that she finds herself stationary at the lowest point of a vertical wheel watching other people stuck at various points on the circumference. On emerging from this experience, the reader will, I think, have no doubt that she has been rotated at high speed.

But, if she has indeed been rotated, then the question arises 'relative to what?' Newton performed a less spectacular but simpler experiment, a version of which may be found in many 'hands on' science museums.

> If a vessel, hung by a long cord, is so often turned about that the cord is strongly twisted, then filled with water, and held at rest together with the water [and then let go] ... while the cord is untwisting itself, the vessel continues for some time in this motion, the surface of the water will at first be plain, as before the vessel began to move; but after that, the vessel, by gradually communicating its motion to the water, will make it begin sensibly to revolve, and recede little by little from the middle, and ascend to the sides of the vessel, forming itself into a concave figure (as I have experienced), and the swifter the motion becomes, the higher will the water rise.

How does the water in Newton's bucket know that it is rotating and should therefore form itself into a concave figure?

One way to start answering these questions is to imagine yourself at the North Pole. If you look up at the starry sky, you see the starry heavens appear to rotate once every 24 hours. Are the stars rotating round the earth or are the stars still and the earth rotating? Newton's laws tell us that if we suspend a small weight from a string and let it go then the resulting pendulum will move back and forth *in a straight line*. Let us set such a pendulum in motion and draw a straight line on the earth's surface in the direction of the first swing. If the starry sky is rotating and the earth is still, then we will see no change of direction in the pendulum's swing with respect to the line marked on the ground. However, if the earth is rotating and the sky is still, then every 24 hours the earth (and with it the line marked on the ground) will perform a full rotation with respect to the line of the pendulum swing and so the line of the pendulum swing will perform a full rotation with respect to the line on the ground.

So far as I know, no-one has performed the experiment at the North Pole, but a similar argument applies at other points on the earth's surface. (The details of the argument are, however, substantially harder.)

Exercise 7.4.1   *(i) Show that (if the sky is fixed and the earth rotates) we will see the same effect at both poles except that the sense of the rotation of the line of the pendulum swing with respect to the line on the ground will be reversed.*

[Hint *Consider a transparent clock face viewed from behind. Do the hands go clockwise or anti-clockwise?*]

(ii) What will happen at the equator?

In order to test what happens, we need a pendulum with a very long string, a very heavy bob and a nearly frictionless support, but the experiment, which may be seen in many science museums, gives the unequivocal answer that the earth rotates and the stars stand still. (The experiment is called 'Foucault's pendulum' after its inventor.)

There is another experiment we could carry out at the North Pole, and that is simply to throw a stone southwards. Newton's laws tell us that the particle will travel directly south whilst earth (as we believe) rotates beneath. The path of the stone will thus appear to drift slightly in a direction opposite to the earth's rotation. (Of course, if the earth were fixed there would be no such effect.) Since the earth rotates so slowly (only once every 24 hours) and the stone is in the air for such a small time, the effect will be tiny but, if we were to fire a long-range gun, the drift would be several tens of metres.

Exercise 7.4.2   *If you have studied the mathematics governing the path of a projectile, you should be able to estimate how long a shell fired at 45° by a gun with a range of 10 kilometres will remain in the air. Use this to check my statement about drift.*

Once again there is a similar effect at other points on the earth's surface (though the magnitude and direction of drift depend not only on where the gun is fired but also on the direction in which it is fired).

Early in the First World War, British and German ships engaged in battle off the Falkland Islands. The British had the advantage of longer-range guns and fought at maximum range. However, they found that their salvos were continually falling 100 metres to the left. It emerged that the gun-sights had been adjusted for the effects of the earth's rotation but this had been done on the assumption that naval battles would take place in the North Sea (at latitude 50° North). The

Falklands lie at latitude 50° South, so the rotational effect is reversed. The resulting double error at extreme range (15 000 metres) accounted for the leftward drift†.

Our answer to the question 'Why do we have to allow for drift in long-range gun-fire?' is thus 'Because the earth is rotating with respect to distant stars and galaxies.' In the same way, the answer to the question 'Why does a child get dizzy when turning round in circles but not when standing still?' is that the earth rotates much more slowly with respect to distant stars than the child does. If we are now asked why the distant stars should decide what makes a child dizzy, we reply that most of the universe is made up of the distant stars and galaxies and very little of the universe is made up of the child. The universe of Galileo's judges is a small one dominated by the earth at its centre — the universe of Galileo is a large one in which nothing is central and no single object dominates.

In Einstein's first paper on special relativity, he remarks that his theory has a 'peculiar consequence' which has come to be known as the 'twin paradox'. Suppose (to vary our metaphor from ships and quaysides) that two twin space travellers, Anne and Caroline, live in a space station (at rest relative to the fixed stars). At time $t = 0$, Anne gets into her spaceship and, after a brief acceleration, travels with velocity u (measured in light speed units) away from the space station for a time $(1 - u^2)T$ given by her watch, then after a brief deceleration and acceleration travels with velocity u towards the space station for a time $(1 - u^2)T$ given by her watch and then briefly decelerates and rejoins her sister Caroline on the space station. Because of time dilation (remember the discussion of the muon in the previous section), the outward and inward journey will each take time T according to Caroline's watch. (We neglect the brief acceleration times.) Thus, when the twins compare watches, Anne's watch will show that a time $2(1 - u^2)T$ has passed and Caroline's watch will show that a time $2T$ has passed. Since special relativity claims to apply to all systems whether mechanical, electromagnetic or whatever, it applies to human beings as well as watches so that Anne has aged by $2(1 - u^2)T$ and Caroline has aged by $2T$.

Many arguments have been urged against this conclusion, but most of them are of the form 'If the earth were round and people lived at the Antipodes they would fall off' or 'If the earth rotated we would all get dizzy.' The most plausible argument urges that everything is relative and that instead of talking about Anne travelling with speed u away from Caroline we could talk about Caroline travelling with speed u away from Anne. However, Anne and Caroline do not have identical

†This anecdote comes, presumably via Blackett, from Littlewood's *Miscellany* [145]. Littlewood worked on ballistics (the mathematics of gun ranging and so on) during that war and recalls a paper he wrote as part of this work which 'ended with the sentence "Thus σ must be as small as possible." This did not appear in the printed [version but] … a speck in a blank space at the end proved to be the tiniest σ I have ever seen (the printers must have scoured London for it).'

experiences. Anne has accelerated three times to move from some frame of reference to another whilst Caroline has remained unaccelerated in one frame of reference. (When the surge of acceleration of a plane taking off pushes us back in our seats, the watchers in the airport lounge do not experience an equal and opposite sensation.) There is no symmetry between Anne and Caroline and so no paradox†.

In 1971 a group of scientists flew an atomic clock round the world (to save the US taxpayer money, it flew economy class by ordinary scheduled flights) and compared it with a twin clock which stayed at home. This experiment did not have the mathematical simplicity of our spaceship experiment since the differing gravitational force at different altitudes means that general relativity must be used to calculate the expected time changes‡. However, the joint predictions of special and general relativity were verified — *one clock twin aged more than the other*.

There are still philosophers, historians and sociologists who believe that the way we view the universe is a matter of convention or social agreement or simply the result of power struggles between competing scientists. All these views contain some elements of truth, but a greater truth is contained in Galileo's counterblast to one of his early opponents:

> Possibly he thinks that philosophy is a book of fiction by some writer, like the *Iliad* or *Orlando Furioso,* productions in which the least important thing is whether what is written there is true. Well … that is not how matters stand. Philosophy is written in that grand book, the universe which stands continually open to our gaze. But the book cannot be understood unless one first learns to comprehend the language and read the letters in which it is composed. It is written in the language of mathematics, and its characters are triangles, circles, and other geometric figures without which it is humanly impossible to understand a word of it; without these one wanders about in a dark labyrinth.

†Not everything that looks symmetric is symmetric. Consider a ski resort full of young ladies looking for husbands and husbands looking for young ladies.

‡For a more detailed discussion as well as a clear description of how theories like general relativity can be put to experimental test see the excellent popular text by C. M. Will *Was Einstein Right?*

CHAPTER 8

A Quaker mathematician

8.1 Richardson

A customer in a British bank would feel perturbed if greeted by a manager wearing jeans, a sweater and shoes which had clearly seen better days. In the same way, the man in the street feels unhappy when introduced to a mathematician who wears a three piece suit. A good mathematician, he feels, should be untidy, absentminded and odd. Since mathematicians are not only permitted but actively encouraged to be eccentric, it is not surprising that a substantial minority are a little peculiar. There are mathematicians who dress like Einstein, others who dress like priests of some Far Eastern religion and some who dress like garden gnomes. There are mathematicians who never open their mail, mathematicians who sleep all day and work during the night, mathematicians who eat yoghurt with a fork, mathematicians who only eat yoghurt, mathematicians who lecture in bare feet, and several who know the railway and long-distance coach timetables for the whole of the British Isles.

This diversity lies mainly on the surface. Most mathematicians share the same kind of mathematical values and pursue similar careers. Lewis Fry Richardson was different — and he was different because he was a Quaker who lived according to the Quaker rules of service and pacifism. He was born in 1881 in Newcastle into a middle class Quaker family. From primary school where his chief enjoyment was Euclid 'as taught by Mr Wilkinson', he moved on to a Quaker boarding school where various masters showed him 'glimpses of the marvels of science', taught him 'how to observe and describe' and convinced him that 'science ought to be subservient to morals'†. He spent two years at the Durham College of Science (a Newcastle offshoot of the University of Durham, later to become the University of Newcastle, but at that time not a degree-granting institution), followed by a three year science degree at

†Here and elsewhere I use the accounts in Gold's obituary notice and Ashford's *Prophet or Professor*?

Cambridge. In the next ten years he occupied a succession of teaching and research jobs.

The more scholarly histories of sonar and radar mention Richardson's 1912 patents for an *Apparatus for Warning a Ship of its Approach to Large Objects in a Fog* and *Apparatus for Warning a Ship at Sea of its Nearness to Large Objects Wholly or Partly under Water* which proposed a system of echo location using high frequency sounds. However, as we saw in our discussion of radar, it makes little sense to try to assign priority of invention to systems which rely on technological advances to implement specific ideas†. I mention it chiefly in order to quote the charming picture of Richardson demonstrating his idea to his son:

> That evening we attached a high-pitched whistle to the focal point of sound of a large golf umbrella and set off looking for buildings about the size of ships. On finding such a building, we pointed the umbrella in various directions (including that of the building), blowing the whistle and timing the echoes with a stop watch. An amused and puzzled crowd gathered including a policeman! My father's answers to questions rapidly turned their mirth and heckling to interest and attention.

One of his posts was with a firm which extracted peat. Here he tackled the following problem 'Given the annual rainfall, how must the drains [in a peat moss] be cut in order to remove just the right amount of water?' The equations involved had no known exact solutions and he had to use approximate methods. When, a few years later, he joined the Meteorological Office it was natural for him to apply similar ideas to weather forecasting. At the time, naturally enough, forecasting was done by looking for past weather patterns sufficiently close to the present pattern that it might be expected to develop in the same way. Richardson proposed to use instead the fundamental physical laws which govern the weather.

To propose such a scheme is easy; to implement it, quite another matter. Some of the equations governing the weather were known, others had to be worked out from first principles. The equations could not be solved exactly and numerical methods for solving them had to be worked out. The general idea of such methods is, again, not hard to describe. We cannot hope to describe the weather at all points, so we only consider points in some regular array — say at various heights above the vertices of some square grid. At a time t we know the temperature, wind speed, pressure, etc. at each of these points. At time $t + \delta t$ these quantities will change but (to a good approximation) the way in which they change will depend in an easily calculated way on the temperature, wind speed, pressure, etc. at neighbouring grid

†The British Royal Commission on Awards to Inventors which sought to assign priority and recommend financial rewards for inventions used during the Second World War was startled to receive from Germany a 1904 Patent for the 'Telemobilscope' of Christian Hüldmeyer which covered the basic idea of radar.

points. Once we have recalculated all the relevant quantities at all the grid points, we can then use these new values in exactly the same way to compute further values at time $t + 2\delta t$ and so on. Notice that if it requires k computations to recompute the 'weather' at one point, it will require Nk computations to recompute the 'weather' at all the N points of our three dimensional grid and MNk computations to 'forecast' the 'weather' at time $M\delta t$. To improve the quality of our forecast we need to make N as large as possible and δt as small as possible — but the larger N is and the smaller δt the more computations we need.

Although this work was additional to his full time job, Richardson managed to work out suitable methods and, in less than three years, by 1916, he had a first draft of a book ready; but publication was to be delayed for six years. There were two reasons for this. The first was that he wished to complete the book with a fully worked out numerical example. The second may be described in his own words.

> In August 1914 I was torn between an intense curiosity to see war at close quarters [and] an intense objection to killing people, both mixed with ideas of public duty, and doubt as to whether I could endure danger. After much difficulty I extricated myself [from the position in the Meteorological Office] in May 1916 and joined the Friends' [Quakers'] Ambulance Unit. In September 1916 I was attached to a motor ambulance convoy lent to the French army. We carried the wounded and sick of the 16-ième division of infantry.

Richardson took his work with him.

> The manuscript was revised and the detailed example ... worked out in France in the intervals of transporting wounded .... During the battle of Champagne in April 1917 the working copy was sent to the rear, where it became lost, to be rediscovered later under a heap of coal.

Cambridge University Press only undertook publication after demanding a subsidy supplied jointly by the Royal Society, the Meteorological Office and Richardson himself. When the book *Weather Prediction by Numerical Process* came out in 1922 it was widely and sympathetically reviewed as an interesting, if quixotic, work but, although only 750 copies were printed, the edition failed to sell out in the next 30 years (so that, even with the subsidy, CUP made a loss). It fulfilled the prediction of one reviewer that it would 'have but a limited number of readers and [would] probably be quickly placed upon a library shelf and allowed to rest undisturbed by most of those who purchase a copy.'

There were two reasons for the relative failure of the book. The first was that the method outlined appeared, by Richardson's own account, totally impractical.

It took me the best part of six weeks to draw up the computing forms and to work out the new distribution in two vertical columns for the first time. My office was a bed of hay in a cold rest billet. With practice the work of an average [arithmetician] might go perhaps ten times faster. If the time step were 3 hours, then 32 individuals could just compute two points so as to keep pace with the weather ...

Richardson calculates that 6400 arithmeticians (he uses the word 'computers' which then meant 'human computers')

... would be needed to race the weather for the whole globe. That is a staggering figure. Perhaps in some years' time it may be possible to report a simplification of the process but in any case the organisation indicated is a central forecast-factory for the whole globe.

He then continues in a much quoted passage:

Imagine a large hall like a theatre, except that the circles and galleries go right round through the space usually occupied by the stage. The walls of this chamber are painted to form a map of the globe. The ceiling represents the north polar regions, England is in the gallery, the tropics in the upper circle, Australia on the dress circle and the antarctic in the pit. A myriad [arithmeticians] are at work upon the weather of the part of the map where each sits, but each [arithmetician] attends only to one equation or part of an equation. The work of each region is coordinated by an official of higher rank. Numerous little 'night signs' display the instantaneous values so that neighbouring [arithmeticians] can read them. Each number is thus displayed in three adjacent zones so as to maintain communication to the North and South on the map. From the floor of the pit a tall pillar rises to half the height of the hall. It carries a large pulpit on its top. In this sits the man in charge of the whole theatre; he is surrounded by several assistants and messengers. One of his duties is to maintain a uniform speed of progress in all parts of the globe. In this respect he is like the conductor of an orchestra in which the instruments are slide-rules and calculating machines. But instead of waving a baton he turns a beam of rosy light upon any region that is running ahead of the rest and a beam of blue light upon those who are behind hand.

... Outside are playing fields, houses, mountains and lakes, for it was thought that those who compute the weather should breathe of it freely.

The second problem lay with the worked example itself. Starting with the state of the atmosphere at 7 am GMT over central Europe on 20

May 1910 'chosen because the observations form the most complete set known to me at the time of writing,' Richardson applied his method to calculate the weather at two points six hours later. The result indicated storms of unprecedented violence on a day which was actually perfectly calm. That Richardson published this result is a testament to his rigid intellectual honesty (he could, quite properly, have chosen to illustrate his method with a more or less artificial example) but hardly helped his case.

In 1920 an administrative reorganisation saw the Meteorological Office placed under the control of the Air Ministry. This would have meant that, if Richardson continued there, he would be an employee of the armed forces. After a struggle with his conscience, he resigned and took up a lectureship at Westminster Training College. Here he prepared students for London external degrees and (rather more unusually) took the exams himself, obtaining first a pass BSc in psychology with pure and applied mathematics and then in 1929 (at the age of 48 and two years after being made a fellow of the Royal Society) the special BSc in psychology.

In 1929 he was appointed Principal of Paisley Technical College where he became a full time teacher and administrator. (In addition to a full day his duties included teaching evening classes in mathematics and physics to London external degree standard.) He devoted what free time he had to psychology, on which he published several papers, and to the study of the causes of war.

After 1945, with the advent of the electronic computer, interest in numerical weather forecasting revived, initially under the impulsion of the great mathematician von Neumann. It turned out that the computational procedures proposed by Richardson required substantial modification to avoid the production of 'artificial storms' caused by the nature of the weather model or the numerical methods used, and bearing no relation to actual events. For example, if we ask how fast disturbances can travel in air, the answer is clearly the speed of sound. But changes in the weather travel much more slowly and computations geared to the speed of sound give very misleading outcomes†. However, as Richardson's description of his forecast-factory shows, numerical meteorology makes an ideal field for the use of electronic computers (and particularly, we may add with an eye on the future, for highly parallel computers). Before his death in 1953 Richardson was able to see the first results which 'although not a great success of a popular sort, [are] anyway an enormous scientific advance on the single, and quite wrong result in which [my book] ended.' In 1965, Richardson's book was reissued and this time sold 3000 copies in less than ten

†There is a discussion of why Richardson's 1922 forecast failed in an interesting article by Platzman entitled *A Retrospective View of Richardson's Book on Weather Prediction.* Needless to say, the reasons are fairly subtle.

years. Since then, faster computers, improved mathematical methods and better observations (particularly using satellites) have gone hand in hand to make Richardson's dream a reality.

8.2 Richardson's deferred approach to the limit

Mathematicians stand on the shoulders of their predecessors. Each generation refines and develops the methods of the previous one with the result that our tools become more powerful but also more difficult to master. This process is inevitable and, on the whole, productive, but may obscure the basic ideas of the subject. One of Richardson's strengths was the ability to look at these underlying ideas with a fresh eye.

The central theme of modern mathematics (at least from 1600 to 1900) is the calculus. Suppose you are given a machine consisting of an input scale which you can set at any value x that you wish and an output scale which registers $f(x)$ the result of your input x. You know that $f(10.02) = 6.04$ and $f(10.08) = 6.07$ and you are now asked to guess a value of y which gives $f(y) = 6.05$. What do you guess? If, as I expect, you guess $y = 10.04$ (which is the guess that I would also make) then you are, probably, assuming that for small changes in the input x the output is approximately linear in x and that

$$f(10.02 + h) \approx 6.04 + 0.5h,$$

at least when h is small. (We saw this used by Blackett in Section 4.2.) The calculus is the study of 'smooth' functions which are locally very well approximated by polynomials. Thus, if f is smooth, we have

$$f(x + h) \approx b_0 + b_1 h,$$

that is f is approximately linear, at least when h is small. Further, there is an even better quadratic approximation

$$f(x + h) \approx b_0 + b_1 h + b_2 h^2,$$

and an even better cubic approximation

$$f(x + h) \approx b_0 + b_1 h + b_2 h^2 + b_3 h^3,$$

at least when h is small, and so on. We write

$$f(x + h) = b_0 + b_1 h + b_2 h^2 + b_3 h^3 + \cdots, \tag{8.1}$$

or, sometimes,

$$f(x + h) = a_0 + \frac{a_1}{1!} h + \frac{a_2}{2!} h^2 + \frac{a_3}{3!} h^3 + \cdots. \tag{8.2}$$

(Of course Formula (8.2) is just Formula (8.1) with $a_r = r! b_r$, but it may be more familiar to any reader who knows about Taylor expansions.)

We shall illustrate our discussion by trying to calculate π. Since the area of a circle of unit radius is π, one way is to observe that the Cartesian equation of the circle with centre the origin $(0,0)$ and radius 1 is

$$x^2 + y^2 = 1.$$

If $y \geq 0$, this equation can be rewritten as

$$y = (1 - x^2)^{1/2},$$

and we see that the area of the first quadrant

$$\{(x,y) : x^2 + y^2 \leq 1, \ x, y \geq 0\},$$

shown as the shaded area in Figure 8.1, is $\pi/4$. If you know a little calculus, you will probably prefer to write

$$\frac{\pi}{4} = \int_0^1 (1 - x^2)^{1/2} \, dx.$$

Although most non-mathematicians would say that the curve $y = (1 - x^2)^{1/2}$ shown in Figure 8.1 is smooth, it is not smooth in the sense used in this chapter since it has a vertical tangent at $x = 1$ (and so does not look like a polynomial near $x = 1$). To get round this problem we look instead at the area Δ^* of the set

$$\{(x,y) : x^2 + y^2 \leq 1, \ 2^{-1/2} \geq x \geq 0, \ y \geq 0\},$$

shown as the shaded area in Figure 8.2(a).

Figure 8.1: The first quadrant of the unit circle.

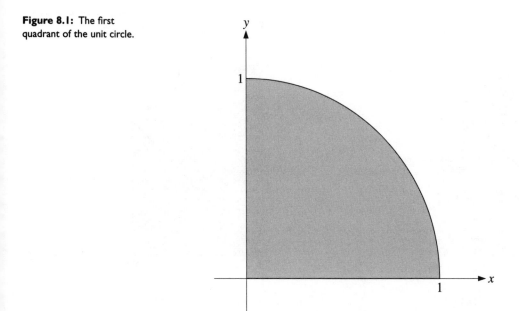

Exercise 8.2.1   *By using the decomposition shown in Figure 8.2(b), or otherwise, show that* $\Delta^* = \pi/8 + 1/4$.

How can we estimate Δ^*? More generally, if f is a well behaved positive function and $b > 0$ how do we estimate the area Δ of the set

$$E = \{(x, y) : f(x) \geq y \geq 0,\ b \geq x \geq 0\},$$

shown as the shaded area in Figure 8.3(a)? (Or, as those who know calculus might write, how do we estimate $\int_0^b f(x)\,dx$?) One natural way is to split the set E into N strips

$$E_r = \{(x, y) : f(x) \geq y \geq 0,\ rb/N \geq x \geq (r-1)b/N\}$$

of area Δ_r $[r = 1, 2, \ldots, N]$ as shown in Figure 8.3(b). Obviously

$$\Delta = \Delta_1 + \Delta_2 + \ldots + \Delta_N$$

In the language of the calculus

$$\int_0^b f(x)\,dx = \sum_{r=1}^{N} \int_{(r-1)b/N}^{rb/N} f(x)\,dx.$$

The advantage of this approach is that (at least when N is large) then the length of the interval between $(r-1)b/N$ and rb/N is small and we can apply the ideas set out in the second paragraph of this chapter. In particular, we recall that f is approximately linear for small changes. Now if f were *exactly* linear on the interval between

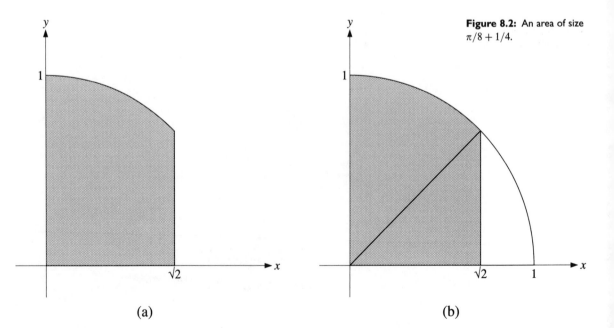

(a)   (b)

Figure 8.2: An area of size $\pi/8 + 1/4$.

$(r-1)b/N$ and rb/N then E_r would be a trapezium with vertices having (x, y) coordinates

$((r-1)b/N, 0)$, $((r-1)b/N, f((r-1)b/N))$, $(rb/N, f(rb/N))$ and $(rb/N, 0)$

and area $\Delta_r = (f((r-1)b/N) + f(rb/N))b/N$. (Figure 8.4(a) illustrates this.) Using this approximation in Formula (8.2) we obtain

$$\Delta \approx (f(0) + 2f(b/N) + 2f(2b/N) + 2f(3b/N) +$$
$$\dots + 2f((N-2)b/N) + 2f((N-1)b/N) + f(b))b/N.$$

This formula is sometimes called the trapezium rule.

How accurate is the trapezium rule? We know that in general E_r is not an exact trapezium because f is only approximately linear even for small changes in the variable. Let us, instead, consider the better quadratic approximation

$$f((r-1)b/N + h) \approx f((r-1)b/N) + A_r h + B_r h^2,$$

which we know will hold provided that h is small enough. If this formula held *exactly* on the interval between $(r-1)b/N$ and rb/N then E_r would be contained within a trapezium G_r with vertices

$((r-1)b/N, 0)$, $((r-1)b/N, f((r-1)b/N + |B_r|(b/N)^2))$,

$(rb/N, f(rb/N) + |B_r|(b/N)^2)$ and $(rb/N, 0)$

Figure 8.3: Estimating area by polygonal approximation.

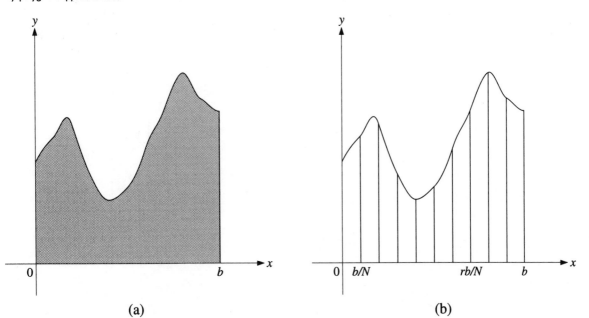

(a)   (b)

and contain a trapezium H_r with vertices

$$((r-1)b/N, 0), \ ((r-1)b/N, f((r-1)b/N - |B_r|(b/N)^2)),$$
$$(rb/N, f(rb/N) - |B_r|(b/N)^2) \text{ and } (rb/N, 0).$$

(See Figure 8.4(b).) In this case

$$\text{area} H_r \leq \text{area} E_r \leq \text{area} G_r$$

and so

$$(f((r-1)b/N) + f(rb/N)b/N - |B_r|(b/N)^3$$
$$\leq \Delta_r \leq (f((r-1)b/N) + f(rb/N)b/N + |B_r|(b/N)^3.$$

In other words

$$|\Delta_r - (f((r-1)b/N) + f(rb/N)b/N| \leq |B_r|(b/N)^3,$$

and our maximum error in estimating the area of E_r in this way is proportional to N^{-3}. Of course, B_r depends on r and N, but it is reasonable to suppose (and, though we shall not prove this, true) that we can find a B independent of r and N such that $B \geq |B_r|$, whatever the values of N and r, and so, from our previous inequality,

$$|\Delta_r - (f((r-1)b/N) + f(rb/N))b/N| \leq B(b/N)^3.$$

Figure 8.4: Estimating the area of a single strip.

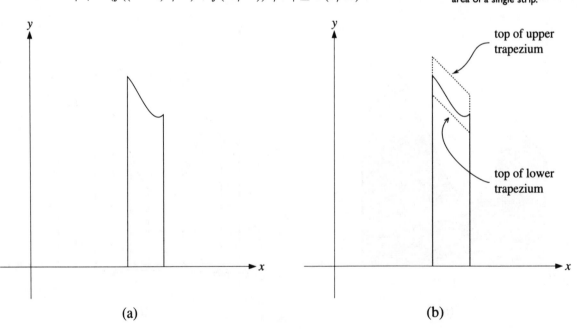

(a)   (b)

Adding up these inequalities for $r = 1, 2, \ldots, N$ we get

$$|\sum_{r=1}^{N} \Delta_r - \sum_{r=1}^{N} (f((r-1)b/N) + f(rb/N))b/N| \leq \sum_{r=1}^{N} B(b/N)^3,$$

and so

$$|\Delta - (f(0) + 2f(b/N) + 2f(2b/N) + 2f(3b/N) +$$
$$\ldots + 2f((N-2)b/N) + 2f((N-1)b/N) + f(b))b/N| \leq Bb^3 N^{-2}. \quad (8.3)$$

The maximum error in estimating each strip is proportional to N^{-3}, but we add N such strips to get the whole area so the maximum total error is N times as large and so proportional to N^{-2}.

(The reader may wonder what would happen if instead of using the quadratic approximation of Equation (8.2) we used a higher order approximation, say the cubic approximation

$$f((r-1)b/N + h) \approx f((r-1)b/N) + A_r h + B_r h^2 + C_r h^3.$$

It is easy to check that the additional error term corresponding to $C_r h^3$ is small compared with that corresponding to $B_r h^2$, at least when h is small, and so our estimates are essentially unchanged.)

The error estimate in (8.3) tells us that the trapezium rule is quite a good way of estimating area. The number of computations required is proportional to N but the error decreases like N^{-2}. Thus (roughly speaking), increasing the work 10-fold decreases the error 100-fold. As an example let us use the trapezium rule to estimate Δ^*, the area we set out to compute originally (see Figure 8.2(a)). Table 8.1 shows N the number of strips, Δ_N^* the estimate using the trapezium rule, and $P_N = 8\Delta_N^* - 2$ the associated estimate for π. (Recall from Exercise 8.2.1 that $\pi = 8\Delta^* - 4$.)

Exercise 8.2.2 *Check the table for $N = 10$ by using a hand calculator. Write a program to check the remaining entries and, if possible, run it.*

Table 8.1. *Estimating π by the trapezium rule.*

N	Δ_N^*	$P_N = 8\Delta_N^* - 2$	error
10	0.6422828269041334	3.1382626152330673	3.3×10^{-3}
20	0.6425949409968993	3.1407595279751943	8.3×10^{-4}
100	0.6426949150737193	3.1415593205897547	3.3×10^{-5}
200	0.6426980400346621	3.1415843202772971	8.3×10^{-6}
1000	0.6426990400320625	3.1415923202564997	3.3×10^{-7}
2000	0.6426990712820574	3.1415925702564591	8.3×10^{-8}

In Richardson's day the only aids available were a hand cranked machine for doing addition, subtraction, multiplication and division together with books of tables of common functions like square roots. (Notice that increasing the accuracy by one decimal place meant either using a table which was ten times as long, or resorting to still more computation.)

The trapezium rule is an obvious way of estimating areas and has been used for centuries. The fact that the error decreases like N^{-2} has also been known for a very long time. The next exercise shows how to establish this rigorously but is not necessary for the understanding of this chapter.

Exercise 8.2.3   *(This is intended for people who have done a substantial amount of calculus.)*

(i) Suppose g is a twice differentiable function, $a > 0$ and $g''(x) \geq 0$ for all $|x| \leq a$. Show that if $g'(-a) > 0$, then $g(-a) < g(a)$ and that if $g'(a) < 0$, then $g(-a) < g(a)$ also.

(ii) Suppose that, in addition to our previous conditions, we know that $g(-a) = g(a) = 0$. Show, by using the result of (i), or otherwise, that there exists a c with $-a \leq c \leq a$ such that $g'(c) = 0$. By considering the sign of $g'(x)$ for $-a \leq x \leq c$, or otherwise, show that $g(x) \leq 0$ for $-a \leq x \leq c$. Show also that $g(x) \leq 0$ for $c \leq x \leq a$, and so $g(x) \leq 0$ for all $-a \leq x \leq a$.

(iii) Suppose F is a twice differentiable function, $a > 0$, $F(-a) = F(a) = 0$ and $|F''(x)| \leq M$ for all $|x| \leq a$. By looking at

$$g(x) = M\frac{x^2 - a^2}{2} - F(x),$$

and using the result of (ii), or otherwise, show that

$$F(x) \geq M\frac{x^2 - a^2}{2}$$

for all $|x| \leq a$. Deduce that

$$\int_{-a}^{a} F(x)\,dx \geq -\frac{2Ma^3}{3},$$

and show that

$$\left| \int_{-a}^{a} F(x)\,dx \right| \leq \frac{2Ma^3}{3}.$$

(iv) Suppose G is a twice differentiable function, $a > 0$ and $|G''(x)| \leq M$ for all $|x| \leq a$. By looking at

$$F(x) = G(x) + Ax + B,$$

for a suitable choice of A and B, or otherwise, show that

$$\left| \int_{-a}^{a} G(x)\, dx - a(G(a) + G(-a)) \right| \leq \frac{2Ma^3}{3}.$$

(v) Suppose f is a twice differentiable function, h > 0 and $|f''(x)| \leq M$ for all $b \leq |x| \leq b + h$. Show that

$$\left| \int_{b}^{b+h} f(x)\, dx - h(f(b) + f(b+h)) \right| \leq \frac{Mh^3}{12}.$$

(vi) Suppose f is a twice differentiable function, N is a positive integer and b > 0 and $|f''(x)| \leq M$ for all $0 \leq |x| \leq b$. Show that

$$\left| \int_{0}^{b} f(x)\, dx - (f(0) + 2f(b/N) + 2f(2b/N) + 2f(3b/N) + \right.$$

$$\left. \ldots + 2f((N-2)b/N) + 2f((N-1)b/N) + f(b))b/N \right|$$

$$\leq Bb^3 N^{-2} \leq \frac{Mb^3}{12N^2}.$$

(vii) You ought to have drawn diagrams at each stage of the argument in order to help you understand what was going on. If you have not drawn such diagrams, do so now.

(viii) Suppose h > 0 and M > 0. Write down a twice differentiable function f such that

$$\left| \int_{b}^{b+h} f(x)\, dx - \frac{h}{2}(f(b) + f(b+h)) \right| = \frac{Mh^3}{12}.$$

Check by direct calculation that it has the property.

[Hint. If you cannot guess f directly, go through the argument of parts (i)–(v) looking for the 'worst case' function.]

However, Richardson introduced a new idea. In many ways it is more natural to think in terms of the 'step length' $\eta = b/N$ rather than N itself. The trapezium rule gives the estimate

$$\Delta(\eta) = (f(0) + 2f(\eta) + 2f(2\eta) + 2f(3\eta) +$$

$$\ldots + 2f((N-2)\eta) + 2f((N-1)\eta) + f(N\eta))\eta$$

for Δ. What Richardson did was to consider $\Delta(\eta)$ as a function of η. Since f is 'well behaved' and since the trapezium formula is a 'well behaved formula' it seems plausible that $\Delta(\eta)$ will be a well behaved function of η. In particular, since it is 'well behaved', we have, in the spirit of Equation (8.1),

$$\Delta(\eta) = \Delta(0 + \eta) = c_0 + c_1\eta + c_2\eta^2 + c_3\eta^3 + \ldots .$$

We recall that this means, for example, that when η is small the approximation

$$\Delta(\eta) \approx c_0 + c_1\eta + c_2\eta^2 \qquad (8.4)$$

is very good.

As we make η closer and closer to zero, the left hand side $\Delta(\eta)$ of Equation (8.4) approaches the required area Δ, and the right hand side $c_0 + c_1\eta + c_2\eta^2$ approaches c_0. It follows that $\Delta = c_0$ and so Equation (8.4) can be rewritten as

$$\Delta(\eta) \approx \Delta + c_1\eta + c_2\eta^2. \qquad (8.5)$$

We can, however, say more than this. If $c_1 \neq 0$, then (at least when η is very small) $c_2\eta^2$ will be very small in magnitude compared with $c_1 \neq 0$. Thus

$$\Delta(\eta) - \Delta \approx c_1\eta,$$

when η is very small, and the magnitude of the error $|\Delta(\eta) - \Delta|$ in approximating the area by the trapezium rule will be proportional to η. But we know from our previous discussion (see Equation (8.3)) that the maximum possible error is proportional to N^{-2} and so to η^2. The only way to reconcile this more rapid decrease with the rest of this paragraph is to conclude that, in fact, $c_1 = 0$ and so

$$\Delta(\eta) \approx \Delta + c_2\eta^2. \qquad (8.6)$$

We expect the approximation given by Equation (8.6) to be very good (at least when η is very small). Let us suppose for the moment that it is *exact* and so

$$\Delta(\eta) = \Delta + c_2\eta^2. \qquad (8.7)$$

We wish to know Δ, but we know neither Δ nor c_2. Thus, if we only know $\Delta(\eta)$ for one value of η, we cannot find Δ. However, if we know $\Delta(\eta)$ for two values of η, then we can. In particular, since

$$\Delta(\eta) = \Delta + c_2\eta^2$$

and

$$\Delta(\eta/2) = \Delta + c_2\eta^2/4,$$

we have

$$\Delta = (4\Delta(\eta/2) - \Delta(\eta))/3.$$

Exercise 8.2.4 *(i) Check this.*

(ii) If you know Simpson's rule check that the expression we have obtained for Δ *is that given by Simpson's rule†.*

†Very little in mathematics is *mere coincidence* and this result is no exception, However, it is, I think, correct to say that Richardson's idea is a general strategy which happens to give Simpson's rule in this case. If the reader wishes to engage in the game of 'hunt the forerunner' she could look at the 'rectification method' used by the seventeenth century Japanese mathematician Seki Kōwa for the calculation of π among other things.

This last equation has been obtained under the assumption that Equation (8.7) holds exactly for all η. Since, even when η is small, it is, in fact, only an approximation, we conclude that

$$\Delta \approx (4\Delta(\eta/2) - \Delta(\eta))/3. \qquad (8.8)$$

In our derivation of Equation (8.8) we have skated over some very thin ice. But, instead of rejecting it out of hand, try it out on our main example with the results shown in Table 8.2. (Remember that it was a brave man who first ate an oyster.) The result is striking. It appears that using this simple idea we can achieve a 30-fold decrease in computational effort for essentially the same accuracy.

When I was an undergraduate, Cambridge was served by one central computer. By dint of massive and continuous effort, it was kept running day and night. Since computation time was easiest to come by at night, anyone who needed to do serious computation rapidly developed nocturnal habits. Nowadays, if you walk into any office in any university, you will see a desk computer much more powerful than any 1965 machine *and it will usually be switched off.* In the 1960s, most computer users were limited by the available power of their machines and were forced to be as economical as possible. Today most users are under no such compulsion. There remains, however, a minority of 'serious users' whose demand for computing power is essentially unlimited.

As a very rough rule, computing speed available (measured in calculations per second) for the same cost has increased 10-fold every 10 years. Thus (for serious users) a 100-fold decrease in computational effort is equivalent to leapfrogging 20 years of technological development. (It is also possible that, because of the constraints imposed by

Table 8.2. *Estimating* π *by the trapezium rule modified according to Richardson's ideas.*

N	$(4\Delta_{2N}^* - \Delta_N^*)/3$	$Q_N = 8(4\Delta_{2N}^* - \Delta_N^*)/3 - 2$	error
10	0.6426989790278212	3.1415918322225700	8.2×10^{-7}
100	0.6426990816883097	3.1415926535064775	8.3×10^{-11}
1000	0.6426990816987224	3.1415926535897789	1.4×10^{-14}

the laws of physics, we may be reaching the end of this particular technological road and that, after another 100-fold speed-up, it will become much harder to push up the speed of our machines.) Many of the problems which make serious demands on computing power have a 'typical grid length'. For example, numerical weather forecasting still follows the pattern first laid out by Richardson and described in the fourth paragraph of Section 8.1. The smaller the grid length, the more accurate the result, but the more computations that are needed. If we could work out the result first with grid length η, then the result with grid length $\eta/2$ and then, somehow, combine the two results in a way paralleling our derivation of Equation (8.8), then we could obtain substantial gains in speed and accuracy. Richardson called this strategy a 'deferred approach to the limit' though it might be better named 'extrapolation to the limit'. (His work on this topic was done in collaboration with a younger friend, J. R. Grant, while the younger friend was still a Cambridge mathematics undergraduate!)

Unfortunately, there are quite a lot of things that can go wrong with the Richardsonian approach. (This mattered less for hand computation. A whiff of danger helped keep the human computer awake during what would otherwise be a long and tedious sequence of routine calculation. However, no electronic computer, and only the rare, canny user of such machines, possesses a sense of danger so the methods now used must have safety built in.) Because of this, Richardson's methods were not much used until the 1960s. Since then, however, they have been intensively developed and form the basis of the method of choice for several important problems.

Swift devoted one of Gulliver's voyages to satirizing the mathematicians of his day. Elsewhere he makes clear his opinion

> that whoever could make two ears of corn, or two blades of grass to grow upon a spot of ground where only one grew before would deserve better of mankind, and do more essential service to his country than the whole race of politicians put together.

Better weather forecasting means better managed, and so more productive, harvests. Faster computation means better weather forecasting. Richardson and his successors have, indeed, given us two ears of corn where we had only one before.

Exercise 8.2.5   (*In this exercise we return to the dimensional analysis of the simple pendulum considered in Section 6.1.*)

(*i*) *In Figure 8.5, O is the centre of a circle of radius R passing through A and B. The circular arc AB has length s. Convince yourself (if you need* convincing) that we can measure angle by defining the angle $\theta = \angle AOB$ to be

$$\theta = \frac{s}{R} = \frac{\text{arc length}}{\text{radius}}.$$

(If we do this then the angle is said to be measured in radians.) Deduce that angle is a dimensionless quantity.

(ii) Suppose that the key quantities governing the period t of a simple pendulum are its length l, the mass m of the weight on the end, g the acceleration due to gravity and θ the angle the string makes with the downward vertical when the pendulum is let go. Use dimensional analysis to show that

$$t = F(\theta)\sqrt{\frac{g}{l}},$$

where F is some unknown function. Since t depends on θ we write $t = t(\theta)$ and our formula becomes

$$t(\theta) = F(\theta)\sqrt{\frac{g}{l}}.$$

(iii) We expect F to be well behaved and so, in the spirit of this section, to have a very good quadratic approximation

$$F(h) \approx b_0 + b_1 h + b_2 h^2$$

when h is small. Continuing in the spirit of this section, we shall temporarily assume the approximation is an exact equation so that

$$F(h) = b_0 + b_1 h + b_2 h^2$$

and

$$t(\theta) = (b_0 + b_1\theta + b_2\theta^2)\sqrt{\frac{g}{l}}.$$

Figure 8.5: Angle as a ratio of lengths.

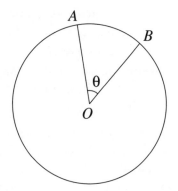

Explain why $t(\theta) = t(-\theta)$ and deduce that $b_1 = 0$. Remembering, once again, that we have to deal with an approximation, we obtain

$$t(\theta) \approx (b_0 + b_2\theta^2)\sqrt{\frac{g}{l}}.$$

Thus when the amplitude is small (i.e. when θ is small) the period is very close to $b_0(g/l)^{1/2}$.

(iv) (This part of the question requires elementary mechanics.) Choose a horizontal x axis and a vertical y axis. Consider a particle of mass m moving under gravity on the path $y = k|x|$ where $k > 0$. (You should assume that energy is conserved.) If the particle is released from rest at $x = a$, find the time $t = t(a)$ required for the particle to return to its starting point. Is the result of the form that you would have expected from the earlier parts of the question? If not, what do you think accounts for the difference?

8.3 Does the wind have a velocity?

Before continuing my account of Richardson's work I shall digress for a page or so to give Kac's view of the relation between physics and mathematics. At the end of the autobiographical note which introduces his *Selected Papers*, Mark Kac recalls how, as a young mathematics student, he tried to learn thermodynamics and statistical mechanics (the study of heat and the kinetic theory of gases) and found it

> ... a particularly disquieting experience because I found myself constantly on the outer edge of understanding, and every line I read raised new questions and doubts. Thermodynamics was especially difficult because it is conceptually subtle while at the same time technically rather trivial. In kinetic theory volumes Δv 'small enough to be taken as elements of integration yet large enough to contain many particles' rendered the subject unpalatable and even repulsive to a young mind already conditioned to look for clarity and rigour.

After a distinguished pre-war career as a pure mathematician with a particular interest in probability, Kac moved, mainly as a result of his war work, and, in particular, his association with the mathematical physicist Uhlenbeck, towards physics and he wrote deep papers on statistical mechanics — the very subject that had so puzzled him as a student. He recounts his early experience

> ... to underscore the differences between the acts of *understanding* in mathematics and in physics. Physics, to the extent that it is at all deductive, is only *locally* deductive; and since its concepts are not fixed but undergo constant evolution it is impossible to understand it

in mathematical terms alone. In mathematics ambiguity is a sin, and contradiction a disaster. In physics a certain amount of ambiguity is a fact of life, and a contradiction may well be the beginning of an important new development. Mathematics, according to the wonderful *bon mot* of Niels Bohr, is concerned with 'ordinary truths', that is statements whose negations are false; physics, on the other hand, deals with 'profound truths', that is statements whose negations are also profound truths. It is because of all this that learning physics by oneself is essentially impossible, or to quote Uhlenbeck, 'one must follow a master'. There are of course parts of physics whose foundations are so clearly formulated that many problems become almost entirely mathematical. It is not likely that the solutions of such problems will be decisive in unravelling deep mysteries of nature.

Everybody knows what is meant by velocity. We drive in cars at a speed of 60 kilometres per hour and ride in planes at speeds of 900 kilometres per hour. We know what is meant by wind velocity and, at a pinch, we can even devise instruments for measuring it. To achieve Richardson's aim of numerical weather forecasting he needed to measure wind velocity at various heights above the ground and developed a typically Richardsonian method for doing so by firing ball bearings upwards from a gun. ('For steel spheres 0.4 cm in diameter it is sufficient to wear a thick cloth cap with a good brim over the eyes. For steel balls of the pea size (0.8 cm in diameter) the observer was sometimes protected by a thick overcoat and a military steel helmet .... For the cherry size of steel ball (1.8 cm in diameter) a substantial fixed roof is necessary.')

However, Richardson begins one of his major papers (bearing the off-putting title *Atmospheric diffusion shown on a distance–neighbour graph* and reprinted in his collected paper with a section entitled *Does the wind possess a velocity?*):

> This question, at first sight foolish, improves on acquaintance. A velocity is defined, for example, in Lamb's 'Dynamics' to this effect : Let Δx be the distance in the x direction passed over in a time Δt, then the x-component of the velocity is the limit of $\Delta x/\Delta t$ as $\Delta t \to 0$. But for an air particle it is not obvious that $\Delta x/\Delta t$ attains a limit as $\Delta t \to 0$.

Consider a car travelling along a straight road so that it is a distance $x(t)$ kilometres from its starting point after a time t hours. To find its speed (in kilometres per hour) when $t = 0$ we might start by measuring the distance it travels in an hour and write

$$\text{speed} \approx x(1) - x(0).$$

However, during that hour the car speeds up and slows down, so it is obviously better to measure the distance it travels in a tenth of an hour and write

$$\text{speed} \approx \frac{x(10^{-1}) - x(0)}{10^{-1}}.$$

Even better, since even in a tenth of an hour conditions will vary, we might measure the distance it travels in a hundredth of an hour and write

$$\text{speed} \approx \frac{x(10^{-2}) - x(0)}{10^{-2}}.$$

Thus we expect

$$\frac{\triangle x}{\triangle t} = \frac{x(\triangle t) - x(0)}{\triangle t}$$

to get closer and closer to 'the true speed v at time $t = 0$'. (This is the same as saying, in the language of the previous section that

$$x(h) \approx x(0) + vh$$

where the approximation gets better as h gets smaller.)

Richardson asks whether what we expect to happen will happen and points out that Weierstrass had constructed a function for which this expectation failed completely. In the exercise that follows I give a similar but simpler construction along lines given by Van der Waerden.

Exercise 8.3.1 *(Do not worry if you cannot get to the end of this exercise, but do try to get as far as the end of (vi).)*

(i) Sketch the graph of the function g given by the condition

$$g(x + k) = |x| \quad \textit{if k is any integer and } -1/2 < x \le 1/2.$$

(ii) Let $g_n(t) = 2^{-n} g(8^n t)$. Sketch the graphs of g_1, g_2 and g_3.

(iii) Let $x_n(t) = g_1(t) + g_2(t) + \ldots + g_n(t)$. Sketch the graphs of x_1, x_2 and x_3.

(iv) Suppose that $x_3(t)$ represents the position of a particle (moving along the x axis) at time t. Suppose that we wish to measure its 'velocity' at time t by considering

$$\frac{x(t + h) - x(t)}{h},$$

for some suitable h. Explain (without doing any calculations) why the 'velocity' will depend crucially on the value of h.

(v) It is also clear, I think, that if things are bad for x_3 they will be still worse for x_4 and so on. On the other hand x_4 is close to x_3 and x_5 is even closer to x_4.

(a) Show that $0 \leq g(t) \leq 1/2$ for all t.

(b) Show that $0 \leq g_n(t) \leq 2^{-n-1}$ for all t.

(c) Show that if $M \geq N + 1$, then

$$x_M(t) - x_N(t) = g_{N+1}(t) + g_{N+2}(t) + \ldots + g_M(t)$$

and use part (b) to show that

$$0 \leq x_M(t) - x_N(t) \leq 2^{-N} + 2^{-N-1} + \ldots + 2^{-M-1} < 2^{1-N}.$$

It is thus extremely plausible (and can be proved using ideas from a first or second year university course) that, as $n \to \infty$, x_n approaches a continuous function x. It is also plausible that x will inherit the difficulties concerning 'velocity' that we saw with x_3, x_4 and so on. The rest of the exercise is concerned with proving this.

(vi) Show that $|(g(t+h) - g(t))/h| \leq 1$ for all t and all $h \neq 0$. Deduce that $|(g_n(t+h) - g_n(t))/h| \leq 4^n$ and hence that

$$\left| \frac{x_n(t+h) - x_n(t)}{h} \right| \leq 4^1 + 4^2 + \ldots + 4^{n-1} + 4^n < \frac{4^{n+1}}{3}$$

for all t, all $h \neq 0$ and all integers $n \geq 1$. Conclude that

$$\left| \frac{x_{n-1}(t+h) - x_{n-1}(t)}{h} \right| < \frac{4^n}{3}$$

for all t, all $h \neq 0$ and all integers $n \geq 2$

(vii) Let r be an integer. Show that $g_n(r8^{-n}) = 0$ and find $g_n((r + \frac{1}{2})8^{-n})$. Show that

$$\left| \frac{g_n((r + \frac{1}{2})8^{-n}) - g_n(r8^{-n})}{\frac{1}{2}8^{-n}} \right| = 4^n.$$

Combine this with the final result of (vi) to obtain

$$\left| \frac{x_n((r + \frac{1}{2})8^{-n}) - x_n(r8^{-n})}{\frac{1}{2}8^{-n}} \right| \geq 4^n - \frac{4^n}{3} = \frac{2}{3}4^n.$$

(viii) Show that

$$x_n(r8^{-n}) = x_{n+1}(r8^{-n}) = x_{n+2}(r8^{-n}) = \ldots$$

and conclude that $x(r8^{-n}) = x_n(r8^{-n})$. Show similarly that $x((r+\frac{1}{2})8^{-n}) = x_n((r + \frac{1}{2})8^{-n})$. Use the result of (vii) to conclude that

$$\left| \frac{x((r + \frac{1}{2})8^{-n}) - x(r8^{-n})}{\frac{1}{2}8^{-n}} \right| \geq \frac{2}{3}4^n. \qquad (*)$$

Thus if we measure the 'velocity' v at time $t = r8^{-N}$ by considering

$$\frac{x(t+h) - x(t)}{h},$$

with $h = \frac{1}{2}8^{-n}$ we will find $|v| \geq \frac{2}{3}4^n$. When n is large this will give an extremely curious view of velocity.

Show also that

$$\left| \frac{x((r+1)8^{-n}) - x((r+\frac{1}{2})8^{-n})}{\frac{1}{2}8^{-n}} \right| \geq \frac{2}{3}4^n.$$

(ix) Now consider any fixed t. For each integer $n \geq 1$ show that we can either find an integer r with $r8^{-n} \leq t < (r+\frac{1}{2})8^{-n}$ or we can find an integer r with $(r+\frac{1}{2})8^{-n} \leq t < (r+1)8^{-n}$. Suppose the first case holds. Using the fact that

$$(x((r+\tfrac{1}{2})8^{-n}) - x(t)) + (x(t) - x(r8^{-n})) = x((r+\tfrac{1}{2})8^{-n}) - x(r8^{-n})$$

together with the inequality (*) in *(viii)*, show that

$$\max(|x((r+\tfrac{1}{2})8^{-n}) - x(t)|, |x(t) - x(r8^{-n})|) \geq \frac{2^{-n}}{6}$$

and by choosing $h = ((r+\frac{1}{2})8^{-n}) - t$ or $h = t - r8^{-n}$ show that there exists an h with $|h| \leq \frac{1}{2}8^{-n}$ such that

$$\left| \frac{x(t+h) - x(t)}{h} \right| \geq \frac{\frac{2^{-n}}{3}}{\frac{1}{2}8^{-n}} = \frac{4^n}{3}.$$

Prove that the same result holds in the second case.

By taking n larger and larger we see that any attempt to measure a consistent 'velocity' is bound to fail. In this case it is not true 'that $\triangle x / \triangle t$ attains a limit as $\triangle t \to 0$.'

Richardson continues:

In view of the following considerations, let us not think of velocity, but only of various hyphenated velocities, such as the one-minute-velocity, or the six-hours-velocity, the words attached by the hyphen indicating the value of $\triangle t$.

When the first Tay bridge was built, the expert consulted gave wind speeds of the many-minutes-velocity type, neglecting the possibility of gusting (bursts with much higher one-minute-velocity). Gusting during a ferocious storm helped bring the bridge down.

Not only does wind speed and direction measured at some fixed point vary erratically, but the direction and speed of the wind measured at some fixed time varies erratically from place to place. In his book *Weather Prediction by Numerical Process* Richardson writes that

> [c]onvectional motions are hindered by the formation of small eddies resembling those due to dynamical instability. Thus C. K. M. Douglas writing of observations from aeroplanes remarks: 'The upward currents of large cumuli give rise to much turbulence within, below, and around the clouds, and the structure of the clouds is often very complex.' One gets a similar impression when making a drawing of a rising cumulus from a fixed point; the details change before the sketch can be completed. We realise thus that: big whirls have little whirls that feed on their velocity, and little whirls have lesser whirls and so on to viscosity — in the molecular sense.

It is well known that the weather moves from West to East, in some sense, blown by the prevailing winds. If it were not for the big and small whirls, weather forecasting would be much easier. We call the phenomenon 'turbulence'.

The study of turbulence is one of the themes of twentieth century physics, but at least one earlier student of nature fell under its spell. Commentators on Leonardo da Vinci's art tend to regret

> ... what can almost be regarded as an obsession .... Something in the movement of water, its swirls and eddies, corresponded to some deep-seated twist in his nature. All his life he studied these movements. The writings in which he sought to codify the waves and reduce the eddies to a geometrical system form one of the largest sections of his voluminous writings.

†Both drawings are well known. They form Plates 281 and 290 in Popham's collection of Leonardo's drawings.

Figure 8.6 shows some of his studies of water. Figure 8.7† shows horsemen caught by a storm. Note the wind gods riding the storm and, more relevant to our present discussion, how

> Leonardo has filled the atmosphere with the cascades and currents which he had studied in moving water.

We now return to Richardson's argument. One way of studying the whirls is to release a collection of balloons and see what happens to them (of course such an experiment would not detect eddies much smaller than the balloons). However, if we measure how far they move from a fixed point in a given time, the main contribution will come from the big eddy in which they all find themselves. Richardson suggests that we look instead at the distance *between pairs* of balloons. More precisely, he suggests that we should look at $d(t)^2$, the square of the distance between them after a time t. (We suggest a reason for looking at the square of the distance in Exercises 8.3.2 and 8.3.3 but it is the natural choice for many reasons.) Of course, for any particular pair, the value of $d(t)^2$ depends on which particular gust catches them

Figure 8.6: A page from Leonardo. (The Royal Collection © Her Majesty The Queen.)

and which eddies separate them, so we consider the *average* value $\langle d(t)^2 \rangle$ obtained by averaging over many pairs of balloons. When Richardson wrote there was already a substantial amount of theoretical and experimental evidence to show that $\langle d(t)^2 \rangle$ is linear in t, that is

$$\langle d(t)^2 \rangle = Kt + d(0)^2,$$

where K is a constant and $d(0)$ is the initial distance apart.

Exercise 8.3.2 *(The drunkard's walk: this exercise requires very elementary probability.)*

(i) Consider a drunkard wandering along the x axis starting at the origin at time $t = 0$. Each minute he takes a step of one unit to the left or the right with equal probability independent of anything he has done before. Let us set $U_i = -1$ if he moves to the left at step i and $U_i = +1$ if he moves to the right. If the man is at X_n at time n, show that

$$X_n = U_1 + U_2 \ldots + U_n.$$

Figure 8.7: Storm. (The Royal Collection © Her Majesty The Queen.)

(ii) Show that $\mathbb{E}U_i = 0$ *and* $\mathbb{E}U_i^2 = 1$. *(Recall that* $\mathbb{E}Z$ *is the expectation or average value of* Z.) *Show that, if* $i \neq j$, *then*

$$\Pr\{U_iU_j = 1\} = 1/2$$
$$\Pr\{U_iU_j = -1\} = 1/2$$

and deduce that $\mathbb{E}(U_iU_j) = 0$. *What happens if* $i = j$?

(iii) Show that

$$X_n^2 = \sum_{i=1}^{n}\sum_{j=1}^{n} U_iU_j$$

and so

$$\mathbb{E}X_n^2 = \sum_{i=1}^{n}\sum_{j=1}^{n} \mathbb{E}(U_iU_j).$$

By using (ii) show that $\mathbb{E}X_n^2 = n$. *Thus, for our drunkard, moving at random like a balloon caught in the wind, the average of the square of the distance travelled from his starting place varies directly with the time taken.*

(iv) Suppose the drunkard now wanders about the two dimensional plane so that his position at time n *is* (X_n, Y_n) *with* $(X_n, Y_n) = (0,0)$. *Let us write* $U_n = X_n - X_{n-1}$ *and* $V_n = Y_n - Y_{n-1}$. *Suppose further that*

$$\Pr\{U_n = 1, \ V_n = 0\} = \Pr\{U_n = -1, \ V_n = 0\} = \Pr\{U_n = 0, \ V_n = 1\}$$
$$= \Pr\{U_n = 0, \ V_n = -1\} = 1/4.$$

Describe in words the nature of the path.

Show that $\mathbb{E}U_i = 0$, $\mathbb{E}(U_iU_j) = 0$ *if* $i \neq j$ *and* $\mathbb{E}U_i^2 = 1/2$. *Show that* $\mathbb{E}X_n^2 = n/2$. *What is* $\mathbb{E}Y_n^2$? *Show that*

$$\mathbb{E}(X_n^2 + Y_n^2) = n$$

and so, as in the one dimensional case, the average of the square of the distance travelled from his starting point varies directly as the time taken.

(v) Extend the result to three dimensions†.

†One of the readers of the manuscript comments: 'You may need to replace alcohol with other substances.'

Exercise 8.3.3   *(i) Suppose g is a function with*

$$g(s + t) = g(s) + g(t)$$

for all real s and t. By setting $s = t = 0$, *or otherwise, show that* $g(0) = 0$. *Deduce that* $g(-t) = -g(t)$.

If n is a positive integer show that $g(nt) = ng(t)$. *Use the previous result to deduce that if m is an integer,* $g(mt) = mg(t)$. *Show that if m* *and n are integers with $n \geq 1$, then $g(1/n) = g(1)/n$ and*

$$g\left(\frac{m}{n}\right) = g(1)\frac{m}{n}.$$

Provided that g is reasonably well behaved, it follows that

$$g(t) = g(1)t$$

for all t.

   (ii) Suppose $g(t)$ is only defined for $t \geq 0$ and

$$g(s+t) = g(s) + g(t)$$

for all $s, t \geq 0$. Show that $g(0) = 0$. Show that if we define $\tilde{g}(t) = g(t)$ for $t \geq 0$ and $\tilde{g}(t) = -g(-t)$ for $t \leq 0$, then

$$\tilde{g}(s+t) = \tilde{g}(s) + \tilde{g}(t)$$

for all real s and t. Use (i) to deduce that

$$g(t) = g(1)t$$

for all $t \geq 0$.

   (iii) (This requires slightly more knowledge of probability than the previous exercise.) Suppose a particle is moving randomly along the x axis in such a way that its position $X(t)$ is continuous. Thus we see, as it were, a particle tossed hither and thither by random gusts of wind. Since $X(t)$ is random, we may talk about expectations. We assume that:

   (a) $\mathbb{E}(X(t) - X(s)) = 0$, that is the average change in position between time s and time t is zero $[t > s]$.

   (b) If $t > s > u$, then $X(t) - X(s)$ and $X(s) - X(u)$ are independent, so what happened to the particle between time u and time s has no effect on what happens to the particle between time s and time t.

   (c) The rules governing the behaviour of the particle do not change with time or position.

   Show using (a) and (b) that if $t > s > u$

$$\mathbb{E}(X(t) - X(s))(X(s) - X(u)) = 0.$$

Explain why (c) implies

$$\mathbb{E}(X(t+s) - X(s))^2 = \mathbb{E}(X(t) - X(0))^2$$

whenever $t, s > 0$.

Now let $g(t) = \mathbb{E}(X(t) - X(0))^2$. *By writing*

$$g(s + t) = \mathbb{E}((X(s + t) - X(s)) + (X(s) - X(0)))^2,$$

and expanding, show that

$$g(s + t) = g(s) + g(t)$$

if $s, t \geq 0$ *and deduce, using part (ii), that* $g(t) = Kt$ *for all* $t \geq 0$ *and some constant* K. *Conclude that if* $t \geq s$

$$\mathbb{E}(X(t + s) - X(s))^2 = K(t).$$

Thus under quite general conditions the average of the square *of the distance travelled varies directly as the time taken.*

(iv) Suppose that another particle is moving randomly along the x axis according to the same rules as the particle in (iii) but completely independently. If the position of this particle at time t is $Y(t)$ *and* $Z(t) = Y(t) - X(t)$, *show that if* $t \geq s$

$$\mathbb{E}(Z(t + s) - Z(s))^2 = 2K(t).$$

8.4 The four-thirds rule

Richardson notes that though several authors have equations of the type

$$\langle d(t)^2 \rangle = Kt + d(0)^2$$

[t]he measured values of K have been found to be 0.2 cm^2 sec^{-1} in capillary tubes, ... 10^5 cm^2 sec^{-1} when gusts are smoothed out of the mean wind, ... 10^8 cm^2 sec^{-1} when means extend over times comparable with 4 hours [and] 10^{11}cm^2 sec^{-1} when the mean is taken to be the general circulation characteristic of the latitude .... Thus the so-called constant K varies in a ratio of 2 to a billion [modern trillion].

He therefore suggests that the 'constant K' is not, in fact, constant but depends on the distance apart of the two particles. Here he describes how mixing takes place in the atmosphere.

Suppose that we were to let loose a sphere 0.01 cm in diameter of acetylene, which has much the same density of air. The sphere contains about 10^{13} molecules. For the first few hundredths of a second its rate of diffusion will be the molecular one $K = 0.2$; then micro-turbulence will spread it less slowly; then after a few seconds part may get caught in one of those gusts such as are shown by a pressure tube manometer, while another part may remain in a lull, so that it is torn asunder and gusts scatter it, K being 10^4. Next squalls of several minutes' duration scatter it more rapidly. Its rate of diffusion is now measured by $K = 10^8$. Then one part gets into a cyclone and another remains behind in an anticyclone, and its rate of diffusion is measured by Defant's value $K = 10^{11}$. Finally, it is fairly uniformly spread throughout the earth's atmosphere at the rate of about one molecule of acetylene for every cube of surface air 70 metres in the edge.

Most of the paper is taken up with detailed mathematics in which he seeks to model his new picture. He shows (as is intuitively plausible) that the principal agent in separating two particles a distance l apart are eddies of size comparable with l (or, as we shall say, of typical size l). Larger eddies sweep both particles along in much the same way and smaller eddies sweep particles back and forth randomly with little total effect. Thus if $d(0) \approx l$ our equation becomes

$$\langle d(t)^2 \rangle = K(l)t + d(0)^2$$

and the equation will only remain valid if t is sufficiently small that $K(l)t$ is not much bigger than l. (The reader will hardly need to be informed that Richardson's treatment of the problem is rather more sophisticated.)

How does $K(l)$ vary with l? Richardson supplemented earlier measurements of $K(l)$ by studying the fall of volcanic dust and the records of competitions in which balloons were released and the winner was the competitor whose balloon travelled furthest. (A later note included observations on motes in a sunbeam and the release of dandelion seeds†.)

The simplest natural way in which $K(l)$ might depend on l would be via an equation of the form

$$K(l) = Cl^\alpha$$

with C and α constants. Just as in Section 5.2 we take logarithms, obtaining

$$\log K(l) = \log C + \alpha \log l.$$

If we plot $\log K$ against $\log l$, then, if such a relation does indeed hold, we should see a straight line of slope α. (Recall, for example, Figure 5.1 in Section 5.2.) When Richardson plotted his data he obtained the result shown in Figure 8.8. It will be seen that the straight line $\log K(l) = \log 0.2 + \frac{4}{3} \log l$ corresponding to the relation $K(l) = 0.2l^{4/3}$ fits the data quite well.

G. I. Taylor wrote a nice summary of the development of the theory of atmospheric turbulence in which he describes Richardson as 'a very interesting and original character who seldom thought on the same lines as his contemporaries and often was not understood by them,'

†'Two slender bamboo poles were fixed vertically 30 cm apart and steadied by strings. To the top of each pole was fixed a small pair of scissors. A seed of dandelion (*Taraxacum officinale*), with its fluffy parachute attached was inserted in each pair of scissors and held there gently so as not to be cut. ... By a sudden jerk of a string, both pairs of scissors were closed, and both pieces of fluff set free, at the same instant as the starting of a stop watch.'

and says that Richardson's paper initiated the modern approach to the subject of turbulence†. Discussing Figure 8.8 he remarks that

> Since the curve here seems to contain all the observational data that Richardson had when he announced his remarkable law, it reveals a well-developed physical intuition that he chose as his index $\frac{4}{3}$ instead of, say, 1.3 or 1.4 but he had the idea that the index was determined by something connected with the way energy was handed down from larger to smaller and smaller eddies. He perceived that this is a process which, because of its universality, must be subject to some simple universal rule.

The words 'well-developed physical intuition' correspond to the murmur that runs round the audience when a snooker player brings off a shot in which skill and luck are inextricably mixed. Richardson recalled that '$\frac{4}{3}$ was chosen partly as a rough mean and partly because it simplified some integrals.' If we are to consider $\frac{4}{3}$ as merely a lucky accident, then we should also remember that when choosing generals Napoleon would always ask, 'Is this man lucky?'

The paper I have just described was Richardson's last major work on meteorology. The only large group in Britain interested in his researches were the meteorologists attached to the Chemical Warfare Experimental Station at Porton, where one of them wrote:

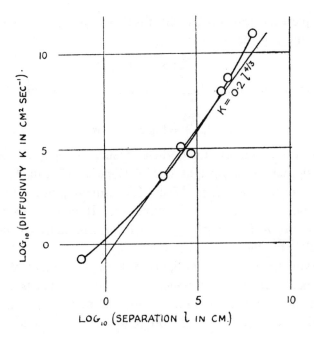

Figure 8.8: Richardson's 4/3 graph.

†Richardson would have been pleased with praise from such a source. An important non-dimensional constant governing the onset of turbulence in the atmosphere is now called the Richardson number *Ri*. Called on to comment on a paper *Note on the use of the Richardson number in meteorological problems* and, referring to wartime rationing, he wrote 'I deeply appreciate the honour of having a number called after me. Yet it makes what should be an objective study, for me embarrassingly personal. From the point of view of rationing G. I. Taylor deserves to have lots of numbers called after him, say Ta_1, Ta_2, Ta_3 ... .' (The problem of how to refer to a mathematical object named after oneself has been solved in various ways. One German Professor talked of 'The function whose name I have the honour to bear.' Dirac created endless confusion by always referring to 'Dirac's equation' as 'Equation 4.2'. J. F. Adams spoke of the 'So called Adams Spectral Sequence' until he overheard his research students referring to a certain 'Professor So Called Adams'.)

As far as the War Office was concerned, the primary aim of the researches was the production of reliable 'range tables' for chemical weapons, but it was recognised that before this could be done there had to be a much deeper understanding of the dynamics and physics of the lower atmosphere. ...

We meteorologists were not really 'poison gas' experts. Our concern was with atmospheric turbulence, and we did not bother very much with chemical weapons. We rarely used war gases in our trials, preferring, for obvious reasons, harmless substitutes. But we could not have done the work without calling on resources and skills assembled for defence purposes. As gas was not employed in World War II, the range-tables we built up so carefully were never put to use, but there has since been a notable pay-off in atmospheric pollution studies, the assessment of evaporation and in the meteorology of the lower atmosphere in general. If Richardson could have been persuaded to join in this work he would have had, for the first time, means whereby some of his most imaginative hypotheses could have been subjected to experimental tests carried out with real precision and accuracy. The gain to meteorology would have been considerable.

†Those who, like myself, are inclined to find this attitude wholly admirable, should bear in mind that Hitler was only restrained from the use of nerve gases in the Second World War by the erroneous belief that the Allies must have similar terrible weapons.

Delicate approaches were made to Richardson and, according to his wife, 'there came a time of heart-break when those most interested in his "upper air" researches proved to be the "poison gas" experts. [He] stopped his meteorological researches, destroying such as had not been published. What this cost him none will ever know.'† For this and other reasons he shifted his interests first to the applications of mathematics and physics to pure psychology and then to the mathematical study of war. Meteorology was from now on a temptation to be resisted.

Richardson retired early from his position at Paisley and refused an offer of a professorship elsewhere because, he wrote, 'I feel that I must make time to prosecute thoroughly researches on the Insta- bility of Peace.' This involved a substantial loss of income and from now on the Richardsons lived even more frugally than before. Even so Richardson could not resist a brief return to his old intellectual passion. When the young American oceanographer Henry Stommel wrote asking whether he might visit, Richardson wrote back 'Come but bring some golf balls'. Before he had succeeded in finding any (luxuries such as these were in short supply after the war), a telegram arrived with the message 'Forget the golf balls they all sink.' Richard- son wished to see if the 4/3 rule for atmospheric turbulence applied to the turbulent flow in the sea lochs he saw every day from his win- dows. The end result was a joint paper in the *Journal of Meteorology* beginning

We have observed the relative motion of two floating pieces of parsnip, and have repeated the observation for many such pairs at different initial separation.

The parsnips also satisfied the 4/3 rule. The paper ends with a note

> After this manuscript was submitted the writers have read two unpublished manuscripts by C. L. von Weizsäcker and W. Heisenberg in which the problem of turbulence for large Reynolds number is treated deductively with the result that they arrive at the 4/3 rule.

In fact, the 4/3 rule had been derived theoretically in 1941. The great Russian mathematician Kolmogorov worked in such a large number of distinct mathematical fields that eleven experts were required to describe his work for his London Mathematical Society obituary and he touched nothing that he did not adorn. He shot to fame at the age of 19 with a paper describing 'an almost everywhere divergent Fourier series'† but the result was just the first note in the symphony of his mathematical life. Kolmogorov was one of the founders of modern probability theory and it was thus natural that he should try his hand at the application of probabilistic techniques to turbulence.

In a later paper Kolmogorov says that his work 'was based on the pictorial idea of Richardson that in a turbulent flow there exist eddies of all possible scales $l < r < L$ between an 'external scale' L and an 'internal scale' and a unified mechanism of energy transfer from large-scale eddies to small-scale eddies.' (However, it seems clear that Kolmogorov's knowledge of Richardson's work was second-hand. In particular, Kolmogorov knew nothing of Richardson's paper on the 4/3 rule.) In his first paper on the subject Kolmogorov summarises the process in two paragraphs which I now paraphrase:

> Turbulent flow of the type considered can be represented in the following way. The mean flow is accompanied by 'first order fluctuations' imposed on it, consisting of chaotic motions (relative to one another) of individual fluid volumes with diameters of the order of l_1 and relative velocities of the order of v_1. In turn, the first order fluctuations are unstable and there are 'second order fluctuations' imposed on the former having typical length scales $l_2 < l_1$ and typical relative velocities $v_2 < v_1$. This process of successive refinement of turbulent fluctuations goes on until, for fluctuations of sufficiently high order, $l_n v_n$ turns out to be sufficiently small that the effect of viscosity on the nth order fluctuations is appreciable and prevents the formation of $(n + 1)$th order fluctuations imposed on the former.
>
> From the energetic standpoint, the process of turbulent mixing can be thought of in the following manner: the first-order fluctuations absorb

†He says 'I know nothing about the order of magnitude of the coefficients.' The present author is inordinately proud of having been able to say something about this.

the energy of the mean motion and transfer it subsequently to higher order fluctuations; as to the energy of the smallest fluctuations, it is dissipated and transferred into thermal energy due to viscosity.

Now, in steady state, as much energy must leave the rth-order fluctuations as enters. The whole process is thus governed solely by the rate ϵ, at which energy is removed (per unit mass, say) from the system by viscosity, and ϵ will also be the rate at which energy is transferred from the rth order fluctuations to the $(r+1)$th order fluctuations and the rate at which energy enters the system. It is natural to think of the system as a *cascade* whose length (the number of stages before viscosity prevents further fluctuations) is governed by the viscosity. By using simple and general arguments† based on elementary statistical and dimensional ideas, Kolmogorov obtained the equivalent of the 4/3 rule in a context which allowed substantial further theoretical investigation.

G. I. Taylor points out that, had Richardson used the same idea from dimensional analysis as did Kolmogorov, he would have obtained a good theoretical reason for choosing $\frac{4}{3}$ rather than some other index. (The reader may wish to consult Section 6.1 before proceeding.) Remember that Richardson started from the equation

$$\langle d(t)^2 \rangle = Kt + d(0)^2,$$

where $\langle d(t)^2 \rangle$ is the average of the square of the increase in distance between two particles at time t. Thus K has dimensions $L^2 T^{-1}$. Richardson now suggests that, although the equation is valid over periods of time when the distance l between the particles does not change too much, we must modify it by taking K to be a function of l. If we look for a relation of the form

$$K = Al^{\alpha},$$

then A has dimensions $L^{2-\alpha} T^{-1}$. If we look at the Richardson cascade in which energy is handed down from larger eddy to smaller eddy and eventually dissipated by viscosity, we see that the only relevant quantity governing A is ϵ, the rate at which energy is dissipated per unit mass. (The viscosity determines the *length* of the cascade but not its *structure*.) Now ϵ has dimensions $L^2 T^{-3}$ and we wish to form a dimensionless quantity from A and ϵ. Since $A^{\lambda} \epsilon^{\mu}$ has dimensions $L^{\lambda(2-\alpha)+2\mu} T^{-\lambda-3\mu}$ the only way that this can be done is to take $\alpha = \frac{4}{3}$.

Exercise 8.4.1 *Check the statements about the dimensions of K, A and ϵ. Check that (if λ and μ are not both zero) $A^{\lambda} \epsilon^{\mu}$ can be dimensionless only if $\alpha = \frac{4}{3}$. If $\alpha = \frac{4}{3}$ what value of λ/μ makes $A^{\lambda} \epsilon^{\mu}$ dimensionless?*

†This phrase troubles me as a perfect example of 'Math-speak'. On the one hand, I think that any mathematician who has had Kolmogorov's argument explained to them would call it 'simple and general' in the sense that once the ideas are understood the mathematics 'writes itself'. On the other hand, Barenblatt says that 'Kolmogorov's two papers on isotropic turbulence may now seem to be absolutely transparent but in the late forties even his own students found them difficult to comprehend. Accordingly, G. K. Batchelor's paper on Kolmogorov's theory of locally isotropic turbulence came to play an extremely important role in disseminating Kolmogorov's ideas not only in the West, but also in the USSR itself and amongst Kolmogorov's closest associates. This happened despite the fact that Komogorov's students were unable to read English — they used a Russian translation of Batchelor's paper that has been carefully treasured ever since.'

Richardson kept a 'revision file' for his book on numerical weather forecasting. One note read:

> When we were schoolboys and the answer to an arithmetical problem came out as a whole number, we felt sure we had got it right. The idea that simplicity implies rightness is not confined to schoolboys.
>
> Einstein has somewhere remarked that he was guided towards his discoveries partly by the notion that the important laws of physics were really simple. R. H. Fowler has been heard to remark that, of two formulae the more elegant is likely to be true. Dirac very recently sought an explanation alternative to that of spin in the electron because he felt that Nature could not have arranged things in such a complicated way . . . .
>
> [If these mathematicians] would condescend to attend to meteorology the subject might be greatly enriched. But I suspect they would have to abandon the idea that the truth is really simple.

Under these circumstances, it was hardly to be expected that Kolmogorov's work would mark the end of the tale†. On the one hand, the 4/3 rule applies (at least approximately) over a much larger range than that to which Kolmogorov's theoretical reasoning applies, and on the other hand, even where Kolmogorov's theoretical reasoning does apply, the experimental evidence suggests that turbulence has a more complex structure than the simple assumptions of Kolmogorov would suggest. 'The Richardson 4/3 diffusion law continue[s] to intrigue physicists and mathematicians; but in the meantime the [formula] provide[s] the essential basis for calculating many natural and industrial flows.'

In 1953 Richardson saw an announcement of some fellowships at his old college for which he might be eligible. After receiving a moderately encouraging response (he was now 71), he made a formal application. He wished to continue his research on the causes of wars, writing that he had already analysed the larger wars since 1820 'but the list of the smaller pugnacious incidents, in each of which about 1000 people were killed, is far from complete; and the search for these small incidents requires a library larger than those in Glasgow. The resources of the Cambridge University Library would help fill the gaps.' Under the heading 'Qualifications' he wrote:

> Broadly speaking, my qualification is a tendency to find out, to arrange and to publish; when that is achieved, I soon forget. Reputations are usually out of date: mine is connected with activities that I no longer engage in, and results that I have partly forgotten.

A few days later, he died quietly in his sleep. His wife wrote to the Provost of King's saying how

†There is an old story of a distinguished scientist who is asked which two questions he would most like to ask God. 'Oh, I would ask Him to explain the theory which links quantum mechanics and general relativity.' 'And the second question? Would you ask Him to explain turbulence?' 'No, I don't wish to embarrass Him.'

[h]e loved his college with so real a love; and often regretted that he had to give up or refuse scientific appointments wherein he might have brought honour to his college, because he feared his researches might be used in war. ... He was dreaming of King's College the night before he died and dreamed he was welcomed there by old friends and colleagues.

CHAPTER 9

Richardson on war

9.1 Arms and insecurity

During the last 25 years of his life, Richardson's main scientific interest was the study of the causes of war. To most people, any attempt to apply the methods of mathematics to such a complex social phenomenon appears doomed from the start. The two books which he wrote on the subject failed to find a publisher during his lifetime and it was not until seven years after his death that they were published with the help of a subsidy from the Littauer Foundation. (In fact, both books had to be reprinted and the royalties more than covered the subsidy.)

The first, entitled *Arms and Insecurity*, is an attempt to produce a mathematical theory of arms races. The mathematics used is not very hard and can be explained to anyone who has done a couple of years of calculus. (If you have not, just skip this part of the discussion.) Consider two nations who spend at a rate $x(t)$ and $y(t)$ (measured, for example, in dollars per year) on armaments at time t. The first nation will tend to increase its expenditure in response to the perceived threat of the expenditure of its rival, but will be restrained by the burden that its own expenditure represents. The greater $y(t)$ is, the greater the rate of increase of $x(t)$ but the greater $x(t)$ is, the smaller the increase. We may seek to model these statements by the differential equation

$$\frac{dx}{dt} = ky - \alpha x,$$

where α is a positive constant representing the fatigue and expense of armaments and k is a positive constant (called the 'defence coefficient' by Richardson) representing the response to threat. As it stands, the equation implies that if $x = y = 0$ at some time then there would be no tendency to arm, so we insert an additional term g 'to represent grievances and ambitions, provisionally regarded as constant' obtaining

$$\frac{dx}{dt} = ky - \alpha x + g. \tag{9.1}$$

In the same way, we describe the behaviour of the second nation by

$$\frac{dy}{dt} = lx - \beta x + h. \tag{9.2}$$

The full analysis of these equations is not hard and will be found in Richardson's book. A particularly simple case occurs when the two nations are sufficiently similar that we may take $k = l$ and $\alpha = \beta$, obtaining

$$\frac{dx}{dt} = ky - \alpha x + g, \tag{9.3}$$

$$\frac{dy}{dt} = kx - \alpha y + h. \tag{9.4}$$

If we add the two equations we get

$$\frac{dx}{dt} + \frac{dy}{dt} = k(x+y) - \alpha(x+y) + g + h, \tag{9.5}$$

so that the total military expenditure of both sides $z(t) = x(t) + y(t)$ satisfies the differential equation

$$\frac{dz}{dt} = (k - \alpha)z + (g + h). \tag{9.6}$$

It is easy to solve Equation (9.6), obtaining the solution

$$z(t) = A \exp((k - \alpha)t) - (g + h)/(k - \alpha), \tag{9.7}$$

where A is a constant

Exercise 9.1.1   *Verify the computations above.*

If $k - \alpha > 0$ (and $A > 0$) then $z(t)$ increases exponentially. If the equations formed a complete model, then within a fairly short time $z(t)$ would exceed the economic capacity of the nations involved but, before this can happen, war breaks out. So far, the theory looks like the kind of thing one might dream up in the bath and we could dismiss it out of hand if it were not for one thing. Equation (9.6) says, in effect, that over a small time Δt the increase Δz in $z(t)$ is given to a high degree of accuracy by

$$\frac{\Delta z}{\Delta t} = (k - \alpha)z + (g + h). \tag{9.8}$$

Richardson took the defence budgets of the Franco-Russian and Austro-German alliances in the years 1909–14 and obtained the remarkable graph reproduced in Figure 9.1. (Our z is his $u + v$, his monetary units are millions of pounds sterling.)

If the experimental test of a physical theory produced such a graph, we would consider it a remarkable confirmation. Richardson goes on to consider disarmament after the First World War and rearmament in the years 1929–39, in each case presenting striking numerical evidence. However, as Richardson himself indicates, there are several problems. The first is that only a small minority of wars appear to have been preceded by an arms race. Richardson argues that this is because arms races are a modern phenomenon (only modern war requires a lengthy mobilisation of resources and only a modern state can support such a mobilisation), but his analysis of three major wars (the Franco-Prussian, Turko-Russian and Russo-Japanese) in the second half of the nineteenth century and the very early twentieth century gives no evidence of preceding arms races. The second, which Richardson could not have discussed, is that the years 1945–88, which appear to have been characterised by high levels of armaments and by deep mistrust between opposing blocs, did not give rise to an arms race ending in war†.

The third problem is this: Richardson studied war in the hope that greater knowledge would aid its prevention. But, if a Richardsonian arms race follows a pre-determined path, then how can we as individuals hope to influence it? To this objection Richardson argues that once we recognise that conditions for an arms race exist, we may be able to prevent one starting. If, as Figure 9.1 suggests, the total $z = u + v$ defence budgets of the rival blocs before the First World War were actually governed by the equation

$$\frac{\Delta z}{\Delta t} = 0.73(z - 194),$$

then if, instead of spending a total of £199 million on defence in 1909,

†References to empirical studies are given in §4.7 of Sandler and Hartley's book *The Economics of Defence* which contains much else of interest. Their general conclusion is that pure Richardsonian arms races did not occur in the post-War period.

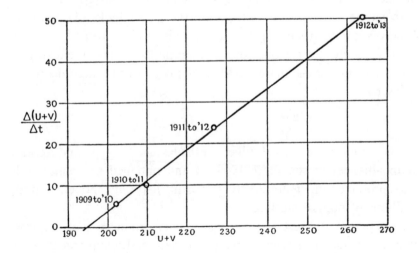

Figure 9.1: Total defence budgets of the competing alliances before the First World War.

the countries had spent £194 million, no arms race and so, presumably, no war would have occurred. The sceptical reader may observe that this equation predicts that, had they spent any less than £194 million, there would have been a disarmament race in which z decreased at an ever accelerating pace eventually becoming negative.

Exercise 9.1.2   *Briefly discuss the analogous behaviour for Equation (9.7).*

This does not invalidate Richardson's argument, since, just as we expect our model to break down for very large z, so we would expect it to break down for z too small. (For example, as a major colonial power, France required a large army and navy to control its empire, whatever the state of her relations with Germany.)

The specific example of defence spending in 1909 may be too simple-minded but, more generally, Richardson argues that

> before a situation can be controlled, it must be understood. If you steer a boat on the theory that it ought to go to the side to which you move the tiller, the boat will seem uncontrollable. 'If we threaten,' says the militarist, 'they will become docile.' Actually, they become angry and threaten reprisals. He has put the tiller to the wrong side. Or, to express it mathematically, he has mistaken the sign of the defence coefficient.

There is something very strange to us in the spectacle of one of the foremost meteorologists of his day spending 20 years seeking a mathematical theory governing the outbreak of wars. To believe, as Richardson evidently believed, that the causes of war are knowable and that, if these causes were known, nations would act on that knowledge to prevent war seems absurd. To most of us, the outbreak of the First World War appears both accidental and unavoidable — for Richardson it was not accidental and was therefore avoidable.

> It seemed that out of battle I escaped
> Down some profound dull tunnel, long since scooped
> Through granites which titanic wars had groined.
> Yet also there encumbered sleepers groaned,
> Too fast in thought or death to be bestirred.
> Then, as I probed them, one sprang up, and stared
> With piteous recognition in fixed eyes,
> Lifting distressful hands as if to bless.
> And by his smile, I knew that sullen hall,
> By his dead smile, I knew we stood in Hell.
> With a thousand pains that vision's face was grained;
> Yet no blood reached there from the upper ground,
> And no guns thumped, or down the flues made moan.

'Strange friend,' I said, 'here is no cause to mourn.'
'None', said the other, 'save the undone years,
The hopelessness. Whatever hope is yours,
Was my life also; I went hunting wild
After the wildest beauty in the world,
Which lies not calm in eyes, or braided hair,
But mocks the steady running of the hour,
And if it grieves, grieves richlier than here.
For by my glee might many men have laughed,
And of my weeping something had been left,
Which must die now. I mean the truth untold,
The pity of war, the pity war distilled.
Now men will go content with what we spoiled.
Or, discontent, boil bloody, and be spilled.
They will be swift with swiftness of the tigress,
None will break ranks, though nations trek from progress.
Courage was mine, and I had mystery,
Wisdom was mine and I had mastery;
To miss the march of this retreating world
Into vain citadels that are not walled.
Then, when much blood had clogged their chariot-wheels,
I would go up and wash them from sweet wells,
Even with truths that lie too deep for taint.
I would have poured my spirit without stint
But not through wounds; not on the cess of war.
Foreheads of men have bled where no wounds were.
I am the enemy you killed, my friend.
I knew you in this dark: for so you frowned
Yesterday through me as you jabbed and killed.
I parried; but my hands were loath and cold.
Let us sleep now ... '

9.2 Statistics of deadly quarrels

In addition to the theoretical work described in the previous section, Richardson set out to collect data on all wars between 1820 and 1949 in as systematic and as objective a way as possible. The results are set out in tabular form in the first part of his book *Statistics of Deadly Quarrels*.

A typical entry is shown in Figure 9.2. The number 6 in the upper left hand corner indicates that there were about 10^6 dead. (The reader who may not even have heard of this war should be told that Paraguay lost an incredible 56% of its total population and 80% of its men of military age.) We see that the Uruguayan Blancos fought the Brazilians from October 1864 to March 1870. The notation $A|B|C$ is used to separate factors A which are mentioned by historians as reducing enmity, factors B which were conspicuously present but not mentioned as causes by the historians consulted, and factors C which are mentioned as causes. Thus Richardson records no factors as reducing enmity between the Blancos and the Brazilians. He notes the factors g (similar religions) and I (different languages) as present but not mentioned by historians as influencing the outbreak of conflict. The notation M_{13} means that the two groups had fought previously and that their last conflict ended 13 years before the outbreak of this one. Finally the notation $\eta\rangle$ tells us that enmity was increased by the sympathy of the Brazilians for those under the control of the Blancos.

There are many problems involved in drawing up such a list and one of the chief pleasures in reading Richardson lies in watching a really clever person dealing with these difficulties. One problem is the definition of what constitutes a war. Here Richardson avoids some of the difficulties by considering instead 'deadly quarrels' where

> [by] deadly quarrel is meant any quarrel which caused death to humans. The term thus includes murders, banditries, mutinies, insurrections, and wars small and large.

(In his discussion of murders, Richardson estimates the number of murders committed in the world between 1820 and 1945 as 6×10^6. He believes that this estimate is unlikely to be in error by more than a factor of 3.) A second problem is that the number of deaths caused by a war is never known exactly. Richardson remarks that three sources

Figure 9.2: Richardsonian history.

Great War in La Plata 1865–70

	Uruguayan Blancos suspicious of greater neighbours.	Paraguayans led by López against Brazilian interference in Uruguay.	*References*
6			Calogeras. Levene. H **23**, 660, 616, 620.
Uruguayan Colorados[1]	— | g i | M$_{0\cdot2}$ 1864.vii–70.iii	— | g i | $\phi\rangle$ T$\rangle$ $\omega\rangle$ 1865.ii.–70.iii	E **17**, 259. E **14**, 387.
Brazilians[2] against cattle-raiding by Uruguayans.	— | g I M$_{13}$ | $\eta\rangle$ 1864.x–70.iii	— | g I | $\phi\rangle$ $\omega\rangle$ 1864.xi–70.iii	
Argentines[2] led by Mitre, a Unitarian.	— | g i M$_{13}$ | — 1865.iii–70.iii	— | g i | $\phi\rangle$ T$\rangle$ $\omega\rangle$ 1865.iii–70.iii	

Result. López killed. Paraguay depopulated.

[2] All against the insolence of the Paraguayan dictator.

state the number of dead in the Union Army in the American Civil War
as 359,258 or 279,376 or 166,623; each, if seen alone, apparently accurate
to a man. The corresponding logarithms [to base 10] may be rounded
off to 5.6, 5.4 [and] 5.2.

For this reason, Richardson uses the logarithmic scale when giving the
number of deaths.

Exercise 9.2.1   *Richardson indicates that*

> *For a few wars the magnitude appears to be known to within 0.04, for many
> wars uncertainties of 0.1 to 0.5 occur and for a few we may be in doubt by
> as much as 0.8.*

*If the number of deaths in a war has magnitude 6 ± 0.3 find the upper
and lower estimates $10^{6.3}$ and $10^{5.7}$ for the death toll. Do the same for
the various error ranges given by Richardson.*

Richardson's psychological studies gave him another reason to choose
a logarithmic scale. Our response to stimuli like light and sound tends
to be proportional to the logarithm of the magnitude of the stimulus.
   Another problem is to decide when a list of quarrels of given magnitude is complete.

The best evidence is provided by the progress search. ... At first my
collection grew rapidly, then slowly [see Table 9.1†] .... It is seen
that there has been a sort of convergence, which makes the present
numbers worthy to be discussed. For smaller incidents in the range
$2.5 < \mu < 3.5$, my collection has continued to grow. [Richardson found
184 such incidents between 1820 and 1929 and thought that this might
represent about half of those that occurred.] For magnitudes between
0.5 and 2.5 the information is scrappy and unorganised; what there is
of it suggests that such small fatal quarrels were too numerous to be
systematically recorded as history, and yet too large and too political to
be recorded as crime.

†The increase in quarrels
with magnitude $6 \pm 1/2$
represents Richardson's
changed estimates for
deaths in the civil war
following the Russian
Revolution.

Table 9.1. *Number of fatal quarrels between 1820 and 1929 in
Richardson's list.*

Range of magnitude	$7 \pm 1/2$	$6 \pm 1/2$	$5 \pm 1/2$	$4 \pm 1/2$
Nov. 1941	1	3	16	62
Dec. 1948	1	3	20	60
Aug. 1953	1	4	19	67

Exercise 9.2.2 *(i) How would you go about compiling such a list?*
(ii) Richardson identified 24 fatal quarrels between 1820 and 1929 of
magnitude greater than 4.5 (that is resulting in more than 30 000 deaths).
How many can you identify?

Once statistics have been gathered, we can test various theories
concerning the causes of war. For example, the experiences of the first
half of this century lead one to classify Germany as a militaristic power.
Is this correct? Richardson worked out the proportion of fatal quarrels
of each range of magnitude involving each nation, obtaining Table 9.2.
(Since Germany did not exist over the whole period 1820–1945, Prussia
is counted before 1870 and Germany thereafter.)

Even if Table 9.2 does not force us to change our views it makes us
aware that we must present them more carefully. It certainly supports
Richardson's conclusion that

> aggression was so widespread that any scheme to prevent war by re-
> straining any one named nation is not in accordance with the history of
> the interval A.D. 1820 to 1945.

(Richardson's discussion was written just after a great war led by Britain
and the United States to destroy German and Japanese militarism

Table 9.2. *Appearances of some belligerents during the interval AD*
1820–1945 in Richardson's list.

Range of magnitude	7.5 to 6.5	6.5 to 5.5	5.5 to 4.5	4.5 to 3.5	3.5 to 2.5	Average number of frontiers of named belligerent
Total no of fatal quarrels in the whole world	2	5	24	63	94	
	Number of fatal quarrels in which the named belligerent took part					
Britain	2	0	1	25	28	22.5
France	2	0	4	15	18	15
Russia	2	0	6	10	18	9.5
Turkey	1	0	8	6	15	8.7
China	1	2	4	7	14	10
Spain	0	1	2	8	11	4.9
Germany or Prussia	2	0	5	3	10	10.6
Italy or Piedmont	2	0	3	5	10	6.1
Austria in var. combinations	2	0	2	5	9	9.3
Japan	2	0	2	5	9	1.6
U.S.A.	2	1	2	4	9	3.3
Greece	2	0	1	3	6	2.7

for ever. In 1991 the Governments of Britain and the United States complained bitterly that Germany and Japan refused to engage in a war against Iraq.)

The Greeks are a brave and martial people. Why then did they engage in so few wars? One natural explanation is that as a poor nation they could only fight with their neighbours. On the other hand Britain as a rich imperial sea power could be and was embroiled in wars all over the world. Stripped of its cynicism (which is mine and not Richardson's), this explanation suggests that the number of external wars a nation engages in might be proportional to the number of different nations with which it shares a frontier. Table 9.2 does seem to show such a tendency, and Richardson certainly believed that his statistical analysis confirmed it — but I suspect that, if the two main colonial empires, Britain and France, who both had many different frontiers and fought many wars, are excluded, the pattern becomes less clear.

Another collection of thought-provoking data comes from *Arms and Insecurity* and is presented in Figure 9.3. The justifications usually given for high defence expenditure in peacetime are that being well armed 'ei-

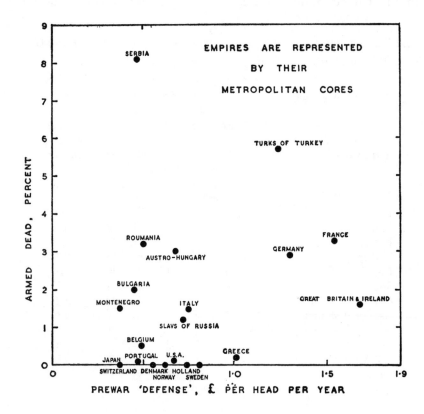

Figure 9.3: Did pre-war defence expenditure decrease casualties in World War I?

ther keeps a nation out of war or diminishes its losses should it become involved.' Richardson's plot of pre-war defence expenditure against the percentage of population killed in World War I reveals no such simple relation. (Equally, as Richardson remarks, the figure fails to support the opposite proposition that the greater the armament the greater the losses.) The reader may feel (and I sympathise with her feeling) that this is not a fair test. However, she must then ask herself what would constitute a fair test. If it is true that armaments increase security (or if the reverse is true), then that truth must be testable. It would surely be very strange if we simply took such important statements on trust.

Exercise 9.2.3 *(i) In your opinion (not mine, not Richardson's), do armaments increase or decrease the security of a nation? Why?*

(ii) What evidence (statistical or other) would make you change your mind?

Is there any pattern to the time of outbreak of wars (or, indeed, of peace)? Richardson tabulates the number $f(0)$ of years between 1820 and 1929 in which there were no outbreaks of deadly quarrels of magnitude between 3.5 and 4.5 and then the number $f(1)$ of years in which there were exactly 1, the number $f(2)$ in which there were exactly 2 and so on. He then does the same thing for outbreaks of peace. The outbreak of war between two countries is a rare event. For the purposes of comparison Richardson considers another kind of rare event, to wit the appearance of the sequence 111 as the last three figures of an entry in a table of seven-figure logarithms. Just as there are many pairs of countries between whom war may break out, so there are many entries on each page of the table. He tabulates the number $h(x)$ of times a page containing exactly x occurrences of the sequence 111 as the last three figures of an entry in pages 6–105 in a book of seven-figure logarithms. The results are given in Tables 9.3–9.5.

First year university mathematics shows that events of the kind recorded in Table 9.5 are governed by the Poisson law of probability. Suppose that you play a fruit machine repeatedly. The probability η of winning a jackpot in any one go is very small, but if you play

Table 9.3. *Outbreaks of deadly quarrels of magnitude 3.5 to 4.5 between 1820 and 1929.*

x =no. of outbreaks of war in a year	0	1	2	3	4	> 4	Total
$f(x)$ =no. of such years	65	35	6	4	0	0	110

repeatedly very many times (N times, say) you will have a reasonable chance of winning several times. The probability of winning exactly n times is given (to a very good approximation) by $k(n) = \lambda^n e^{-\lambda}/n!$ (where $\lambda = N\eta$). Suppose, as an illustration, that we take $\lambda = 59/110$ (there is no need to know the actual values of η and N to apply the formula). We then obtain Table 9.6. If we go on M such gambling sprees, we will (on average) expect to win exactly n jackpots in about $Mk(n)$ of our outings. If we set $M = 110$ we obtain Table 9.7.

Tables 9.3–9.5 and Table 9.7 look very similar. In the case of Tables 9.5 and 9.7 it is easy to account for the resemblance. Looking up one entry in the table of logarithms corresponds to having one go on the fruit machine. The unlikely event that the last three figures of the entry are 111 corresponds to winning the jackpot. Looking at all the entries on a page corresponds to going on a gambling spree. Table 9.5 records the result of 100 such sprees.

Exercise 9.2.4 *(i) Would you expect the same kind of results if, instead of using seven-figure logarithms, Richardson had used a table of nine-figure logarithms? What if he had used a table of four-figure logarithms?*

(ii) Had Richardson been writing today he would have used a computer to simulate a Poisson process. Write a program using the random number generator on your machine to simulate the gambling sprees described above with $\eta = 10^{-4}$, $N = 5636$ and $M = 110$. Tabulate your results in the manner of Tables 9.3–9.5 and Table 9.7. (Note that $59/110 \approx 0.5636$.)

(iii) Train crashes are, fortunately, rare. Would you expect the number of deaths per year due to this cause in England to follow a Poisson distribution? Why?

Table 9.4. *Endings of deadly quarrels of magnitude 3.5 to 4.5 between 1820 and 1929.*

x =no. of outbreaks of peace in a year	0	1	2	3	4	> 4	Total
$g(x)$ =no. of such years	63	35	11	1	0	0	110

Table 9.5. *Occurrences of a specified pattern of least significant digits in a table of logarithms.*

x =no. of occurrences on a page	0	1	2	3	4	> 4	Total
$f(x)$ =no. of such pages	56	37	3	4	0	0	100

In the same way, Table 9.3 resembles the pattern that would emerge if, every day, during each year the god of war has a go on a fruit machine. The jackpot, for him, is represented by the outbreak of a deadly quarrel, the outcome for each year corresponds to the outcome of one gambling spree and the table records the result of 110 such sprees.

Exercise 9.2.5 *(This exercise is intended only for readers who have done a first course in statistics.) Use a χ^2 test to examine Richardson's assertion that Tables 9.3–9.5 are well explained by assuming an underlying Poisson distribution in each case. (Remember to think about degrees of freedom.) The inventor of the χ^2 test was Karl Pearson and it is an indication of the coherence of Richardson's intellectual life that he could record that 'in 1906 ... I sold my physics books in order to raise money to go and see Professor Karl Pearson and learn about statistical proof.'*

The fruit machine model has a grim plausibility when applied to mutual nuclear deterrence for which, each day, there is a very small chance of war due to accident. Table 9.3 seems to show that the model applies more generally. If so, then, as Richardson remarks,

> ... [this] statistical and impersonal view of the causation of wars is in marked contrast with the popular belief that a war can usually be blamed on one or two named persons. But there are similar contrasts in other social affairs; the statistics of marriage for example are in contrast with any biographer's account of the incidents that led two named people to marry each other.

The outbreak of the First World War was preceded by a series of crises each of which *could* have precipitated a European war but did not,

Table 9.6. *Probability of n jackpots in one gambling spree.*

x =no. of jackpots won	0	1	2	3	4	> 4	Total
$k(x)$ =prob. of winning	0.5848	0.3137	0.0841	0.0150	0.002	0	1

Table 9.7. *Expected number of occasions with n jackpots in 110 gambling sprees.*

x =no. of jackpots won	0	1	2	3	4	> 4	Total
$Mk(x)$ =av. no. of occasions	64.3	34.5	9.3	1.7	0.2	0	110

and the assassination of Archduke Franz Ferdinand which led to the crisis which *did* was an affair of peculiar coincidences†. The institution of slavery in the United States was bound to produce conflict of some kind ('I tremble for my country when I reflect that God is just'), but the date and the form of that conflict were the result of the interaction of many unpredictable events and decisions.

Naturally, Richardson sought factors that reduce or increase the risk of deadly quarrels between groups. One of Richardson's brothers was an enthusiast of Ido (an artificial language derived from Esperanto‡) and one of the driving forces behind the promotion of such 'international languages' was the belief that common languages promote understanding and thus peace. A fellow member of Richardson's ambulance corps recalled that Richardson

> had learnt to speak Esperanto as he thought it would bring people closer together. I remember on one occasion we were in a dressing station just behind the lines when some German prisoners were brought in for treatment; he tried to question them in Esperanto and was disappointed when they just stared back at him.

However, his analysis of wars revealed no overall tendency for a common language to decrease (or increase) the number of deadly quarrels between groups. (There were variations between languages; Spanish speakers tended to fight more and Chinese less.)

Richardson's book contains several other thought provoking studies. (Does a common religion decrease the risk of war? Are wars becoming larger? How important are economic causes?) The interested reader should obtain Richardson's books and judge for herself. It does, however, seem to me that the only major force for peace to emerge from his studies is time. In Table 9.8 Richardson records the time since the previous conflict between the same pair of opponents for major deadly quarrels between 1820 and 1929§.

Although other interpretations are possible, the simplest interpreta-

Table 9.8. *Time since previous war between the same pair of belligerents.*

Years between wars	0 to 10	10 to 20	20 to 30	30 to 40	40 to 50	50 to 60	60 to 70	70 to 80	80 to 90	90 to 100	Above 100	Total
Number of times observed	32	20.5	9.5	10	7	6	1	0	0	0	3	98

† A few years ago, the three surviving members of the group who carried out the assassination were asked if they regretted it. Two of them said that, in view of the consequences, they felt that they had been mistaken. The third said that he would do it again.

‡ In the language Ido, the word 'ido' means 'offspring'.

§ Richardson writes that 'The table is for wars of magnitude greater than 3.5 between 1820 and 1929 (with a backward search to 1750 for previous wars). The .5 occurs because of observed intervals that lay on the boundary between adjacent decades. The three cases of chronic warfare have been omitted. The number of pairs of opponents who had not previously fought depends on the notion of an independent belligerent but lies between 97 and 118.'

tion of Table 9.8 is that the longer two nations have been at peace the more likely they are to stay at peace. On this optimistic note, let me conclude my account of Richardson's book.

Exercise 9.2.6 *(This exercise is intended only for readers who have done a first course in statistics.) Richardson suggests that Table 9.8 is well explained by assuming an underlying geometric probability distribution*

$$\Pr(Decades \ since \ last \ war = n) = \frac{r^n}{1-r}.$$

Use a χ^2 test to examine Richardson's suggestion. (You will need to think about the size of cells.)

Richardson's work on war is novel and remarkable. Is it successful? There are many problems with his approach. From the statistical point of view, the number of wars of magnitude greater than 3.5 since 1820 is quite small and the wars themselves do not have the homogeneity that we usually require of events subject to statistical enquiry. This reduces our confidence in the significance of observed effects. We can only extract a limited amount of information from a limited amount of data.

One of the first things that strikes the reader is Richardson's refusal to assign blame. Since it is usual for participants in a deadly quarrel (or indeed in more minor affairs) to accuse the other side of starting it, this is, perhaps, the only scientific way to proceed and it certainly makes a refreshing change. However, it leaves me uneasy on two points. Richardson approaches war in the same way that a medical researcher approaches illness, that is as a pathology of the system to be understood and, some day, cured. He does not investigate the possibility that war may, under certain circumstances, represent rational state policy. It is, of course, unlikely that both sides in a war will benefit, and it often happens that both sides end up worse off, but this does not appear always to have been the case. It could be argued that for states like Bismarck's Prussia, Great Britain from 1660 to 1914 and the United States from its foundation to the present day, the costs of wars have (on average) been outweighed by their benefits. It would be difficult but interesting to use Richardsonian methods to find the proportion of wars which left one side better off.

More seriously, Richardson's methods leave out everything that cannot be counted, everything that is special to one particular war and everything that is a matter of personal opinion. It is, surely, part of our knowledge of the Mexican War of 1846 that General Grant considered it 'one of the most unjust ever waged by a stronger against a weaker nation.' There is something inadequate about a system of classification which draws no moral distinction between the War of Greek Independence and the First Opium War. It seems to me that the historians with their emphasis on the specificity of historical events see a much greater part of the truth. But, although Richardson saw only a small part of the truth, it was a *new* part and that is what makes his work so exciting and, in my view, so valuable.

In the next section, I give part of a paper of Richardson so that the reader may hear his authentic voice. We have seen that Table 9.2 suggests that nations with many frontiers fight more wars than those with few frontiers. This suggests that a world consisting of a few large federations might be more peaceful than one of many sovereign states. That this possibility was important to Richardson is shown by the following extract from the Socratic dialogue with which he begins his book.

FEDERATOR [When we have] a world government elected by the people, affairs will go much better.

CRITIC We shall merely have most of the old troubles under new names. There will be wars, but they will be called rebellions or civil wars.

FEDERATOR But history shows that peoples whom habit has joined seldom split asunder. Civil wars have been rarer than international wars.

CRITIC That assertion would be important if it was known to be true. But it needs to be tested by statistics.

In order to compare the opportunities for civil and foreign war, Richardson sought to divide the world into cells of equal population which should be compact (in a sense that he discusses) and respect natural and national boundaries. He wrote a long paper which was published after his death in Volume VI of *The Yearbook of the Society for General Systems Research* — a title unknown to the mathematical community and sounding (to the uninitiated, at least) like the house journal of a snake-oil manufactury. Section 7 of his paper is entitled *Lengths of Land Frontiers or Seacoasts* and forms the next section of this chapter.

9.3 Richardson on frontiers

[The contents of this section are taken directly from Richardson's paper.]

In the previous section [of Richardson's paper] integrals were taken around simple geometric figures, as a preliminary to taking them around frontiers shown on political maps. An embarrassing doubt arose as to whether actual frontiers were so intricate as to invalidate that otherwise promising theory. A special investigation was made to settle this question. Some strange features came to notice; nevertheless an overall general correction was found possible. The results will now be reported. ...

At first I tried to measure frontiers by rolling a wheel of 1.8 centimetres diameter on maps; but there is often fine detail, which the wheel cannot follow; some convention would be needed as to what detail should be ignored and what retained: considerable skill would be needed to guide the wheel in accordance with any such decision; and in practice the results were erratic.

Much more definite measurements have been made by walking a pair of dividers along a map of the frontier so as to count the number of equal sides of a polygon, the corners of which lie on the frontier. In this respect the polygon resembles one 'inscribed' to a circle, but some sides may lie outside the frontier, and the polygon need not be closed. Its total length, Σl, has been studied as a function of the length, l, of its side. This process comes down to us from Archimedes, and is standard in pure mathematics. For perfection the dividers should be stepped along a map on a globe; but only plane maps were available. To avoid most of the errors caused by the projection of the globe onto a plane, the investigation has been restricted to moderately small portions of the earth's surface, the largest being Australia. Also, for definiteness, the maps have been specified. Usually l was fixed in advance, and a fractional side was estimated at the end of the walk. If it was desired to have a whole number of sides, then l had to be adjusted by successive approximations. The main purpose was to study the broad average variation of Σl with l. But some of the incidental details are so interesting that they deserve mention; for they are in marked contrast with the properties of the smooth curves which have their lengths integrated in text books. The moving spike of the dividers describes a circle which may intersect the frontier in more than one point; if so the intersection to be chosen is that one which comes next in order forward along the frontier. This obvious rule has surprising consequences; it sometimes prevents the polygon from having a whole number of sides.

Table 9.9. *The west coast of Britain.*

Start	N or S	N	S	S	S	S	S
Length of side, km	971	490	-	200	100	30	10
Number of sides	1	2	2	5.9	15.4	69.1	293.1
Total length, km	971	980	-	1180	1540	2073	2931

As an explanation of how chance can arise in a world which he regarded as strictly deterministic, Henri Poincaré [183] drew attention to insignificant causes which produced very noticeable effects. Seacoasts provide an apt illustration. For the spike of the dividers may just miss, or just catch, a promontory of land or the head of a loch; so that an insignificant change in l may alter Σl noticeably.

The west coast of Britain from Land's End to Duncansby Head was chosen as an example of a coast that looks more irregular than most other coasts in an atlas of the world. The quality and scale of the maps of Britain were such that the irregularities of the coast greatly predominated over any errors that are likely to have occurred in the processes of drawing, printing, or reading the maps. For the longer steps the *Times Atlas 1900* was used. The 10 kilometre steps were counted on the *British Isles Pocket Atlas 1935* by John Bartholemew FRGS. Narrow waters were regarded as barred by any bridges shown in the latter atlas. The Mersey tunnel was also regarded as a barrier, but the Severn tunnel was disregarded. These rules were kept the same for all lengths of step. The over-all length in one step was found to be 971 km. The attempt to make exactly two equal steps is worth stating in detail because it illustrates principles and peculiarities. The map was page 15 of the *Times Atlas* dated 1900. A circle with centre at Duncansby Head and radius 490 km cuts the coast of Cumberland

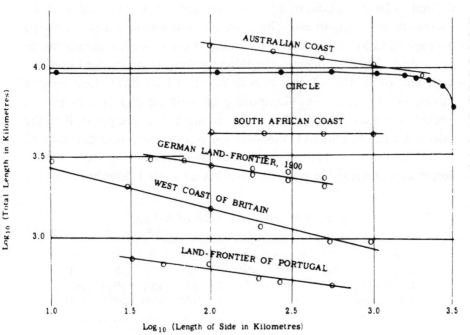

Figure 9.4: Measurements of curves by way of polygons (not shown) which have equal sides and have their corners on the curve. The slope of a graph shows how the total length of the polygon increases as its side becomes shorter. The convergence for the circle contrasts with the behaviour for frontiers.

at a point P near Silecroft. There are other intersections, but they are further forward along the coast, and therefore must be ignored. A circle with centre P and the same radius cuts the coast in about ten points in the north of Scotland; these, being backward along the coast, must be ignored. The first forward intersection is at Land's End. So there are two steps with a total length of 980 km on the Southward journey. The Northward journey is quite different. There is a point Q on Morecambe Bay which is 498 km from either Land's End or Duncansby Head. A circle with centre at Land's End and radius 498 km cuts the coast in several points, but Q comes first along the coast. A second circle with centre Q and the same radius cuts the coast in several points, of which the next forward along the coast is near Cape Wrath. So we cannot arrive at Duncansby Head. Moreover a search for some other midpoint, which would allow the northward journey to be in two equal steps, failed to find any. The northward journey is impossible according to the rule of 'next intersection forward along the frontier'; yet this rule seems too reasonable to be abandoned. The strange impossibility arises from the attempt to hit off a whole number of sides; it does not occur, if a fractional side is estimated at the end. The results are summarised [in Table 9.9]. As to how the total length Σl may be expected to vary with the length l of the side, I have no theory. Quite empirically the logarithms of these variables were plotted against one another [in Figure 9.4]; and a straight line was drawn through the points. More evidence would be needed before one could say whether the deviations from the straight lines are of any interest. I am inclined to regard them as random. The important feature for present purposes is that the slope of the graph is only moderate even for such a ragged line as the western shore of Great Britain. *On the straight line in Figure 9.4 the total length varies inversely, as the fourth root of the side, that is*

$$\Sigma l \propto l^{-0.25}. \tag{9.9}$$

As an example of a less indented coast, that of Australia was selected, in particular as shown on Map 30 of the *Times Atlas* dated 1900. The starting point was always at the southernmost point, and the motion was always counterclockwise. [The results are shown in Table 9.10.] The logarithms are plotted in Figure 9.4. The points lie near a straight line of slope -0.13. This is about half the slope for the west coast of Britain. For comparison with the Australian coast, a circle containing a plane area of 7.636×10^6 km^2, which is given as the area of the Australian mainland, was inscribed with regular polygons, and their calculated perimeters are shown on the diagram.

The coast of South Africa was studied because it is exceptionally smooth over a long length. The starting point was Swakopmund, and the end point was Cape Station Lucia. The map was page 118 of the *Times Atlas, 1900*. [The results are given in Table 9.11.] The same type of empirical formula is again found to be suitable but with a smaller parameter so that

$$\Sigma l \propto l^{-0.02}. \tag{9.10}$$

The land-frontier of Canada is in part defined by a meridian and a parallel of latitude which are so far definite in length; but many other land-frontiers follow winding rivers. It is not surprising therefore to find disagreements between official statements about the lengths of frontiers. Here [in Table 9.12] are some collected from amongst otherwise concordant data in *Armaments Year-Book 1938*. There is evidently no official convention which can hamper the intellectual discussion of the lengths of frontiers.

Each of these frontiers is irregular. That between Spain and Portugal was selected for measurement, and was more precisely defined as extending from a railway bridge at the northwest to the tip of an estuary in the south as shown in the *Times Atlas* dated 1900, Map 61. Some of the results depend on whether the polygon is started at the S or the NW end, as may be seen from [Table 9.13]. When an attempt is made to construct a polygon of three sides starting at the northwest, a peculiar difficulty occurs. If the side≥206 km, it is much too short; if the side>206 km, it is much too long; so that there is no solution. The critical length, 206 km, is that at which the point of the dividers just grazes the north-east shoulder of Portugal. In spite of this peculiarity the general run of the points on the diagram is tolerably straight.

As an example of a variegated land-frontier bounded partly by rivers,

Table 9.10. *The coast of the Australian mainland.*

Length of step, km	2000	1000	500	250	100
Number of steps	4.7	11.04	23.9	52.8	144.2
Total length, km	9400	11040	11950	13200	14420

Table 9.11. *The coast of South Africa.*

Side in km	1000	500	250	100
Number of sides	4.12	8.31	19.78	43.34
Total length, km	4120	4155	4263	4334

partly by mountains, and elsewhere by neither, I chose that of Germany as shown in the *Times Atlas* dated 1900, pp.39–40. Since then the frontier has been much altered, but previously it had persisted from 1871. [The results are given in Table 9.14]. Again the total length depends on whether the start is at the east or west. Nevertheless, for a conspectus, a straight line on the graph may serve.

9.3.1 Conclusions on polygons of equal sides inscribed to frontiers

1   The fitting of a whole number of sides can only be done by troublesome successive approximations.

2   It is occasionally impossible to fit a whole number of sides.

3   The possibility of fitting a whole number of sides occasionally depends on which end of the frontier is taken as the starting point.

4   For the above three reasons it is preferable to allow the final side to be an estimated fraction of the standard side.

5   The total polygonal length, including the estimated fraction, usually depends slightly on the point that is taken as the start.

6   It is doubtful whether the total polygonal length of a seacoast tends to any limit as the side of the polygon tends to zero.

7   To speak simply of the 'length' of a coast is therefore to make an

Table 9.12. *Land-frontiers.*

Land-frontier between	Kilometres as stated by:	
	the former country	the latter country
Spain and Portugal	987	1214
Netherlands and Belgium	380	449.5
USSR and Finland	1590	1566
USSR and Romania	742	812
USSR and Latvia	269	351
Estonia and Latvia	356	375
Yugoslavia and Greece	262.1	236.8

Table 9.13. *The land-frontier between Spain and Portugal.*

Start	S or NW	S or NW	S	NW	S	NW	S	NW	S	NW
No. of sides	1	2	3	—	7.05	7.07	13.10	13.06	27.2	27.05
Side, km	543	285	201	—	100	100	56.29	56.36	30	30
Total, km	543	570	603	—	705	707	737	736	816	812

unwarranted assumption. When a man says that he 'walked 10 miles along the coast' he usually means that he walked 10 miles *near* the coast.

8   Official statements of the lengths of some land-frontiers disagree with one another in ratios such as [1.1 or 1.2].

9   Although the phenomena mentioned in 2, 3, 5 and 6 are peculiar and disconcerting, yet coexisting with them are some broad average regularities which are summarised by the useful empirical formula

$$\Sigma l \propto l^{-\alpha}, \tag{9.11}$$

where Σl is the total polygonal length, l is the length of the side of the polygon, and α is a positive constant, characteristic of the frontier.

10   The constant α may be expected to have some positive correlation with one's immediate visual perception of the irregularity of the frontier. At one extreme $\alpha = 0$ for a frontier that looks straight on the map. For the other extreme the west coast of Britain was selected, because it looks one of the most irregular in the world; for it α was found to be 0.25. Three other frontiers which, judging by their appearance on the map were more like the average of the world in irregularity, gave: $\alpha = 0.15$ for the land-frontier of Germany in about AD 1899: $\alpha = 0.14$ for the land-frontier between Spain and Portugal; and $\alpha = 0.13$ for the Australian coast. A coast selected, as looking one of the smoothest in the atlas, was that of South Africa, and for it $\alpha = 0.02$.

11   The relation $\Sigma l \propto l^{-\alpha}$ is in marked contrast with the ordinary behaviour of smooth curves for which

$$\Sigma l = A + Bl^2 + Cl^4 + Dl^6 + \cdots, \tag{9.12}$$

where A, B, C, D,... are constants. This property is used in the 'deferred approach to the limit' [discussed in Section 8.2].

[The remainder of this part of Richardson's paper considers how

Table 9.14. *The land-frontier of Germany.*

Start	E	W	E	W	E	W	E	W	W	W
Side in km	500	500	300	299	180	180	100	99.6	66.0	40.5
No. of sides	4.02	4.52	7.62	7.85	13.75	14.92	28.05	28.1	44.45	74.9
Total, km	2010	2260	2286	2353	2475	2686	2805	2799	2934	3053

earlier formulae in his paper must be changed in the light of Formula (9.11).]

9.4 Why does a tree look like a tree?

Pure mathematicians have investigated the properties of curves of infinite length since the end of the nineteenth century. A particularly simple and elegant example is the 'snowflake curve' of von Koch. We construct it in stages as shown in Figure 9.5. At the first stage we have a straight line segment AB of unit length. In the second stage we divide the segment into three equal parts AX, XY and YB and form an equilateral triangle XZY. The four line segments AX, XZ, ZY, YB form the second stage of the snowflake E_2 say. To form the third stage E_3 we repeat the construction for each of the four line segments of the third stage as shown in the figure. The fourth stage E_4 is obtained by repeating the construction for each of the sixteen line segments of the third stage and so on.

Exercise 9.4.1 *Show that the nth stage E_n of the von Koch snowflake has 4^{n-1} line segments each of length 3^{-n+1}. Show that the total length of the nth stage is $(4/3)^{n-1}$.*

It is extremely plausible (and, in fact, true) that the sequence of snowflake curves E_1, E_2, ... approach a final intricate curve E which constitutes the von Koch snowflake. (Compare Exercise 8.3.1. Von Koch intended his snowflake to be a geometric, and therefore more easily grasped, analogue of the kind of example given there.)

What happens if we try Richardson's procedure of 'walking a pair of dividers' along the curve E? If we use any of the sequence of curves E_1, E_2, ... first set the dividers at a distance 1 and set one leg at A then the second will come to rest at B. It seems reasonable to suppose that the same will be true for the final curve E. Now suppose that we set the dividers at a distance $1/3$. If we walk along E_1, we will mark out A, X, Y and B giving a total of 3 steps but if we walk along any of the curves E_2, E_3, ..., we will mark out A, X, Z, Y and B giving a total of 4 steps and it seems reasonable to suppose that the same will be true for the final curve E. In the same way if we set the dividers at a distance 3^{-2}, then it will require 4^2 steps to walk them along any of the curves E_3, E_4, ..., and so we expect the same to be true for E. The pattern is now clear and we set it out in Table 9.15. When we plot the results logarithmically in Figure 9.6 (corresponding to Figure 9.4 in Richardson's paper) we obtain a line of slope $\log 4 / \log 3 \approx 1.26$ and

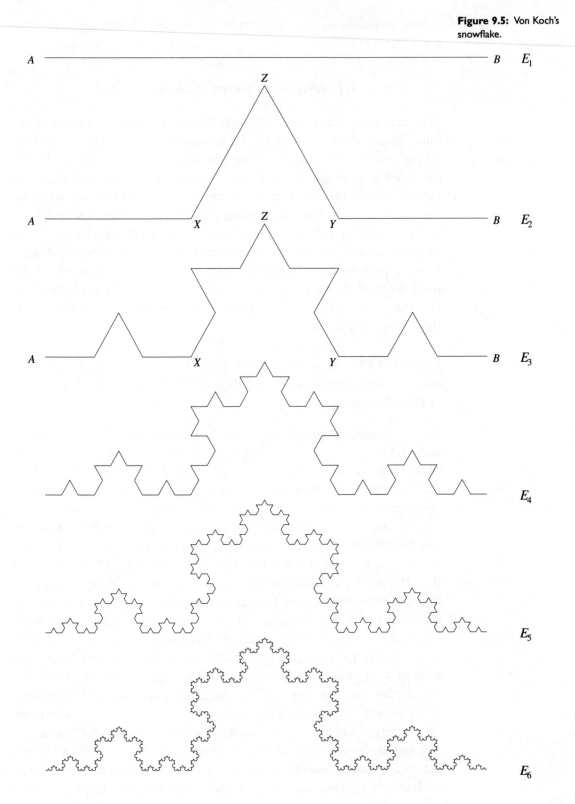

see that Richardson's formula (9.11)

$$\Sigma l \propto l^{-\alpha}$$

applies with $\alpha = \log 4 / \log 3$. In what follows we shall refer to the α in Richardson's formula as the 'Richardson number' and say that von Koch's snowflake has Richardson number $\log 4 / \log 3$.

A properly suspicious reader will remark that we have made life easy for ourselves by always starting at A and by setting our dividers at 3^{-n}. She is quite right. It is quite hard to show that Richardson's procedure

Table 9.15. *The snowflake curve.*

Start	A	A	A	A	A	A
Length of side	1	3^{-1}	3^{-2}	3^{-3}	3^{-4}	3^{-5}
Number of sides	1	4	4^2	4^3	4^4	4^5
Total length	1	$(4/3)$	$(4/3)^2$	$(4/3)^3$	$(4/3)^4$	$(4/3)^5$

Figure 9.6: Snowflake measurements corresponding to coastline measurements.

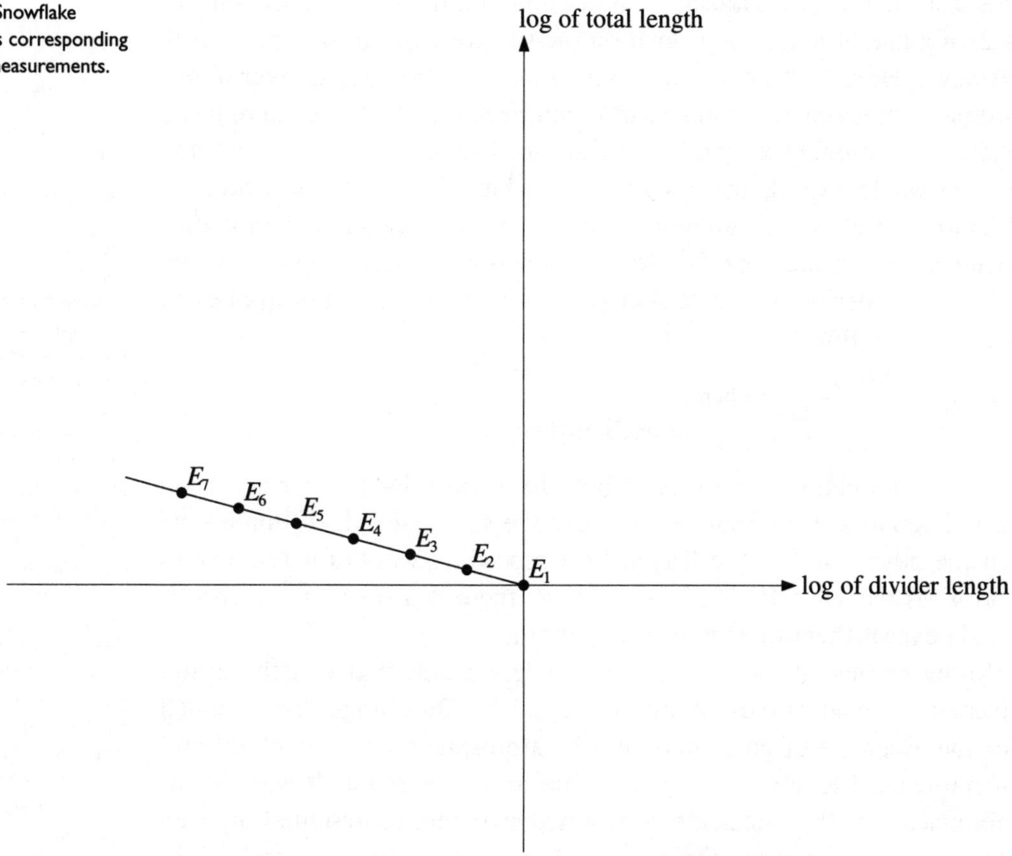

will work wherever we start and whatever sequence of divider settings we choose. (Remember that we do not expect

$$\Sigma l \propto l^{-\alpha}$$

to hold exactly but that we do expect the error to decrease towards zero as the length of the divider settings decreases towards zero.) However, in this case Richardson's procedure does indeed work and gives the stated Richardson number.

In the construction of von Koch's curve we always chose the added triangle so that it pointed 'outwards'. If, instead, we toss a coin to decide whether each new triangle points 'outwards' or 'inwards' we obtain the kind of 'random snowflake' illustrated in Figure 9.7. Notice that Richardson's method will give the same results for the 'random' and the 'non-random' snowflake so the random snowflake will also have Richardson number $\log 4/\log 3 \approx 1.26$.

Mathematicians have long been fascinated by the idea of dimension and in 1919 Hausdorff invented a new kind of dimension. Hausdorff dimension corresponds to ordinary dimension for 'normal objects' so that, for example, the Hausdorff dimension of a sphere is 3, of its surface is 2, of a line of longitude drawn on the surface is 1 and of a point is 0. However, Hausdorff's definition applies to any subset whatsoever of our ordinary three dimensional space. It can be shown that if a curve has a Richardson number α then it has Hausdorff dimension α but there are, as one would expect, many curves for which Richardson's procedure fails to produce any meaningful result. The pure mathematician is thus inclined to consider the Hausdorff dimension, which applies to any curve, as superior to the Richardson number, which only applies to some curves. But

> When every one is somebodee,
> Then no one's anybody.

Just as a doughnut and a bicycle both have Hausdorff dimension 3, so very different kinds of curves can have the same Hausdorff dimension. On the other hand, if the Richardson procedure works for two curves and gives the same Richardson number, there is a sense in which one would expect them to resemble one another.

Many mathematicians were more or less aware that coastlines and rivers resembled curves of infinite length†. Our lungs are intended for the exchange of gases between the atmosphere and our blood and therefore need to have as large a surface area as possible. It was a commonplace that their intricate convoluted structure represented nature's close approach to a surface of infinite area. However, mathematicians

†Thus, for example, the 1950 English edition of Steinhaus's *Mathematical Snapshots* contains a clear statement of the 'paradox of length' near the end of the chapter entitled *Tessellations, Mixing of Liquids, Measuring Areas and Lengths* (repeated in later editions). Another mathematical 'paradoxical problem', which circulates in the same kind of way, asks what happens if you drop a perfectly flexible piece of string on a table. This question was asked by the eminent theoretical physicist J. L. Synge and has been discussed at length by several other eminent mathematicians and physicists but I am not sure if it has a really satisfactory answer.

Figure 9.7: A random snowflake.

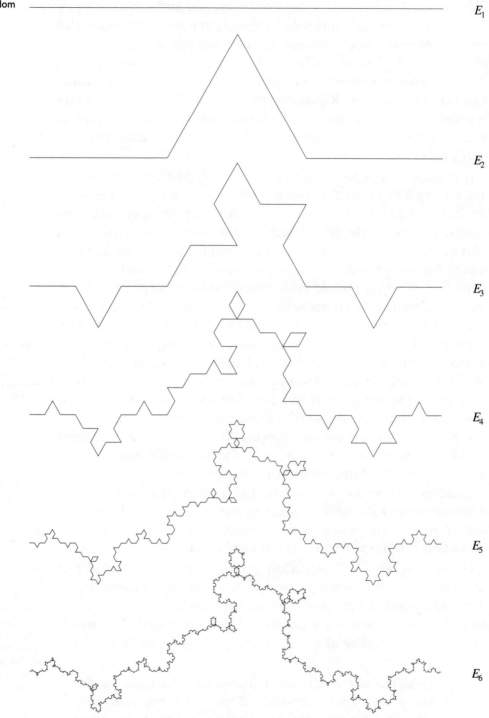

E_1

E_2

E_3

E_4

E_5

E_6

did not ask themselves what kind of curves and surfaces nature produced in this way and, in particular, no-one appears to have expected them to be well enough behaved for anything like Richardson's procedure to work. The Richardson of Section 8.4 was prepared to look for order where others only saw formlessness. Figure 9.4 is the intellectual twin of Figure 8.8. Richardson who was much more interested in explaining than describing and much more interested in the causes of war than the structure of coastlines, made no effort to draw attention to his discovery.

It is hard to imagine a mathematician more different from the shy and retiring Richardson than the man who was to interpret his work on the British coastline. Mandelbrot was an intensely ambitious and competitive product of the élite French mathematical educational system who at the age of 20 'told anybody who would listen that my ambition was to find a corner of science, not necessarily very extensive or even significant, of which I would know enough to be its Kepler or even its Newton.' Entering middle age with a substantial reputation he was still prey to a 'devouring yet ill defined ambition ... to be the first to find order where everybody else had seen chaos'†. Before disposing of a series of journals which were rarely consulted, a librarian asked Mandelbrot to glance through them. One was the issue of the *General Systems Yearbook* containing the paper from which Section 9.3 was copied. It was not hard for a mathematician with Mandelbrot's background to recognise that the α in Richardson's formula $\Sigma l \propto l^{-\alpha}$ was simply related to the Hausdorff dimension. However, Mandelbrot did more. Instead of saying that the British coastline looked like a random‡ curve of Richardson number α_B, he demonstrated by drawing such curves *that a random curve of Richardson number α_B looks like the British coastline!* (Although our 'random snowflake' is insufficiently random to carry real conviction, it is not totally unlike a stylised coastline.)

In the same way, although many people must have observed that some trees consist of a 'simple tree' from which grow smaller 'simple trees' from which in turn grow even smaller 'simple trees', very few can have thought of a tree as a Richardson–Kolmogorov cascade in which, to adapt the description of turbulent flow given in Section 8.4,

> ... a tree can be represented in the following way. The trunk has 'first order trees' growing from it, consisting of branches of lengths about l_1. In turn, the first order trees have 'second order trees' growing from them having typical length scales $l_2 < l_1$. This process of successive refinement goes on until for branches of sufficiently high order l_n turns out to be sufficiently small that leaves grow from them.

†The reader who wishes to know more about the character and career of this remarkable man should read the interview in *Mathematical People* [3].

‡Here and in what follows, I duck the important question of what random means in the context. Another important contribution of Mandelbrot concerns what random ought to mean here. Incidentally, Mandelbrot gave a simple explanation of why different countries give different lengths for their joint frontier. The larger the country the smaller-scale map it uses to show its frontiers. The results of Table 9.12 illustrate this.

We would expect the scaling between trees of various orders to depend on some scaling process such as that revealed in the 'mouse-to-elephant curve' (Figure 5.1). What Mandelbrot saw was that a random Richardson–Kolmogorov cascade of the appropriate type with the correct scaling would look like a tree.

Mandelbrot's observation is one of the very few contributions mathematics has made to the visual arts. (Perspective is another, but beyond that the list is short and controversial.) It is no wonder that when he published his ideas in *The Fractal Geometry of Nature* they met with instant and well deserved popular and academic acclaim. Since there are now many fine books devoted to Mandelbrot's 'fractals', I shall not continue this discussion much further, though I shall give a note on further reading at the end of the chapter.

I would, however, point out that the observation that some object is 'fractal' leaves unanswered the question of why it is fractal. We are in the position of having Richardson's 4/3 rule without Kolmogorov's explanation. I may know that the 'fractal structure' of a tree makes it look like a tree but I do not know why the tree has the fractal structure it does.

A good example of what might constitute such an explanation is given by our circulatory system where each artery branches repeatedly into smaller arteries. For each level of this 'arterial cascade' we know that during a reasonable period of time as much goes in as comes out. On the other hand, the fourth power law which we obtained in Section 6.1 in the paragraph labelled *Flow in a circular pipe* shows that some sort of scaling will be required if the system is to work efficiently. It turns out that the required scaling is given by Murray's law: 'The cube of the radius of the parental vessel equals the sum of the cubes of the radii of the daughter vessels.'

Exercise 9.4.2 *(i) If the daughter vessels divide into granddaughter vessels show that the cube of the radius of the grandparental vessel equals the sum of the cubes of the radii of the granddaughter vessels.*

(ii) If after repeated division a single vessel has split into n vessels of equal size show that each of the new vessels will have radius $n^{-1/3}$ of the original but the total cross-sectional area will be multiplied by $n^{1/3}$. Give explicit values when $n = 2$, $n = 10$ and $n = 100$.

(iii) Suppose that after repeated division a single vessel of radius r has split into n vessels of sizes $r_1, r_2, \ldots, r_n$ and that $r_j < \delta$ for each j. By observing that $r_j^3 < \delta r_j^2$, or otherwise, show that the total cross-sectional area will be multiplied by at least δ^{-1}. Think what this means about your circulatory system.

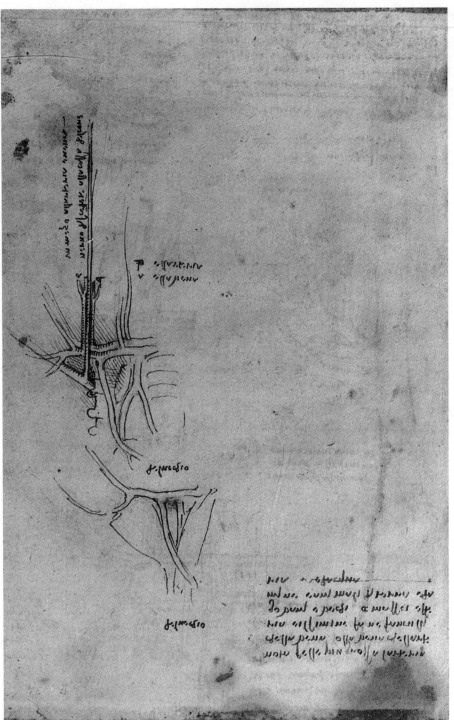

Figure 9.8: Anatomical
drawings by Leonardo. (The
Royal Collection © Her
Majesty The Queen.)

Exercise 9.4.3 *(Murray's derivation of Murray's law. This requires a little calculus towards the end.) Murray starts out by asking how much it costs the body to maintain circulation at a rate Q (measured in volume per unit time) through a particular blood vessel. For simplicity he considers the vessel as a simple cylindrical pipe of radius a. (Measurements of actual flow show that the simplification is a reasonable one.)*

(i) Murray measures the cost in energy used per unit time. Do you think this is reasonable and why?

(ii) If W_p is the energy used per unit time per unit length of a tube to propel a liquid of viscosity η at a rate Q through a tube of radius a show that, in the notation of Section 6.1,

$$[W_p] = MLT^{-3}.$$

Use a dimensional argument along the lines of that given in the paragraph labelled Flow in a circular pipe *to show that*

$$W_p = \frac{K_p Q^2 \eta}{a^4},$$

for some constant K_p. (The same result could be obtained directly from our previous rule connecting pressure, flow, viscosity and radius by thinking of the relation between pressure, flow and work done.)

(iii) Murray then points out that the cost of pumping blood round is not the only cost associated with the system. So far as the body is concerned blood is, itself, an expensive commodity. He points out that in addition to the small amount of energy consumed directly by the blood cells, there is the cost of regular replacement of the constituents, the cost of the containing vessels and 'the burden placed on the body in general by the mere weight of blood'. The energetic cost per unit time of a certain volume of blood is presumably proportional to that volume. In particular, the energetic cost W_b per unit time per unit length of the blood required to fill the blood vessel is proportional to the cross-sectional area and so

$$W_b = K_b a^2$$

for some constant K_b. The total cost involved in operating the blood vessel is thus

$$W(a) = \frac{K_p Q^2 \eta}{a^4} + K_b a^2.$$

If Q is fixed, find the value of a which minimises $W(a)$. Deduce that the optimum radius for a vessel carrying blood at a rate Q is given by

$$Q = Ca^3,$$

where C is a constant depending on K_p, K_b and η.

(iv) Suppose a vessel of radius a carrying blood at a rate Q splits into n vessels with the jth vessel having radius a_j carrying blood at a rate Q_j. If

the vessels have optimum radius, use the fact that $Q = Q_1 + Q_2 + \ldots + Q_n$
to deduce Murray's law

$$a^3 = a_1^3 + a_2^3 + \ldots + a_n^3.$$

(v) It is possible to compute K_p *or to find it by experiment, and it is possible to find* η *by experiment, but our theory seems to contain an unknown constant* K_b. *Murray shows that* K_b *can be found as follows. Returning to the calculations which concluded part (iii) suppose* Q *is fixed and we write* $W_p(a)$ *for the pumping cost,* $W_b(a)$ *for the blood cost and* $W(a)$ *for the total cost with given radius a. Show that* $\tilde{a}$, *the value of a which minimises* $W(a) = W_p(a) + W_b(a)$, *satisfies the equation*

$$2W_p(\tilde{a}) = W_b(\tilde{a}). \qquad (*)$$

Thus, if as we expect, the blood vessel is of optimum size the energy cost of the blood filling the vessel is twice the energy cost of pumping it through the vessel. *Since this is true for each part of the circulatory system it must be true of the whole.* Thus the energy cost of the blood filling the circulatory system is twice the energy cost of pumping it through the system. *But all the work of pumping is done by the heart and biologists have measured the energy per unit used by the heart in pumping blood. If Murray's theory is correct, we thus know the total energy cost of our blood per unit time and simple arithmetic now gives the cost per unit volume of our blood (and, if we want, the value of* K_b). *The cost of blood given by Murray's calculations is of the same order of magnitude as the cost of other parts of our body.*

Murray sums up his position as follows. 'Were blood a cheaper material we might expect all arteries to be uniformly larger than they are, thereby greatly reducing the burden on the heart. (For example, doubling the radii of all arteries, the lengths of which are constant, would mean a fourfold increase in volume, and a reduction of the work of the heart to one-sixteenth.)' Conversely if blood were more expensive we would expect a 'scant-blooded, large-hearted race'. Check that you understand Murray's numerical example.

(vi) Equation (*) in part (v) is not entirely a stroke of luck. If $A, B > 0$
and $n, m > 0$ *show that, if* $f(x) = Ax^n$ *and* $g(x) = Bx^{-m}$, *then the positive value* $\tilde{x}$ *of x which minimises* $f(x) + g(x)$ *satisfies the equation*

$$nf(\tilde{x}) = mg(\tilde{x}).$$

What goes wrong if $A > 0 > B$?

In his splendid exposition of our *Vital circuits* Vogel says that Murray's law holds (to the accuracy that it is reasonable to expect a biological law to hold) for a wide range of organisms. It even works for certain sponges which take in water through tiny pores, separate microedibles from it, and then pass the water through pipes of increasing size until finally disgorging it. He goes on:

> If this were as far as we could push Murray's law, it would certainly be a fine generalisation, another satisfying application of theory ... and another bit of evidence that physics matters. Recently, though, some tantalising concomitants have emerged. It has been found that if the flow is experimentally reduced in a blood vessel, the inside diameter of the vessel decreases until it stabilises at a new and smaller size. The reduction in diameter is just about what one would expect if the system were rearranging itself to keep the speed gradient at the wall unchanged.

Although dimensional analysis cannot be expected to reveal how the liquid actually flows, the more detailed direct discussion to be found in most introductory texts on hydrodynamics† reveals that the rate of increase of the velocity of the fluid as we move away from the wall (what Vogel calls the 'speed gradient') is the same for all pipes satisfying Murray's law.

†Look in the index of such a text under Poiseuille flow.

> The mechanism for the corrective adjustment is becoming clear. Without going into the details, it looks as if the cells lining blood vessels can quite literally sense changes in the speed gradient next to them. An increase in gradient stimulates cell division, which would increase vessel diameter as appropriate to offset the faster flow. Neither change in blood pressure nor cutting the nerve supply makes any difference — this is apparently a direct effect of the gradient on synthesis of some chemical signal by the cells. Perhaps the neatest feature of the scheme is that a cell needn't know anything about the size of the vessel of which [it is] part. As a consequence of Murray's law it can be given the same specific instruction wherever it might be located, a command telling it to divide when the speed gradient exceeds a specific value.

Vogel points out how neatly this allows for our growth from baby to adult and asks how we could recover from even the most minor injury or surgery without such self-adjusting mechanisms.

If I have let myself be carried away in my account of Richardson's work, I can only plead that he has long been a hero of mine. Coming across his two books on the mathematics of conflicts in the public library of my home town (how they got there I do not know) was one of the factors that made me choose mathematics as my degree subject. In any case, it is good for us to be reminded that:

> If a man does not keep pace with his companions perhaps it is because he hears a different drummer. Let him step to the music he hears, however measured or far away.

A note on books

In the last 25 years there have been three related sets of mathematical ideas which have caught the public imagination: catastrophe theory, fractals and chaos. Faced with these sudden upsurges of, often not very well informed, public interest many mathematicians have simply echoed Newton.

> Now is this not very fine? Mathematicians that find out, settle and do all the business must content themselves with being nothing but dry calculators and drudges and another that does nothing but to pretend and grasp at all things must carry away the invention as well of those that were to follow him as of those that went before.

I must confess to a sneaking sympathy with this view since, although my heart sinks slightly when conversations begin 'Ah, you're a mathematician. I never could do mathematics at school,' it sinks much deeper when they begin 'Ah, you're a mathematician. Then I can talk to you about the fractal structure of Tristram Shandy.'

(Since mathematicians are generally considered incapable of making conversation†, let me digress from the matter in hand to offer a little advice. Some day the reader will have a conversation with a courteous old gentleman, probably speaking with a slight central-European accent. When he leaves, the reader will have the impression of having taken part in one of the most interesting half-hours of her life. Only by an effort of profound concentration will she be able to reconstruct a conversation that actually went as follows:

†*Question* How do you tell whether a mathematician is an extrovert or an introvert? *Answer* Extrovert mathematicians look at *your* feet when they talk to you.

C. O. G. Tell me, what do you do for a living?

SELF Actually I work as a 'White Fish Quality Controller'.

C. O. G. That is remarkable. I have always admired the work of White Fish Quality Controllers. Tell me, how do you go about your work?

SELF Well there are actually three sides to our work ... (*Talks non stop for 5 minutes.*)

C. O. G. It all seems very difficult. You must need a great deal of training before you can do it.

SELF (*Modestly.*) Well it's partly a gift, of course, but we all do at least ten weeks training. And, of course, you continue to learn on the job ... (*Talks on for another 5 minutes.*)

C. O. G. Fascinating. But there must be many problems facing White Fish Quality Controllers. Which do you think are the most important?

And so on. Take the courteous old gentleman as a model and you cannot go far wrong.)

Fortunately, other mathematicians have taken a more generous view and sought to inform the public of the ideas behind the slogans. There are thus several excellent books on each topic (I might mention in particular those of Ian Stewart). If asked to choose *one book only* on each topic I would choose the following.

- *Catastrophe Theory,* V. I. Arnol'd. A beautiful little book by a great mathematician.
- *Fractals,* H. Lauwrier. Explains the mathematics simply and has a genuine feel for the aesthetics of fractals.
- *The Essence of Chaos,* E. N. Lorenz. A little drier than some other accounts but straight from the horse's mouth†.

†It reveals that Lorenz's numerical experiments, which must count among the most influential ever performed, were made on a Royal-MacBee LGP-30, a machine which I thought existed only in hacker mythology. (See page 406 of Raymond's *The New Hacker's Dictionary*.)

However, if asked to choose the book with the most breath-taking pictures, I would, like everybody else, nominate *The Beauty of Fractals* by H.-O. Peitgen and P. H. Richter.

PART III

The pleasures of computation

CHAPTER 10

Some classic algorithms

10.1 These twice five figures

If you are a shepherd counting sheep there is a natural way of keeping tally. Each sheep is recorded by a simple stroke I. When you have ten strokes IIIIIIIIII (check by counting on your fingers) cross them out and make a cross sign X. Now start again with strokes I, crossing out when you reach ten and making another X and so on. In this way if you have forty-three sheep you will end up with four Xs and three Is i.e. XXXXIII. If you have a large flock you may reach ten Xs in which case you cross them out and make a C (from 'centum', the Latin for one hundred). Thus CCXIIIIIIII represents two hundred and eighteen. To make numbers easier to read the Romans used V for five and L for fifty giving CCXVIII for two hundred and eighteen and CCLXXXI for two hundred and eighty-one†.

†Later the Romans modified this scheme still further so that the order in which the letters were written mattered and, for example, IX meant nine. For simplicity I shall ignore this complication.

As society became more complex (a Roman legion under Julius Caesar consisted of six thousand men, the ordinary soldiers being paid two hundred and twenty-five denarii a year with the cost of rations being deducted) the Romans added the symbols M for one thousand and D for five hundred. Since we are not used to working with Roman numerals, we are apt to exaggerate the difficulty of calculating with them but, for example, addition is no more difficult with Roman numerals than with our own. The sum

$$\text{MCCCXVIIII} + \text{MCVII}$$

is done by noting that we have six Is giving a V and an I. We now have three Vs giving an X and a V. This gives in turn two Xs, four Cs and two Ms so our answer is MMCCCCXXVI.

Exercise 10.1.1 *Find* MMCCCCLXVI+CCLXXXXVIIII *in the same way.*

Subtraction is no more difficult, and multiplication by small integers can be done by repeated addition. More complicated calculations might be harder, but surely not too hard for someone with experience and a collection of 'tricks of the trade'. I doubt whether the chief problems of a Roman tax collector, grain importer or army paymaster were arithmetical.

In any case, such a person would probably use an abacus. In this instrument, beads or similar counters are laid out in rows; the beads in the first row representing units, in the second tens, in the third hundreds and so on. The abacus was used in modern times in China and Japan, and a skilled operator could work faster than a mechanical calculator. (The counter in the Roman abacus was called a 'calculus', that is a pebble, and from this was derived the verb 'calculare' to calculate.)

The abacus is of very ancient origin and was found throughout the ancient world. Some time around 500 AD, Indian clerks hit upon a new way of recording the contents of an abacus. Let us denote the numbers one to nine by symbols **a,b,c**, etc. (thus **c** is three, **f** is six and so on). We can then record the number of counters in each column starting from the highest and ending with the units. Thus two thousand five hundred and seventeen would be recorded as **beag**. This works well unless one of the intermediate columns is empty (as occurs with two hundred and five, for example). The problem was solved by the invention of a special symbol **Z**, say, for an empty column. Thus one thousand and twenty five (one pebble in the thousand column, no pebbles in the hundred column, two in the ten column and five in the one column) becomes **aZbe** or, in our usual notation, 1025.

Around 800 AD, the Arab empires formed the most advanced civilisation of the world. The caliphs of Bagdad, in particular Harun al-Rashid (a major character in the Arabian Nights) and his son al-Mâmûn were major patrons of learning. The mathematician al-Khowârizmî worked at the court of Bagdad. He wrote two works on algebra and arithmetic in which he brought together Greek and Indian ideas (and, probably, some of his own).

The title of the first text was 'Hisâb al-jabr w'almuqâbah' or 'The Science of Reuniting and Cancelling'. When, after 300 years, his work reached Europe it created such a sensation that the word 'al-jabr' passed into all European languages as 'algebra', 'algèbre', etc.. His second text dealt with computation using the new Indian notation, and his own name, latinised as 'Algoritmi' and later further distorted as Algorithm, came to mean 'computation with positional notation'. One thirteenth century French monk was so carried away with enthusiasm that he wrote a 'Song of the Algorithm' to help his pupils learn the new system.

> Here begins the algorithm
> This new art called the algorithm, in which
> Out of these twice five figures
> 0 9 8 7 6 5 4 3 2 1
> Of the Indians we derive such benefit.

Eventually algorithm came to mean 'a standard method of computation' and I shall devote this chapter and the next to describing some typical algorithms. The problems people had with the new Indian notation, and in particular with the idea of zero (representing an empty column) have left traces in the English language. The Indians called their new symbol 'śūnya', i.e. void; this became 'ṣifra' in Arabic; and the Arabic word gave rise both to 'zero' and 'cipher'. If we say that someone is a cipher we mean that he is a nonentity, a null; if we say that something is in cipher we mean that it is in code, and so incomprehensible to ordinary people; if we say that someone learnt ciphering we mean that they learnt arithmetic (but this usage is now old-fashioned).

Why did the new system eventually oust the old? Historians of science have pointed out that systems close to the Indian appear to have been invented at other times and other places without catching on. The new system was undoubtedly superior to the old in representing very large numbers like 6 530 461 (which under the Roman system would require new symbols for 5 000 000, 1 000 000, 500 000 and 100 000) but such a number was then unlikely to occur in real life (now, of course, such numbers occur in the accounts of even quite small firms). It also, though this could hardly be anticipated from its origin, lent itself to what we would call pencil and paper computation, without using an abacus. This discovery could not have been exploited without reasonably plentiful and cheap writing materials, but the arrival of the Indian numbers in Europe coincided with that of paper (originally invented in China and transmitted via the Arabs), the first cheap method for the permanent storage of information.

The ability to make and record long computations was particularly appealing to merchants engaged in more and more complicated transactions and there seems to have been a symbiotic relation between the rise of the new merchant class and the rise of the new arithmetic. The new arithmetic helped the merchants with their trade, and they in turn insisted that their children be taught the new arithmetic.

Since anything the new system could do could also be done by the old (though sometimes, as with long division, in a more cumbersome manner) the two systems coexisted for three or four centuries. The invention of decimals by Stevin, followed by that of logarithms by Napier, dealt the death blow to the old methods since these concepts are natural with Indian numerals and not with Roman ones.

In view of what we have seen, it is natural to ask whether our present system of writing numbers could not be improved. The fact that we count in units, tens, hundreds and so on is presumably due to the accident of our having ten fingers. Various proposals have been made for recording numbers to base b as

$$\sum_{n=0}^{N} a_n b^n$$

with $0 \le a_n < b$. The most interesting choices seem to be:

$b = 12$. The advantage claimed here is that 12 has more factors than 10.

Exercise 10.1.2 (i) *Find the decimal expansion of $1/n$ in base 10 (i.e. with our usual base) for $n = 1, 2, 3, \ldots, 9$.*

(ii) *Find the 'decimal' expansion† of $1/n$ in base 12 (use A to represent ten and B to represent eleven) for $n = 1, 2, 3, \ldots, 8, 9, A, B$.*

†More correctly the 'duodecimal' expansion.

$b = 11$. Bertrand Russell wrote:

> Experience has taught me a technique for dealing with [cranks]... I counter the devotees of the Great Pyramid by adoration of the Sphinx; and the devotee of nuts by pointing out that hazelnuts and walnuts are just as deleterious as other foods and only Brazil nuts should be tolerated by the faithful .... On this basis a pleasant and inconclusive argument becomes possible.

On this principle, mathematicians, faced with enthusiasts for $b = 12$, have argued that choosing a base with no non-trivial factors would lead to a more uniform treatment of reciprocals.

Exercise 10.1.3 *If you like this sort of thing, find the 'decimal' expansion‡ of $1/n$ in base 11 (use A to represent ten) for $n = 1, 2, 3, \ldots, 8, 9, A$.*

‡More correctly the 'undecimal' expansion. At least, that is what various classicists assure me.

$b = 2$. As everybody knows, the modern electronic computer stores numbers in binary (and is, in part, a large, fast, electronic abacus). Moreover, if the reader takes the trouble to practice a bit, she will find that computation in binary is very easy for humans. However, this number system is unsuitable as a system of record (for human beings). The most famous date in English history becomes

10000101010 and whilst one object of this form can be memorised, I doubt whether, even with practice, most people could manage a long list.

$b = 8$. This is a rather attractive choice. It is perfectly acceptable as a system of record (the famous date mentioned above becomes 2052) and numbers in this base (octal) can be easily translated into binary and vice versa.

Exercise 10.1.4 *Give a rule for translating numbers in base 8 to base 2 and a rule for translating numbers in base 2 to base 8. Check your rules on the date given above.*

There is another proposal of a different type, which was explained to me at school, and which seems to have been repeatedly rediscovered during the last thousand years. The idea here is to use negative integers as well as positive integers and record numbers as

$$\sum_{n=0}^{N} a_n 10^n$$

with $-5 \le a_n \le 5$. We write $\bar{n} = -n$ so that, for example, $\bar{3} = -3$. Using our standard positional notation we have, again for example,

$$3\bar{4}25\bar{2}\bar{1} = 3 \times 10^5 + (-4) \times 10^4 + 2 \times 10^3 + 5 \times 10^2$$
$$+ (-2) \times 10^1 + (-1) \times 10^0$$
$$= 2 \times 10^5 + 6 \times 10^4 + 2 \times 10^3 + 4 \times 10^2 + 7 \times 10^1 + 9 \times 10^0$$
$$= 262479$$

Exercise 10.1.5 (i) *Translate $45\bar{1}43$ and $30\bar{2}3\bar{3}$ into the usual form. Translate 2773 and 10926 from the usual form to our new form.*

(ii) *What integer is represented by $\bar{4}3$? If n is written in the new form, give a simple rule for obtaining $-n$ in the new form.*

Addition with the new notation follows exactly the same pattern as with the old except that we must be prepared to carry negative amounts as well as positive.

Example 10.1.6 *Noting that $6 = 1\bar{4}$, $\bar{6} = \bar{1}4$ and $\bar{5} = \bar{1}5$, we have the following sum.*

$$3\bar{4}42$$
$$1\bar{2}34+$$
$$\overline{3344}$$

Exercise 10.1.7   *Choose three six-figure numbers written in the new style and add them. Check your answer by translating all the numbers back into the old style.*

Subtraction in the new system reduces to addition using the observation of Exercise 10.1.5(*ii*).

Multiplication of two single digits is easy. For example $4 \times 4 = 16 = 2\bar{4}$, $4 \times \bar{3} = -12 = \bar{1}2$, and $\bar{2} \times \bar{3} = 6 = 1\bar{4}$. Multiplying a multi-digit number by a single digit number in the new system is done as in the old except that (just as with addition) we must be prepared to carry negative amounts as well as positive. The reader will check that

$$4 \times 3\bar{3}2\bar{4} = 11\overline{1}4\overline{4}$$

and

$$\bar{3} \times 3\bar{1}2 = \bar{1}124.$$

Long multiplication is now done according to the standard pattern:

$$
\begin{array}{r}
4\bar{3}2 \\
1\bar{4}2 \times \\
\hline
\bar{1}3\overline{4}4 \\
\bar{2}512 \\
4\bar{3}2 \\
\hline
22\overline{4}2\overline{4}
\end{array}
$$

Exercise 10.1.8   (i) *Translate the problem just done into old notation, do it by long multiplication (no calculators!) and check that the two answers coincide.*

(ii) *Choose two four-figure numbers written in the new style and multiply them as in the way just shown. Check your answer by translating all the numbers back into the old style and doing a long multiplication (no calculators).*

What is the advantage of the new scheme? If the reader has done the calculations suggested (or if she thinks for a moment), she will see that the new style multiplication only requires us to use multiplication tables for $n \times m$ with $0 \le n, m \le 5$ instead of $0 \le n, m \le 10$ together with the simple rules for multiplying signed numbers. Since there is good evidence that the speed with which we can perform mental tasks depends on the number of different things we have to keep in mind, this means that we should, after sufficient practice, be able to do arithmetic faster using the new system than the old. (In addition, if the idea were generally adopted, children would have to learn by heart smaller and much simpler tables than at present.)

However, I dare say, not one reader in a hundred will be prepared to give the new system a fair trial. What, after all, is the point of spending a lot of time learning a new system if all the results have to be translated back into the old? If the reader considers why she will not adopt the new system, she will understand better why the Indian numbers took so long to oust the Roman ones.

Of course, the system described works to any base $b \geq 3$.

Exercise 10.1.9 (i) *Describe the system in the case $b = 3$. This is an extremely attractive system. According to Knuth, it was seriously considered as an alternative to the binary system for early computers. 'Perhaps' he says 'the symmetric properties and simple arithmetic of this system will prove to be quite important one day — when the "flip-flop" is replaced by the "flip-flap-flop".'*

(ii) *What happens when $b = 2$?*

It seems to me that some of the systems described above (in particular working to base 8) are superior to our present one, but that none of them are so much better as to make worthwhile the agony of a change-over. But perhaps there is a system which is that much better? We know that Archimedes, the greatest mathematician of antiquity, was interested in such matters, since he wrote on a possible notation for large numbers in 'The Sand Reckoner'. (Since this is written like a very good *Scientific American* article the reader may wish to read it for herself.) If even Archimedes could miss a path which now seems so obvious, perhaps we too may have overlooked some important point.

10.2 The good old days

When politicians and grandparents talk about the good old days when schools taught the basic skills, they have in mind something like the following (see, for example, Chapter XXII of Durell's excellent textbook *General Arithmetic for Schools*). (The monetary units are those used in Great Britain until 1971.)

Example 10.2.1   *A man borrows £132 17s $6\frac{3}{4}$d at compound interest of $2\frac{3}{4}\%$ payable yearly. How much will he owe after 2 years correct to the nearest farthing?*

To appreciate this example fully the reader needs various items of information.

1   The sum of money £*A* B*s* C*d* is to be read as *A* pounds, *B* shillings and *C* pennies (or pence). The d is said to be short for denarius, a small Roman coin and the £ sign to come from the libra, a Roman unit of weight (in Anglo Saxon times, a pound was the value of a pound weight, in Latin 'pondo libra', of silver coins). I do not know why the £ sign precedes and the s and d follow the figures to which they refer.

2   One pound was worth 20 shillings, one shilling was worth 12 pennies (or pence). One quarter of a penny was called a farthing. Farthing coins were still used in the 1950s. (The coinage consisted of farthings, half-pennies, pennies, three-penny pieces, six-penny pieces, shillings, florins worth two shillings and half-crowns worth two shillings and six-pence (six pennies). There was no crown. Professional people like lawyers charged in guineas; there was no guinea coin or note but the guinea, being worth one pound and one shilling, was felt to be more gentlemanly.)

3   In compound interest at *x*% payable yearly, *x*% of what you owed at the beginning of the year is added to your debt at the end of the year. (Monetary units may change but compound interest never goes out of fashion.)

4   When I did this sort of question as a schoolboy, pocket-calculators did not exist.

To solve this problem we work in farthings.

£132	is	2640s
£132 17s	is	2657s
£132 17s	is	31884d
£132 17s 6d	is	31890d
£132 17s 6d	is	127560 farthings
£132 17s $6\frac{3}{4}$d	is	127563 farthings

The next stage is the computation of the compound interest in farthings. We work to two places of decimals in order to preserve sufficient accuracy. In order to understand why the calculation proceeds as it does, the reader should constantly recall that we are doing everything without the aid of machinery.

Initial sum	is	127563.00 farthings
$\frac{1}{4}\%$ of initial sum	is approx.	318.91 farthings
$\frac{3}{4}\%$ of initial sum	is approx.	956.73 farthings
2% of initial sum	is	2551.26 farthings
First year sum	is approx.	131070.99 farthings
$\frac{1}{4}\%$ of first year sum	is approx.	327.68 farthings
$\frac{3}{4}\%$ of first year sum	is approx.	983.04 farthings
2% of first year sum	is approx.	2621.42 farthings
Second year sum	is approx.	134675.45 farthings

Of course, if we have a calculator we simply enter our instructions to obtain

Second year sum is approx. $127563 \times 1.0275 \times 1.0275 = 134675.43$ farthings.

Although we lost a little accuracy in working to two places of decimals, both methods give the amount owed after two years as 134675 farthings correct to the nearest farthing. All that remains is to convert back to pounds, shillings and pence.

134675 farthings	is	$33668\frac{3}{4}$d
33668d.	is	2805s 8d
2805s	is	£140 5s

so the sum owing after 2 years is £140 5s $8\frac{3}{4}$d.

The reader can now try some examples for herself.

Exercise 10.2.2 *A man borrows £119 12s $8\frac{1}{2}$d at compound interest of $3\frac{3}{4}\%$ payable yearly. How much will he owe after two years correct to the nearest farthing? Start the calculations without using a calculator. If things get too much for you, check your calculations so far and complete them using a calculator. If you manage to finish without using a calculator, congratulations! Now check your result using a calculator.*

Exercise 10.2.3 *Do Example 10.2.1 by first converting the sum of money from pounds, shillings and pence to pounds (rather than farthings; make sure you hold sufficiently many decimal places), calculating compound interest and then converting back to pounds, shillings and pence.*

Since the mathematics teachers at my school were extraordinarily good, it is worth asking why they set questions like this. They were not under the impression that the commercial world was full of people seeking loans of £213 14s $7\frac{1}{2}$d, and, even in my childhood, interest rates of $2\frac{3}{4}\%$

per year were not readily obtainable. (We did not complain; a certain lack of realism was preferable to calculating with $5\frac{7}{8}\%$.)

In fact, we were the inheritors of a proud tradition of long, accurate calculations going back to Kepler and passing through Newton and Gauss. If a mathematician wished to know the orbit of a planet or the values of a newly invented function, there was no alternative to computing it by hand. We were being taught the art of organising long computations clearly and concisely. With the coming of the computer the skills we acquired have become obsolete, but they were necessary and useful in their time.

On the other hand, the system of currency (for which my schoolmasters were not responsible) was simply ridiculous. If the reader does the next example, with or without a calculator, she will see how much of the difficulty of the previous problems is due to the monetary units involved.

Exercise 10.2.4 *A man borrows £172.34 at compound interest of $2\frac{3}{4}\%$ payable yearly. How much will he owe after two years correct to two decimal places?*

Associated with the old British currency was an equally peculiar system of weights and measures. It is said that the Swedes have no sense of humour, but I have reduced a roomful of them to hysterics by reciting the British system of lengths.

12 inches	make	1 foot
3 feet	make	1 yard
$5\frac{1}{2}$ yards	make	1 rod, pole, or perch
4 poles	make	1 chain
10 chains	make	1 furlong
8 furlongs	make	1 mile

An acre is 10 square chains.

Exercise 10.2.5 *(i) How many acres to a square mile?*
(ii) What is the length of the side of a square of area 1 acre in yards, feet and inches correct to the nearest inch? (Use a calculator.)

One of the joys of British householding was meeting problems like the following in real life.

Exercise 10.2.6 *A man and his wife wish to carpet a rectangular room which measures 17 feet 4 inches by 14 feet 3 inches. If carpet costs 17 shillings and 4 pence a square yard, how much will this cost them?*

As an encore I would give the British system of weights.

16 ounces	make	1 pound
14 pounds	make	1 stone
2 stones	make	1 quarter
4 quarters	make	1 hundredweight
20 hundredweight	make	1 ton

†Naturally. Recall the 'pondo libra', *pondo* means 'by weight' and *libra* is the 'pound' quantity. The pound sign £ is a decorated L.

The abbreviation for ounce was oz, for pound was lb†and for hundredweight cwt.

It might be supposed that the British groaned under a system which made calculation so unnecessarily difficult. On the contrary, they were so deeply attached to it that it required a 100 years to persuade them to decimalise their currency. The process of changing to metric weights and measures is still not complete in 1996; one of the first actions of the incoming Conservative government in 1979 was to abolish the body in charge of the change-over‡.

‡Napoleon would have agreed with this decision. In exile on Saint Helena he indulged in a tirade against the metric system starting with the folly of consulting mathematicians 'upon a question which was wholly within the province of the administration. ... Nothing can be more contrary to the organisation of the mind, the memory, and the imagination [than the metric system]. ... The new system of weights and measures will be the subject of embarrassment and difficulties for several generations .... Thus are nations tormented about trifles.'

Why was this? First, I think, we must remember that, for most adults, the change-over involved *extra* work. The old system may have been hard to learn, but they had already learnt it, the new system easy to learn, but they had to learn it. The difficulties they may have had as school children were subsumed under the vague heading 'I was never very good at mathematics at school'. (Moreover, whatever vicissitudes adults may suffer, childhood is not one of them.)

Second, very few people are active calculators. If you bought 2 pounds of apples at 1 shilling and 8 pence per pound it was rarely the first time that you had done so. From experience you knew both what 2 pounds of apples looked like and roughly what it should cost. If the sum charged was a bit over 3 shillings most people would be satisfied. (Do most people know in advance what their supermarket bill will be? Do you?) If units are changed, we lose all the benefit of experience in the old units.

The reader may ask why I have not mentioned the economic costs and benefits. It might be possible, at least in principle, to estimate some costs like that of new arithmetic text-books (but remember that such text-books are replaced regularly in the normal course of events) and some benefits (weighing machine manufacturers can produce the same model for the home and export market). Other aspects are controversial. Most economists would expect substantial benefits from increased competition as a result of the removal of artificial trade barriers; a minority would be more sceptical. Certainly we could expect no agreement on the degree to which the use of different units of measurements hampered international trade.

Finally, if the reader still cannot understand how there could be any opposition to such reforms, she should recall that our units of time and angle remain undecimalised. (The division of units into 60 sub-units goes back to the Babylonian astronomers of 1000 BC.) She should try to make out a case for a day consisting of 10 new hours, each divided into 100 new minutes, each in turn divided into 100 new seconds.

I suspect that scientists and engineers are happy to work in seconds, leaving minutes and hours as units of record rather than calculation. Since the radian is the natural unit of angular measure for the mathematician, the pressure for decimalising the degree is much reduced. (However, my scientific calculator has a mode marked 'grad' in which the right angle is divided into 100 units.)

As I indicated at the beginning of this section, many people regret the changes in the school syllabus consequent on the metrication of weights and measures and the introduction of the hand calculator. 'What would today's student do if deprived of her calculator?' they ask. But the only way I could be denied the use of a calculator would be if I were marooned on a desert island and, in such a case, I doubt if I should spend much of my time calculating compound interest. Mathematics should be taught because it is useful or amusing. Long-hand calculation was never very amusing; it is no longer useful. We should have no regrets at its disappearance.

10.3 Euclid's algorithm

One of the most marvellous algorithms known to mathematicians is also among the oldest. Euclid's algorithm is over 2300 years old (it appears in Euclid's *Elements*), yet Knuth devotes more than 40 pages of his work *The Art of Computer Programming* to discussing it.

The reader will probably have met the notion of the greatest common divisor (sometimes called the highest common factor) in school. If m and n are strictly positive integers, the largest integer which divides both m and n is called their *greatest common divisor*. Thus, for example, the greatest common divisor of 12 and 30 is 6, the greatest common divisor of 15 and 30 is 15 (since 15 divides 15) and the greatest common divisor of 17 and 30 is 1. For the sake of brevity I shall sometimes write $\gcd(m, n)$ for the greatest common divisor of m and n.

If the two integers m and n are small we may be able to spot their greatest common divisor on sight. However, even for relatively small pairs of integers like 1890 and 7623, a more systematic approach is needed.

When I was at school we were taught a method of finding the greatest common divisor which involved factoring both integers. Thus, for example,

$$1890 = 2 \times 3^3 \times 5 \times 7$$
$$7623 = 3^2 \times 7 \times 11^2$$

and so, by inspecting the common prime factors,

$$\gcd(1890, 7623) = 3^2 \times 7 = 63.$$

This method only carries us a limited way forward since it is not easy to factorise even moderately sized numbers. (My school only wished to make factorising more interesting for us.)

Example 10.3.1 *Choose two three-figure numbers and try to find their greatest common divisor. Repeat this with two others. Now do the same with a couple of randomly chosen pairs of four-figure numbers. If you found that easy, repeat the exercise with five-figure numbers (if not, do not bother).*

Euclid's algorithm is based on the following easy observation.

Lemma 10.3.2 *Suppose m and n are integers with $m \geq n \geq 1$. Suppose that n goes into m q times with remainder r. (More formally suppose*

$$m = qn + r$$

with q and r integers and $n > r \geq 0$.) Then

$$\gcd(m, n) = \gcd(n, r).$$

Proof Observe that if the integer u divides n and r, then it divides $qn + r = m$. Thus the greatest common divisor of n and r divides m and so divides n and m. Thus

$$\gcd(n, r) \leq \gcd(m, n).$$

Conversely if u divides m and n, then it divides $m - qn = r$. Repeating our previous argument, we now obtain

$$\gcd(m, n) \leq \gcd(n, r)$$

and so combining our two results we have $\gcd(m, n) = \gcd(n, r)$ as stated. ∎

[Some readers may need to be told that 'lemma' is another word for 'theorem'. Mathematicians usually call their central results 'theorems' and the results used to establish these central results 'lemmas'† but they rarely maintain complete consistency of usage. The symbol ∎ was invented by Halmos to indicate the end of a proof.]

†Swinnerton-Dyer observes that, as a consequence, anyone who can prove a theorem can have it named after them, but 'It is the height of distinction to have a lemma named after you'.

Let us apply this observation to the pair of integers 270 and 80. We note that

$$270 = 3 \times 80 + 30 \quad \text{and so} \quad \gcd(270, 80) = \gcd(80, 30),$$
$$80 = 2 \times 30 + 20 \quad \text{and so} \quad \gcd(80, 30) = \gcd(30, 20),$$
$$30 = 1 \times 20 + 10 \quad \text{and so} \quad \gcd(30, 20) = \gcd(20, 10).$$

But 10 divides 20 so $\gcd(20, 10) = 10$ and it follows that $\gcd(270, 80) = 10$.

This does not look very impressive, but we can apply the same method to two more or less randomly chosen seven-figure integers 2 064 135 and 1 515 562. Simple computations give

$$2\,064\,135 = 1 \times 1\,515\,562 + 548\,573, \quad 1\,515\,562 = 2 \times 548\,573 + 418\,416,$$
$$548\,573 = 1 \times 418\,416 + 130\,157, \quad 418\,416 = 3 \times 130\,157 + 27\,945,$$
$$130\,157 = 4 \times 27\,945 + 18\,377, \quad 27\,945 = 1 \times 18\,377 + 9\,568,$$
$$18\,377 = 1 \times 9\,568 + 8\,809, \quad 9\,568 = 1 \times 8\,809 + 759,$$
$$8\,809 = 11 \times 759 + 460, \quad 759 = 1 \times 460 + 299,$$
$$460 = 1 \times 299 + 161, \quad 299 = 1 \times 161 + 138, \quad 161 = 1 \times 138 + 23.$$

Thus since 23 divides 138, the same reasoning as before shows that the greatest common divisor of 20 644 135 and 1 515 562 is 23.

Example 10.3.3 *There must always be a suspicion that any example selected by the author has been chosen to show his favourite method in a particularly favourable light. Redo the examples you chose in Example 10.3.1 using the method of the two previous paragraphs. Now choose a pair of seven-figure numbers for yourself and try out the method on them.*

(To find the remainder when 652 divides 526 290 using my hand calculator I first compute 526 290/652. My calculator shows 807.19325515 so I see that 652 goes 807 times into 526 290 leaving a remainder between 0 and 806. Since (using my calculator again)

$$526\,290 - 807 \times 652 = 126$$

the required remainder is 126.)

We summarise our method as follows.

Algorithm 10.3.4 (Euclid) *Suppose a_1 and a_2 are integers with $a_1 > a_2 \geq 1$. We define a sequence of integers $a_1 > a_2 > a_3 > \ldots$ by the rule that a_{j+2} is the remainder when we divide a_j by a_{j+1}. (More formally,*

$$a_j = q_{j+1}a_{j+1} + a_{j+2}$$

with q_{j+1} and a_{j+2} integers and $a_{j+1} > a_{j+2} \geq 0$.) *We stop the process at the first point that* a_{k+2} *divides* a_{k+1}. *The integer* a_{k+2} *is the greatest common divisor of* a_1 *and* a_2.

The alert reader may ask how we can be sure that there is *any* k such that a_{k+2} divides a_{k+1}. But if not, we would get an endless sequence of positive integers, each one strictly smaller than the one before and this is clearly impossible.

Any reader who has tried a few examples along the lines suggested in Example 10.3.3 will have few doubts that Euclid's algorithm often works extremely well. Can we always guarantee, not simply that, as we have already proved, it will get the right answer, but also that it will get that answer quickly? The answer is yes and we show this by a simple sequence of observations.

Lemma 10.3.5 *(i) Suppose* $m = qn + r$ *with* q *and* r *integers and* $n > r \geq 0$. *Then* $m/2 > r$.

(ii) If $a_1, a_2, a_3, \ldots$ *is the sequence in Algorithm 10.3.4 then* $a_r/2 > a_{r+2}$ *for all* r *with* $1 \leq r \leq k$.

(iii) If $a_1, a_2, a_3, \ldots$ *is the sequence in Algorithm 10.3.4 and* $a_1 \leq 2^N$ *for some positive integer* N *then* $k < 2N$.

Proof *(i)* Observe that, since $n > r$ and $q \geq 1$,

$$m = qn + r \geq n + r > 2r.$$

(ii) This follows at once from *(i)*.

(iii) From *(ii)* we see that $a_3 < 2^{-1}a_1$, that $a_5 < 2^{-1}a_3 < 2^{-2}a_1$, and, more generally, that $a_{2l+1} < 2^{-l}a_1$ for all integers l such that $1 < 2l + 1 \leq k + 2$. Since $a_{2l+1} \geq 1$ and $a_1 \leq 2^N$, it follows that

$$1 \leq a_{2l+1} < 2^{-l}a_1 \leq 2^{-l}2^N = 2^{N-l}$$

and so $N - l > 0$, i.e. $N > l$, for all integers l with $1 < 2l + 1 \leq k + 2$.

If k is odd, we take $l = (k + 1)/2$ to obtain $N > (k + 1)/2$ and, if k is even, we take $l = k/2$ to obtain $N > k/2$. In either case $k < 2N$ as claimed. ∎

Since the integer k represents the number of steps required by Euclid's algorithm, we may restate part *(iii)* of the previous lemma as follows.

Lemma 10.3.6 *If* m, n *and* N *are integers with* $2^N \geq m, n \geq 1$, *then Euclid's algorithm takes at most $2N$ steps to find the greatest common divisor of* m *and* n.

Proof Use Lemma 10.3.5 *(iii)*. ∎

As every computer buff† knows,

$$2^{10} = 1024 > 1000 = 10^3.$$

Thus Lemma 10.3.6 tells us that if m and n are less than 1000 (and so if m and n are three-figure numbers) Euclid's algorithm requires at most 20 steps. This may not strike the reader as very impressive but, by the same token, since $2^{30} = (2^{10})^3 > (10^3)^3 = 10^9$, we know that Euclid's algorithm applied to two nine-figure integers will require at most 60 steps. In particular the reader armed with an accurate ten-figure calculator will feel confident of doing the next exercise in considerably less than an hour.

†And educated musician. The near equality $1024 \approx 1000$ gives $(5/4)^3 \approx 2$ and shows that 3 major thirds equal about one octave. Schroeder's charming book *Number Theory in Science and Communication* gives a little more detail. See the chapter on temperament in any book on 'physics for musicians' or 'music for physicists' for a substantial discussion.

Example 10.3.7   *Choose two nine-figure integers at random (or get someone else to choose them for you) and find their greatest common divisor.*

The example chosen by the reader may well take rather less than 60 steps. In order to *guarantee* that our estimate will always work, we have made rather pessimistic assumptions. We return to this point in the next chapter.

Now suppose we have a modern desk computer programmed to work to the required degree of precision. If we have two 150-figure integers, Euclid's algorithm will require at most 1000 steps. The machine will return the greatest common divisor much faster than I could type in the integers.

Exercise 10.3.8   *(Bezout's theorem. We shall make use of this result in Exercise 16.1.1 which describes a modern method for enciphering secret messages.) Suppose we use Euclid's algorithm to find the greatest common divisor w of two integers $u = a_1$ and $v = a_2$ with $u \geq v \geq 1$. We obtain a sequence of equations*

$$a_1 = q_2 a_2 + a_3 \qquad (1)$$
$$a_2 = q_3 a_3 + a_4 \qquad (2)$$
$$\vdots \qquad\qquad \vdots$$
$$a_j = q_{j+1} a_{j+1} + a_{j+2} \qquad (j)$$
$$\vdots \qquad\qquad \vdots$$
$$a_{n-2} = q_{n-1} a_{n-1} + a_n \qquad (n-2)$$
$$a_{n-1} = q_n a_n + a_{n+1} \qquad (n-1)$$
$$a_n = q_{n+1} a_{n+1}. \qquad (n)$$

The algorithm tells us that $w = a_{n+1}$.

If we look at Equation $(n-1)$ we see that

$$w = a_{n-1} - q_n a_n,$$

and so

$$w = k_{n-1} a_{n-1} + l_{n-1} a_n, \qquad (n-1)'$$

for suitable integers k_{n-1} and l_{n-1}. By Equation $(n-2)$ we know that

$$a_{n-2} = a_n - q_{n-1} a_{n-1},$$

so, substituting for a_n in Equation $(n-1)'$, we obtain

$$w = k_{n-2} a_{n-2} + l_{n-2} a_{n-1}, \qquad (n-1)'$$

for suitable integers k_{n-2} and l_{n-2}. In general, once we have obtained

$$w = k_{j+1} a_{j+1} + l_{j+1} a_{j+2}, \qquad (j+1)'$$

we know from Equation (j) that

$$a_{j+2} = a_j - q_{j+1} a_{j+1},$$

so, substituting for a_{j+2} in Equation $(j+1)'$ we obtain

$$w = k_j a_j + l_j a_{j+1}, \qquad (j)'$$

for suitable integers k_j and l_j. Continuing in this way we eventually obtain

$$w = k_1 a_1 + l_1 a_2, \qquad (1)'$$

for suitable integers k_1 and l_1. In other words

$$w = ku + lv$$

for suitable integers k and l.

(i) In the discussion so far we have assumed that $u \geq v \geq 1$. Show that, given any non-zero integers u and v with greatest common divisor w, there exist integers k and l such that

$$w = ku + lv.$$

(ii) If u and v are non-zero integers with greatest common divisor w, and r is any integer show that there exist integers k and l such that

$$r = ku + lv,$$

if and only if r is divisible by w.

(iii) The preliminary discussion may seem a bit abstract so here is a specific example. Let us find the greatest common divisor of 538 and 191.

Euclid's algorithm gives

$$538 = 2 \times \mathbf{191} + \mathbf{156}$$
$$\mathbf{191} = 1 \times \mathbf{156} + \mathbf{35}$$
$$\mathbf{156} = 4 \times \mathbf{35} + \mathbf{16}$$
$$\mathbf{35} = 2 \times \mathbf{16} + \mathbf{3}$$
$$\mathbf{16} = 5 \times \mathbf{3} + \mathbf{1}$$
$$\mathbf{3} = 3 \times \mathbf{1}.$$

The use of bold numbers is intended to help the reader follow the calculations but otherwise has no further significance†.

$$\mathbf{1} = \mathbf{16} - 5 \times \mathbf{3}$$
$$= \mathbf{16} - 5 \times (\mathbf{35} - 2 \times \mathbf{16}) = 11 \times \mathbf{16} - 5 \times \mathbf{35}$$
$$= 11 \times (\mathbf{156} - 4 \times \mathbf{35}) - 5 \times \mathbf{35} = 11 \times (\mathbf{156} - 49 \times \mathbf{35}$$
$$= 11 \times (\mathbf{156} - 49 \times (\mathbf{191} - \mathbf{156}) = 60 \times \mathbf{156}$$
$$= 60 \times (\mathbf{538} - 2 \times \mathbf{191}) - 49 \times \mathbf{191} = 60 \times \mathbf{538} - 169 \times \mathbf{191}.$$

† I took the idea from Childs' excellent book *A Concrete Introduction To Higher Algebra* which any reader interested in the kind of thing discussed here will find it well worth her while to consult.

Verify by direct computation that

$$1 = 60 \times 538 - 169 \times 191.$$

   Choose a pair of three-figure numbers u and v at random. Find their greatest common divisor w and find integers k and l such that

$$w = ku + lv.$$

Verify your answer by direct computation.

   Choose another pair of three-figure numbers u' and v'. Set $u = 3u' + 9$, $v = 9v'$ and repeat the process of the previous paragraph.

   (iv) Find the greatest common divisor w of 92 and -16. Find integers k and l such that

$$w = k \times 92 + l \times (-16).$$

Exercise 10.3.9   *Euclid's algorithm has been known for over 2000 and the reasons for its excellence have been well understood for over a 100 years. It thus came as a surprise to me to learn that it now has a genuine competitor, invented by J. Stein in 1961. Here it is. Suppose we wish to find the greatest common divisor of integers a and b with $a, b \geq 1$.*

STEP 1 *Set $a_1 = a$, $b_1 = b$, $c_1 = 1$.*
STEP n *We have integers a_n, b_n, c_n with a_n, $b_n \geq 1$.*
   (i) If $a_n = b_n$ stop. The greatest common divisor is $a_n c_n$.

(ii) If a_n and b_n are both even, set $a_{n+1} = a_n/2$, $b_{n+1} = b_n/2$ and $c_{n+1} = 2c_n$.

(iii) If a_n is even and b_n is odd, set $a_{n+1} = a_n/2$, $b_{n+1} = b_n$ and $c_{n+1} = c_n$.

(iv) If a_n is odd and b_n is even, set $a_{n+1} = a_n$, $b_{n+1} = b_n/2$ and $c_{n+1} = c_n$.

(v) If a_n and b_n are both odd, set $a_{n+1} = |a_n - b_n|$, $b_{n+1} = \min(a_n, b_n)$ and $c_{n+1} = c_n$.

Now continue to step $n+1$.

(a) Before proceeding to the rest of the question try the algorithm out on a few cases (for example $a = 2152$ and $b = 764$) to see how it works.

(b) Show that (if the algorithm does not stop before the nth step)

$$c_{n+1} \gcd(a_{n+1}, b_{n+1}) = c_n \gcd(a_n, b_n),$$

and that (if the algorithm does not stop before the $(n-1)$th step)

$$a_{n+2}b_{n+2} \leq a_n b_n/2.$$

Hence show that, if $a, b \leq 2^N$, the algorithm will stop after at most $4N$ steps. Explain how to obtain $\gcd(a,b)$ from the final values of a_n, b_n and c_n.

(c) By considering

$$a = 3(2^{N-2} + 2^{N-3} + \ldots + 2^2 + 2 + 1), \quad b = 3,$$

show that, if $N \geq 5$, there exist integers $a, b \leq 2^N$ for which the algorithm takes at least N steps.

The Stein algorithm is thus guaranteed to work in roughly the same number of steps as Euclid's algorithm. However, Euclid's algorithm requires a 'long division' at each step whereas the Stein algorithm only requires division by 2 which is a simple operation in binary arithmetic (why?). Thus if every microsecond counts and we are prepared to program in machine language, the Stein algorithm may run faster†.

The Stein algorithm is not a world shattering discovery, but the pleasure it has given me to understand it and write it down here makes me echo the words of Poincaré:

> ... adepts find in mathematics delights analogous to those that painting and music give. They admire the delicate harmony of numbers and forms; they are amazed when a new discovery discloses for them an unlooked-for perspective; and the joy they thus experience, has it not the aesthetic character although the senses take no part in it? Only the privileged few are called to enjoy it fully, it is true; but is it not the same with all the noblest arts?

†The mathematicians who read my manuscript all wrote long essays in the margin discussing the extent to which the Stein algorithm is novel and the degree to which the advantages of 'easy division' are outweighed by the increase in 'branching operations'. Clearly there is more to be said.

10.4 How to count rabbits

The man who did most to introduce the new Indian numerals and their associated methods of computation to Europe was Leonardo Fibonacci. The text-book which he wrote in 1202 was a storehouse of computational methods and remained a standard work for two centuries.

Like most good text-book writers, Fibonacci borrowed examples freely from his predecessors (including al-Khowârizmî). Indeed, one of his problems which ran:

> There are seven old women on the road to Rome, each woman has seven mules, each mule carries seven sacks, each sack contains seven loaves, with each loaf are seven knives and each knife is in seven sheaths. Women, mules, sacks, loaves, knives and sheaths, how many are there on the road to Rome?

has been traced back to around 1650 BC when something very similar occurs as a teaching example in the Rhind Papyrus. Most English children still know a variant, with a trick answer, which begins:

> As I was going to St Ives I met a man with seven wives ...

But Fibonacci also added new examples to the common stock.

> 'How many pairs of rabbits,' he asks, 'can be produced from a single pair in a year if every month each pair begets a new pair which from the second month on becomes productive?'

Let us write D_r for the total number of non-breeding pairs at the end of the rth month, E_r for the total number of breeding pairs at the end of the rth month and F_r for the total number of pairs. A reasonable interpretation of Fibonacci's problem gives Table 10.1. Thus, if $r \geq 1$,

$$F_r = E_r + D_r$$

and, if $r \geq 2$,

$$E_r = E_{r-1} + D_{r-1} \text{ and } D_r = E_{r-1}.$$

Table 10.1. *Fibonacci's rabbits.*

Month	r	1	2	3	4	5	$\cdots$	r	$\cdots$
Non-Breeding	D_r	1	0	1	1	2	$\cdots$	E_{r-1}	$\cdots$
Breeding	E_r	0	1	1	2	3	$\cdots$	$D_{r-1} + E_{r-1}$	$\cdots$
Total	F_r	1	1	2	3	5	$\cdots$	$D_r + E_r$	$\cdots$

Combining these results, we see that

$$E_r = F_{r-1} \text{ for } r \geq 2 \text{ and } D_r = E_{r-1} = F_{r-2} \text{ for } r \geq 3,$$

and so, if $r \geq 3$

$$F_r = E_r + D_r = F_{r-1} + F_{r-2}.$$

The conditions

$$F_r = F_{r-1} + F_{r-2}, \quad F_1 = F_2 = 1,$$

define the famous Fibonacci numbers F_n.

Exercise 10.4.1 *Compute F_r for $r = 1, 2, \ldots, 15$. [As a check, $F_{15} = 610$.]*

Fibonacci numbers are involved in many beautiful results and turn up in unexpected and important places throughout mathematics†. As an example of an elegant formula let me cite Cassini's identity which dates back to 1680.

Theorem 10.4.2 *If $n \geq 2$ then*

$$F_{n+1}F_{n-1} - F_n^2 = (-1)^n.$$

Proof Using the definition $F_r = F_{r-1} + F_{r-2}$ we obtain

$$F_{n+1}F_{n-1} - F_n^2 = (F_n + F_{n-1})F_{n-1} - F_n^2 = F_{n-1}^2 + F_n(F_{n-1} - F_n)$$
$$= F_{n-1}^2 - F_nF_{n-2} = -(F_nF_{n-2} - F_{n-1}^2).$$

Thus, if we set $G_n = F_{n+1}F_{n-1} - F_n^2$ we obtain

$$G_n = -G_{n-1} = G_{n-2} = -G_{n-3} = \ldots = (-1)^{n-2}G_2.$$

But $G_2 = F_3F_1 - F_2^2 = 2 - 1 = 1$, so $G_n = (-1)^{n-2} = (-1)^n$ and this is the required result. ∎

Cassini's identity is the basis for one of Lewis Carroll's favourite puzzles in which a square of side F_n units is cut up as shown in Figure 10.1 and apparently reassembled to form an F_{n+1} by F_{n-1} rectangle. Cassini's identity tells us that we have gained or lost one square unit of area. In the figure, $n = 6$ so we cut up an 8×8 square and obtain a 13×5 rectangle. An area of 64 units has been converted into an area of 65 units!

Exercise 10.4.3 *Show that if n is even the dissection actually reassembles as shown in a very exaggerated form in Figure 10.2. Here ABC and DEF are straight lines, $ACFE$ is a rectangle and $AHFG$ is a parallelogram. What is the length of AF? What is the area of the triangle AHF? What* is the length l_n of the perpendicular from H onto AF? Compute l_6. Why is the illusion so good? What happens if n is odd?

Exercise 10.4.4   *(i) Write a computer program to invert a 2×2 matrix,*

(ii) Find the determinant and the inverse of the matrix

$$E_n = \begin{pmatrix} F_n & F_{n+1} \\ F_{n-1} & F_n \end{pmatrix}$$

for all $n \geq 2$.

(iii) Use your computer program on a real computer (or programmable calculator) to invert E_n for various values of n (you could start with $n = 10, 20, 30, \ldots$ and then fill in the gap which looks most interesting). For which value of n does your program first show signs of distress? What happens as you increase n beyond this point? Why could this have been foreseen? If your pocket calculator has a matrix inversion program try the same problem on it.

(iv) What connections can you see between this exercise and the previous one on Carroll's puzzle?

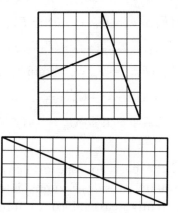

Figure 10.1: Carroll's puzzle with $n = 6$.

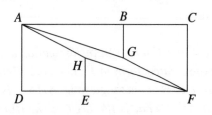

Figure 10.2: Spot the extra area.

Turning to more serious matters, we note the following nice formula for F_n.

Lemma 10.4.5

$$F_n = \frac{1}{\sqrt{5}} \left(\left(\frac{1+\sqrt{5}}{2} \right)^n - \left(\frac{1-\sqrt{5}}{2} \right)^n \right).$$

Proof Let us write

$$u = \left(\frac{1+\sqrt{5}}{2} \right) \quad \text{and} \quad v = \left(\frac{1-\sqrt{5}}{2} \right).$$

The reader may easily check that

$$u^2 = u + 1 \text{ and } v^2 = v + 1,$$

and so, if we write

$$G_n = \frac{1}{\sqrt{5}} (u^n - v^n),$$

it follows that

$$
\begin{aligned}
G_{r-1} + G_{r-2} &= \frac{1}{\sqrt{5}} (u^{n-1} - v^{n-1}) + \frac{1}{\sqrt{5}} (u^{n-2} - v^{n-2}) \\
&= \frac{1}{\sqrt{5}} (u^{n-1} + u^{n-2}) - \frac{1}{\sqrt{5}} (v^{n-1} + v^{n-2}) \\
&= \frac{1}{\sqrt{5}} u^{n-2}(u+1) - \frac{1}{\sqrt{5}} v^{n-2}(v+1) \\
&= \frac{1}{\sqrt{5}} u^{n-2} u^2 - \frac{1}{\sqrt{5}} v^{n-2} v^2 \\
&= G_n.
\end{aligned}
$$

But, as the reader may easily check, $G_1 = 1$ and $G_2 = 1$. Thus G_n satisfies the same defining relations as F_n and so $F_n = G_n$ as stated. ∎

Exercise 10.4.6 *Check the things that 'the reader may easily check' in the proof just given.*

Exercise 10.4.7 *We may generalise Lemma 10.4.5 as follows. Suppose that U_n is defined by the conditions*

$$U_r = aU_{r-1} + bU_{r-2}, \quad U_1 = u_1, \quad U_2 = u_2,$$

and that p and q are distinct roots of the equation

$$t^2 = at + b$$

(i.e. $p^2 = ap + b$, $q^2 = aq + b$). Show that if C and D satisfy

$$Cp + Dq = u_1$$
$$Cp^2 + Dq^2 = u_2,$$

then

$$U_r = Cp^r + Dq^r.$$

Lemma 10.4.5 has a remarkable consequence. Using a pocket calculator it is easy to check that

$$0 > \left(\frac{1 - \sqrt{5}}{2}\right) > -0.62 \text{ and } 0.45 > \frac{1}{\sqrt{5}} > 0.$$

Thus

$$\left| \frac{1}{\sqrt{5}} \left(\frac{1 - \sqrt{5}}{2}\right)^n \right| < 0.3$$

for all $n \geq 1$. Lemma 10.4.5 thus gives the following theorem.

Theorem 10.4.8 F_n *is the closest integer to* $\dfrac{1}{\sqrt{5}} \left(\dfrac{1 + \sqrt{5}}{2}\right)^n$.

Exercise 10.4.9   *(i) Check the inequalities just given.*

(ii) Compute F_{16} exactly using the definition. Use your pocket calculator to compute $\dfrac{1}{\sqrt{5}} \left(\dfrac{1 + \sqrt{5}}{2}\right)^{16}$. *Do you expect* $\dfrac{1}{\sqrt{5}} \left(\dfrac{1 + \sqrt{5}}{2}\right)^n$ *to become an ever better approximation to F_n as n increases? Why?*

Now let us return to the study of Euclid's algorithm started in the last section. In order to refresh her memory, the reader should do the following exercise.

Exercise 10.4.10   *(i) Pick two five-figure numbers at random and apply Euclid's algorithm to them.*

(ii) Compute F_{20} and F_{19} using Theorem 10.4.8 and a pocket calculator. Apply Euclid's algorithm to these two numbers. What do you notice? What is the greatest common divisor of F_n and F_{n-1} for general $n \geq 2$? How many steps does Euclid's algorithm require to find it?

Exercise 10.4.10 strongly suggests that the Fibonacci series gives the 'worst cases' for Euclid's algorithm. Let us see if we can prove this. Suppose that a and b are integers with $a > b \geq 1$ and that Euclid's algorithm requires exactly n steps to find their greatest common divisor *c.* If we write $a_{n+3} = a$, $a_{n+2} = b$ and $a_2 = c$, then we can set out the n stages of the algorithm as follows.

1st step	$a_{n+2} = q_{n+1}a_{n+1} + a_n,$
2nd step	$a_{n+1} = q_n a_n + a_{n-1},$
$\vdots$	$\vdots$
$(n+2-k)$th step	$a_{k+1} = q_k a_k + a_{k-1},$
$\vdots$	$\vdots$
$(n-1)$th step	$a_4 = q_3 a_3 + a_2,$
nth step	$a_3 = q_2 a_2.$

Here a_{k+1}, a_k, a_{k-1} and q_k are integers with $a_{k+1} > a_k > a_{k-1} > 0$ (i.e 'a_k goes into a_{k+1} q_k times with a_{k-1} left over') for $3 \le k \le n+1$ and a_3, a_2 and q_2 are integers with $a_3 > a_2$ (i.e 'a_3 goes into a_2 exactly q_2 times'). Notice that, since $a_{k+1} > a_k$ we must have $q_k \ge 1$ for $3 \le k \le n+1$ and $q_2 \ge 2$.

If we now apply Euclid's algorithm to F_{n+2} and F_{n+1}, we get an instructive parallelism

$$a_{n+2} = q_{n+1}a_{n+1} + a_n, \quad F_{n+2} = F_{n+1} + F_n,$$
$$a_{n+1} = q_n a_n + a_{n-1}, \quad F_{n+1} = F_n + F_{n-1},$$
$$\vdots \qquad\qquad \vdots$$
$$a_{k+1} = q_k a_k + a_{k-1}, \quad F_{k+1} = F_k + F_{k-1},$$
$$\vdots \qquad\qquad \vdots$$
$$a_4 = q_3 a_3 + a_2, \qquad F_4 = F_3 + F_2,$$
$$a_3 = q_2 a_2, \qquad F_3 = 2 = 2 \times 1 = 2F_2.$$

If we recall the facts about q_k given in the last sentence of the previous paragraph, we get the even more instructive parallel

$$a_{n+2} \ge a_{n+1} + a_n, \quad F_{n+2} = F_{n+1} + F_n,$$
$$a_{n+1} \ge a_n + a_{n-1}, \quad F_{n+1} = F_n + F_{n-1},$$
$$\vdots \qquad\qquad \vdots$$
$$a_{k+1} \ge a_k + a_{k-1}, \quad F_{k+1} = F_k + F_{k-1},$$
$$\vdots \qquad\qquad \vdots$$
$$a_4 \ge a_3 + a_2, \qquad F_4 = F_3 + F_2,$$
$$a_3 \ge 2a_2 \qquad F_3 = 2F_2.$$

The reader may already see how the argument ends but, in case she does not, we write the last set of equations in reverse order and observe that $a_2 = c \geq 1$, $F_2 = 1$ to obtain

$$a_2 \geq F_2, \qquad\qquad\qquad\qquad (1)$$
$$a_3 \geq 2a_2, \qquad F_3 = 2F_2, \qquad (2)$$
$$a_4 \geq a_3 + a_2, \qquad F_4 = F_3 + F_2, \qquad (3)$$
$$\vdots \qquad\qquad \vdots \qquad\qquad \vdots$$
$$a_{k+1} \geq a_k + a_{k-1}, \qquad F_{k+1} = F_k + F_{k-1}, \qquad (k)$$
$$\vdots \qquad\qquad \vdots \qquad\qquad \vdots$$
$$a_{n+1} \geq a_n + a_{n-1}, \qquad F_{n+1} = F_n + F_{n-1}, \qquad (n)$$
$$a_{n+2} \geq a_{n+1} + a_n, \qquad F_{n+2} = F_{n+1} + F_n. \qquad (n+1)$$

We see at once that

from (1) we have $\qquad\qquad\qquad\qquad a_2 \geq F_2, \qquad (1')$

from (1') and (2) we have $\qquad\qquad a_3 \geq F_3 \qquad (2')$

from (1'), (2') and (3) we have $\qquad a_4 \geq F_4 \qquad (3')$

from (2'), (3') and (4) we have $\qquad a_5 \geq F_5 \qquad (4')$

$$\vdots \qquad\qquad\qquad\qquad \vdots \qquad \vdots$$

from $((k-2)')$, $((k-1)')$ and (k) we have $a_{k+1} \geq F_{k+1} \qquad (k')$

$$\vdots \qquad\qquad\qquad\qquad \vdots \qquad \vdots$$

from $((n-2)')$, $((n-1)')$ and (n) we have $a_{n+1} \geq F_{n+1} \qquad (n')$

from $((n-1)')$, (n') and $(n+1)$ we have $\quad a_{n+2} \geq F_{n+2} \qquad ((n+1)')$

We have thus proved the following remarkable result.

Theorem 10.4.11 *Let a and b be integers with $a > b \geq 1$. If Euclid's algorithm requires n steps to find their greatest common divisor, then $a \geq F_{n+2}$ and $b \geq F_{n+1}$.*

Another way of putting our results is as follows.

Theorem 10.4.12 *Let a and b be integers with $a > b \geq 1$. If $a < F_{n+2}$ or $b < F_{n+1}$ then Euclid's algorithm requires less than n steps to find their greatest common divisor. If $a = F_{n+2}$ and $b = F_{n+1}$ then Euclid's algorithm requires exactly n steps to find their greatest common divisor.*

Combining Theorem 10.4.8 with Theorem 10.4.12 we get

Lemma 10.4.13 *If m, n and N are integers with*

$$\frac{1}{\sqrt{5}} \left(\frac{1 + \sqrt{5}}{2} \right)^N > m, n \geq 1$$

then Euclid's algorithm takes at most N steps to find the greatest common divisor of m and n.

Exercise 10.4.14 *(i) Write down the first 15 Fibonacci numbers and the first few powers of 2 (i.e. $2 = 2^1$, $4 = 2^2$, $8 = 2^3$, ...) and use it to compare the bounds on the number of steps required for Euclid's algorithm given by by Lemma 10.4.13 and by Lemma 10.3.6.*
(ii) Use the fact that

$$\frac{1 + \sqrt{5}}{2} < 2$$

to show that, if Lemma 10.4.13 shows that no more than N steps are required to find the greatest common divisor of n and m, then Lemma 10.3.6 will show that no more than 2N steps are required to find the greatest common divisor of n and m.
(iii) (This probably requires some facility with logarithms.) Show that, if N is very large and Lemma 10.4.13 shows that no more than N steps are required to find the greatest common divisor of n and m, then Lemma 10.3.6 will show that not more than roughly $1.4 \times N$ steps are required to find the greatest common divisor of n and m.

As the preceding exercises show, Lemma 10.4.13 is not much better than Lemma 10.3.6, but when we proved Lemma 10.3.6 we had no idea how much it could be improved. With Lemma 10.4.13 we know we have the best possible result.

Some modern algorithms

11.1 The railroad problem

In 1859, France and Austria went to war in Italy. As the troops marched along the roads, they waved cheerfully to the trains which ran on the railways alongside. It was war in an old tradition, fought for limited objectives by small professional marching armies, with battles which began in the morning and ended in the evening. The American Civil War, which began two years later, was a new kind of war coloured by ideology, fought for unconditional surrender, between great conscript armies rushed across vast distances by rail to fight battles of ever-increasing length over ever-extending spaces.

The elder Moltke, Chief of the Prussian and then of the German Military Staff†, did not think much of the qualities of the generalship exhibited at the beginning of the Civil War (he spoke of the opposing armies as 'mobs of armed men chasing each other round the country-side'). However, he was quick to draw the lessons of this conflict in which the only way of avoiding stalemate was to 'get there first with the most' and the method of achieving this was the railway.

†Victor in wars against the Danes, the Austrians and the French. It is said that he only smiled twice, once when his mother-in-law died and once when he saw the defences of Copenhagen.

> 'Build no more fortresses, build railways,' ordered the elder Moltke who had laid out his strategy on a railway map and bequeathed the dogma that railways are the key to war. In Germany the railway system was under military control with a staff officer assigned to every line; no track could be laid or changed without permission of the General Staff. Annual mobilisation war games kept railway officials in constant practice and tested their ability to improvise and divert traffic by telegrams reporting lines cut and bridges destroyed. The best brains produced by the War College, it was said, went into the railway section and ended up in lunatic asylums.

In 1870 the newly united Germany had beaten France and the German military staff had insisted on the seizure of Alsace-Lorraine to increase military security against a possible revenge by France. Of course, the loss of Alsace-Lorraine ruled out any permanent reconciliation with France and, whilst Germany was strong enough to defeat France if she fought alone, this might not be the case if she found allies. At the time Russia was allied with Germany and her railway system was underdeveloped but, if she changed sides and developed her railway system and industry, she would represent a formidable challenge. In order to slow down this process, the German government closed the German capital markets to Russia, thus driving her into the arms of France. The Franco-Russian alliance now made it imperative for Germany to plan for a war on two fronts, and the plan developed required a knock-out blow against France while the Russian armies were still mobilising, the armies being rushed back after victory to confront the Russian attack.

As each year passed and the Russian railway system grew, the time available for the decisive victory over France decreased. To gain this rapid victory, mobilisation would have to flow seamlessly into attack. The trains and armies could not stop at the French frontier while diplomats negotiated, and the technical requirements of the planned victory would require an advance through neutral Belgium bringing Britain into the war against Germany.

Amidst the mounting diplomatic tension that followed the murder of Archduke Franz Ferdinand in 1914, the Russians began to mobilise. For Russia this was just another card in the diplomatic game, but for the German army each day that the Russians were allowed to mobilise meant a day lost in the conflict with France. The French, anxious to show Britain that, if war broke out, it was not of their choosing, withdrew their forces ten kilometres behind their frontiers. The German general staff decided to accept the odium of unprovoked aggression rather than accept the military risks of delay. The first troops crossed into neutral Luxembourg to seize railway junctions. At this point the Kaiser panicked. Would it not be possible to proceed against Russia without involving France? The troops were ordered back and it was explained that 'a mistake had been made'. The Commander of the German armies now told the Kaiser that it was impossible to change plans. 'It cannot be done†. The deployment of millions cannot be improvised. ... Those arrangements took a year to complete ... and once settled cannot be altered.' The troops reentered Luxembourg and the First World War had begun‡.

Railways played a key role in the two World Wars which followed and in the plans for the third formulated in the succeeding Cold War. The United States armed forces commissioned research to discover ways in which the newly invented electronic computer could be used to solve

†When the advice given to the Kaiser was revealed after the war, the Chief of the German Army's Railway Division was so incensed by this slur on his professional competence that he wrote a book replete with charts and graphs to show that it could have been done and that he could have done it.

‡In the course of this book the reader has been presented with three possible causes of the First World War: that it was caused by the murder of Archduke Franz Ferdinand, that it was caused by an arms race and that it was the consequence of a railway timetable. These three putative causes are not mutually exclusive.

problems of transport and supply. One of the problems was the 'railroad problem' of getting the maximum number of trains from A to B along a given railway network. This was solved by Ford and Fulkerson. As often happens, they and others later found simpler and more appealing solutions and it is one of these that I shall present. (Pioneering work is clumsy and so, according to Besicovitch, mathematicians' reputations rest on the number of bad proofs they have given.)

First we need to formulate the problem mathematically. We suppose that we have n towns which we label $1, 2, \ldots, n$. The railway line between towns i and j can carry a maximum of c_{ij} trains an hour from i to j. (If there is no railway between i and j then we set $c_{ij} = 0$.) Our method does not assume that $c_{ij} = c_{ji}$ so that the number of trains that can be carried per hour in one direction is the same as that in the other, though this will often be the case. We write x_{ij} for the actual number of trains per hour that run from i to j under our plan. We take $c_{ii} = x_{ii} = 0$ (trains running from town i and back again do not interest us). Clearly

$$0 \leq x_{ij} \leq c_{ij},$$

for all $1 \leq i, j \leq n$. If we wish to move trains from town 1 to town n then as many trains will have to leave intermediate towns as enter them. Thus

$$\sum_{j=1}^{n} x_{ij} = \sum_{j=1}^{n} x_{ji}$$

for $2 \leq i \leq n-1$ and the number F of trains per hour leaving 1 and arriving at n (the *flow*) is given by

$$F = \sum_{j=1}^{n} x_{1j} = \sum_{j=1}^{n} x_{jn}.$$

We wish to maximise the *flow value F*.

In order to gain some insight into the problem, divide the set of towns $\{1, 2, \ldots, n\}$ into two subsets S_1 and S_2, the first containing 1 (the source) and the second n (the destination). (Formally, we demand $S_1 \cup S_2 = \{1, 2, \ldots, n\}$, $S_1 \cap S_2 = \emptyset$, $1 \in S_1$ and $n \in S_2$. We call the decomposition $\{S_1, S_2\}$ a *cut*.) Suppose now we place observers beside each track from towns in S_1 to towns in S_2. The total number of trains they observe each hour will be greater than or equal to the total transfer per hour from S_1 to S_2 (some trains may run from S_2 to S_1) and this in turn will be equal to the flow value F of trains per hour going from 1 to n. Formally

$$\sum_{i\in S_1, j\in S_2} x_{ij} \geq \sum_{i\in S_1, j\in S_2} x_{ij} - \sum_{i\in S_1, j\in S_2} x_{ji} = F.$$

(Here $\sum_{i\in S_1, j\in S_2} x_{ij}$ means the sum of all x_{ij} with $i \in S_1$ and $j \in S_2$.) Recalling that $x_{ij} \leq c_{ij}$ (the number of trains per hour cannot exceed the capacity of the track) we see that

$$\sum_{i\in S_1, j\in S_2} c_{ij} \geq \sum_{i\in S_1, j\in S_2} x_{ij} \geq F.$$

If we call $\sum_{i\in S_1, j\in S_2} c_{ij}$ the *cut value C* of the cut $\{S_1, S_2\}$, we obtain

the cut value $C \geq$ the flow value F.

Since we did not specify any particular flow or cut in our argument, we have, in fact, shown

any cut value $\geq$ any flow value,

and so

the minimum cut value $\geq$ the maximum flow value.

Further, if we can find a flow x_{ij} and a cut $\{S_1, S_2\}$ for which the flow value and the cut value are equal, then the previous equation shows that there is no flow with larger flow value and no cut with smaller cut value.

We summarise our conclusions in a theorem.

Theorem 11.1.1   *For any network*

the minimum cut value $\geq$ the maximum flow value

and, if we can find a flow and a cut such that

cut value = flow value,

then that flow is the best possible.

This theorem in turn forms the basis for an algorithm to find a best possible flow. (There may be many different flow patterns giving the same maximum flow value.)

We assume that all the c_{ij} are integers. On the face of it this looks like a very restrictive assumption but it is not. If one of the tracks has a capacity of $3\frac{1}{2}$ trains an hour and another a capacity of $4\frac{2}{3}$, it suffices to work in units of $\frac{1}{6}$ of a train per hour to restate the problem as one involving integers. Our algorithm requires:

An initial integer flow We need a possible flow pattern with all the x_{ij} integers. One such flow is obvious since we could use $x_{ij} = 0$ for all

i, j (that is no trains running at all) but, in practice, we may start with a much better guess at an optimal flow.

Once we have an integer flow we can apply:

The central improving step Define

$$A_0 = \{1\},$$
$$A_1 = A_0 \cup \{j : c_{1j} > x_{1j} - x_{j1}\},$$
$$A_2 = A_1 \cup \{j : c_{ij} > x_{ij} - x_{ji} \text{ for some } i \in A_1\},$$

and, more generally,

$$A_{r+1} = A_r \cup \{j : c_{ij} > x_{ij} - x_{ji} \text{ for some } i \in A_r\}.$$

Thus A_0 consists just of the source town 1. The set A_1 consists of the source town 1 together with all those towns j which are undersupplied from 1; that is such that the total number of trains $x_{1j} - x_{j1}$ transferred from 1 to j each hour is less than the capacity c_{1j} of the track from 1 to j. Once the set A_r of towns has been found, we take A_{r+1} to be the set consisting of all the towns in A_r together with all those towns j which are undersupplied from A_r; that is, such that there exists a town i in A_r for which the total number of trains $x_{ij} - x_{ji}$ transferred from i to j each hour is less than the capacity c_{ij} of the track from i to j.

At each stage, either $A_{r+1} = A_r$ or A_{r+1} contains at least one more town than A_r. Since there are only n towns, it follows that, for some $r \leq n$, we arrive at a position where either

(I) $n \in A_r$, or

(II) $n \notin A_r$ but $A_{r+1} = A_r$.

The moment we reach such a position we stop. If (II) holds, we move to the *final step* given below. If (I) holds, we proceed as follows.

Write $i(r) = n$. Since $n = i(r)$ was in A_r but not in A_{r-1} it must be undersupplied from a town $i(r-1)$ in A_{r-1} but not in A_{r-2}. In turn the town $i(r-1)$ which is in A_{r-1} but not in A_{r-2} must be undersupplied from a town $i(r-2)$ in A_{r-2} but not in A_{r-3}. Continuing in this way we find a chain of towns $i(s) \in A_s$ each of which is undersupplied from the previous town i_{s-1}, and so satisfy

$$c_{i(s-1)i(s)} > x_{i(s-1)i(s)} - x_{i(s)i(s-1)}$$

for $1 \leq s \leq r$. Since $i(0) \in A_0$ and A_0 consists just of the source town 1 we have $i(0) = 1$ and so we have a chain of undersupplied towns stretching from the source town 1 to the destination town n. We now do the obvious thing by increasing the number of trains per hour along this chain. Set

$$u = \min_{1 \leq s \leq r} (c_{i(s-1)i(s)} - (x_{i(s-1)i(s)} - x_{i(s)i(s-1)}))$$

(thus u is the least underflow along the chain). By increasing $x_{i(s-1)i(s)}$ or decreasing $x_{i(s)i(s-1)}$ (or both) we can find new integer values $x'_{i(s-1)i(s)}$ and $x'_{i(s)i(s-1)}$ still satisfying

$$0 \leq x'_{i(s-1)i(s)} \leq c_{i(s-1)i(s)} \text{ and } 0 \leq x'_{i(s)i(s-1)} \leq c_{i(s)i(s-1)},$$

but with

$$x'_{i(s-1)i(s)} - x'_{i(s)i(s-1)} = x_{i(s-1)i(s)} - x_{i(s)i(s-1)} + u$$

for each $1 \leq s \leq r$, so that the number of trains per hour increased by u along the whole chain. If we now set $x'_{ij} = x_{ij}$ for all other pairs i, j (thus leaving the rest of the traffic flow unaltered), then we have a new integer flow whose flow value is u trains per hour larger than the one with which we started. (Since u is a strictly positive integer we have thus increased the flow value by at least 1.)

We now subject our new flow to the *central improving step*. This will either lead us to Case (II) and the *final step* or produce another flow with a flow value increased by at least 1 which in turn can be subjected to the *central improving step* and so on. This process must terminate because we know that the flow value cannot exceed any cut value and each application of the *central improving step* increases our flow value by at least 1. The only way the process can terminate must be by arriving at Case (II) and the *final step*.

The final step Looking back at how we arrived via Case (II) of the *central improving step* we see that we have a set A_r containing 1 but not containing n such that

$$A_r = A_{r+1} = A_r \cup \{j : c_{ij} > x_{ij} - x_{ji} \text{ for some } i \in A_r\},$$

and so

$$c_{ij} = x_{ij} - x_{ji} \text{ for all } i \in A_r \text{ and all } j \notin A_r.$$

Since $x_{ji} \geq 0$ and $c_{ij} \geq x_{ij}$ this last remark can be strengthened to give

$$c_{ij} = x_{ij} \text{ and } x_{ji} = 0 \text{ for all } i \in A_r \text{ and all } j \notin A_r.$$

If we write $S_1 = A_r$ and take S_2 to be the set consisting of the remaining towns, then the previous paragraph tells us that we have a cut such that

$$c_{ij} = x_{ij} \text{ and } x_{ji} = 0 \text{ for all } i \in S_1 \text{ and all } j \in S_2,$$

and no trains travel from S_2 to S_1 whilst all the tracks from S_1 to S_2 are used to full capacity. Thus, for the flow and cut under discussion,

$$\text{cut value} = \sum_{i \in S_1, j \in S_2} c_{ij} = \sum_{i \in S_1, j \in S_2} x_{ij} = \text{flow value},$$

and so by Theorem 11.1.1 our flow is the best possible.

We illustrate the procedure in Figure 11.1. The reader who wishes to try her hand at the Ford–Fulkerson algorithm will find a collection taken from Cambridge mathematics examination papers in Figure 11.2. (Note that the arrows show the permitted directions. If no line is shown between i and j then $c_{ij} = 0$, if a line but no arrow is shown then $c_{ij} = c_{ji}$.)

Figure 11.1: The Ford–Fulkerson algorithm in action.

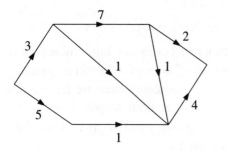

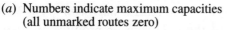

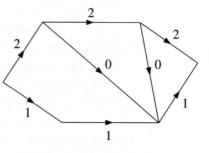

(*a*) Numbers indicate maximum capacities (all unmarked routes zero)

(*b*) A possible flow

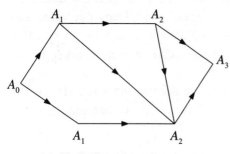

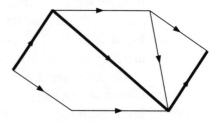

(*c*) Underflow towns

(*d*) An underflow path

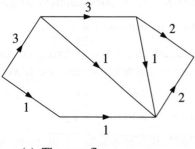

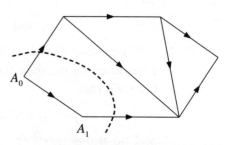

(*e*) The new flow

(*f*) Underflow towns for the new flow, and a cut proving that the flow is maximal

Exercise 11.1.2 *Find the optimal flows for the systems shown in Figure 11.2. In each case, check your solution by finding a cut for which the flow value equals the cut value.*

Figure 11.2: Exercises on the Ford–Fulkerson algorithm.

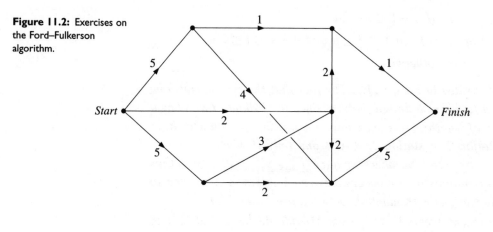

(a)

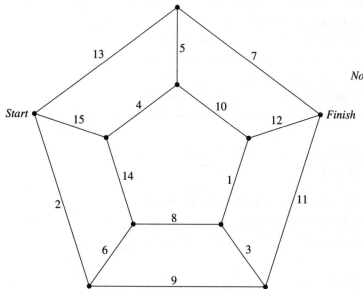

Note: All edges are bidirectional; that is, they allow flow in either direction.

(b)

Exercise 11.1.3 *(i) The Titipu metro system has $2n + 2$ stations numbered 0 to $2n + 1$ and the capacity (in trains per quarter hour) of the railway from i to j is c_{ij} with*

$$c_{0j} = n \quad if \ 1 \le j \le n$$
$$c_{i\,2n+1} = n \quad if \ n+1 \le i \le 2n$$
$$c_{ij} = 0 \ or \ c_{ij} = 1 \quad if \ 1 \le j \le n \ and \ n+1 \le i \le 2n$$
$$c_{ij} = 0 \quad otherwise.$$

The system is illustrated in Figure 11.3. The fact that the system runs one way only is not a fault but a design feature, since the Mikado (who has a car) is in favour of healthy exercise. Show that n trains per quarter hour can run from station 0 to station $2n + 1$ if and only if whenever S is a subset of $\{1, 2, \ldots, n\}$ with k members we have $\sum_{i \in S} \sum_{j=n+1}^{2n} c_{ij} \ge k$ (thus each collection of stations with numbers between 1 and n is connected to at least as many stations with numbers between $n + 1$ and $2n$).

(ii) After the recent events in Titipu the Mikado declares that 'Young people should get married with less song and dance.' At a given date each of the n eligible little maids from school must hand in a list of those eligible young men whom she is prepared to marry. The Lord High Everything Else (who, incidentally, planned the metro) must then draw up a list, marrying each little maid to one of those young men whom she is prepared to consider. (By a strange coincidence there are exactly n eligible young men in Titipu.) The penalties for failure are those usual to Titipu. Show that the Lord High Everything Else can succeed if and only if every collection of k maids names at least k possible swains (this is called 'Philip Hall's marriage lemma' and turns out to be an extremely useful result in a broad range of mathematical disciplines). Give an algorithm to enable the Lord High Everything Else to complete his task if it is possible.

How would your answers change if the number of eligible young men in Titipu was not specified to be n and why?

Just as the maximal flow need not be unique so there may be be more than one 'critical cut' for which the flow value equals the cut value. However, if we have a critical cut (S_1, S_2) with $1 \in S_1$, $n \in S_2$, we know that:

1   Increasing the capacity c_{ij} of the route from i to j will not increase the maximum total flow unless i is in S_1 and j is in S_2. (Since all the lines from S_1 to S_2 are working to capacity, there is no point in trying to increase the flow of trains elsewhere.) The critical cut is a 'bottleneck'.

2   If the route from a town i in S_1 to a town j in S_2 is damaged so that the capacity c_{ij} is reduced, then the cut value of (S_1, S_2) is reduced and so (since the maximum flow is no more than the value of any cut) maximum total flow is reduced. Thus, in a military context, you should concentrate on protecting routes across the cut, and your opponent should concentrate on destroying them.

If the reader has not been carried away by the beauty of the mathematics, she may retain some doubts about its applicability to real railways running real trains. Surely there is much more to a railway than a number c_{ij} — there are railway workers who get tired and who make mistakes; there are trains which break down; even in peacetime there are a hundred and one things that can go wrong and in the fog and chaos of war there are a thousand. Such objections apply, I think, if we treat the computer as an infallible oracle but not if we treat it as an extra advisor suggesting plans to be tried and modified in the light of other information not treated in the algorithm.

The Ford–Fulkerson algorithm is, of course, applicable to everything from telephone and computer networks to pipelines. It was the first of an ever-expanding set of network algorithms including, for example, timetabling for complex projects. In the next section we look at one

Figure 11.3: The Titipu metro.

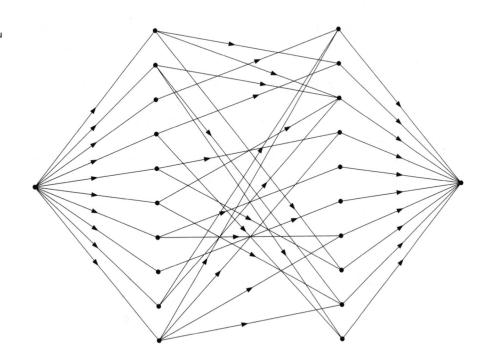

of the many problems which arise in managing large communication networks.

Exercise 11.1.4 *Suppose you are given n towns 1, 2, ..., n together with a list which, for each pair of towns i, j with i ≠ j, tells you either that there is no direct link between the two towns or that the shortest direct link has length l_{ij}.*

(i) Give an algorithm to tell whether it is possible to get from town 1 to town n.

(ii) If there is a route from town 1 to town n give an algorithm to find the shortest route from 1 to n.

[There are many different possible algorithms but you should try for something significantly faster than 'examine all possible routes'. Part (ii) is harder than Part (i). If you have no ideas for Part (ii), draw a reasonably complex road map and colour the shortest routes from each of the towns 1, 2, ... n − 1 in red. What do you notice about the pattern?]

11.2 Braess's paradox

Mathematicians like to prove things if they can, partly because they enjoy exact reasoning for its own sake, but also because experience has shown them that what is 'obvious' is not always true. In 1988 Braess produced the following example.

Consider car drivers driving from town A to town B during a rush hour. There are two routes, one via X and one via Y. As more cars use a road the average speed of each car drops. In our example we shall suppose that if p thousand cars per hour use road AX then the time taken by each car to travel from A to X is $10p + 10$ minutes. The times for all four roads are shown in Table 11.1 and in Figure 11.4.

Suppose now that $2n$ thousand cars per hour travel from A to B (the factor 2 reduces the number of fractions in our calculations but has no other significance) with x thousand cars per hour using the route through X and $y = 2n - x$ thousand cars per hour using the route through Y. The time taken to travel via X will be

$$t_X = (10x + 10) + (x + 60) = 11x + 70,$$

Table 11.1. *Times for Braess's route system.*

Route	AX	XB	AY	YB
Minutes taken	$10p + 10$	$p + 60$	$p + 60$	$10p + 10$

and the time taken via Y

$$t_Y = 11y + 70.$$

If the drivers are allowed to choose their route then, presumably, they will choose the one taking the shortest time. But as more choose this route, it will get more crowded and the time it takes will rise whilst as fewer take the other route, the time it takes will fall until the time taken by the two routes is equal and there will be no incentive for drivers to take one route rather than the other. The system will settle down to the state when both routes take the same time and so

$$11x + 70 = t_X = t_Y = 11y + 70.$$

Thus if drivers make their own decisions we will have $x = y$ and so $x = y = n$. The journey time for each driver will be $70 + 11n$ minutes.

Now suppose we decide that, instead of allowing free choice, we direct drivers in such a way as to minimise the *average* journey time $A(x, y)$. It is easy to see that

$$A(x, y) = (2n)^{-1}(xt_X + yt_Y) = (2n)^{-1}(x(11x + 70) + y(11y + 70))$$
$$= (2n)^{-1}(11(x^2 + y^2) + 70(x + y)),$$

and so, since $x + y = 2n$,

$$A(x, y) = A(x) = 11(2n)^{-1}(x^2 + (2n - x)^2) + 70$$
$$= 11(2n)^{-1}(2x^2 - 4nx + 4n^2) + 70.$$

Figure 11.4: Braess's route system. Before.

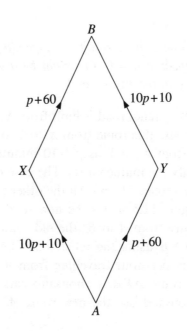

We must choose x so as to minimise $A(x)$. Many of my readers will know how to do this.

Lemma 11.2.1 *If a, b and c are real and $a > 0$ then*

$$at^2 + bt + c \geq c - \frac{b^2}{4a}.$$

The inequality becomes an equality at the unique value $t = -b/2a$.

Proof Observe that

$$at^2 + bt + c = a\left(t + \frac{b}{a}\right)^2 + c - \frac{b^2}{4a}.$$

(This formula is said to be obtained 'by completing the square'.) Since

$$a\left(t + \frac{b}{a}\right)^2 \geq 0,$$

with equality at the unique value $t = -b/2a$, the stated result follows. ∎

Lemma 11.2.1 tells us that $A(x) \geq 70 + 11n$ and that the minimum occurs uniquely when $x = y = n$. Thus our complicated calculations† for the greater good of society turn out to give the same result as the individual selfish decisions of the motorist. When this kind of thing occurs, economists look smug and refer to Adam Smith's metaphor of 'the invisible hand'.

† It is possible to reduce the work by exploiting symmetry. During a lecture, Einstein informed his audience that 'There is a clever way of Minkowski for doing this calculation. However, chalk is cheaper than brains so we shall do it the stupid way.'

Exercise 11.2.2 *State and prove a result corresponding to Lemma 11.2.1 when $a < 0$. What happens if $a = 0$. (It might be a good idea to graph the function $at^2 + bt + c$ in the two cases.)*

Now suppose a one-way relief road is built from X to Y and that if p thousand cars per hour use this route from X to Y then the time taken by each car to travel from X to Y is $p + 10$ minutes. (We make the road one-way to simplify the mathematics. The case of a two-way road is left to the reader in Exercise 11.2.6.) All the other parts of the system remain unchanged. Figure 11.5 shows the new road system. There are now three possible routes from A to B: the old routes AXB and AYB and the new route $AXYB$ using the relief road. Let us suppose that of the $2n$ thousand cars per hour travelling from A to B, x thousand cars per hour use the route AXB, y thousand cars per hour use the route AYB and $2z$ thousand use the new route $AXYB$. We observe that $x + y + 2z = 2n$.

Since the cars using the route AXB share the road AX with cars using the route $AXYB$, the time taken to travel by the route AXB is

$$t_X = (10(x + 2z) + 10) + (x + 60) = 11x + 20z + 70.$$

Similarly the time taken to travel by the route AYB is

$$t_Y = 11y + 20z + 70$$

and the time taken to travel the new, three-road route $AXYB$ is

$$t_Z = (10(x+2z)+10)+(2z+10)+(10(y+2z)+10) = 10(x+y)+42z+30.$$

If the motorists are free to choose their own routes then, if $x < y$, we will have $t_X < t_Y$ and motorists will switch from the route AYB to the quicker route AXB. Thus $x \geq y$ and similarly $y \geq x$ so $x = y$. It follows that $x = y = n - z$,

$$t_X = t_Y = 11n + 9z + 70 \text{ and } t_Z = 20n + 22z + 30.$$

Thus

$$t_X - t_Z = 40 - 9n - 13z. \qquad (*)$$

Since $z \leq n$, it follows that $t_X - t_Z \geq 40 - 22n$ and so, if $n \leq 20/11$, the new route will be quicker than the old routes. The motorists will all choose the new route so $z = n$, $x = y = 0$ and each car will take

Figure 11.5: Braess's route system. After.

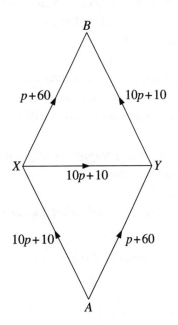

$42n + 30$ minutes. Thus

time before new road built − time after new road built
$$= (11n + 70) − (42n + 30) = 40 − 31n,$$

and, provided that $n \leq 40/31$, motorists will bless the builders of the new road. However, when n increases past $40/31$ motorists will have longer journeys than they did before the new road was built. *Building a new road has increased journey times!* The time added to the journey increases as n increases until, when $n = 20/11$, each motorist's journey time has increased by $180/11$ minutes (over a quarter of an hour).

Now suppose that traffic builds up so that $40/9 > n > 20/11$. If $z = n$ then equation (∗) tells us that $t_X − t_Z < 0$ so motorists should move from the new route to the old ones, that is choose $z < n$. On the other hand if $z = 0$ then (since $40/9 > n$) equation (∗) tells us that $t_X − t_Z > 0$ so motorists should move from the old routes to the new ones. Thus we will have $n > z > 0$ and both the old and new routes will be used. Since all three routes are in use, they must take the same time (for otherwise drivers will transfer from the slower to the faster route). Thus $t_X = t_Z$ and, from (∗), $z = (40 − 9n)/13$, $x = y = n − z = (22n − 40)/13$. Now

time before new road built − time after new road built
$$= (11n + 70) − t_X = (11n + 70) − (11n + 9z + 70)$$
$$= −9z = −9(40 − 9n)/13$$

so the loss of time for each motorist caused by building the new road falls from its maximum at $n = 20/11$ to zero when $n = 40/9$. The fact that adding extra links to a congested system can make matters worse is known as Braess's paradox†.

† This paradox may have not surprised engineers familiar with the slogan 'Local strength means general weakness' as much as the rest of us.

Exercise 11.2.3   *(i) Check the calculations above.*

(ii) Show that if $n > 40/9$ and motorists are allowed free choice, no motorist will use the new route and traffic will revert to the old pattern $z = 0$, $x = y = n$.

(iii) Graph the time taken to complete a journey as a function of n for the old and the new road patterns.

Exercise 11.2.4   *It is natural to ask what pattern of flow will give the* least *average time*
$$A(x, y, z) = (2n)^{-1}(xt_X + yt_Y + 2zt_Z).$$

(i) Show that
$$2nA(x, y, z) = 11(x^2 + y^2) + 4z^2 + 80(n − 1)z + 140n.$$

(ii) Show that, if z is kept fixed, then $A(x, y, z)$ attains a minimum at the unique point $x = y = n - z$.

(iii) Show that $2nA(n - z, n - z, z) = B(z)$, where

$$B(z) = 26z^2 + (36n - 80)z + (140n + 22n^2)$$

$$= 26 \left(z - \frac{20 - 9n}{13} \right)^2 + 140n + 22n^2 - \frac{2(20 - 9n)^2}{13}.$$

(iv)(a) If $n \leq 10/11$, show that $B(z)$ decreases as z increases from 0 to n. Deduce that $A(x, y, z)$ is a minimum when $z = n$, $x = y = 0$. Thus the invisible hand works and the selfish individual decisions of the drivers give the least average journey time.

(b) If $10/11 \leq n \leq 20/9$, show that $B(z)$ decreases as z increases from 0 to $(20 - 9n)/13$ and then increases as z increases from $(20 - 9n)/13$ to 1. Deduce that $A(x, y, z)$ is a minimum when $z = (20 - 9n)/13$, $x = y = (22n - 20)/13$. Thus we can make use of the new road to reduce the average journey time (but some drivers will have a longer journey than others; for what happens if drivers are prepared to cooperate but not to accept worse journey times than other drivers see Exercise 11.2.5).

(c) If $20/9 \leq n$ show that $B(z)$ increases as z increases from 0 to n. Deduce that $A(x, y, z)$ is a minimum when $z = 0$, $x = y = n$. The invisible hand fails to find this minimum when $20/9 \leq n \leq 40/9$ but (see Exercise 11.2.3(ii)) succeeds for $40/9 \leq n$.

(v) Graph how the best average time varies with n.

Exercise 11.2.5   *The reader may ask what happens if instead of seeking to minimise the average time $A(x, y, z)$ spent by a motorist (as in the previous exercise), we seek to minimise the longest journey time $M(x, y, z)$ spent by a motorist.*

(i) Write

$$S_X = T_X \text{ if } x \neq 0, \quad S_X = 0 \text{ if } x = 0,$$
$$S_Y = T_Y \text{ if } y \neq 0, \quad S_Y = 0 \text{ if } y = 0,$$
$$S_Z = T_Z \text{ if } z \neq 0, \quad S_Z = 0 \text{ if } z = 0.$$

Explain why $M(x, y, z) = \max(S_X, S_Y, S_Z)$.

(ii) Show that $M((x + y)/2, (x + y)/2, z) \geq M(x, y, z)$ and deduce that we can confine our search to the case $x = y$ and the function

$$N(z) = M(n - z, n - z, z).$$

(iii) Show that $M(x, y, z)$ is minimised by taking $x = y$ and

$$z = n \qquad \text{if } 0 < n \le 20/11,$$

$$z = \frac{40 - 9n}{13} \qquad \text{if } 20/11 \le n \le 40/31,$$

$$z = 0 \qquad \text{if } 40/31 < n.$$

(iv) Graph the time taken to complete a journey as a function of n. Graph the optimal z as a function of n. Verify that, for certain ranges of values of n to be specified, the minimum longest journey time solution differs both from the invisible hand and from the minimum average (Exercise 11.2.4) solutions.

Exercise 11.2.6 *Most of my readers will feel they have done enough exercises by now. However, some readers may like an exercise which is not broken down into bite-size hints and others may not be prepared to take my word that Braess's paradox is unaffected if the road XY is made two-way. Both sets of readers should consider Braess's network as shown in Figure 11.5 but with XY allowing two-way flow and such that, if p thousand cars per hour travel from X to Y (respectively from Y to X), the time taken by each car to travel from X to Y (respectively from Y to X) is $10p + 10$ minutes†. Find the invisible hand, least average time and minimum longest journey time for all $n > 0$.*

There has, of course, been a great deal of discussion as to whether Braess's paradox actually occurs in road systems. It is certainly true that road improvements sometimes fail to produce their stated objectives. More generally, mathematicians and engineers tend to be more sceptical of Adam Smith's invisible hand than economists or, in other contexts, ecologists, but no one knows whether the complex structures of natural systems or modern economies are close to optimal — or indeed what optimal means in such contexts.

However, if we move from roads and economies to telecommunications networks we reach a context in which Braess's paradox and related problems are known to occur in practice. The reader may not see why this is a problem. Computers and telephone exchanges are not motorists — surely they can cooperate for the common good? But in order to cooperate they must communicate and because, at best, signals between computers can only travel at the speed of light, there is not sufficient time for such communication before decisions on routing must be taken. (The same kind of problems occur *within* a single computer. In order to reduce communication times, the core of the machine must be as compact as possible, with the attendant problem

†This is about the most complicated such network which can be solved with our bare hands. (For mathematicians the phrase 'with bare hands' carries both good and, except among the Narodniks of North Oxford [82], bad overtones. Compare 'The author does not use Lagrange multipliers but carries out the calculations bare handed' with 'He did not know how to operate a crane so he built the dam with his bare hands'.)

of heat production and cooling. Thus a 1980s supercomputer consisted of a small box and a roomful of refrigeration equipment to deal with the heat produced in the small box.)

To grasp the problem, consider Braess's network with cars travelling both ways along the roads and going between any of the four towns. Suppose, further, that the number of cars entering the system varies in an erratic manner and that *the only way of sending messages between the towns is by car.* How should traffic managers at the various towns allocate cars to routes? How can we construct local rules for decision making which will combine to give good global decisions. (In parallel computers where many separate 'sub-computers' try to work on the same problem, how and when should they communicate with each other? If we give poor rules they will spend all their time talking to each other rather than working.) If the reader wishes to appreciate the mixture of hard modern mathematics and old-fashioned rule-of-thumb which is used to attack such problems, she could consult Kelly's paper *Network Routing*, particularly since the paper ends with the description of a commercially successful application.

11.3 Finding the largest

We have now seen two important algorithms — Euclid's algorithm and the Ford–Fulkerson algorithm. Looking back at them, we can see various important questions that must be asked about any algorithm.

1 *Can my valet use it?* That is are the instructions clear enough that they can be used without special knowledge? In less picturesque language, can they be translated into a computer program?

2 *Does it start?* The Ford–Fulkerson algorithm works by improving a given flow. We must give it an initial flow (fortunately the zero flow can always be used for this). Anyone who has used a computer seriously will have produced programs which fail to start.

3 *Does it stop?* Anyone who has used a computer seriously will also have produced programs which fail to stop†. If you look back at my accounts of Euclid's Algorithm and the Ford–Fulkerson Algorithm you will see that I was very careful to prove that they stopped.

4 *Does it stop at the right answer?* This requires no comment.

5 *Is it fast?* We considered this question for Euclid's algorithm (see the discussion at the end of Section 10.3 and Lemmas 10.3.6 and 10.4.13), but not for the Ford–Fulkerson algorithm.

†It is traditional for books on algorithms to have the index entries '*Loop* See Cycle' and '*Cycle* See Loop'.

The study of sorting algorithms illustrates the general nature of Question (5) very clearly. We are all familiar with the ordered list given by a dictionary. (Although a crossword-hater by inclination, I also possess an anagrammatic dictionary and a dictionary ordered by word length n and then by the letters in the kth position $[1 < k \leq n]$, just for the pleasure of ownership.) The computer system I am using to write this book will, if I wish, produce a concordance (ordered list) of all the words in it. Later I shall use sorting systems to put my index and bibliography into alphabetical order. Both human beings and computer systems find it easier to search and work with ordered lists, so the problem of sorting is an important one.

The third volume of Knuth's as yet unfinished epic *The Art of Computer Programming* devotes 388 fascinating pages to sorting algorithms. In his introduction he writes:

> Computer manufacturers estimate that over 25 percent of the running time on their computers is currently being spent on sorting, when all their customers are taken into account. There are many installations in which sorting uses over half the computing time. From these statistics we may conclude either that (i) there are many important applications of sorting, or (ii) many people sort when they shouldn't, or (iii) inefficient sorting algorithms are in common use. The real truth probably involves some of all three alternatives.

Partly as a result of Knuth's attack on (iii), the figure of 25% is probably now a substantial overestimate but sorting remains a major task.

One way of measuring the efficiency of a sorting algorithm is to ask how many comparisons it requires.

Exercise 11.3.1 *The following exercise may help the reader appreciate the kind of problem involved. Take a cheap pack of cards and choose some order on the cards (for example ace beats king, king beats queen, etc and, if two cards have the same face value, clubs beat spades, spades beat hearts and hearts beat diamonds). Shuffle the pack well and deal out ten cards face down in a line on the table. You are allowed to turn over any two cards, inspect them and then either interchange them or leave them alone before turning them face down again. You repeat this as many times as you need to be sure that they are in the correct order. You are not allowed to use information remembered from earlier goes or to use 'extraneous information' (for example that the ace of clubs is the highest card).*

If you find this too easy, do the same exercise with 20 or 40 cards. You may find it useful to try out the algorithms that follow in the same way.

Let us begin with the simpler problem of finding the largest of a set of n unequal numbers $x_1, x_2, \ldots, x_n$. Here is one method.

First step Set $y_1 = x_1$.

Second step Compare y_1 and x_2. If $y_1 < x_2$ set $y_2 = x_2$; if $y_1 > x_2$ set $y_2 = y_1$.

Third step Compare y_2 and x_3. If $y_2 < x_3$ set $y_3 = x_3$; if $y_2 > x_3$ set $y_3 = y_2$.

jth step Compare y_{j-1} and x_j. If $y_{j-1} < x_j$ set $y_j = x_{j-1}$; if $y_{j-1} > x_j$ set $y_j = x_j$.

nth step Compare y_{n-1} and x_n. If $y_{n-1} < x_n$ set $y_n = x_{n-1}$; if $y_{n-1} > x_n$ set $y_n = x_n$. Record y_n and stop.

Exercise 11.3.2 *Apply the algorithm to $x_1 = 3$, $x_2 = 1$, $x_3 = 4$, $x_4 = 6$, $x_5 = 5$, $x_6 = 7$ and $x_7 = 2$.*

At the beginning of this section we posed five questions to be asked about any putative algorithm. I hope that the reader will accept that the answer to Question (1) ('Can my valet use it?') is yes. The answers to Questions (2) and (3) ('Does it start and stop?') are also yes. What about Question (4)? Does it stop at the right answer? The reader is probably already convinced by doing Exercise 11.3.2 that it does. If not, it suffices to observe that $y_j = \max(x_1, x_2, \ldots, x_j)$. (The method is called 'bubble max' with y_j as the rising bubble.) We are left with Question 5 ('Is it fast?'). Observe that 'bubble max' requires $n - 1$ comparisons to find the largest elements. Could we do it with less?

Lemma 11.3.3 *Any algorithm for finding the largest of a set of n unequal numbers requires at least $n - 1$ comparisons.*

Proof Each comparison tells us that one element is less than another. If we make r comparisons, then there will be at most r elements which have been shown to be less than some other and so at least $n - r$ which have not been shown to be less than some other element and each of which could be the largest element. ■

We have thus shown that bubble max is the fastest method of finding the largest element (at least if we measure speed by the number of comparisons). It is not, however, the unique fastest method. Consider what we shall call the 'knock-out' method for finding the largest of 2^N distinct real numbers $x_1, x_2, \ldots, x_{2^N}$.

First step Set $x_r(1) = x_r$ for $1 \leq r \leq 2^N$.

Second step For each $1 \leq p \leq 2^{N-1}$ compare $x_{2p-1}(1)$ and $x_{2p}(1)$. If $x_{2p-1}(1) > x_{2p}(1)$ set $x_p(2) = x_{2p-1}(1)$; if $x_{2p}(1) > x_{2p-1}(1)$ set $x_p(2) = x_{2p}(1)$.

Third step For each $1 \leq p \leq 2^{N-2}$ compare $x_{2p-1}(2)$ and $x_{2p}(2)$. If $x_{2p-1}(2) > x_{2p}(2)$ set $x_p(3) = x_{2p-1}(2)$; if $x_{2p}(2) > x_{2p-1}(2)$ set $x_p(3) = x_{2p}(2)$.

jth step For each $1 \leq p \leq 2^{N-j+1}$ compare $x_{2p-1}(j-1)$ and $x_{2p}(j-1)$. If $x_{2p-1}(j-1) > x_{2p}(j-1)$ set $x_p(j) = x_{2p-1}(j-1)$; if $x_{2p}(j-1) > x_{2p-1}(j-1)$ set $x_p(j) = x_{2p}(j-1)$.

Nth step Compare $x_1(N-1)$ and $x_2(N-1)$. If $x_1(N-1) > x_2(N-1)$ set $x_1(N) = x_1(N-1)$; if $x_2(N-1) > x_1(N-1)$ set $x_1(N) = x_2(N-1)$. Record $x_1(N)$ and stop.

Exercise 11.3.4 *Apply the algorithm to $x_1 = 5$, $x_2 = 1$ $x_3 = 4$, $x_4 = 6$, $x_5 = 8$, $x_6 = 7$, $x_7 = 3$ and $x_8 = 2$.*

What about our five questions? The fact that international committees can actually organise knock-out competitions shows that the answer to Question (1) ('Can the algorithm be operated without intelligent intervention?') is yes. The answers to Questions (2) and (3) ('Does it start and stop?')are also yes. What about Question (4)? Does it stop at the right answer? Again the fact that knock-out competitions are acceptable to players and supporters in so many sports is strong presumptive evidence but we now require proof. To prove the correctness of the algorithm observe that at each step (round) we remove half the remaining elements (teams) but that at no step can we remove the greatest element (best team) since only elements which are smaller than some other element (beaten by another team) are removed. After $N-1$ steps (rounds) only one element (team) remains and since the remaining elements (teams) always include the largest element (best team), the single remaining element (team) must be the largest (best). Turning to Question (5) we remark that each team, except the ultimate winner, loses exactly once (each element except the largest is the smaller in exactly one comparison) and each game has exactly one loser (each comparison reveals that exactly one of the two elements compared is the smaller) so that

$$\text{number of games} = \text{number of losers} = \text{number of teams} - 1 = 2^N - 1$$

i.e.

number of comparisons

$$= \text{number of elements found smaller than some other}$$
$$= \text{number of elements} - 1 = 2^N - 1.$$

Setting $n = 2^N$, we see that the knock-out method is exactly as fast as bubble max if we measure speed by the number of comparisons involved.

We have only described the knock-out method when there are 2^N teams (elements). Unless the reader has led a very isolated life, perhaps tending sick penguins in the Antarctic, she will know that to arrange a knock-out between n teams with $2^{N-1} + 1 \leq n \leq 2^N$ it is usual to give $2^N - n$ teams a bye into the second round. (Such an arrangement is shown for 12 competing teams in Figure 11.6.)

Exercise 11.3.5 *(i) Give a description of a knock-out method (with byes into the second round) for finding the largest of n distinct real numbers $x_1, x_2, \ldots, x_n$ along the same lines as the one given above for the special case $n = 2^N$.*

(ii) Answer Questions (1)–(5) for your method.

(iii) Apply your method to the set of numbers given in Exercise 11.3.2

Although it is usual to run knock-out competitions by giving byes to the second round, other arrangements are possible. An alternative knock-out arrangement is shown for 12 teams in Figure 11.7.

Indeed, even bubble max can be arranged as a knock-out as shown in Figure 11.8. However, whereas the two arrangements shown in Figures 11.6 and 11.7 only require 4 rounds to produce a winner from 12 teams, bubble max requires 11 rounds. In general, if we wish to find the best team (largest element) out of n with $2^{N-1} + 1 \leq n \leq 2^N$ both bubble max and 'knock-out with byes into the second round' require

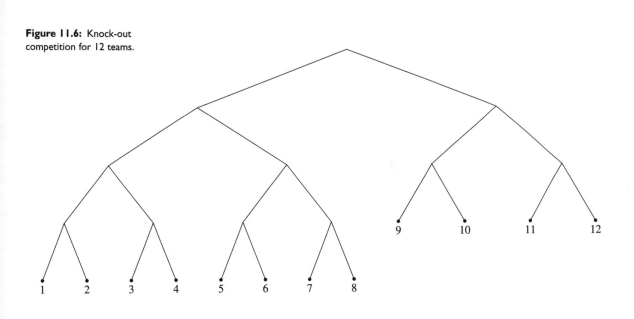

Figure 11.6: Knock-out competition for 12 teams.

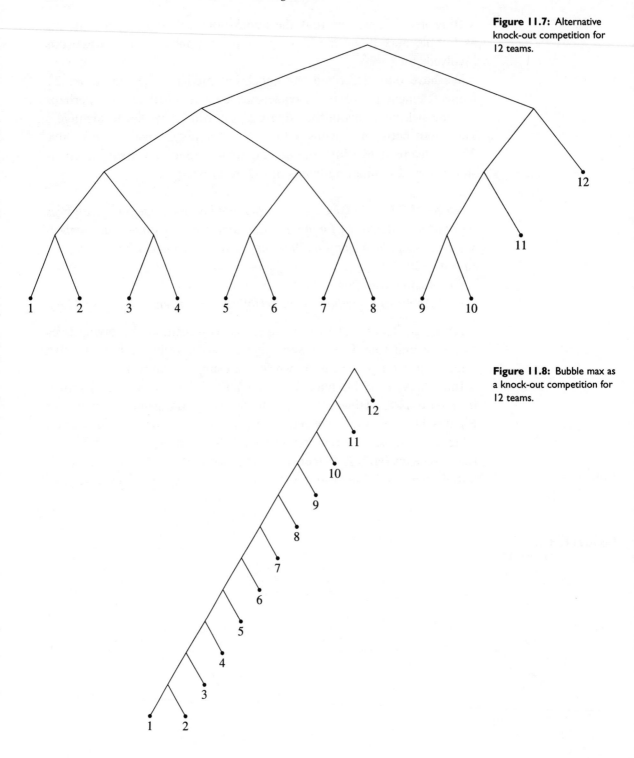

Figure 11.7: Alternative knock-out competition for 12 teams.

Figure 11.8: Bubble max as a knock-out competition for 12 teams.

$n-1$ games, but bubble max requires $n-1$ rounds whilst 'knock-out with byes into the second round' requires only N rounds. The greater economy in rounds turns out to have important consequences which we shall discuss in the next section.

Exercise 11.3.6 *(A computation used in the next section.) There is a well known story, repeated, with his usual trimmings, by Bell in his* Men of Mathematics, *that when Gauss was ten his teacher, Bütner, seeking an hour's repose, set his pupils the 100 term sum*

$$81297 + 81495 + 81693 + \cdots + 100899.$$

The teacher had barely finished stating the problem when, to quote Bell:

> ... *Gauss flung his slate on the table: 'There it lies,' he said — 'Ligget se' in his peasant dialect. Then for the remaining hour, while the other boys toiled, he sat with his hands folded, favoured now and then by a sarcastic glance from Bütner, who imagined the youngest pupil in his class was just another blockhead. At the end of the period, Bütner looked over the slates. On Gauss's there appeared but a single number. To the end of his days Gauss loved to tell how the one number he had written was the correct answer and all the others were wrong.*

(A more restrained account of Gauss's early life, and a more sympathetic estimate of Bütner will be found in Bühler's excellent biography[22].)
(i) Verify that

$$\begin{aligned}
1 + 2 + 3 + 4 + 5 + 6 + 7 &= (1 + 7) + (2 + 6) + (3 + 5) + 4 \\
&= (4 + 4) + (4 + 4) + (4 + 4) + 4 \\
&= 7 \times 4 = 28
\end{aligned}$$

and

$$\begin{aligned}
1 + 2 + 3 + 4 + 5 + 6 &= (1 + 6) + (2 + 5) + (3 + 4) \\
&= (3\tfrac{1}{2} + 3\tfrac{1}{2}) + (3\tfrac{1}{2} + 3\tfrac{1}{2}) + (3\tfrac{1}{2} + 3\tfrac{1}{2}) \\
&= 6 \times 3\tfrac{1}{2} = 21.
\end{aligned}$$

(ii) Work out Bütner's sum.
(iii) Show more generally that

$$a + (a + d) + (a + 2d) + \cdots + (a + (n-1)d) = na + \tfrac{1}{2}n(n-1)d.$$

In my opinion, there is something basically wrong with an educational philosophy which includes results like that of the previous exercise in a 'formula booklet'. The reason generally given for such an inclusion is that memorising formulae is not part of mathematics. It is true that many mathematicians do not *remember the formula*, but all mathematicians *know how to derive it*. The student who memorises the formula and the one who looks it up are both seeking to pacify remote and incomprehensible forces by uttering meaningless incantations. The student who can work it out for herself is following, however distantly, in the footsteps of Gauss. A method can be extended (see, for example, Exercise 11.4.13); a formula can only be applied.

11.4 How fast can we sort?

In the previous section I explained the importance of the sorting problem but we only tackled the problem of finding the maximum. The most obvious way of tackling the full sorting problem is by successive extraction of the maximum.

To be more explicit, consider the problem of sorting of a set of n unequal numbers $x_1, x_2, \ldots, x_n$. Here is the method known as 'bubble sort'.

First step Write $x_1 = x_1(1)$, $x_2 = x_2(1)$, $\ldots$, $x_n = x_n(1)$. Use bubble max to find the largest element of $x_1(1), x_2(1), \ldots, x_n(1)$. Call it y_1 and call the remaining $n-1$ elements $x_1(2), x_2(2), \ldots, x_{n-1}(2)$.

Second step Use bubble max to find the largest element of $x_1(2), x_2(2)$, $\ldots, x_{n-1}(2)$. Call it y_2 and call the remaining $n-2$ elements $x_1(3), x_2(3)$, $\ldots, x_{n-2}(3)$.

jth step Use bubble max to find the largest element of $x_1(j), x_2(j)$, $\ldots, x_{n-j+1}(j)$. Call it y_j and call the remaining $n-j$ elements $x_1(j+1)$, $x_2(j+1), \ldots, x_{n-j}(j+1)$.

nth step There is one remaining element $x_1(n)$. Write $y_n = x_1(n)$ and stop.

It is, I think, reasonably clear that the algorithm produces the list y_1, $y_2, \ldots, y_n$ of the elements in decreasing order.

Exercise 11.4.1 *Apply the algorithm to $x_1 = 3$, $x_2 = 1$, $x_3 = 4$, $x_4 = 6$, $x_5 = 5$, $x_6 = 7$ and $x_7 = 2$.*

How many comparisons does bubble sort take? We saw in the previous section that bubble max requires $m-1$ comparisons to find the largest of m unequal numbers. Thus, in the algorithm just described, the first step requires $n-1$ comparisons, the second $n-2$, the jth $n-j$ and so on. The total number of comparisons required is thus

$$(n-1) + (n-2) + (n-3) + \ldots + 3 + 2 + 1 = \tfrac{1}{2}n(n-1).$$

(See Exercise 11.3.6 if you need help with the equality.) Whether the algorithm is practical depends on the speed of the computer and the size of n. If $n = 1000$ we require about $500\,000$ comparisons and a modern computer will take the twinkling of an eye. If $n = 1\,000\,000$ we require about $500\,000\,000\,000$ comparisons which (at least by 1996 standards) represents a substantial computational task. What is more worrying is that every time we increase n by a factor of 10, the number of comparisons increases by a factor of 100. (As we noted in Section 8.2, there is a rule-of-thumb that machine speed increases by a factor of 10 each decade.)

Exercise 11.4.2 *What is the rough size of the list which could be sorted by your personal computer (or your school's computer) in an hour using the method above? What is the rough size of the list which could be sorted by the fastest machine that you know?*

Can we do better than this? The surprising answer is yes, if, instead of using methods based on bubble max, we use methods based on knock-out competition. Reverting to the sporting analogy, suppose we have arranged an N round knock-out between n teams. After the competition is over (but fortunately before the winner's cup has been presented), the winning team is disqualified. Do we have to replay all the matches to find a new winner? Surely not. The results of matches not involving the disqualified team are unaffected, so we need only consider those teams who have lost against the disqualified team.

Suppose that the disqualified team first played in the kth round (because of the possibilities of byes, we cannot assume that $k = 1$), eliminating team x_k in the kth round, team x_{k+1} in the $(k + 1)$th round, team x_{k+2} in the $(k + 2)$th round and so on. All we need do is readmit team x_k into to the $(k + 1)$st round (a 'walk-over' since it won its previous rounds and its opponents in the kth round have been disqualified), where it plays team x_{k+1} and the winner y_{k+1} goes on into the $(k + 2)$nd round where it plays team x_{k+2} to produce a winner y_{k+2} which, in turn, goes on into the $(k + 3)$th round to challenge team x_{k+3} and so on. Combining the results already obtained of the games not involving the disqualified teams with these new games, we obtain the results of an N (or, if there was only one game in the first round and it involved the disqualified team, an $N - 1$) round knock-out between $n - 1$ teams. The new winner is the best of the remaining $n - 1$ teams (and so the second best of the initial n teams). (We observe that the new winner is one of the teams eliminated by the old winner, since only the best team can beat the second best.) We note that only $N - k + 1$ new games had to be played and that this number is less than or equal to N.

Now suppose that the new winners are disqualified in their turn. Repeating the procedure of the first paragraph, we see that at most N

new games need be played to produce the result of an N (or fewer) round knock-out between the remaining $n-2$ teams. The new winner is the best of these $n-2$ teams (and so the third best of the initial n teams). Continuing in this manner we see that, after the jth disqualification, at most N new games need be played to produce the result of an N (or fewer) round knock-out between the remaining $n-j$ teams. The new winner is the best of these $n-j$ teams (and so the $(j+1)$th best of the initial n teams). After $n-2$ disqualifications there will only be two teams left and these teams will be the two weakest. The winner of the final match will be the weakest but one and the loser the weakest. I give an example in Figure 11.9.

Exercise 11.4.3 *Apply the knock-out method to rank the x_j given in Exercise 11.4.1.*

We observe that, after the knock-out competition has been played once to determine the best team (this requires $n-1$ games) we need only $n-2$ disqualifications (each requiring at most N new games) to rank all the teams in order. Thus

$$\text{total number of games required} \le (n-1) + (n-2)N$$
$$\le N + (n-2)N = (n-1)N$$

(The reader will probably be able to improve this estimate but it is not necessary.) We saw in the previous section that if $2^{N-1} + 1 \le n \le 2^N$, a knock-out between n teams requires only N rounds. Thus, if we have to rank n teams and $2^{N-1} + 1 \le n \le 2^N$, then

$$\text{total number of games required} \le (n-1)N < N2^N.$$

Dropping the sporting metaphor, we see that, if $n \le 2^N$, the 'knock-out method' requires less than $N2^N$ comparisons to sort n unequal numbers into order.

Bearing in mind that $2^{10} = 1024 > 1000$, we see that when $n = 1000$ the knock-out method requires less than $10\,000$ comparisons (50 times fewer than bubble sort). If $n = 1\,000\,000$ the knock-out method requires less than $20\,000\,000$ comparisons ($250\,000$ times fewer than bubble sort and an easy task for a small 1996 desk-top computer). Moreover, every time we increase n by a factor of 10, the number of comparisons increases by a factor of only just more than 10.

Exercise 11.4.4 *What is the rough size of the list which could be sorted by your personal computer (or your school's computer) in an hour using the knock-out method? What is the rough size of the list which could be sorted by the fastest machine that you know?*

Figure 11.9: The knock-out algorithm applied to ranking six teams.

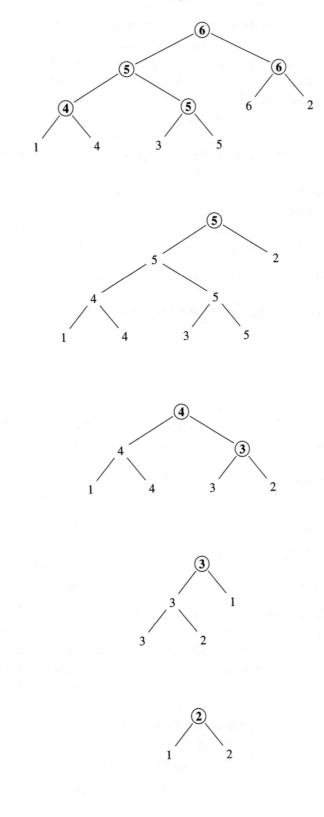

Can we do still better? Not much, since we can now sort 2^N elements with no more than $N2^N$ comparisons, and we saw in Lemma 11.3.3 that we always require at least $2^N - 1$ comparisons merely to find which is the largest. Still, we might be able to do a little better. In order to discuss this possibility I need to introduce the idea of a 'yes–no question' (or to give it a more formal name 'binary chop').

It is said that a few years ago Polish radio ran a 'Twenty questions' competition along the following lines. The game-show host thought of a word and the competitors were allowed to ask up to twenty questions to which the host replied either yes or no. This worked well until a team of mathematics students turned up with a dictionary and a line of questioning that ran, essentially, as follows. (We assume that the dictionary had 2^{20} words in it.)

STUDENTS Is the word the mth in the Dictionary with $1 \le m \le 2^{19}$?
HOST No.
STUDENTS Is the word the mth in the Dictionary with $2^{19} + 1 \le m \le 2^{19} + 2^{18}$?
HOST Yes.
STUDENTS Is the word the mth in the Dictionary with $2^{19} + 1 \le m \le 2^{19} + 2^{17}$?

Exercise 11.4.5 *(i) Write out explicitly the algorithm that the students employed. Show that M questions can determine a word in a dictionary of 2^M words.*
[The algorithm can, of course, be expressed more or less elegantly. One way of obtaining a neat formulation is to use the binary expansion

$$m - 1 = \epsilon_{M-1}2^{M-1} + \epsilon_{M-2}2^{M-2} + \epsilon_{M-3}2^{M-3} + \dots + \epsilon_2 4 + \epsilon_1 2 + \epsilon_0$$

with ϵ_j taking the value 0 or 1.]
(ii) Presumably the host could not instantly locate the 2^{19}th word in the dictionary. What questions do you think the students actually asked?

Can we do better than the Polish students? Suppose we wish to find a particular element x (known to the 'host' but not to us) in a set A containing M elements. If we are allowed just one yes–no question then all we can do is pick some subset B of A and ask 'Does x belong to B?'. If the answer is 'Yes' then we know that x belongs to B and if the answer is 'No' we know that x belongs to $C = A \setminus B$, the complement of B in A (i.e. the set of elements of A which do not belong to B). We observe that at least one of B and C contains at least $M/2$ elements. Symbolically, if we write $|E|$ for the number of elements in E,

$$M = |A| = |B| + |C| \le 2\max(|B|, |C|), \text{ and so } \max(|B|, |C|) \ge M/2.$$

Thus the answer to our question need only restrict our search to a set A_1 of size at least $M/2$. In other words, a single yes–no question can, at best, halve the size of the set in which the unknown x lies. It follows that two yes–no questions can, at best, restrict our search to a set A_2 of size at least $M/4$, that three yes–no questions can, at best restrict our search to a set A_3 of size at least $M/8$, and so on. Since we can only be certain of what x is if we have reduced the size of the set A_k containing it to 1 element, it follows that, if we can guarantee to find x with k questions $M/2^k \le 1$. We summarise our conclusions in a lemma†.

†An alternative approach uses the pigeonhole principle. There are 2^k possible patterns of answers to k yes–no questions. If we have more than 2^k members of the set then two of them must elicit the same pattern of replies and so be indistinguishable.

Lemma 11.4.6 *If $M > 2^k$, then at least $k + 1$ yes–no questions are required to locate an unknown element in a set of size M.*

What does this have to do with the problem of sorting? We make the following two observations.

Lemma 11.4.7 *There are $n! = n \times (n-1) \times (n-2) \times \ldots \times 3 \times 2 \times 1$ different ways of arranging n things in order.*

Sketch proof We can choose the first object in n ways. There then remain $n-1$ objects so we can choose the second object in $n-1$ ways. There then remain $n-2$ objects so we can choose the third object in $n-2$ ways, and so on. ∎

Lemma 11.4.8 *If we wish to sort a set of n unequal numbers $x_1, x_2, \ldots , x_n$, then comparing two of them is a yes–no question about their order.*

Proof The question 'Is $x_i > x_j$?' is a yes–no question. ∎

Combining Lemmas 11.4.6, 11.4.7 and 11.4.8 gives us a lower bound on the number of comparisons required to sort n numbers.

Theorem 11.4.9 *If we can sort a set of n unequal numbers $x_1, x_2, \ldots ,x_n$ using k comparisons then*

$$n! \le 2^k.$$

Proof There are $n!$ possible orderings of $x_1, x_2, \ldots , x_n$. Each comparison is a yes–no question about the possible orderings. Since k such questions suffice to determine a unique ordering, Lemma 11.4.6 tells us that $n! \le 2^k$. ∎

We already know

Theorem 11.4.10   *If $n \leq 2^N$, the 'knock-out method' requires less than $N2^N$ comparisons to sort a set of n unequal numbers $x_1, x_2, \ldots, x_n$.*

To compare Theorem 11.4.9 with Theorem 11.4.10 we require estimates for $n!$.

Lemma 11.4.11 (A simple Stirling's estimate)

(i) $n^n \geq n! \geq (n/2)^{(n-2)/2}$.
(ii) If $N \geq 3$ then $2^{N2^N} \geq 2^N! \geq 2^{(N-1)2^{N-2}}$.

Proof   (i) Since $n \geq r$ for each $n \geq r \geq 1$ we have

$$n^n = n \times n \times n \times n \times \ldots \times n \times n \times n$$

$$\geq n \times (n-1) \times (n-2) \times (n-3) \times \ldots \times 3 \times 2 \times 1 = n!.$$

Further there are at least $(n-2)/2$ integers r with $n \geq r \geq n/2$ and so, writing K for the least integer greater than or equal to $n/2$,

$$n! = n \times (n-1) \times (n-2) \times (n-3) \times \ldots \times 3 \times 2 \times 1$$

$$\geq n \times (n-1) \times (n-2) \times (n-3) \times \ldots \times k$$

$$\geq \frac{n}{2} \times \frac{n}{2} \times \frac{n}{2} \times \frac{n}{2} \times \ldots \times \frac{n}{2}$$

$$\geq \left(\frac{n}{2}\right)^{(n-2)/2}.$$

(ii) Setting $n = 2^N$ in (i) we get

$$2^{N2^N} = (2^N)^{2^N} = n^n \geq n!$$

and

$$n! \geq (n/2)^{(n-2)/2} = (2^{N-1})^{(2^N-2)/2} \geq (2^{N-1})^{2^{N-2}} = 2^{(N-1)2^{N-2}},$$

as stated.   ■

Combining this estimate with the two theorems we get the required comparison.

Theorem 11.4.12   *If $2^N \geq n \geq 2^{N-1}$ and $N \geq 4$, any method of sorting a set of n unequal numbers $x_1, x_2, \ldots, x_n$ requires at least $N2^N/16$ comparisons and the 'knock-out method' requires less than $N2^N$ comparisons.*

Proof   By Theorem 11.4.9, if we can sort the set using k comparisons, then

$$2^k \geq n!.$$

By Lemma 11.4.11 (ii)

$$n! \geq 2^{N-1}! \geq 2^{(N-2)2^{N-3}},$$

and so

$$k \geq (N-2)2^{N-3} \geq N2^{N-4} = N2^N/16.$$

The result for the 'knock-out method' is copied directly from Theorem 11.4.10. ∎

Thus no method of sorting can improve on the knock-out method by more than a factor of 16. (As the reader may remark, the factor 16 is a substantial overestimate but our result is spectacular enough as it stands.) Of course, the speed of a sorting algorithm does not only depend on the number of comparisons that the computer makes and, in practice, people use a closely related method called 'heap sort' which is easier to program and requires less 'administrative work' from the machine. (For more details and for a description of competing methods such as 'quicksort' see Chapter 5 of Knuth's marvellous *Art of Computer Programming*.) As a simple practical application, the program *MakeIndex* which I am using to index this book announces, with modest pride, that it required 11 036 comparisons to sort my 1030 entries.

It is difficult to think of a more satisfactory result than Theorem 11.4.12. We set out to find a fast method of sorting and we have ended up with a method that is provably the fastest (to within a constant factor). Unfortunately this kind of outcome is exceptional. Mathematicians have algorithms for many tasks but very few are known to be optimal. In general, we have an algorithm which, when applied to n pieces of data, is guaranteed to work in less than $f(n)$ steps and a proof that no algorithm can work in less than $g(n)$ steps, but $f(n)$ grows much faster than $g(n)$ as n increases. Can we narrow the gap by finding an algorithm with a smaller f or a lower bound with a larger g? Outsiders often ask what mathematicians will do when they run out of problems and are sceptical when told that new problems are forever turning up. Here is a new field, suggested and nurtured by the growth of electronic computing, with hundreds of new and difficult problems (some of whose solutions would be of great practical importance if they could be found) to keep mathematicians busy for decades to come. We shall return to this question in Section 16.1.

In fact, the situation is even more complicated than at first appears. Many algorithms for which the best provable number of steps is $f(n)$, say, turn out to work much faster in practice. Sometimes this is because the guaranteed running time represents the worst case and we can show that the average running time (for a suitable meaning of average) is less, sometimes there is a big gap between what we can prove about the algorithm and what is actually true and sometimes both reasons apply. Since the prospective user cannot wait a few years for mathematicians to sort the matter out, competing algorithms are usually tested on a collection of typical problems to see which runs faster. (Though, even here, there is the non-trivial difficulty of deciding what is a typical problem.)

Exercise 11.4.13 (A neater Stirling's estimate)   *The result of Lemma 11.4.11 can be improved by using Gauss's idea from Exercise 11.3.6.*

(i) Suppose $a > 0$. Sketch the function $f(x) = x(a - x)$. Show that $f(x) = a^2/4 - (x - a/2)^2$ and deduce, or prove otherwise, that, if $0 \le b \le a/2$, then

$$b(a - b) \le x(a - x) \le a^2/4$$

for all x with $b \le x \le a - b$.

(ii) By matching terms as follows

$$n! = 1n \times 2(n - 1) \times 3(n - 3) \times \ldots$$

and using (i) show that

$$(n/2)^n \ge n! \ge n^{n/2}.$$

Deduce, in particular, that

$$2^{(N-1)2^N} \ge 2^N! \ge 2^{N2^{N-1}}$$

for all $N \ge 1$.

Exercise 11.4.14 (A closer Stirling's estimate)   *We can obtain even better estimates for $n!$ by using calculus. (The reader will need to have done about two years of calculus for this question. British readers may be accustomed to the provincial habit of writing $\ln$ for what I call $\log$†.) Our method is illustrated by Figure 11.10.*

(i) Show that $\log r \le \log x$ for all x with $r \le x \le r + 1$ and deduce that

$$\log r \le \int_r^{r+1} \log x \, dx$$

for all integer r with $R \ge 1$.

(ii) By summing the results of (i) show that

$$\log n! \le \int_1^{n+1} \log x \, dx.$$

(iii) By integrating $\int_1^{n+1} \log x \, dx = \int_1^{n+1} 1 \log x \, dx$ by parts show that

$$\log n! \le (n + 1)\log(n + 1) - n,$$

†In the days when I was at school we used logarithms to the base 10 for calculation so we had to distinguish $\log_{10}$ from $\log_e$. Since then tables of logarithms to the base 10 have been relegated to the museum. (A reader points out that ln is used in *Concrete Mathematics* [74]. This just goes to show that nobody and no book is perfect.)

and so

$$n! \leq e^{-n}(n+1)^{n+1}.$$

(iv) By modifying the ideas above along the lines suggested by Figure 11.11, show that

$$n! \geq e^{-n+1}(n)^{n}.$$

(v) We have thus shown that $a_n \geq n! \geq b_n$ where $a_n = e^{-n}(n+1)^{n+1}$ and $b_n = e^{-n+1}n^n$. Using the approximation $(1+n^{-1})^n \approx e$ show that for large n

$$\frac{a_n}{b_n} \approx n$$

so that we have obtained very close upper and lower bounds for $n!$.

(vi) Can you do still better?

It is possible by refining the calculations above and using a further idea to obtain Stirling's formula

$$n! \approx 2\pi e^{-n}n^{n+1/2}.$$

We cannot prove it here but the reader may care to verify the excellence of the approximation for the largest n that her calculator can handle.

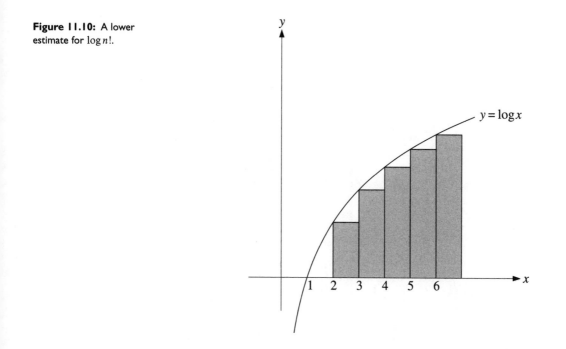

Figure 11.10: A lower estimate for $\log n!$.

11.5 A letter of Lord Chesterfield

The study of algorithms is part of the constant search for better ways
of doing things just like the search for ways of preventing cholera or
hunting submarines. But the task does not end there. Having found
a better way you must persuade others to adopt it. Learning how to
do this can only come from long and painful experience. (Experience
is an excellent teacher but her lessons are expensive.) Until the reader
has experienced the shock of having her well thought-out and carefully
argued proposal thrown out for the most fatuous of reasons, she will
not understand why I write this section or why, before any difficult
committee meeting, it is important to recite the two phrases: 'You can
only resign once' and 'Those who fight too long with dragons become
dragons themselves'†.

When a government committee was set up in 1978 to consider mathematics teaching in English and Welsh schools, their first step was to set
up a survey to find out the opinions and mathematical needs of a representative sample of adults. However, the survey ran into an unexpected
difficulty when many of those approached refused to be interviewed.

> Both direct and indirect approaches were tried, the word 'mathematics'
> was replaced by 'arithmetic' or 'everyday use of numbers', but it was clear

†He who fights with
monsters must take care
lest he thereby become a
monster. And if you gaze
too long into an abyss, the
abyss gazes also into you.

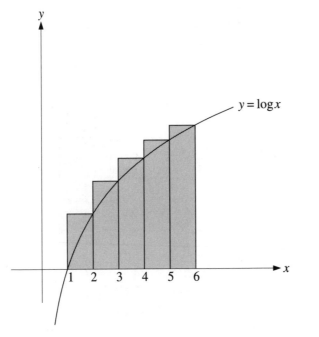

Figure 11.11: An upper
estimate for $\log n!$.

that the reason for people's refusal to be interviewed was simply that the subject was mathematics .... Several personal contacts pursued by the enquiry officer were also adamant in their refusals. Evidently there were some painful associations which they feared might be uncovered. This apparently widespread perception amongst adults of mathematics as a daunting subject pervaded a great deal of the sample selection; half of the people approached as being appropriate for inclusion in the sample declined to take part.

Even among those who agreed to take part,

> The extent to which the need to undertake even an apparently simple and straightforward piece of mathematics could induce feelings of anxiety, helplessness, fear and even guilt in some of those interviewed was, perhaps, the most striking feature of the study.

There did not appear to be any connection between mathematical competence and occupational group. However,

> The feelings of guilt to which we referred earlier appeared to be especially marked among those whose academic qualifications were high and who, in consequence of this, felt they ought to have a confident understanding of mathematics, even though this was not the case.

In view of this, it is clear that mathematicians should not refer to mathematics in advocating a given course of action†. By itself this is not a grave disadvantage. Darwin's *On the Origin of Species* is a marvellous example of sustained book-length argument without any recourse to mathematics. Unfortunately, most people are also unwilling to follow sustained argument‡.

Often there is good reason to distrust the 'real life' application of the long chains of reasoning beloved of mathematicians. In real life it is not usually true that A always produces the effect B. The most we can say is that A usually produces the effect B. The mathematical argument:

A_1 always produces the effect A_2, A_2 always produces the effect A_3, ... , A_{n-1} always produces the effect A_n, so A_1 always produces the effect A_n,

is valid however large n is. The political argument:

A_1 usually produces the effect A_2, A_2 usually produces the effect A_3, ... , A_{n-1} usually produces the effect A_n, so A_1 usually produces the effect A_n,

begins to look distinctly shaky even for $n = 3$.

Mathematics proceeds from the simple to the complex, but real life is rarely simple. Goethe says that 'Mathematicians are like Frenchmen: whatever you say to them they translate into their own language and

†Since many people impute infallibility, impartiality and omniscience to computers, some reference to 'computer simulation' or even a computer display with lots of arcade-style action might be helpful — but I wish to discuss the principles of argument not those of advertising.

‡In Anglo-Saxon countries, the word 'intellectual' which the dictionary defines as 'a person possessing, or supposed to possess, superior powers of intellect; given to pursuits that exercise the intellect' is, in fact, an insult, even among intellectuals. In British society, clever men waste a great deal of time pretending to be stupid. (In France the situation is reversed.)

forthwith it is something entirely different'†. A related sentiment is expressed in the engineer's joke about the hot-air balloonist caught up in a great storm and blown miles off course. For a moment the wind slackens and through a gap in the clouds she sees a man walking his dog. 'Where am I?' she shouts down. The man scratches his head, reflects for a time and finally shouts back 'In a balloon.' 'Are you a mathematician?' she yells. 'As a matter of fact I am, how did you guess?' And, as the clouds come down again and the wind carries the balloon away, the balloonist has just time to yell, 'Because your answer took a long time to arrive and when it came was perfectly accurate and perfectly useless.'

There are no fixed rules for persuading people; what you say or write must depend on who they are, who you are (some bodies can be moved by witty speeches, but if you are not witty you should not try) and what you wish to get done. However, here are some suggestions for you to adapt in the light of your own experience.

1   Do not try to argue for two things at the same time; your audience will become confused as to which argument supports which course of action. Decide what is your most important object and argue for it alone.

2   Do not offer your audience a series of possible actions. Most bodies are incapable of deciding between three alternatives. (This is not as cynical a remark as it may seem. There exists a substantial body of mathematics showing that decisions genuinely become harder to make as the number of alternatives increases.) Choose your preferred action and argue for it.

3   State what the problem is. Explain what your proposal is. Explain why your proposal solves the problem. Restate your proposal. Stop.

4   You should answer any objections to your proposals when they are made but not before. Again this may seem cynical advice (and there are times when it is your duty to put both sides of the case), but the natural mode of discussion is for *A* to propose, *B* to oppose, *A* to answer *B*'s objections and so on. If you attempt to play the parts of *A* and *B* simultaneously you will probably only confuse the issue.

5   Bidders at auctions are sometimes carried away by the excitement into paying much more than their purchase is worth. The same can happen in argument. Do not invest more hope or more energy in an issue than it is worth. Do not treat technical issues as moral ones (or moral issues as technical ones).

†*Question* How do you tell that you are in the hands of the Mathematical Mafia? *Answer* They make you an offer you can't understand.

6   Remember that honest men may disagree. Remember that few people change their opinions as the immediate result of discussion, but that if your case is good and well argued it may go on working in their minds for a long time. How many great ideas have begun as the lunacy of one, become the heresy of a few, the opinion of a minority, the view of the majority and finally the common sense of all?

It is well known that the year (more precisely the 'solar year') is not a simple multiple of days (more precisely the 'solar day') in length. Many hundreds of years of observations gave

$$1 \text{ year} = 365.2422 \text{ days.}$$

Even this statement must be qualified in various ways (for example, by replacing 'year' by 'average year' and 'day' by 'average day'), but we shall accept it as a basis for discussion. In the Julian calendar, introduced by Julius Caesar, the year is taken to have 365 days except that all years divisible by 4 have 366 days (leap years).

Exercise 11.5.1   *Show that the Julian calendar will have one day too many roughly every 128 years. (Thus after 128 years an astronomical event which should occur near midnight on 2 January will occur near midnight on 1 January.)*

†The dates of the main Christian festivals were not fixed for the first three centuries of the Christian era so the 'zero' with respect to which discrepancy was measured was displaced correspondingly.

By 1588 the discrepancy had grown to 10 days† and Pope Gregory XIII introduced a new calendar. Ten days were omitted altogether (so that Thursday, 4 October 1582 was followed by Friday, 15 October 1582). The new Gregorian calendar differed from the Julian calendar in that years divisible by 100 (centennial years) were to be leap years only if divisible by 400.

Exercise 11.5.2   *Show that the Gregorian calendar will have one day too many roughly every 3333 years.*

The Gregorian system was adopted all over Europe but not in Britain, with the result that, by 1750, the day which in France or the Netherlands was called 16 August was called 5 August in Britain. The following is the first half of a letter, dated London, March 18, 1751, from Lord Chesterfield to his son:

My Dear Friend,

I acquainted you in a former letter that I had brought a bill into the House of Lords, for correcting and reforming our present calendar which is the Julian, and for adopting the Gregorian. I will now give you a more particular account of that affair, from which reflections will naturally occur to you that I hope may be useful, and which I fear you have not made. It was notorious that the Julian calendar was erroneous, and had overcharged the solar year with eleven days. Pope Gregory XIII corrected this error; his reformed calendar was immediately received by all the Catholic Powers of Europe, and afterwards adopted by all the Protestant ones, except Russia, Sweden, and England. It was not, in my opinion, very honourable for England to remain in a gross and avowed error, especially in such company; the inconvenience of it was likewise felt by all those who had foreign correspondences, whether political or mercantile. I determined, therefore, to attempt the reformation; I consulted the best lawyers, and the most skillful astronomers, and we cooked up a bill for that purpose. But then my difficulty began; I was to bring in this bill, which was necessarily composed of law jargon and astronomical calculations, to both of which I am an utter stranger. However, it was absolutely necessary to make the House of Lords think that I knew something of the matter, and also to make them believe that they knew something of it themselves, which they did not. For my own part, I should just as soon have talked Celtic or Sclavonian to them as astronomy, and they would have understood me full as well; so I resolved to do better than speak to the purpose, and to please instead of informing them. I gave them, therefore, only an historical account of calendars, from the Egyptian down to the Gregorian, amusing them now and then with little episodes; but I was particularly attentive to the choice of my words, to the harmony and roundness of my periods, to my elocution, to my action. This succeeded and ever will succeed; they thought I informed, because I pleased them; and many of them said, that I had made the whole very clear to them, when, God knows, I had not even attempted it. Lord Macclesfield, who had the greatest share in forming the bill, and who is one of the greatest mathematicians and astronomers in Europe, spoke afterwards with infinite knowledge, and all the clearness that so intricate a matter would admit of; but as his words, his periods, and his utterance were not near so good as mine, the preference was most unanimously, though most unjustly, given to me.

[From Lord Chesterfield's *Letters to his Son*]

Exercise 11.5.3 *In 1923 the Soviet Union adopted a new 'Soviet' calendar which follows the Julian except that a centennial year is a leap year only if it is $n \times 100$ where n leaves a remainder 2 or 6 when divided by 9.*

(i) Show that the Soviet calendar will have one day too many roughly every 45 000 years.

(ii) What is the first year when the Soviet and Gregorian calendars will differ?

The following problem is taken from early editions of Rouse Ball's *Mathematical Recreations.* (In preparing this section I used [212]. If the reader wishes to become a wizard at finding the day of the week from the date then the quickest method I know is due to Conway and is given in volume II of *Winning Ways* [14].)

Exercise 11.5.4 *(i) Show that (for the Gregorian calendar) the first or last day of every alternate century must be a Monday.*

(ii) Show that (for the Gregorian calendar) the thirteenth day of each month is more likely to be a Friday than to be any other day of the week. (Since this is essentially just a hard slog you may wish to use a computer.)

CHAPTER 12

Deeper matters

12.1 How safe?

In the early days of railways, there was no means of communicating between different parts of the line and trains were run on a time-interval basis. After the passage of a train, a danger signal would be shown for five minutes (say) to prevent another train following too closely. For the next five minutes a caution signal was displayed to indicate that any following train should slow down and finally a clear signal would be shown. If a train had to stop between signalling points, the guard was supposed to run back to warn the following train. (The book from which this account is taken remarks that 'accidents were not as frequent as might be expected'.)

Gradually, as experience was gained and technology permitted, improvements were made. Points and signals were mechanically interlocked so that the signals could only show clear if the points were correctly set. The invention of the electric telegraph allowed signalmen to communicate and led to the replacement of time-interval by space-interval working under the block system. The line was divided into blocks and when a train passed the signalman at the beginning of a block, he placed his signals at danger until informed by the signalman at the other end that the train had left the block. Since only one train can be in any one block at any one time, collisions are, in some sense, 'impossible'.

Actually, the system as described has various weaknesses

1   The signals may fail to operate. In 1876 a signal frozen in the 'safe position' caused a major accident. Up to this time, as we have described above, signals were normally set at 'safe', only being placed at 'danger' when there was a train in the succeeding block. Since then, signals have been set at 'danger', only being set at 'safe'

to permit a train to pass (or some more complex form of 'fail safe' device is used).

2   The signalman at the end of a block may wrongly record a train as leaving his block. To prevent this, an interlocking device was introduced which only allowed him to record clear when the train operated a treadle at the end of the section.

3   The signalman at the beginning of a block may wrongly allow a train to enter the block even though the previous train has not been signalled as having left the block. This was prevented by interlocking signals at the beginning and the end of the section.

4   The train driver may not observe the signals. This requires more advanced technology, but nowadays (at least on major lines) a train passing a danger signal is automatically halted.

  Have we now, by eliminating all scope for signal failure and human error, produced a perfectly safe railway? No, because by abstracting the notions of signal, section and train we have hidden one of the most common causes of accidents.

5   A train is not a single object. It may shed carriages. The fact that part of a train has left a section is no guarantee that all of it has. This problem was overcome by devices for counting the number of axles entering and leaving a section, or by track-circuiting in which the presence of a train on a given section is detected electrically.

Even now our problems are not over. If a signal fails, then since it 'fails safe', it will set permanently at danger. If no further steps are taken, all traffic on the line will come to a stop until the signal is repaired. It is, of course, true that a railway with no moving trains will have no collisions, but such a railway is unacceptable to its users. There must therefore be provision to override safety features and this introduces new risks ... †.

†The plans for the San Francisco metro BART called for the trains to be fully automated but to carry a conductor in case of emergencies. It was pointed out that the kind of person who would be happy to ride back and forth without doing anything for nine years would be precisely the person least capable of coping with an emergency in the tenth.

Today we must deal with problems which are much more complicated than keeping two trains off the same section of track. Air traffic control involves many objects moving in three dimensional space. Atomic power stations require decisions to be made faster than mere human beings can think. Fortunately we have computers capable of dealing with masses of data practically instantaneously (and public relations men ready to reassure us that when trains collide, ferries sink, planes crash into mountains and atomic power stations release radioactivity, these are 'one-off' accidents which will never happen again). It is clear that, without computers, many activities of modern life would be either prohibitively expensive or prohibitively dangerous. It is less immediately clear why computers cannot make these activities perfectly safe rather than just acceptably safe.

Suppose we have a computer program to control some complex process. How do we know that it will always do the right thing under every possible circumstance? One way is to run through all the possibilities and check that the program always gives the right answer. There are two problems with this. One is that there may be too many possibilities to check. The other, which is far more important, is that we may not know all the possibilities. (Remember how we neglected the possibility that a train might divide.) It is significant that many railway safety routines are named after the accident which showed they were needed.

A cautionary tale (among many such tales) is provided by the computer-aided dispatch system for the London Ambulance Service. In 1991, the Conservative Government, anxious for reelection, issued guidelines under which ambulances should reach 95% of emergencies within 15 minutes of being called. The LAS management decided on a deadline of 14 months to get a computer-aided dispatch scheme up and running†. Such a short time-scale left too little time for testing (though trials did reveal that the software ignored every 53rd vehicle in the 776-strong fleet) and staff training. The ambulance crews felt, correctly, that the new system was being imposed from above without consultation. Under the old system they were in direct voice contact with the dispatchers and there was scope for local initiative. The new system treated them as cogs in a machine (typically, the management's reaction to the discontent among the crews was to worry about possible sabotage). Low morale meant that once things started to go wrong any confidence in the new system among the ambulance crews would vanish and this attitude would feed back to the control room. When the manual system was replaced by a semi-computerised system, the number of emergencies reached within 15 minutes dropped from 65% to 30%. Undeterred, the management moved over to a fully computerised system without paper records.

Immediately, things began to go wrong. Whenever the system lost radio contact with an ambulance it generated an 'error message' to warn the operator. If the operator was too busy to answer, the system generated an error message about the error message, and then an error message about the new error message and so on until the operator could no longer work because the screen was full of them. In addition, the machine would dispatch a new ambulance to replace the one it had lost. As the system dispatched more and more ambulances, it also spent more and more time deciding which of its remaining ambulances

† All countries have technological disasters, but each country adds its own national ingredient — obsessive secrecy in the ex-USSR, bribery in Italy, low technology failures in high technology projects in the USA. In Britain we have the doctrine that 'A manager's job is to manage'.

was closest to each reported incident. The staff knew that things were going seriously wrong, but without paper records they were unable to find out what was happening. As the day continued, the strain on the system was further increased by a rising volume of calls to the control room from people worried by the non-arrival of ambulances. Six hours after the change-over, the control room was taking 10 minutes to answer incoming phone calls. During the first day fewer than 20% of emergencies were reached within 15 minutes and the proportion fell again on the next day. (Figure 12.1 taken from [223] shows the process in graphic detail.)

At this point the management decided to return to the semi-computerised system. There was some improvement, but a week later the system crashed completely. A piece of program left over from some earlier 'patching' of the system ensured that every time an ambulance was dispatched, a small piece of memory was used and then not freed. After three weeks the bug had used up all the memory. (The fallback procedures failed because they had been designed for a paperless system; they were, however, untested even for that.) At this point, the 1.5 million pound project was abandoned and the ambulance dispatch service returned to its previous methods and efficiency. One member of the resulting enquiry estimated that it would take five years to introduce the system the management had envisaged to allow for proper testing and staff training.

The lesson from successful and unsuccessful schemes† alike is the need for a long testing period, measured in years rather than months, but here we run into an interesting, though not fatal, problem. The lifetime of a computer system is now about five years, after which it is so behind the times that the manufacturer loses all interest in it. This means that the advanced machines the system was designed to use will be obsolete by the time that it is running smoothly‡. Computer languages have a longer lifetime — perhaps 10 or 15 years. But most major works of mankind, from atomic power stations to space shuttles and from air traffic systems to warships, take years to design and build, and are then expected to last at least 20 years. Towards the end of their useful life, they will thus be controlled by computers out of a science museum, running programs written in a language no one uses, by people who have long since departed.

†At the end of 1994 the *Scientific American* reported that 'Studies have shown that for every six new large-scale software systems put into operation two others are cancelled. The average software development overshoots its schedule by half; larger projects generally do worse. And some three quarters of all large systems are "operating failures" that either do not function as intended or are not used at all.'

‡There is a cottage industry building obsolete computers for precisely these purposes.

Exercise 12.1.1 *Ask one of your friends to give you a long computer program they have written without telling you what it is supposed to do. Try to work out, from the program alone, what its purpose is and how it achieves it.*

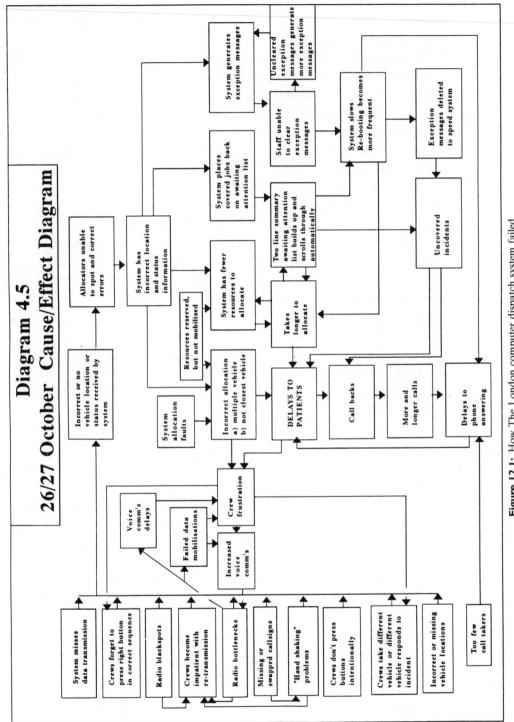

Figure 12.1: How The London computer dispatch system failed.

One suggested way of improving matters is to look at how an aeroplane is designed. The designer does not design the engines or the instruments but takes them as given with certain guaranteed specifications. It is up to the engine and instrument builders to make sure that the parts for which they are responsible meet the specifications. In the same way, it is suggested, large computer programs should be assembled from smaller components, each of which is guaranteed to work in a specified way. (Note that this reflects the railway safety systems described above. One system ensures that the exit signalman cannot report a train as leaving the block unless it actually has. Another system ensures that the entry signalman cannot reset the entry signal from danger unless the exit signalman has reported the block clear and a third system ensures that a train cannot pass a signal at danger.)

We have already produced such guaranteed components. Our algorithms for finding maximum flows, greatest common divisors and so on come with proofs (that is guarantees) that they will always work. (Though, of course, we must still check that the program implements the algorithm.) Unfortunately, any large program will contain many such components and, as the reader will probably agree, it is likely to be quite a long process to prove that any particular component is correct. The obvious way out is to check correctness mechanically, that is to write a program to check the correctness of each component (algorithm). (We would have to check the correctness of the checking program (algorithm), but this would be a once-and-for-all effort.) Does there exist such a universal checking program (algorithm)? The negative answer was given by a young Cambridge mathematician, Alan Turing, in 1937. (This does not mean that the ideas outlined in the two previous paragraphs are useless, but it does mean that they cannot provide a universal panacea.)

The statement that an algorithm (program) is correct is a complex one to analyse, so Turing concentrated on one aspect of correctness. In order to be useful an algorithm (program) must terminate so Turing asked: does there exist a terminating algorithm to check that any given algorithm terminates? In Section 12.3 I shall sketch a proof that the answer is no. Before embarking on the proof I shall outline some of its background.

The piece of background we consider in the remainder of this section concerns how this book is produced. Few mathematics texts have been as influential as the three volumes of Knuth's *The Art of Computer Programming*. (The reader will have no difficulty in detecting its influence on this book.) Knuth's plans called for a seven-volume work but, in the middle of the project, Knuth decided to write a program to typeset his own books (and another to design his own fonts). The result was one of the largest programs ever successfully completed†, a glorious revolution in mathematical publishing and the non-appearance (at least up to now) of any further volumes.

Knuth's program is called TEX (pronounced 'tech' as in technology) and this book is written in a version called LATEX (pronounced 'lay-tech'). I type out on my home computer something like:

```
If $\alpha>0$ then $x^{\alpha}$ grows faster than $\log x$
as $x\rightarrow\infty$.
```

To translate this, note that $ means 'mathematical formula starting or finishing', \ means 'instruction follows', ^ means 'superscript enclosed in braces { and } follows'. The instruction alpha means 'print small Greek alpha' and so on†. (You will see that what I have written is a set of instructions, i.e. a program.) The end result I hope for should look something like:

If $\alpha > 0$ then x^α grows faster than $\log x$ as $x \to \infty$.

Exercise 12.1.2 *Explain why if I type*

```
If $alpha>0$ then $x^{\alpha}$ grows faster than $\log x$
as $x\rightarrow\infty$.
```

I get

If $alpha > 0$ then x^α grows faster than $\log x$ as $x \to \infty$.

The TEX program ignores spaces in mathematical formulae. Explain why if I type

```
If $\alpha>0 then x^{\alpha}$ grows faster than $log x$
as $x\rightarrow\infty$.
```

I get

If $\alpha > 0 then x^\alpha$ grows faster than $log x$ as $x \to \infty$.

Since my home computer cannot process LATEX, I transfer what I have written to disk. What is written on the disk looks like a meaningless jumble, since the computer has converted my text to ASCII in which, essentially, each possible key stroke of my key pad is assigned a number between 0 and 127. (Thus the disk carries a very long number which can be translated back into a series of key strokes.) I transfer the contents of the disk to my office machine. A program in that machine then converts my program (which was written in LATEX) into a program

†When the Cambridge Mathematics Departments chose their first mathematical word processing system, it was assumed that Knuth would never finish the job.

†Any TEXnician who reads this should remember the Master's words. 'The author feels that [the] technique of deliberate lying ... actually make[s] it easier ... to learn the ideas.'

written in plain TEX. Knuth's program now takes over and rewrites and extends my program adding instructions on spacing, line-breaking, page-breaking and all the other things necessary to produce a printed page. At the end, it outputs a detailed set of instructions to the laser printer telling it where to put every dot on the page.

Knuth gives the full details of his program in a 524-page book (Volume B of *Computers and Typesetting†*). Knuth wrote his program in a computer language called WEB. A computer program translated it into another language called PASCAL. When Knuth's program is installed on a new machine, another program rewrites the PASCAL program as a machine code program which is the one actually obeyed by the machine.

I give these details not primarily to sing the glories of TEX, glorious though it is. (Most innovations begin among graduate students and rise through the mathematical community with those graduates, only arriving at the professoriate 20 years later when the graduates become professors. TEX and its offspring have been almost universally adopted within ten years‡.) Instead, I wish to emphasise how much of computing involves programs acting on other programs. If I had to explain a computer to mathematicians who had never seen one, I suspect that one of the hardest tasks would be to explain how this is done. 'How,' they would ask, 'does the computer know what is data and what is program?' It is not hard to propose mechanisms for this, but it is only easy to convince oneself that they can work if you have actually seen it done.

† When I was younger, I believed that authors actually read through the books they cited. This is definitely not the case here. However, if you want to see the highest standards of programming and documentation for long programs in action, the book is well worth a look.

‡ The interested reader will find a few more remarks at the end of Appendix A1.1.

12.2 The problems of infinity

In Sections 5.1 and 5.2. I quoted from Galileo's *Dialogue Concerning Two New Sciences*. The *Dialogue* deals mainly with what we now call dynamics and material science, but in the following passage the friends digress to discuss the problems associated with the notion of infinity.

SIMPLICIO Here a difficulty presents itself which appears to me insoluble. Since it is clear that we may have one line greater than another, each containing an infinite number of points, we are forced to admit that, within one and the same class, we may have something greater than infinity, because the infinity of points in the long line is greater than the infinity of points on the short line. This assigning to an infinite quantity a value greater than infinity is beyond my comprehension.

SALVIATI This is one of the difficulties which arise when we attempt, with our finite minds, to discuss the infinite, assigning to it those properties which we give to the finite and limited: but this I think is wrong, for we cannot speak of infinite quantities as being the one greater or less than or equal to another. To prove this I have in mind an argument which, for the sake of clearness, I shall put in the form of questions to Simplicio who raised the difficulty.

I take it for granted that you know which of the numbers are squares and which are not.

SIMPLICIO I am quite aware that a squared number is one which results from the multiplication of another number by itself: thus 4, 9, etc., are squared numbers which come from multiplying 2, 3, etc., by themselves.

SALVIATI Very well; and you also know that just as the products are called squares, so the factors are called sides or roots; while on the other hand those numbers which do not consist of two equal factors are not squares. Therefore, if I assert that all numbers, including both squares and non-squares, are more than the squares alone, I shall speak the truth, shall I not?

SIMPLICIO Most certainly.

SALVIATI If I should ask further how many squares there are, one might reply truly that there are as many as the corresponding number of roots, since every square has its own root and every root its own square, while no square has more than one root and no root more than one square.

SIMPLICIO Precisely so.

SALVIATI But if I enquire how many roots there are, it cannot be denied that there are as many as there are numbers because every number is the root of some square. This being granted, we must say that there are as many squares as there are numbers because they are just as numerous as their roots, and all the numbers are roots. Yet at the outset we said that there are many more numbers than squares, since the larger proportion of them are not squares. Not only so, but the proportionate number of squares diminishes as we pass to larger numbers. Thus up to 100 we have 10 squares, that is the squares constitute 1/10 part of all the numbers; up to 10 000 we find only 1/100 part to be squares; and up to a million only 1/1000 part: on the other hand in an infinite number, if [some]one could conceive of such a thing, he would be forced to admit that there are as many squares as there are numbers taken together.

SAGREDO What then must we conclude under these circumstances?

SALVIATI So far as I can see we can only infer that the totality of all numbers is infinite, and that the number of squares is infinite, and that the number of their roots is infinite; neither is the number of squares less than the totality of all numbers, nor the latter greater than the former: and finally, the attributes 'equal', 'greater' and 'less' are not applicable to infinite, but only to finite, quantities. When, therefore, Simplicio introduces several lines of different lengths and asks me how it is possible that the longer ones do not contain more points than the shorter, I answer him that one line does not contain more or less or just as many points as another but that each line contains an infinite number. ...

In modern terms, Galileo considers the set $N = \{1, 2, 3, \dots\}$ of strictly positive integers and the subset $S = \{1, 4, 9, \dots\}$ of the squares of positive integers. He observes that there is a correspondence

$$n \leftrightarrow n^2$$

between N and S such that each member of N corresponds to exactly one member of S and each member of S corresponds to exactly one member of N. We call such a correspondence a 'one-to-one' correspondence†. If we take

$$[0, 1] = \{x \in \mathbb{R} : 0 \le x \le 1\}$$

to be the set of real numbers between 0 and 1 inclusive and

$$[0, 2] = \{x \in \mathbb{R} : 0 \le x \le 2\}$$

to be the set of real numbers between 0 and 2 inclusive, then

$$x \leftrightarrow 2x$$

gives a one-to-one correspondence between $[0, 1]$ and $[0, 2]$.

†In more advanced work what we have called a 'one-to-one' correspondence is called a 'bijective' correspondence. This is because the traditional terminology involves distinguishing 'one-to-one' (bijective) and 'one-one' (injective) correspondences and no-one can remember which is which.

Exercise 12.2.1 *If we take*

$$[\alpha, \beta] = \{x \in \mathbb{R} : \alpha \le x \le \beta\}$$

and $a < b$, $c < d$ show that there is a one-to-one correspondence between $[a, b]$ and $[c, d]$.

Thus we have established a one-to-one correspondence between 'lines of different lengths'.

Galileo's main concern in this dialogue is to defend the ideas of what we would now call the calculus and he is quite content to have shown the problems associated with the traditional views of infinity. He has, however, left one question unanswered: Can any two infinite sets be put into one-to-one correspondence? This question was raised and answered by Cantor.

Theorem 12.2.2 (Cantor) *There is no one-to-one correspondence between the positive integers and the real numbers.*

Proof Suppose that there was a one-to-one correspondence between the positive integers and the real numbers. Let $x(n)$ be the real number corresponding to the positive integer n. We know that we can write

$$x(n) = a(n) + y(n)$$

where $a(n)$ is an integer and $0 \le y(n) < 1$ and that we can write $y(n)$ as a decimal

$$y(n) = 0.y_1(n)y_2(n)y_3(n)\ldots.$$

(Formally

$$y(n) = \frac{y_1(n)}{10} + \frac{y_2(n)}{10^2} + \frac{y_3(n)}{10^3} \cdots,$$

or more briefly

$$y(n) = \sum_{r=1}^{\infty} \frac{y_r(n)}{10^r},$$

where y_r is an integer with $0 \le y_r \le 9$; but the formal description tells us nothing we do not already know.)

We now write down a real number x which will turn out *not to be in the list* $x(1), x(2), x(3), \ldots$. We define the nth decimal place x_n of x by

$$x_n = y_n + 3 \quad \text{if } 0 \le y_n(n) \le 4,$$
$$x_n = y_n - 3 \quad \text{if } 5 \le y_n(n) \le 9,$$

and set

$$x = 0.x_1x_2x_3\ldots,$$

or, more formally,

$$x = \sum_{r=1}^{\infty} \frac{x_r}{10^r}.$$

Consider the nth number x_n on our list; by definition

$$|x_n - y_n(n)| = 3$$

and so x and $x_n = a(n) + y(n)$ differ by 3 in the nth decimal place. Thus $x_n \ne x$ and, as we claimed at the beginning of the paragraph, x is not in the list $x(1), x(2), x(3), \ldots$. Since x is missing, our list fails to give a one-to-one correspondence between the positive integers and the real numbers.

The argument works for any putative one-to-one list of the real numbers so Cantor's theorem follows. ∎

Exercise 12.2.3 *Suppose the first few x_n in the proof of Theorem 12.2.2 are given by*

$$x_1 = 12.5346723\ldots$$
$$x_2 = 0.5000000\ldots$$
$$x_3 = -5.350000\ldots$$
$$x_4 = 3.879989\ldots$$
$$x_5 = 8.000000\ldots$$

Find the first few decimal places of x.

Cantor's original proof of his theorem was different, but depended on the same *diagonal argument* (so called because we change the nth 'diagonal element' $y_n(n)$ to obtain the nth 'element' x_n for x). Cantor's diagonal argument was an entirely new method of mathematical argument. Many mathematicians have proved important theorems — very few have introduced new modes of argument into mathematics. In the next chapter we shall give another spectacular example of its use.

Exercise 12.2.4 *Our proof of Cantor's theorem contains the following statement ' ... x and $x_n = a(n) + y(n)$ differ by 3 in the nth decimal place. Thus $x_n \neq x \ldots$'. A particularly cautious student might wish to see a proof of this. (If you are not ultra-cautious, do not bother to read further.)*

(i) Recall that $a(n)$ is an integer and

$$y(n) = \sum_{r=1}^{\infty} \frac{y_r(n)}{10^r}, \quad x = \sum_{r=1}^{\infty} \frac{x_r}{10^r}.$$

Show that

$$Y_n = 10^n \left(a(n) + \sum_{r=1}^{n} \frac{y_r(n)}{10^r} \right) \text{ and } X_n = 10^n \left(\sum_{r=1}^{n} \frac{x_r}{10^r} \right)$$

are integers and that

$$|X_n - Y_n| \geq 3.$$

(ii) If we replaced the condition 'x and $x_n = a(n) + y(n)$ differ by 3 in the nth decimal place' by the condition 'x and $x_n = a(n) + y(n)$ differ by 8 in the nth decimal place' would it follow that

$$|X_n - Y_n| \geq 8?$$

Why?

(iii) Returning to the argument of (i) set

$$\mathcal{Y}(n) = 10^n \left(\sum_{r=n+1}^{\infty} \frac{y_r(n)}{10^r} \right) \text{ and } \mathcal{X}(n) = 10^n \left(\sum_{r=n+1}^{\infty} \frac{x_r}{10^r} \right).$$

Show that

$$0 \le \mathcal{Y}(n) \le 1 \text{ and } 0 \le \mathcal{X}(n) \le 1$$

and deduce, using (i), that

$$|(X_n + \mathcal{X}(n)) - (Y_n + \mathcal{Y}(n))| \ge 2.$$

Hence show that

$$|x - x_n| \ge 2 \times 10^{-n}$$

and so $x \ne x_n$ as required.

In order to prepare the way for the arguments of the next section, let me give another example of a diagonal argument. Suppose I have a computer and a suite of programs P_1, P_2, P_3, ... such that when I apply program P_j to the integer k, the result $P_j(k)$ is always 0 or 1 [$1 \le j,k$]. Then I claim that there is a sequence $P(1)$, $P(2)$, $P(3)$, ... consisting of zeros and ones which is not the same as the sequence $P_j(1)$, $P_j(2)$, $P_j(3)$, ... for any j. (Thus there is a sequence of zeros and ones which is not produced by any of our programs.) To illustrate the proof of this consider Table 12.1 showing the first few terms of the first few sequences $P_j(k)$.

In order to create our new sequence $P(1)$, $P(2)$, $P(3)$, ... we systematically alter the diagonal elements (underlined in the diagram) by setting

$$P(k) = 1 - P_k(k)$$

(that is $P(k) = 1$ if $P_k(k) = 0$ and $P(k) = 0$ if $P_k(k) = 1$). Since the sequence $P(1)$, $P(2)$, $P(3)$, ... and the sequence $P_j(1)$, $P_j(2)$, $P_j(3)$, ... differ in the jth place (that is $P(j) \ne P_j(j)$), they are different sequences for each $j \ge 1$, just as we required.

Table 12.1. *Machine output for a suite of programs.*

$k =$	1	2	3	4	5	...
P_1	1	0	1	0	1	...
P_2	1	1	1	1	1	...
P_3	1	0	0	0	1	...
P_4	0	0	1	1	1	...
P_5	0	1	0	1	0	...

12.3 Turing's theorem

†It is reprinted in various anthologies and in the collected edition of Turing's work.

When Turing wrote his paper bearing the rather off-putting title *On computable numbers with an application to the Entscheidungsproblem* † in 1937, there were no electronic computers. He therefore begins by describing how such a machine might work. His model is a mathematician with a notebook (since the mathematician can always buy more paper, the notebook can be made as large as needed). Each page is either empty or has one of a finite number of symbols on it (can you distinguish an infinite variety of symbols?) and the mathematician is in one of a finite number of states of mind. (You may object that a human being can have an infinite number of states of mind, but you must surely admit that a machine can have only a finite number of states.) According to her state of mind and what is written on the page, the mathematician does some or all of the following:

1   She erases what is written on the page and replaces it with one of a finite collection of possible symbols (or leaves it empty).
2   She changes her mind to one of its other finitely many possible states.
3   She moves n pages forward or back where n is an integer bounded by some fixed N. (You cannot decide to move a million pages at a time.)
4   She starts off with certain symbols written on certain pages (the program and data) and may (if the program does what is intended) halt after a time with certain other symbols written on certain other pages (the answer).

Part of the paper is devoted to arguing that this is indeed a realistic model of what a mathematician does. (This part of the argument is well summarised near the end of Chapter II of Hodge's splendid biography of Turing.)

It is not hard to convince oneself that particular computations can be done provided our machine has enough states and can print enough symbols. Naïvely we might suppose that ever more complicated computations would require machines ever richer in states and with ever-growing repertoires of symbols. In the second part of his argument Turing shows that this is not the case. He constructs what is now called a *universal Turing machine* and shows that any computation performed by some Turing machine can be performed by a universal Turing machine.

My desk-top computer will run programs written for it, but not programs written for the incompatible machines produced by IBM.

However, I can buy a program which enables my machine to simulate (pretend to be) an IBM machine and so run IBM programs. In the same way, though at a time when IBM was just a tabulator manufacturer, Turing showed how his universal Turing machine could simulate any given Turing machine and so perform any computation that the latter could perform. Nowadays, this conclusion strikes us as obvious. My desk-top can do anything a super-computer can do; it just takes longer to do it. The super-super-computer, which is now just a twinkle in its designer's eye, may be 10 or a 100 times faster than today's super-computer but that will be all. Electronic computers are all, we think, universal machines differing only in speed†.

Turing devotes 25 pages to the arguments described above and less than 2 pages to his diagonal argument which he clearly and correctly considered would be the easiest part of his paper for his readers. I shall now try to give a version of his argument, arguing in terms of electronic computers rather than Turing machines.

I program my computer by typing in at a keyboard. Since there are only a limited number M, say, of key strokes, I can designate each one by a unique integer m with $1 \leq m \leq M$. If my program is typed in by pressing keys numbered $a_0, a_1, a_2, \ldots, a_n$, then I can represent my program by the single integer

$$\sum_{r=0}^{n} a_r N^{n-r},$$

(where N is any previously chosen integer with $N \geq M+1$). Table 12.2 shows one of the 'keyboards' (actually 'font tables') used in printing this book. (The entries are numbered from 0 to 127 rather than, as we have done, from 1 to 128. If the reader wonders about the effect of symbols like 11 and 12 she will find that the question answers itself.)

Exercise 12.3.1   *Table 12.3 shows Knuth's typewriter font. Here $\sqcup$ (symbol number 32) represents a space. Suppose that we ignore the character Γ, so that the remaining symbols are numbered 1 to 127, and suppose that we represent the succession of key strokes $a_0, a_1, a_2, \ldots, a_n$ by the single integer*

$$\sum_{r=0}^{n} a_r 1000^{n-r}.$$

(i) What is represented by the number

$$87\,101\,108\,108\,032\,100\,111\,110\,101\,033?$$

† It can be proved that these *obvious* statements are also *true*. A desk-top computer with an infinite supply of disks and someone to feed it those disks on command is a universal machine in the sense of Turing. However, the passage of time which has made the fact obvious has not made the proof trivial.

(ii) Find the number representing

> `'* Solve 2x=4.'`

(iii) Show that any program that can be written by a given typewriter can be written using a typewriter with only two keys.
[Hint: Think binary or think Morse.]

Up to now, I have been a little coy as to what a program actually is. The simplest definition, and the one that I shall use here, is that a program is any sequence of key strokes†. (In the days when computers

†In Germany, it used to be said, everything which is not permitted is forbidden. In England everything which is not forbidden is permitted. In Ireland, who cares?

Table 12.2. *Font table for the main font used in this book.*

	0	1	2	3	4	5	6	7	8	9
0	Γ	Δ	Θ	Λ	Ξ	Π	Σ	Υ	Φ	Ψ
10	Ω	ff	fi	fl	ffi	ffl	ı	ȷ	`	´
20	ˇ	˘	¯	°	¸	ß	æ	œ	ø	Æ
30	Œ	Ø	´	!	''	#	$	%	&	'
40	(	)	*	+	,	-	.	/	0	1
50	2	3	4	5	6	7	8	9	:	;
60	¡	=	¿	?	@	A	B	C	D	E
70	F	G	H	I	J	K	L	M	N	O
80	P	Q	R	S	T	U	V	W	X	Y
90	Z	[	"	]	^	.	'	a	b	c
100	d	e	f	g	h	i	j	k	l	m
110	n	o	p	q	r	s	t	u	v	w
120	x	y	z	–	—	"	~	¨		

Table 12.3. *Font table for Knuth's typewriter font.*

	0	1	2	3	4	5	6	7	8	9
0	Γ	Δ	Θ	Λ	Ξ	Π	Σ	Υ	Φ	Ψ
10	Ω	↑	↓	'	ı	¿	ı	ȷ	`	´
20	ˇ	˘	¯	°	¸	ß	æ	œ	ø	Æ
30	Œ	Ø	␣	!	"	#	$	%	&	'
40	(	)	*	+	,	-	.	/	0	1
50	2	3	4	5	6	7	8	9	:	;
60	<	=	>	?	@	A	B	C	D	E
70	F	G	H	I	J	K	L	M	N	O
80	P	Q	R	S	T	U	V	W	X	Y
90	Z	[	\	]	^	_	'	a	b	c
100	d	e	f	g	h	i	j	k	l	m
110	n	o	p	q	r	s	t	u	v	w
120	x	y	z	{			}	~	¨		

were programmed using punched cards (one card per order) the operators used to amuse themselves by collecting up old punched cards, shuffling them and feeding them into the machine. Usually the machine did nothing. Sometimes it went into endless loops. Sometimes it would print out something like *! and stop. The game quickly palled.) The reader is welcome to investigate other possible definitions but she will find, I think, that they result only in more work for no additional reward. With the scheme outlined above, every program is represented by a positive integer, though not every positive integer represents a program.

Exercise 12.3.2   *Suppose we represent programs in the manner of Exercise 12.3.1. Which of the following numbers represent programs in our extended sense:*

$$123\,452\,032, \quad 100\,100\,045, \quad 100\,000\,011?$$

Outline a simple computer program to decide whether a given positive integer is a program in our extended sense.

As Exercise 12.3.2 indicates, it is easy to arrange that our computer will print ILLEGAL EXPRESSION† and stop if we type in a positive number which is not a program in our extended sense. If we do this, every positive integer may be considered as a program (those previously excluded just being programs to print ILLEGAL EXPRESSION and stop). We write $C(n)$ for the program represented by n. In order that the machine knows when we have finished typing we add a special key to the keyboard called 'BEGIN' which we press when we have finished.

If we feed in the program $C(n)$ and then press 'BEGIN', will the computer eventually stop (the algorithm terminates) or will it run for ever (the algorithm does not terminate)? Suppose there is an algorithm which could decide this for every n. Then we can write a program, Q, say, such that if we type in Q followed by $C(n)$ and then press 'BEGIN', the machine will eventually print 0 if $C(n)$ will stop and 1 if $C(n)$ will never stop.

So far, so good. We now observe that if we type in Q followed by $C(j)$ followed by $C(k)$ and then press 'BEGIN', the machine will eventually print a symbol 0 or 1 according as the program consisting of the keystrokes of $C(j)$ followed by the key strokes of $C(k)$ does or does not terminate. Let us write $P_j(k)$ for the result of typing in Q followed by $C(j)$ followed by $C(k)$ and then pressing 'BEGIN'. In Table 12.4 we draw a possible pattern of $P_j(k)$. If it looks familiar to the reader this is because it is familiar, being the same diagram (Table 12.1) as we drew to illustrate the final diagonal argument in the previous section.

†My children have asked me to point out the difference between unlawful and illegal. Unlawful means against the law but an ill eagle is a very sick bird.

Encouraged by this coincidence, we observe that, starting with the program Q, it is easy to construct a program R such that, if we type in R followed by $C(n)$ and then press 'BEGIN' the machine acts as though we had typed in Q followed by $C(n)$ followed by $C(n)$ again (so $C(n)$ is written twice) and then pressed 'BEGIN'. Thus if we type in R followed by $C(n)$ and then press 'BEGIN' the machine will eventually print $P_n(n)$. If $P_n(n) = 0$ this tells us that were we to type in $C(n)$ followed by $C(n)$ again and then press 'BEGIN' the machine would eventually stop; if $P_n(n) = 1$ this tells us that if we performed the same sequence of actions the machine would never stop.

To obtain a contradiction, we now modify R so that instead of printing 0 the machine goes into an endless loop, but its behaviour is otherwise unaltered. Let us call the new program S. If we type in S followed by $C(n)$ and then press 'BEGIN', the machine will eventually stop if $P_n(n) = 1$ but will never stop if $P_n(n) = 0$.

Since S is itself a program, and since every program has an associated integer we must have $S = C(N)$ for some N. What happens if we apply S to itself by typing S (that is $C(N)$) followed by S (that is $C(N)$) again and pressing 'BEGIN'? There are, apparently, two possiblities.

(1) If the machine stops, then, by the definition of $P_n(n)$, we have $P_N(N) = 0$. But, if $P_N(N) = 0$, the definition of S tells us that the machine cannot stop.

(2) If the machine does not stop, then, by the definition of $P_n(n)$, we have $P_N(N) = 1$. But, if $P_N(N) = 1$, the definition of S tells us that the machine must stop.

Since both assumptions lead to a contradiction, there can be no program S with the properties we have described. But we constructed S from the program Q so there can be no program Q with the properties we have claimed for it.

Table 12.4. *Does the program $C(j)$ followed by $C(k)$ terminate?*

$k =$	1	2	3	4	5	...
P_1	1	0	1	0	1	...
P_2	1	1	1	1	1	...
P_3	1	0	0	0	1	...
P_4	0	0	1	1	1	...
P_5	0	1	0	1	0	...
$\vdots$	$\vdots$	$\vdots$	$\vdots$	$\vdots$	$\vdots$	$\vdots$

Theorem 12.3.3 (Turing) *There is no program (algorithm) for checking whether every program (algorithm) stops.*

The reader may object that I did not show explicitly how to construct R from Q and S from R. In order to do this, I would have had to give more details about the programming language, and I believe that if the reader reflects for a bit, she will see that the program construction required is genuinely trivial.

Example 12.3.4 *Suppose that your computer has a sub-routine TUR, say, such that if x is any binary string (i.e. x is any finite sequence of 0s and 1s like 10010001 or 001011) then $\mathrm{TUR}(x)$ is either 0 or 1. Write a computer program in your favourite language which, given a binary string x, goes into an endless loop if $\mathrm{TUR}(xx) = 0$ but stops and prints YES if $\mathrm{TUR}(xx) = 1$. (Here xx is just the binary string x written twice, so if $x = 10011$ then $xx = 1001110011$.)*

If the reader is unconvinced, she should look at the article by Charlesworth ([30] reprinted in Volume III of [27]) where Turing's theorem is proved in detail for a computer using the BASIC programming language.

PART IV

Enigma variations

CHAPTER 13

Enigma

13.1 Simple codes

In 1929 a group of Polish mathematics students at the university of Poznán were invited, under pledge of secrecy, to attend a weekly night course in cryptology (the art of making and breaking codes). Up until then, code-breaking had been the province of the gifted amateur, usually a linguist (the Cambridge University Library still shelves books on cryptology in its palaeography section, sandwiched between Shorthand and Ancient Greek). Recently, however, the German army had switched to a mechanical enciphering method which had resisted all the efforts of the Polish code-breakers. Perhaps mathematical methods might succeed against machines when traditional means failed.

Let us consider messages written in capital letters using X to mark spaces such as, for example,

SENDXTWOXDIVISIONSXTOXPOINTXFIVEXSEVENTEEN.

It is sometimes convenient to associate letters with numbers in the following obvious way

$$A \leftrightarrow 0, \ B \leftrightarrow 1, \ C \leftrightarrow 2, \ldots, \ Z \leftrightarrow 25.$$

The first code most children learn consists in choosing an integer k and replacing the letter associated with r by the letter associated with $r + k$. This recipe is incomplete since $r + k$ may not lie between 0 and 25. To get round this, we first subtract a multiple of 26, ($26j$, say) so that $r + k - 26j$ lies between 0 and 25 and replace the letter associated with r by the letter associated with $r + k - 26j$. Thus if we choose $k = 20$, B which corresponds to 1 goes to the letter V which corresponds to 21 and J which corresponds to 10 goes to the letter D which corresponds to $30 - 26 = 4$. This is called the Caesar code† because it is said to have been used by Julius Caesar.

†Historians and professional cryptologists draw a clear distinction between codes and ciphers. In a code, groups of letters or numbers are substituted for words or phrases according to a 'dictionary' or code book. In a cipher, letters are changed or shuffled according to a fixed set of rules. Thus the Caesar code is actually a cipher. Mathematicians are mainly interested in ciphers and use the words code and cipher interchangeably. I shall follow the lax mathematical usage. (In the British Navy a cryptographic system was a cipher when used by an officer but became a code in the hands of a non-officer.)

Exercise 13.1.1 *Show that the message above becomes*

MYHXRNQIRXCPCMCIHMRNIRJICHNRZCPYRMYPYHNYYH.

If we represent our message by π and our encoded message by $T^k\pi$ then it is easy to see that

$$T^kT^l\pi = T^{k+l}\pi,$$

and that

$$T^l\pi = \pi \text{ whenever } l \text{ is a multiple of 26.}$$

In particular if we successively form $T^k\pi$, $T^1T^k\pi$, $T^2T^k\pi$, ..., $T^{25}T^k\pi$, one of these 26 expressions will be the original message. If we apply this idea to the coded message ρ of Exercise 13.1.1 we get

$\rho = $ MYHXRNQIRXCPCMCIHMRNIRJICHNRZCPYRMYPYHNYYH

$T^1\rho = $ NZIYSORJSYDQDNDJINSOJSKJDIOSADQZSNZQZIOZZI

$T^2\rho = $ OAJZTPSKTZEREOEKJOTPKTLKEJPTBERATOARAJPAAJ

$T^3\rho = $ PBKAUQTLUAFSFPFLKPUQLUMLFKQUCFSBUPBSBKQBBK

$T^4\rho = $ QCLBVRUMVBGTGQGMLQVRMVNMGLRVDGTCVQCTQCRPCL

$T^5\rho = $ RDMCWSVNWCHUHRHNMRWSNWONHMSWEHUDWRDUDMSDDM

$T^6\rho = $ SENDXTWOXDIVISIONSXTOXPOINTXFIVEXSEVENTEEN.

Exercise 13.1.2 *The following has been produced by applying T^k (for some value of k). Find the original message.*

GOVVHNYXO

The Caesar code is easily broken because there are only 25 such codes (or 26 if, as a mathematician would, you count the case $k = 0$).

The next code that most people learn is simple substitution in which each of the 26 letters is replaced by another in such a way that no two letters are replaced by the same one. We give an example in Table 13.1.

Table 13.1. *A simple substitution code.*

Original letter	A	B	C	D	E	F	G	H	I	J	K	L	M
Substituted letter	G	P	H	M	N	F	A	I	Q	S	U	B	O
Original letter	N	O	P	Q	R	S	T	U	V	W	X	Y	Z
Substituted letter	C	L	R	Z	X	Y	D	E	J	T	K	W	V

Exercise 13.1.3   *Show that the message about the two divisions becomes*

YNCMKDTLKMQJQYQLCYKDLKRLQCDKFQJNKYNJNCDNNC.

Since we can choose 26 letters to replace A, 25 letters to replace B (one has already been used), 24 letters to replace C (two have already been used), and so on, there are

$$26 \times 25 \times 24 \times \cdots \times 3 \times 2 \times 1 = 26! = 403\,291\,461\,126\,605\,635\,584\,000\,000$$

different simple substitution ciphers.

Exercise 13.1.4   *Use Exercise 11.4.14 to verify that the value given for 26! is not too far off.*

We cannot solve this cipher by working through all the possibilities. However, as most people know, almost any message longer than a post card which uses this kind of code can be deciphered using statistical properties of the English language.

Take, for example, the following result of using a simple substitution cipher.

```
VARHTAOWAOTAJBCOQCAORTHIQOIAWOMZMQAU
MOWLEEOJKJLIOVAOLIQTHNFBANOVFQOLOQCL
IDOQCJQOJIZCHWOQCATAOLMOIHOQLUAORHTO
NABLXCATLIKOBLXCATMOELDAOQCLMOLIOQCA
ORLAENOVARHTAOJIZOHROFMOCJNOQLUAOQHO
NABLXCATOJOQAEAKTJUOELDAOQCAOAOJUXEA
OKLPAIOFMOQCAOBHUXJIZOVJQQJELHIOJINO
VTLKJNAOWHFENOEHIKOJKHOCJPAOBAJMANOQ
HOAOLMQOLQOCJMOIHOXTJBQLBJEOMLKILRLB
JIBA
```

We tabulate the number of times each letter occurs in the coded message. as Table 13.2

The most common letter in ordinary text is X, (spaces are more frequent than any single letter), followed by E, followed by T, A, I and O (these last four having similar frequencies). Let us write the letters we guess in lower case. Since the most frequent letter in the coded message

Table 13.2. *Frequency count for a coded message.*

Letter	A	B	C	D	E	F	G	H	I	J	K	L	M
Occurrences	37	11	16	3	11	5	6	16	18	22	8	26	12

Letter	N	O	P	Q	R	S	T	U	V	W	X	Y	Z
Occurrences	10	61	2	23	7	10	13	6	6	5	6	0	4

is O and the second most frequent is A we guess that O corresponds to
x and A to e. With these guesses our text becomes

```
VeRHTexWexTeJBCxQCexRTHIQxIeWxMZMQeU
MxWLEExJKJLIxVexLIQTHNFBeNxVFQxLxQCL
IDxQCJQxJIZCHWxQCeTexLMxIHxQLUexRHTx
NeBLXCeTLIKxBLXCeTMxELDexQCLMxLIxQCe
xRLeENxVeRHTexJIZxHRxFMxCJNxQLUexQHx
NeBLXCeTxJxQeEeKTJUxELDexQCexexJUXEe
xKLPeIxFMxQCexBHUXJIZxVJQQJELHIxJINx
VTLKJNexWHFENxEHIKxJKHxCJPexBeJMeNxQ
HxexLMQxLQxCJMxIHxXTJBQLBJExMLKILRLB
JIBe.
```

The three letter word QCe occurs four times in this short passage
so it is a reasonable guess that it represents the (it might, of course,
represent she or ate or some other word, but we must start somewhere
and the occurs much more frequently in English than any of the other
alternatives). With this guess Q corresponds to t and C to h. Our text
now becomes

```
VeRoTexWexTeJBhxthexRTHItxIeWxMZMteU
MxWLEExJKJLIxVexLItTHNFBeNxVFtxLxthL
IDxthJtxJIZhHWxtheTexLMxIHxtLUexRHTx
NeBLXheTLIKxBLXheTMxELDexthLMxLIxthe
xRLeENxVeRHTexJIZxHRxFMxhJNxtLUextHx
NeBLXheTxJxteEeKTJUxELDexthexexJUXEe
xKLPeIxFMxthexBHUXJIZxVJttJELHIxJINx
VTLKJNexWHFENxEHIKxJKHxhJPexBeJMeNxt
HxexLMtxLtxhJMxIHxXTJBtLBJExMLKILRLB
JIBe.
```

The new text contains the word thJt suggesting that J corresponds to
a and the word tH suggesting that H corresponds to o and the words
L and Lt suggesting that L corresponds to i. With these suggested
substitutions we get

```
VeRoTexWexTeaBhxthexRToItxIeWxMZMteU
MxWiEExaKaiIxVexiItToNFBeNxVFtxixthi
IDxthatxaIZhoWxtheTexiMxIoxtiUexRoTx
NeBiXheTiIKxBiXheTMxEiDexthiMxiIxthe
xRieENxVeRoTexaIZxoRxFMxhaNxtiUextox
NeBiXheTxaxteEeKTaUxEiDexthexexaUXEe
xKiPeIxFMxthexBoUXaIZxVattaEioIxaINx
```

```
VTiKaNexWoFENxEoIKxaKoxhaPexBeaMeNxt
oxexiMtxitxhaMxIoxXTaBtiBaExMiKIiRiB
aIBe.
```

Looking at the words iM, VattaEioI, aKo and haPe we guess that M corresponds to s, I to n, K to g and P to v. The message now reads

```
VeRoTexWexTeaBhxthexRTontxneWxsZsteU
sxWiEExagainxVexintToNFBeNxVFtxixthi
nDxthatxanZhoWxtheTexisxnoxtiUexRoTx
NeBiXheTingxBiXheTsxEiDexthisxinxthe
xRieENxVeRoTexanZxoRxFsxhaNxtiUextox
NeBiXheTxaxteEegTaUxEiDexthexexaUXEe
xgivenxFsxthexBoUXanZxVattaEionxanNx
VTiKaNexWoFENxEongxagoxhavexBeaseNxt
oxexistxitxhasxnoxXTaBtiBaExsigniRiB
anBe
```

Exercise 13.1.5 *Complete the decipherment. The passage comes from page 469 of the Penguin translation of Hašek's* The Good Soldier Švejk.

Exercise 13.1.6 *The following has been produced by a simple substitution cipher. Decipher it.*

```
LZKSTFIKSHYRNSPVSOYAZKMSPOOWMTSYRSLZ
KSTZPMLSTLPMYKTSLZKSCPQNSUWCSUDSAPKS
FRNSLZKSNFROYRCSIKRSUDSOPRFRSNPDQKSL
ZKSTFIAQKSPVSLKSLSWTKNSUDSAPKSYTSJWY
LKSQPRCSUWLSLZFLSCYXKRSUDSNPDQKSYTSM
FLZKMSTZPMLSFRNSZPQIKTSYTSYRSIDSPAYR
YPRSQWOHDSLPSTPQXKSYL.
```


If you need more text, use the rest of the message. Otherwise use it as a check.

```
DPWSOFRSTKKSVMPISLZYTSKSFIAQKSLZFLSL
ZKSQPRCKMSLZKSTFIAQKSLZKSKFTYKMSYLSY
TSLPSVYRNSLZKSOPNKSDPWSOFRSFQTPSTKKS
LZFLSYLSZKQATSLPSHRPESLZKSTWUBKOLSEZ
YOZSYTSUKYRCSNYTOWTTKNSRPLYOKSLZFLSY
RSTPQXYRCSLZYTSOPNKSDPWSZFXKSWTKNSHR
PEQKNCKSEZYOZSYLSYTSZFMNSLPSCYXKSLPS
FSIFOZYRK
```

Exercise 13.1.7 *Show that the Caesar code is a simple substitution code and use this fact to give a very quick method of deciphering any long passage encoded by using a Caesar code.*

Can we find a simple code which is difficult to crack using frequency methods? One such code is the simple rotation code R. Let us associate letters with numbers as we did in the discussion of the Caesar code so that

$$A \leftrightarrow 0, \ B \leftrightarrow 1, \ C \leftrightarrow 2, \ldots, \ Z \leftrightarrow 25.$$

If the rth letter of our code is associated with the integer i_r we replace it by the letter associated with $r - 1 + i_r - 26 j_r$ where j_r is the integer such that $0 \leq r - 1 + i_r - 26 j_r \leq 25$.

Exercise 13.1.8 *(i) Show that simple rotation encodes* ROTATION *as*

RPVDXNUU.

(ii) The following short coded message has been obtained using simple rotation

NPVAFFJ.

Decode it.

If we represent our message by π and our encoded message by Rπ we may define new codes R^2, R^3 and so on by taking

$$R^2\pi = R(R\pi), R^3 = R(R^2\pi), \ldots R^{k+1} = R(R^k\pi),$$

for $k \geq 1$.

Exercise 13.1.9 *(i) Describe* R^k *in the same way as we described* R.
(ii) Why is it reasonable to write $R^1 = R$? *Why is it reasonable to write* $R^0\pi = \pi$? *Why is it reasonable to write* $R^{-k} = R^{26-k}$ *if* $1 \leq k \leq 26$? *(All these questions have at least two good answers, but the reader is only asked for one.) Define* R^l *for a general integer* l *in such a way that* $R^l(R^k\pi) = R^{l+k}\pi$.
(iii) Suppose we use simple rotation R *as our coding method. Explain why, in a long message, on average, any given letter* A, *say, in the original message will be coded as any given letter* C *about 1/26th of the time. Conclude that the resulting coded message will show roughly equal frequencies of each letter.*
(iv) Does the conclusion of (iii) hold if we replace R *by* R^k *and* k *is divisible neither by 13 nor 2? Does it hold if* k *is divisible by 13 but not by 2? Without making a detailed investigation, state what you think will happen if* k *is divisible by 2 but not by 13.*

(*v*) *Describe a method for breaking the code* R^k *for unknown k along the lines of our first method for breaking the Caesar code. Use your method to decode* BVTIIPZW.

Here are two coded passages which cannot be read quite so easily. They both come from an essay by James Thurber entitled *Exhibit X* and included in [240]. In it, he recalls his time as a code clerk at the American Embassy in Paris in 1918 using a ' ... new code book [which] had been put together so hastily that the word "America" was left out, and code groups so closely paralleled true meanings that "LOVVE" for example was the symbol for "love".' (Kahn writes that during the 1920s and 1930s it was rare for 'the major codes of the major powers [to be broken] — always with the exception of those of the United States whose cryptograms were as transparent as a fish tank to any competent cryptanalyst'. At the beginning of the Second World War, the British had to indicate to the American government that, whilst they had never attempted and would never, under any circumstances, dream of attempting to break American codes, nonetheless it had been suggested that those codes might, conceivably, be vulnerable.)

Passage 1

```
IBUACXPYRSRDNQEIEQGNZTMFLFGDUAUNWDPRKYBKPKTVFWVIGFLH
XALZDBGYMSENADWIXQGUACWPYFRHXZAWGKGZVWYRYCPQOIMFIFMO
ZJJJVIJDEPWYOCIKAMPCIFSGALSUQCDTJIVXHPYKMJNOZNWCMVMO
KRUYDSGDUETVCEYQNWRTZIRIYMXLEXPGPUPYAKBBQDAPSLWBFBNO
IBUAXTMZNPUV
```

Passage 2

```
WZRFHFZOUIGBRBVGNMJBDRZNKACUUFGFNJAKTPMMQCTPUAGRBUOV
GUIDMBCEPYPESJJUWZVFHFOFLONSRYGVCQLPFJDJODIIOSSRIGQZ
ETZVCPJJXLPWKXTAJEHFHYFNYIHPOSNRNDAKJPMIOTIJKSMRIUIF
LQJOLLRZWLFHAWTAJMUQWPDNDJETAQGPIVXAZDYHKFPEPSRRDWKA
KPQJEPYQLDFESSOYJMUUWAKNNRTPZFQMHXXMCWSJTKTRJCHYMNRJ
ZYADUMRDADTPTJFOJWBCV
```

The two codes have been obtained by combining the fairly weak simple substitution method with the extremely weak rotation method to produce something stronger than either. Because the passages are so short I think that either would be quite hard to break without some hint as to the method involved. (For example, the use of frequency counts reveals nothing.)

Exercise 13.1.10 (*i*) *The first passage π say has been coded by first*

applying a simple substitution cipher S *and then applying the simple ro-*
tation cipher R *to the result. The code could thus be represented as*
$C = R^k S$. *Let us write* $\sigma = C\pi$. *Explain why, if k is not divisible by*
2 or 13, this ensures that the frequency count will normally reveal noth-
ing.

(ii) Does it really matter if k is divisible by 2 (but not by 13)? Does
it matter if k is divisible by 13? Why?

(iii) Once we know how the code was constructed, it is not too hard to
break. Explain why there exists an l with $0 \le l \le 25$ *such that* $R^l C = S$.
Explain why $R^l \sigma$ *can be decoded by the frequency method. At first sight*
it still looks as though we will have to look at $R^p \sigma$ *for* $p = 0, 1, 2, \ldots, 26$
and try to decode each of the 26 possibilities by the frequency method,
but this is not the case. Explain why, if we find the number n_p *of the*
most frequently occurring letter in $R^p \sigma$, *then we would expect l to be the*
value of p for which n_p *is largest.*

(iv) In order to save the reader time, I now reveal that the code
has the form $R^3 \pi$ *or* $R^{-3} \pi$. *Carry out the procedure suggested in (iii)*
to find out which. (If you guess right the first time, you should still
check the other possibility so as to see what happens if you do not guess
right.)

(v) Find the message. Since the message is short, I include two hints.
The first is a note of reassurance: one of the words in the original mes-
sage does, in fact, end in X. The second is to remark that solution of such
puzzles is much aided by the possession of a dictionary in which words
are sorted by length and kth letter so that, for example, all seven-letter
words with fourth letter E are listed together. Such compilations have
been made easy by, but will also in due course be made obsolete by, the
computer. Such aids must be used with care, but mine [153] gives only
two seven-letter words with fourth letter E ending in TH, namely SEVENTH
and BENEATH†.

†But, it has been pointed
out to me, AGREETH is also
a possible word.

The second passage is coded in much the same way *except that*
we reverse the order in which we apply the two codes by first applying
a rotation cipher R^k for some k and then applying a simple substitution cipher S to the result. The code could thus be represented
as $C' = SR$. In my view, $C' = SR$ presents a much severer problem
than RS or even $C = R^k S$. Whether the reader accepts my opinion
or not she should try to decipher the second message without using
any further information. If she finds an easy solution, I should be
glad to see it. If not, she may agree with me that we have a very
nice example in which the order we do things radically affects the
result.

Writing the second passage in a way that reflects the period of 26 is a sensible first step but I suspect that it will remain baffling. (Since the page is not wide enough we split the table in two.)

	0	1	2	3	4	5	6	7	8	9	10	11	12
1	W	Z	R	F	H	F	Z	O	U	I	G	B	R
2	C	U	U	F	G	F	N	J	A	K	T	P	M
3	G	U	I	D	M	B	C	E	P	Y	P	E	S
4	N	S	R	Y	G	V	C	Q	L	P	F	J	D
5	E	T	Z	V	C	P	J	J	X	L	P	W	K
6	H	P	O	S	N	R	N	D	A	K	J	P	M
7	L	Q	J	O	L	L	R	Z	W	L	F	H	A
8	E	T	A	Q	G	P	I	V	X	A	Z	D	Y
9	K	P	Q	J	E	P	Y	Q	L	D	F	E	S
10	T	P	Z	F	Q	M	H	X	X	M	C	W	S
11	Z	Y	A	D	U	M	R	D	A	D	T	P	T

	13	14	15	16	17	18	19	20	21	22	23	24	25
1	B	V	G	N	M	J	B	D	R	Z	N	K	A
2	M	Q	C	T	P	U	A	G	R	B	U	O	V
3	J	J	U	W	Z	V	F	H	F	O	F	L	O
4	J	O	D	I	I	O	S	S	R	I	G	Q	Z
5	X	T	A	J	E	H	F	H	Y	F	N	Y	I
6	I	O	T	I	J	K	S	M	R	I	U	I	F
7	W	T	A	J	M	U	Q	W	P	D	N	D	J
8	H	K	F	P	E	P	S	R	R	D	W	K	A
9	S	O	Y	J	M	U	U	W	A	K	N	N	R
10	J	T	K	T	R	J	C	H	Y	M	N	R	J
11	J	F	O	J	W	B	C	V					

†Many code messages are vulnerable because the first or last words can be guessed. The German naval code-breakers of the Second World War were much aided by the admiral in command at the Canadian port of Halifax whose messages invariably began 'SNO Halifax BREAK GROUP Telegram in [some number *n*] parts FULL STOP Situation.' In order to prevent this sort of attack, the instructions for the German dockyard cipher called for messages to end with an irrelevant word such as *Wassereimer*, *Fernsprechen* or *Kleiderschrank* but some signalmen followed these instructions to the letter. Similar clues were given by the tendency of some code users to foul language. I have been told that, to prevent such indiscretions on the Allied side, code clerks were issued with anthologies of poetry and told to choose their 'padding' from the poem of the day.

Is the code effectively insoluble? Not if we know or guess correctly even a small part of the passage. Suppose that we know that the last word of the passage is 'Thurber'†. Then we know, looking at the last letter of the passage, that r in the 21st column will be encoded as V. It follows that q in the 22nd column will also be encoded as V, as will p in the 23rd column and so on. We have the following list of decodes:

	0	1	2	3	4	5	6	7	8	9	10	11	12
V decodes as	l	k	j	i	h	g	f	e	d	c	b	a	z

13	14	15	16	17	18	19	20	21	22	23	24	25

V decodes as y x w v u t s r q p o n m

Exercise 13.1.11   *Check that the guess that the last word is 'Thurber'*
yields the additional decodes

	0	1	2	3	4	5	6	7	8	9	10	11	12
B	t	s	r	q	p	o	n	m	l	k	j	i	h
C	x	w	v	u	t	s	r	q	p	o	n	m	l
F	h	g	f	e	d	c	b	a	z	y	x	w	v
J	k	j	i	h	g	f	e	d	c	b	a	z	y
O	w	v	u	t	s	r	q	p	o	n	m	l	k
W	i	h	g	f	e	d	c	b	a	z	y	x	w

	13	14	15	16	17	18	19	20	21	22	23	24	25
B	g	f	e	d	c	b	a	z	y	x	w	v	u
C	k	j	i	h	g	f	e	d	c	b	a	z	y
F	u	t	s	r	q	p	o	n	m	l	k	j	i
J	x	w	v	u	t	s	r	q	p	o	n	m	l
O	j	i	h	g	f	e	d	c	b	a	z	y	x
W	v	u	t	s	r	q	p	o	n	m	l	k	j

Check that, after inserting these substitutions, the passage becomes

	0	1	2	3	4	5	6	7	8	9	10	11	12
1	i	Z	R	e	H	c	Z	p	U	I	G	i	R
2	C	U	U	e	G	c	N	d	A	K	T	P	M
3	G	U	I	D	M	o	r	E	P	Y	P	E	S
4	N	S	R	Y	G	g	r	Q	L	P	x	z	D
5	E	T	Z	i	t	P	e	d	X	L	P	x	K
6	H	P	u	S	N	R	N	D	A	K	a	P	M
7	L	Q	i	t	L	L	R	Z	a	L	x	H	A
8	E	T	A	Q	G	P	I	e	X	A	Z	D	Y
9	K	P	Q	h	E	P	Y	Q	L	D	x	E	S
10	T	P	Z	e	Q	M	H	X	X	M	C	x	S
11	Z	Y	A	D	U	M	R	D	A	D	T	P	T

	13	14	15	16	17	18	19	20	21	22	23	24	25
1	g	x	G	N	M	s	a	D	R	Z	N	K	A
2	M	Q	i	T	P	U	A	G	R	x	U	y	m
3	x	w	U	s	Z	t	o	H	m	a	k	L	x
4	x	i	D	I	I	e	S	S	R	I	G	Q	Z
5	X	T	A	u	E	H	o	H	Y	l	N	Y	I
6	I	i	T	I	t	K	S	M	R	I	U	I	i
7	v	T	A	u	M	U	Q	o	P	D	N	D	l
8	H	K	s	P	E	P	S	R	R	D	l	K	A
9	S	i	Y	u	M	U	U	o	A	K	N	N	R
10	x	T	K	T	R	s	e	H	Y	M	N	R	l
11	x	t	h	u	r	b	e	r					

The partial decode of our passage is quite promising. There are plenty of xs and it looks 'quite English'. The zs are a little hard to fit in but they might be part of place names or nonsense words. (It is usual to include a few nonsense words in messages to be encoded in order to complicate the code-breaker's task.) On the third line (columns 14–26) we have

$$\text{xw?s?toxmak?x}$$

which strongly suggests the phrase 'was to make' and gives the possible new substitutions

	0	1	2	3	4	5	6	7	8	9	10	11	12
H	r	q	p	o	n	m	l	k	j	i	h	g	f
L	c	b	a	z	y	x	w	v	u	t	s	r	q
U	p	o	n	m	l	k	j	i	h	g	f	e	d
Z	o	n	m	l	k	j	i	h	g	f	e	d	c

	13	14	15	16	17	18	19	20	21	22	23	24	25
H	e	d	c	b	a	z	y	x	w	v	u	t	s
L	p	o	n	m	l	k	j	i	h	g	f	e	d
U	c	b	a	z	y	x	w	v	u	t	s	r	q
Z	b	a	z	y	x	w	v	u	t	s	r	q	p

With these substitutions the first line of the passage becomes

$$\text{inRenciphIGiRgxGNMsaDesNKA}$$

The first two words must presumably be 'in enciphering' and there is, I think, a plausible guess for the next word.

Exercise 13.1.12   *Complete the decipherment. (If you find this hard do not worry. It is quite hard. If you find it tedious, you may begin to see that code-breaking consists of weeks of tedium and minutes of excitement.)*

Exercise 13.1.13   *If you had to encipher this code many times, what sort of mechanical aids would you ask for? If you had to decipher such codes regularly, what sort of mechanical aids would you ask for?*

Even if we cannot guess any part of the message, the code is vulnerable if we send too long a message. Observe that, if we write out the coded message in 26 columns as we did above, then *in each column* the same letter will always be encoded in the same way. If the message contains, say, more than 2600 letters, then each column will contain at least 100 letters, and it is a good, though not certain, bet that the most frequently appearing letter in a given column corresponds to the most frequently appearing letter in our messages, that is to say x.

Exercise 13.1.14   *Show that if you* **know** *which letter in each column corresponds to* x *then you can decipher the entire message.*

Even if our guess is wrong for a few columns, there will still be enough correct letters in our proposed decipherment to enable us to pass rapidly to the correct solution.

In fact, for sufficiently long messages encoded using methods of this type we do not even need to know the method used. Suppose that all we know is that there is a reasonably small integer such that if we write out the message in n columns then in each column the same letter will always be encoded in the same way. (The smallest such n is called *the period* of the code.) If we try the effect of writing out the message in m columns, then, if m and the period n have no common factor, the coding method will usually have the required effect of smoothing out letter frequencies. However, if $m = n$, the letter frequencies will usually exhibit great disparities of the same kind as we expected in the special case, discussed in the previous paragraph, when $m = n = 26$. If m and n have common factors, then we will see a greater or lesser degree of statistical regularity, but it will usually not be hard to distinguish the particular case when $m = n$. Once we have guessed n, the method of the previous paragraph may be used.

Looked at in this way, it is clear that the vulnerability of the codes discussed in this section to attacks based either on guessing certain words or on statistical methods, is due to their short periods. The natural way forward is to seek codes with longer periods. However, here we run into another problem.

Exercise 13.1.15 *(i) Encode the following message (the first part of the Zimmermann Telegram) using a code of the same type as we used on the second Thurber quotation:*

```
    WE INTEND TO BEGIN UNRESTRICTED SUBMARINE WARFARE
      ON THE FIRST OF FEBRUARY STOP WE SHALL ENDEAVOUR
    IN SPITE OF THIS TO KEEP THE UNITED STATES NEUTRAL
  STOP IN THE EVENT OF THIS NOT SUCCEEDING WE MAKE MEXICO
         A PROPOSAL ON THE FOLLOWING BASIS COLON
      MAKE WAR TOGETHER COMMA MAKE PEACE TOGETHER
```

Decode it. What will be the usual effect of mistakes in encoding the message?

(ii) (Optional but instructive) Try to construct a code which is harder to break than the ones so far discussed. Carry out the encoding and decoding of the message just given. What will be the usual effect of mistakes in encoding the message? (It is worth noting that the retransmission of a garbled message is often as valuable to the enemy code-breakers as the transmission in code of a known message.)

13.2 Simple Enigmas

If a cipher is to be used in battle, messages must be enciphered and deciphered rapidly. Long-period codes of the type discussed at the end of the previous section take a long time to encipher and decipher by hand and are vulnerable to human error. The only way forward along this route is to construct a machine to do the work for us.

Such a machine was invented by a German electrical engineer called Scherbius in 1918. Consider a simple substitution code in which A is encoded by SA, B by SB and so on. It is not difficult to see that the same effect could be obtained by using a 'code wheel' or 'rotor' with 26 electrical contacts on one side of the wheel labelled A to Z and 26 electrical contacts on the other also labelled A to Z and with a wire connecting A on one side to S(A) on the other, B to S(B) and so on. If electric current enters by contact A on the first side, it leaves by contact S(A) on the other side; if it enters by contact B, it leaves by contact S(B) and so on. The code wheel has mechanised the substitution cipher S.

Now suppose the rotor rotates by one step each time we encipher a letter. The reader should convince herself that this device mechanises the code $C' = SR$ which was the hardest discussed in the previous section. As we saw there, the code C' is fine for short messages but, because it has the short period 26, is vulnerable to frequency analysis for long messages. To increase the period the result of the first encoding is fed into a second code rotor *which rotates by one step once every 26 times we encipher a letter.*

Exercise 13.2.1   *Let* D *be the code which sends* A *to* B, B *to* C, ... , Z *to* A. *Let* S_1, S_2 *be two substitution codes. Show that the code defined by sending the* $(k + 26l)$*th letter* α_{k+26l} *of our message to* $S_2 D^{l-1} S_1 D^{k-1} \alpha_{k+26l}$ *corresponds to one sent by the mechanism just proposed. Show, more generally, that the code defined by sending the* $(k + 26l + n)$*th letter* $\alpha_{k+26l+n}$ *of our message to* $S_2 D^{l-1} S_1 D^{k-1} \alpha_{k+26l}$ *corresponds to one sent by the mechanism just proposed started in an appropriate state.*

Barring unlikely coincidences each of the first 26×26 letters of our message will be encoded by a different substitution code corresponding to the 26×26 different arrangements of the rotors before repeating and the period of our code will be $26 \times 26 = 26^2 = 676$.

By feeding the result of the second coding into a third rotor *which rotates by one step once every* 26^2 *times we encipher a letter*, we obtain a code which will, in general, have period 26^3 and so on.

Exercise 13.2.2   *Describe the three-rotor code in the same way that we described the two-rotor code in Exercise 13.2.1.*

The three-rotor machine may be represented in diagrammatic form as shown in Figure 13.1. We refer to the machine a 'three-rotor Primitive Enigma' since modifications of this machine gave rise to the German Enigma machines which it was the task first of the Polish and then of the British code-breakers to crack.

My purpose in what follows is to explain roughly how the Enigma machines worked and why the apparently impossible task of breaking them was in fact possible. This will have several consequences.

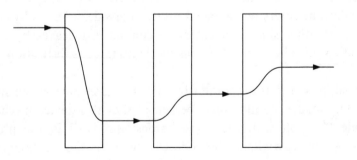

Figure 13.1: A three-rotor Primitive Enigma.

1   I shall not follow the exact path that the Poles and the British took.

2   There were several different versions of Enigma in use at various times with various branches of the German Armed Forces and they were used with various degrees of cryptographic competence. I shall concentrate on the German Naval Enigma, which was the best designed and most carefully used and, consequently, the hardest to break. British memoirs of the First World War had revealed the fact that German Naval codes had been comprehensively broken† and the German Navy was determined not to make the same mistake twice.

3   In the interests of clear exposition‡ I shall slur over or omit details both of the Enigma machine and the methods used for breaking it.

4   The omission of technical and historical detail will make things look a great deal simpler than they were. The reader who is tempted to exclaim when looking at actions of either side, 'Why, even I would have thought of that!' is almost certainly mistaken.

†The ill effects of these revelations should be borne in mind when considering the long British silence after the Second World War. The most important British coup of the First World War was the decoding of the Zimmermann Telegram which finally brought the USA into the war.

‡And for the less easily admitted reason of sheer ignorance.

The reader who wishes to go further will find dozens of books and hundreds of articles on Enigma. The chief sources known to me are the accounts of Hinsley [95], Kozaczuk [132], Welchman [253] and those collected by Hinsley and Stripp in [109]. The best place to start reading is probably Kahn's *Seizing The Enigma* which also has an extensive bibliography. The journal *Cryptologia* acts in part as a newsletter for Enigma buffs.

For the rest of this section, we shall discuss the three-rotor Primitive Enigma, since many of the issues involved in dealing with the actual Enigma machines appear in this simpler context. However, the reader should keep in mind that, for reasons to be discussed later, the machine the Poles had to deal with was a much more formidable device.

At first sight, the Primitive Enigma appears totally impregnable. First, the enemy has to guess the exact nature of the coding device. (Even minor variations in design such as stepping the second rotor after 23 rather than 26 letters have been enciphered might well prove completely baffling. The pre-war British attack on Enigma was stymied by incorrect assumptions about the wiring of the typewriter keys to the first input plate.) Once he has done that, he must discover the wiring of the three rotors. When we deciphered the second Thurber passage in the previous section, we were in fact finding the wiring of the rotor in a one-rotor Primitive Enigma machine. If the reader reflects briefly, she will begin to see the magnitude of the task proposed.

However, the German Navy expected to fight many battles and to lose some of them. Eventually, an Enigma machine and its rotors would be captured†. In previous wars the capture of a code book meant the cracking of the associated code, but now the loss of an Enigma machine still left the code heavily defended. When we encode using Enigma, we start off with the rotors in a certain position and the counting mechanism which turns the rotors in a certain state. There are $26^3 = 17\,576$ different starting positions for the rotors. In addition, although the enemy code-breakers know that the second rotor will move through one step after r_1, $r_1 + 26$, $r_1 + 52$, ... steps of the first rotor, and that the third rotor will move through one step after r_2, $r_2 + 26$, $r_2 + 52$, ... steps of the second rotor for some r_1, r_2 with $1 \le r_1, r_2 \le 26$, they do not know the values of r_1 and r_2 (the state of the counting mechanism). If the enemy wishes to decode an 800-letter message by working through all the possible initial positions of the rotors and all the possible values of r_1 and r_2, then they will need $26^5 = 11\,881\,376$ trial decodes, only one of which will, in general, be correct. A team of 300 cryptographers working three eight-hour shifts a day (with 100 cryptographers on each shift) and each working through each trial in half a minute would take well over a month to work through all these possibilities. It was this new level of security that made Enigma-type machines so attractive.

However, although our Primitive Enigma is indeed hard to break, it is not quite so hard to break as our vivid picture of hundreds of cryptographers working day and night would imply. Battalions of figures, like battalions of men, are not always as strong as they seem. It is not necessary to obtain a complete decode to find the position of the Enigma rotors. For reasons we shall now discuss, a very short snatch of ungarbled prose in the midst of an otherwise useless trial decode gives a sufficient clue to decoding the whole message. The Achilles heel of our Primitive Enigma which, fortunately for the Polish and British code crackers, it shared with the actual German Enigmas was the motion of the first 'fast rotor' relative to the remaining rotors. For 26 encipherments the inner rotors remained stationary while the fast rotor moved forward at a regular pace one step at a time, then the inner rotors changed to new positions and remained stationary whilst, once again, the fast rotor moved round its 26 positions. Thus, if the fragment

ERYUHUBRUNKXTANKERXNEZRTWTYFFFDMNBWUPOIJ

appeared in a trial decode, it would be well worth testing the hypothesis that our rotors were in the correct position during the decode XTANKERX,

†A great deal of thought has gone into the security of the electronic passwords used for automatic 'hole in the wall' banking. However, some criminals use bulldozers to steal the wall together with its automatic bank which they then break open at their leisure. Secure systems require protection against crude methods as well as subtle ones.

but that after XTANKERX, XTANKERXN or XTANKERXNE, either the second rotor of our putative encoder advanced and that of the real one did not, or vice-versa.

The discussion of the previous paragraph shows that our Primitive Enigma has a weakness, but anybody who has tried to read a book for misprints knows that human attention quickly flags and will find the thought of teams of clerks reading through thousands upon thousands of trial decodes in search of a short stretch of meaning fairly unrealistic. A more plausible approach is indicated by our solution of the second Thurber passage which we broke by guessing that it ended with the letters XTHURBER. Routine reports and routine orders tend to be very stereotyped. Welchman recalls that

> [w]e developed a very friendly feeling for a German officer who sat in the Quatra Depression in North Africa for quite a long time reporting every day with the utmost regularity that he had nothing to report. In cases like this, we would have liked to ask the British commanders to be sure to leave our helper alone.

A further aid was provided by the fact that German military men, like military men everywhere, tend to be very punctilious in giving full military titles†. Suppose we guess that a message enciphered on a known three-rotor Primitive Enigma has as its 10th to 16th letters GENERAL, and we know that the enciphered message has as its 10th to 16th letters SDFERTO. We now use six near copies of the Enigma machine E_1, E_2, ..., E_6, each with the same arrangement of rotors as the original but with a different stepping arrangement to be described later.

We start with the two inner rotors of the six mock Enigmas all in the same position. The outer 'fast rotor' of the first Enigma E_1 will be in some position which we call position 1. We set the outer rotor of E_2 one step further on in position 2, the outer rotor of E_3 one further step along in position 3 and so on. We now encipher G using E_1, E using E_2, N using E_3 and so on. If the positions of the rotors in E_1 coincide with those of the original Enigma when it enciphered the 10th letter G of the original message, then E_1 will produce the same encipherment S. Further, *if the inner rotors of the original Enigma did not change between the 10th and the 11th encipherment* then, since the outer rotor of the original Enigma moves forward one step, the position of the rotors in E_2 will coincide with those of the original Enigma when it enciphered the 11th letter E of the original message and so E_2 will produce the same encipherment D. Repeating the argument 4 more times, we see that if the positions of the rotors in E_1 coincide with those of the original Enigma when it enciphered the 10th letter and *if*

†As opposed to academics who like it to be known that they like to be known as Tom but who, in fact, like to be known as Dr Körner.

*the inner rotors of the original Enigma did not change between the 10th
and the 16th encipherment*, the encipherment of GENERAL by our new
arrangement will be SDFERTO, just as it was for the original.

The following objections may occur to the reader:

1   Our new arrangement might have enciphered GENERAL as SDFERTO
     by chance even if the rotor arrangement in the two machines did
     not coincide. It is, I suspect, quite hard to work out the exact
     probability of this happening, but the following argument gives a
     rough estimate. If we encipher a single letter (say G) at random,
     the chance that it will be enciphered as a particular letter (say S)
     is 26^{-1}. Thus if we encipher a sequence of 6 letters (say GENERAL)
     completely at random, the chance that they will be enciphered
     as a particular sequence (say SDFERTO) is 26^{-6}. If we repeat the
     experiment 26^3 times the chance that a particular encipherment
     (say SDFERTO) will come up, is about $26^3 \times 26^{-6} = 26^{-3}$ which is
     quite small. It is, of course, not true that our new arrangement
     works 'at random', but the argument makes it very plausible that
     the rate of 'false alarms' will not be very high. Since routine and
     speedy examination will reveal 'false alarms' for what they are,
     they will not, therefore, cause us any problems.

2   Even if we now know the positions of the rotors of the original
     Enigma at one point, we have little idea which of its 26^2 possible
     states the counting mechanism was in at this point. However, if
     we assume a particular state and decipher the message on that
     assumption, the first point where sense changes to nonsense will
     mark the point where either the second rotor of the actual Enigma
     stepped but the second rotor of the assumed Enigma did not or
     vice-versa. It is thus easy to synchronise the second rotors and,
     when needed, the third ones of the assumed Enigma with the real
     one.

3   Finally the reader will have noted that our method depended on
     the assumption that the inner rotors of the original Enigma did not
     change between the 10th and the 16th encipherment. One possible
     way of dealing with this would be to run the process another five
     times corresponding to the cases when the second rotor steps after
     the first, second, third, fourth or fifth enciphered letter, but this
     will take six times as long. Alternatively, we might note that the
     chance of an inner rotor change during the encipherment is only
     5/26 and if we are not very confident of our guess (that the 10th to
     16th letters of the original message were GENERAL), we are better
     off using the time to test five other guesses.

Exercise 13.2.3 *Suppose we have six guesses, each of which has a chance p of being correct. There is a process A taking a time T which, if the guess chosen is correct, has a probability q of cracking the code but otherwise does nothing. There is another process B taking a time 6T which, if the guess chosen is correct, is certain to crack the code but otherwise does nothing. If we are given a time 6T to crack the code, show that we should apply process A with each of our six guesses rather than process B with one guess provided that*

$$q \geq q_0(p) \text{ where } q_0(p) = \frac{1 - (1-p)^{1/6}}{p}.$$

Compute $q_0(p)$ when $p = 1/4$, when $p = 1/2$ and when $p = 3/4$.

There is another possibility when our guess is quite long. For example, suppose that our guess is 24 letters long. In that case we can split it into two guesses corresponding to the first 12 letters and the last 12 letters and be confident that the inner rotors will not move during the encipherment of at least one of them. It therefore makes sense to concentrate on methods which assume that the inner rotors do not move.

Like our proposed attack on a Primitive Enigma, the British attack on the German Naval Enigma required a mechanised 'brute-force' search through all $26^3 = 17\,576$ starting positions of the three rotors. The first such machines were built at the end of 1938 by the Poles (though they exploited a specific weakness in the German systems which was later removed) who called them *bomby* possibly because of the ice cream sundae (a *bomba*) the mathematicians ate when they discussed the project or possibly because of the regular ticking noise the *bomby* made in operation. Each British *bombe* consisted of the equivalent of 12 mock Enigmas, each of which could, apparently, be driven through the Enigma cycle of 17 576 starting positions in 15 minutes. However, for reasons which will become clear in the next chapter, even when things were going well, it was necessary to run through many Enigma cycles to find the daily German code-setting.

It was the 15-minute cycle time which made it feasible to break Enigma but it was also the 15-minute cycle time which meant that breaking Enigma was only just possible. At some times it took about 24 hours to decipher the day's transmissions for a particular Enigma system. If a change in German procedures required a ten-fold increase in Enigma cycles, then the machines could not be speeded up, since they were already running as fast as possible, nor could the number of machines assigned be increased without taking them from other decoding problems, and, unless the number of machines was increased, the best that might be hoped for was to decode every tenth day's output ten days late. Nor was this best outcome really to be hoped for, since intermittent decodes provide many fewer good initial guesses than can be obtained from a constant supply of decodes.

The great Atlantic convoy battles ran to many time-scales, from the month that it might take a slow convoy to complete its crossing, through to the minutes that a man might survive in the freezing sea. For those who tried to read the U-boat codes there were two time-scales. One was the day — information gained within 24 hours could steer convoys round the waiting submarine packs, information which took 72 hours to obtain usually came too late. The other was the cycle time of their bombes.

13.3 The plugboard

We have described how to encipher messages using our Primitive Enigma, but we have not said how the intended recipient should decipher the message so enciphered. If we describe the effect of the Primitive Enigma in a particular state s by T_s (so that A is enciphered by T_sA and so on), then we seek a deciphering method T_s^{-1} such that $T_s^{-1}(T_s$A$) = $ A, $T_s^{-1}(T_s$B$) = $ B and so on. In principle this is an easy task. If X emerges from the third rotor, then, by reversing the path and finding which letter X$'$ is enciphered by the third rotor as X, which letter X$''$ is enciphered by the second rotor as X$'$, and finally which letter X$'''$ is enciphered by the first rotor as X$''$, we obtain $T_s^{-1}($X$) = $ X$'''$.

If we look again at the diagrammatic version of the Primitive Enigma in Figure 13.2, we see that to decrypt a letter we 'simply reverse the arrows' (or the electric current). It is rather harder to construct a practical device to do what we have so easily described and even with such a device we must either use two machines or run the risk of errors like enciphering a message when the machine is in deciphering mode.

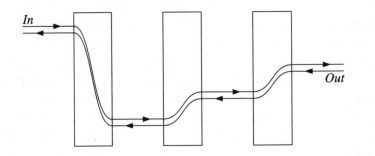

Figure 13.2: In and out of a three-rotor Primitive Enigma.

One of Scherbius's collaborators came up with an ingenious way round the problem. He added a fourth element called a reflector which took the output from the third rotor and fed it back by another path into the same third rotor, with the effect shown in Figure 13.3. If we write Ra for the encipherment of a particular letter a by the reflector acting alone and so on, we see that the effect of following the path given is to encipher a letter b, say, by $T_s^{-1}RT_s b$. Thus, if we write C_s for the effect of our new machine in state s, we obtain

$$C_s = T_s^{-1}RT_s.$$

The reflector did not move, so R did not depend on s.

Since the reflector consisted of 13 wires joining pairs of connectors, R is a simple substitution code with two special properties.

(1) If one wire connected, say, the letters e and q, then Re = q and Rq = e. Mathematicians say that R is self-inverse and non-mathematicians that R both enciphers and deciphers. Both mean that, since

$$R(Re) = Rq = e,$$

applying R twice to any letter leaves it unchanged. Since it is natural for mathematicians to write $R(Re) = R^2 e$ and to write the code which leaves everything unchanged as I (so that $Ie = e$), they would write

$$R^2 e = Ie$$

for every letter E, or still more briefly

$$R^2 = I.$$

(2) R cannot encrypt a letter as itself. (If the wire joined a connection to itself there would be a short circuit.)

It is easy to see by tracing paths through the machine in the manner of Figure 13.3 that the complete encipherment $C_s = T_s^{-1}RT_s$ must also have these properties. The reader may be interested to see this algebraically.

Figure 13.3: A Commercial three-rotor Enigma.

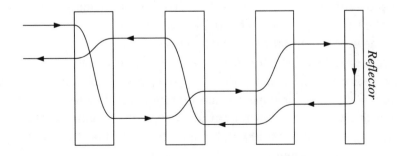

Reflector

To prove that C_s is self-inverse, we note that

$$C_s^2 = C_s C_s = (T_s^{-1}RT_s)(T_s^{-1}RT_s)$$
$$= (T_s^{-1}R)(T_s T_s^{-1})(RT_s) = (T_s^{-1}R)I(RT_s)$$
$$= (T_s^{-1}R)(RT_s) = T_s^{-1}R^2 T_s$$
$$= T_s^{-1}IT_s = T_s^{-1}T_s = I.$$

To show that C_s cannot encrypt a letter as itself, suppose, for example, that $x = C_s x$. Applying the encryption T_s to both sides of our equation, we obtain

$$T_s x = T_s(T_s^{-1}RT_s x) = (T_s(T_s^{-1})(RT_s)x)$$
$$= I(RT_s)x = R(T_s x)$$

so that R encrypts the letter $T_s x$ as itself. Since this is impossible, it follows that our initial assumption that $x = C_s x$ is false.

The three-rotor Enigma with reflector that we have just described corresponds, apart from one further modification to be discussed later, to the Commercial Enigma which Scherbius sought, with limited success, to sell to various businesses. One of the British code-breakers recalls how, whilst the attack on Enigma was Britain's most important secret,

> [a] retired banker named Burberry, living in the [same Bletchley] hotel once ... startled me by describing the Enigma which he had used in his bank. I probably said "fascinating", and raised one eyebrow.

It now had the advantage that, because C_s was self-inverse, the same machine and the same settings could be used for encryption and decryption. The Commercial Enigma was clearly good enough to ensure complete security for normal business transactions, but it is worth asking whether the changes between the Primitive and the Commercial versions made Enigma more or less secure.

Once again we must distinguish between the problems faced by an enemy cryptographer trying to reconstruct the wiring of an unknown machine including the rotors from various clues (capturing the machine by cryptographic means) and those faced by the same enemy who knows the wiring of the machine and its rotors but is trying to find out how it has been set up on a particular day (recovering the daily key). It seems plausible that the extra complication of having to find the wiring of the reflector would make the job of anyone seeking to reconstruct the machine by cryptographic means that much harder, but that argument does not apply to someone who already knows the machine's structure (including the reflector wiring) and is now trying to find the daily key.

At first sight, it might be argued that the extra complexity of the new path through the three rotors, through the reflector and back out through the three rotors in reverse order, must make the task harder, but complexity of method need not produce complexity of outcome. Exactly the same brute-force method that we described at the end of the previous section will (provided, as before, we start with a correct guess of some part of the message) produce the day's setting with the same amount of work. Viewed in this way, the new Enigma is no harder to crack than the old. Viewed in another way, might it not be easier?

A good secret code should have the aspect of a glass wall — vertical, smooth and featureless, presenting no hand-holds for the code-breaker. (As we have seen, one of the attractions of Enigma-type codes is that they automatically flatten letter frequency.) The Commercial Enigma presents two hand-holds (in addition to the Achilles heel of the 'fast rotor' which it inherits from the earlier Primitive version), the first being the, otherwise desirable, property of being self-inverse and the second that it cannot encipher letters as themselves.

One of the Italian Navy systems used an ordinary Commercial Enigma. One day, Mavis Lever, one of the youngest cryptanalysts

> ... sensed something strange in an Italian intercept, realised after a moment that the message had not a single L in it, and knowing that the Enigma never replaced a plaintext letter, concluded that the message was a dummy whose plaintext consisted entirely of *l*'s.

Such windfalls (which might well enable one to capture the whole machine by cryptanalysis) are rare, but the non-self-encipherment property is helpful in any attack on the code by guessing part of an enciphered message (in the jargon of cryptanalysis a 'probable word' method). The problem facing cryptanalysts using this method is that they not only have to guess a part of a message, say `warningxminesxatxsquare` `xsevenxsix`† but they also have to guess *exactly* where the guessed part came in a longer message. If the longer message was encoded as, say,

WSDTCNNXCLKEUFWNPVYQJGIBFE...

then our guessed phrase could not start in the first place because w cannot encipher as W, nor in the third since the fourth letter of the guessed phrase is n which cannot encipher as N the seventh letter of the coded message.

†There was 'a procedure known as "gardening", whereby the RAF laid mines in specified positions solely to generate warning messages. The positions were carefully chosen to avoid numbers, especially 0 and 5, for which the Germans used more than one spelling in their signals.'

Exercise 13.3.1 *Suppose that you have a guessed passage $\omega = w_1 w_2 \ldots w_n$ of length n. Show that the probability that a sequence $\Omega = W_1 W_2 \ldots W_n$ of n letters chosen at random does not coincide with α in any place (in other* *words* $w_j \neq w_j$ *for each* $1 \leq j \leq n$) *is*

$$f(n) = \left(\frac{25}{26}\right)^n = \left(1 - \frac{1}{26}\right)^n.$$

Compute $f(n)$ *for* $n = 6, 12, 18$. *For what value of* n *do we have* $f(n) \approx 0.1$, $f(n) \approx 0.01$, $f(n) \approx 0.001$? *Explain why you would expect to locate the exact position of a phrase of 200 letters 'buried' in a coded message of 400 letters by inspection of matchings, but you would not expect to be able to perform the same trick with a phrase of 6 letters buried somewhere in a coded message of 12 letters.*

There was, however, one further difference between the Primitive and the Commercial Enigmas which greatly strengthened its defences. The rotors were now made removable and interchangeable. Even if only three rotors were available, they could be arranged in any of $3 \times 2 \times 1 = 6$ ways so that even when opponents knew the wiring of each rotor and the reflector, they were, in effect, faced with 6 possible machines. If they then used brute-force search methods, these would take them 6 times as long (or require 6 times as many mock Enigmas) to cope with the machines. During the war, the German Army selected its daily set-up from 5 rotors giving $5 \times 4 \times 3 = 60$ possible machines and the German Navy from 8 rotors giving $8 \times 7 \times 6 = 336$ possible machines.

Exercise 13.3.2 *(i) If it takes a mock Enigma 15 minutes to work through all possible starting positions with a certain combination of rotors, how long will it take to run through 336 combinations? How many mock Enigmas will you require to run through all the combinations in a day?*

(ii) Let u_n *be the number of possible three-rotor machines which can be made with* n *rotors. Find* u_{n+1}/u_n *and tabulate its value for* $n = 3, 4, 5, 6, 7$. *What happens to* u_{n+1}/u_n *as* n *increases? Explain why increasing the number of rotors from which the choice is made beyond a certain point not only makes the system unwieldy but fails to increase its security a great deal.*

The armed forces of several countries adopted Enigma-type machines but most sought to make them more secure than the basic Commercial Enigma. The British Air Force adopted a machine whose name shifted from 'RAF Enigma' to 'Type X' and finally to 'Typex'. Internal memoranda stated frankly that 'The machine ... was copied from the German "ENIGMA" with additions and alterations suggested by the Government Code and Cipher School.' However, 'Difficulty arises in remunerating the patentees, in the near future at any rate,' and the matter of such payment was deferred 'until it is permissible for us to negotiate with them.' The British Army adopted the same system, but the Navy stuck with older and more established methods (see Figure 13.4). It appears that the designers of the Typex machine sought extra security by using many rotors in series (so we might talk of a ten-rotor Enigma) and a more complex stepping system.

How much does the addition of extra rotors add to the security of an Enigma machine? Clearly the answer may depend on the type of machine, so let us ask the question first about our original Primitive Enigma. If we use the method of attack described at the end of the last section, then finding the setting of an n-rotor Primitive Enigma will require a brute-force search through 26^n positions. This suggests that each extra rotor makes the task of our opponent 26 times more difficult.

On the other hand, if we look at our ten-rotor Primitive Enigma as it encodes a 200-letter message, we see the fast outer rotor taking a step with every letter, the more stately second rotor taking a step every 26th letter, and, if we are lucky, we may see the third rotor taking the step it takes every $26^2 = 676$th letter. The remaining rotors will, in all likelihood, imitate the House of Peers which

> ... , throughout the war
> Did nothing in particular
> And did it very well.

Unless we can explain exactly how the remote possibility that the higher

Figure 13.4: Preventing a code book from falling into enemy hands.

order rotors might rotate adds greatly to our security, it seems unwise to take the argument of the previous paragraph at its face value.

There is another problem connected with increasing the number of rotors. Increasing mechanical complexity means decreasing reliability and increasing size, and the new theories of *Blitzkrieg* called for a machine small enough and reliable enough to accompany a general's command car into battle. Some unknown but extremely clever German Army engineer came up with another solution: the 'plugboard' or 'steckel'. On entering and leaving the machine, the current passed through a plugboard which could be set up to interchange any chosen set of non-overlapping pairs of letters and leave the remaining letters unchanged (or 'self-steckered'). The set-up is shown in Figure 13.5.

If the three-rotor Commercial machine has the effect C_s in state s, then adding the plugboard is to produce a machine which has effect

$$E_s = PC_sP$$

where P is the effect of the plugboard alone. Since P interchanges certain pairs of letters and leaves the remaining letters unchanged, the effect of applying P twice is to leave everything unchanged. In other words, P is self-inverse.

$$P^2 = I.$$

It is now easy to check that E_s retains both the desirable property of being self-inverse and the undesirable property of not enciphering letters as themselves.

Exercise 13.3.3 *Prove this by arguments along the same lines as those at the beginning of this section in which we showed that C_s is self-inverse and does not encipher letters as themselves.*

The exciting thing about the plugboard is that it appears to rule out any brute-force attack on Enigma even if the enemy holds copies of

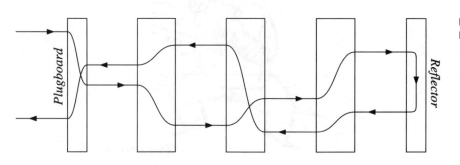

Figure 13.5: A three-rotor Enigma with plugboard.

our machine and its rotors and knows some of our messages, both in encrypted and unencrypted form. In the previous section we talked somewhat glibly about machines stepping through the $26^3 = 17\,576$ different rotor positions. The initial plugboard specification only called for six pairs of letters to be interchanged (leaving 14 self-steckered) but, even with this restriction, there are over 10^{11} different plugboard arrangements — and the plugboard can be changed whenever we wish.

Exercise 13.3.4 *Let v_r be the number of possible arrangements of the plugboard in which $2r$ letters are self-steckered $[0 \le r \le 13]$. Explain carefully why*

$$v_r = \frac{26!}{2^{13-r}(2r)!(13-r)!}.$$

Compute v_{r+1}/v_r. For which values of r do we have $v_{r+1}/v_r \ge 1$? Show that v_r is largest when $r = 2$. Find v_2 and v_3. Verify the statement about 10^{11} plugboard arrangements made just before this exercise.

Later, the number of self-steckered letters was reduced to 6. The exercise just done shows that this is close to optimal, giving about 1.5×10^{14} different plugboard arrangements from which to choose. There is no way we could conduct a brute-force search through this number of possibilities.

Numbers, by themselves, are no guarantee of cryptographic safety, and the German code experts identified one chink in Enigma's armour. Remember that a three-rotor Enigma, once set, runs through a cycle of $26^3 = 17\,576$ steps and then starts again. At each step it acts like a simple substitution cipher acting first, say, as E_1, then, after one step as E_2, after two steps as E_3 and so on. If, every time we use our Enigma, we start at the same point, then the first letter of each message will be enciphered using E_1, the second using E_2 and so on. If we treat the sequence of first letters as a message encoded using the substitution cipher E_1, the simple statistical techniques along the lines used in Section 13.1 (made easier by the fact that E_1 is self-inverse) will usually tell us what E_1 is. If we do the same for E_2 and so on then, just by knowing the E_j, we can now decipher the messages. The situation is, in fact, rather worse since it is quite plausible that possession of sufficiently many E_j will enable the wiring of the rotors to be discovered.

One of the British code-breakers recalls how, when dealing with one small German communication network,

[we] were amazed to be presented with twenty or more messages sent on the same day *with the same [setting]*. Presumably the operator had

not read his manual. This extraordinary lapse enabled us to recover the complete details of an unknown Enigma, using no other data than that the messages were in German.

The Swiss, Spanish, and Italian governments not only used Commercial Enigmas, but enciphered all messages for a given day with the same setting. During the 1930s, the chief British code-breaker 'Dilly' Knox broke both the Spanish and the Italian systems 'by hand'. (The Poles had some trouble with the Swiss system until they realised that the Swiss used German, French and Italian for their messages. The Swiss were warned indirectly that their system was vulnerable.) Although the main Italian Naval codes, which were of the old-fashioned 'book' type, were seldom read, the Italian Navy also used Enigma-type machines for other purposes. Information from this 'secondary' source enabled the British to maintain control of much of the Mediterranean in the face of what could have been a formidable Italian challenge. It also enabled them to locate and sink many of the tankers carrying fuel for Rommel's forces in North Africa, severely limiting his mobility. A secondary but welcome effect of these actions was to help destroy the (already limited) German trust in the reliability of their Italian ally.

In order to avoid this problem we must make sure that different messages start at different stages of the Enigma cycle. Since the Enigma cycle has $26 \times 26 \times 26$ steps, each step can be identified in a unique way by a sequence of three letters like ERT or FKA. We shall call this sequence the 'text setting'. By choosing text settings corresponding to widely spaced stages in the Enigma and using a different text setting each time, we can destroy any prospect of using the methods described in the previous two paragraphs. However, for the receivers (whether friends or foes) to decipher our transmitted message, they will need to know the text setting. How can we ensure that our friends know the text setting but our foes do not?

Consider the organisation of a wolf-pack attack on a convoy. We note the following points:

1   The U-boats need to communicate with each other as well as with High Command.

2   The order in which and the times at which the U-boats communicate cannot be fixed in advance.

3   An individual U-boat may not hear all the messages transmitted. (For example, it might be submerged at the time of a given transmission.) For these reasons the message *itself* must identify the text setting.

Similar considerations apply to the use of Enigma in *Blitzkrieg* operations. To get round the problem, the German Army decided on the following procedure:

1   The operator chose two three-letter sequences at random. The first (say, NDX) we shall call the 'indicator setting' and the second (say, CVE) would act as the text setting.

2   The operator transmitted the indicator setting (in our case NDX) as it stood.

3   Using the indicator setting as a temporary text setting, he encoded the six-letter sequence consisting of the actual text setting repeated twice (in our case, CVECVE) obtaining a new six-letter sequence (say, ERTFYU). He then transmitted the encoded sequence (in our case, ERTFYU).

4   He then used the text setting (in our case, CVE) to encode the rest of the message and transmitted it.

   The recipient then looked at the first nine transmitted letters (in our case, NDXERTFYU). Using the first three letters (in our case, NDX) as a text setting, he decoded the next six letters (in our case, ERTFYU) obtaining a repeated three-letter sequence (in our case, CVECVE). He then used the three-letter sequence (in our case, CVE) as a new text setting and decoded the rest of the message using it.

The repetition of the text sequence guarded against errors in the initial encipherment and transmission. The German Navy used a different procedure.

Exercise 13.3.5   *What weaknesses, if any, can you find in the Army system? If you can find any, how would you exploit them? (An unexploitable weakness is not a weakness.)*

CHAPTER 14

The Poles

14.1 The plugboard does not hide all finger-prints

As an inveterate browser, even in airport and railway bookstalls, I frequently come across books with titles like *Ten Spies Who Changed The World*, but closer inspection reveals that the only common characteristic of the spies described is that their actions failed to change anything. It is, however, just possible that the disaffected employee of the German Cryptographic Agency who sold documents concerning Enigma to the French from the early 1930s onwards did indeed change the course of history. The documents gave the general structure of the Military Enigma, including the existence of the plugboard. They also gave the keys including the plugboard settings for certain periods (in effect, allowing a code-breaker to reduce the Military Enigma to a Commercial Enigma for those periods). What they did not give was the internal wiring of the Enigma and its rotors.

Unable to make much of this material, the French offered it to their British and Polish allies. The British, who did not yet see Germany as a major threat, seem to have given it a fairly cursory inspection before deciding that they too could make nothing of it. The Poles gave the material to one of their new 'mathematical' code-breakers named Rejewski who did what the French and British experts had considered impossible and recovered the complete wiring of the Enigma and its rotors: 'a stunning achievement [and] one that elevates him to the pantheon of the greatest cryptanalysts of all time.' However, this removed only the outer defences of the plugboard Enigma, leaving the inner citadel of the daily settings untouched. In this section we discuss the elementary mathematical idea which gave Rejewski his first handhold on the plugboard Enigma, and in the next section I shall show how Rejewski and his colleagues captured the daily settings. Since we shall be discussing points that escaped the experts of the German Army, the reader must expect to use paper, pencil and hard thought to follow the arguments of the next few sections.

Rejewski's first insight was that the plugboard, complicated though it is, does not hide the inner Commercial Enigma entirely from view. To see why this might be so, let us consider a much simpler situation. Let S and T be substitution codes and T^{-1} the substitution code which recovers the message π from the encoded message $T\pi$. Thus, in the notation introduced in the previous section, $T^{-1}T = I$. We consider the new substitution code $D = TST^{-1}$.

Exercise 14.1.1   *(i) Explain why $TT^{-1} = I$.*
(ii) Explain why D defined above is a substitution code.
(iii) If T is self-inverse show that $D = TST$.

Can we tell anything about S by examining D? To mathematicians a substitution code is a special case of what they call a 'permutation'. The study of permutations goes back to the investigations into the solubility of quintic and other polynomial equations by radicals (now called Galois theory) and is the subject of group theory. The most obvious way of writing a substitution code S is in the form of a table.

ABCDEFGHIJKLMNOPQRSTUVWXYZ
UGCJWNKSLQFTOZPVDBYREXHMAI

Group theorists have another way of writing permutations. Observe that S takes A to U, U to E, E to W, W to H, H to S, S to Y and Y back to A. We thus have a cycle (AUEWHSY). In the same way, starting at B we obtain the cycle (BGKFNZILRTR). Starting at C we return immediately to C giving the cycle (C) and starting at D we obtain the cycle (DJQ). The letters E, F, G, H, I, J, K, L, M and N occur in cycles we have already found, but O does not. Starting at O we obtain the cycle (OPVXM) and this exhausts the letters of the alphabet. We write S in *cycle form* as

(AUEWHSY)(BGKFNZILTR)(C)(DJQ)(OPVXM).

Notice that we can interchange the order of the cycles without changing the code, so that, for example

(AUEWHSY)(BGKFNZILTR)(C)(DJQ)(OPVXM)

$$= (DJQ)(OPVXM)(BGKFNZILTR)(AUEWHSY)(C)$$

and that we can rotate each individual cycle round, so that, for example

(OPVXM) = (PVXMO) = (VXMOP) = (XMOPV) = (MOPVX).

We say that a cycle containing n letters is a cycle of length n so that, for example, (AUEWHSY) is a cycle of length 7.

Exercise 14.1.2   *(i) Express in cycle form the substitution code which has the table*

$$\text{ABCDEFGHIJKLMNOPQRSTUVWXYZ}$$
$$\text{XCDEBHGFPQVTYSZRWJKLMUOANI}$$

(ii) Find the table for the substitution code which has the cycle form

$$\text{(AFYZ)(BLCQHI)(D)(ER)(G)(JK)(MOVXW)(NTS)}.$$

Now suppose that S is given the table

Original letter	A B C D E F G H I J K L M ...
Substituted letter	A'B'C'D'E'F'G'H'I'J'K'L'M' ...

and we wish to find a table for $D = TST^{-1}$. The easiest way to go about this is not to ask for the effect of D on A, B, ... but to ask for the effect of D on TA, TB, ... (which is just a rearrangement of the alphabet A, B, ... We observe that

$$D(TA) = TST^{-1}(TA)$$
$$= TS(T^{-1}TA)$$
$$= TSA = TA',$$

and, similarly $D(TB) = TB'$ and so on, giving a table for $D = TST^{-1}$ in the following form.

Original letter	TA TB TC TD TE TF TG TH TI TJ TK TL TM ...
Substituted letter	TA'TB'TC'TD'TE'TF'TG'TH'TI'TJ'TK'TL'TM' ...

Looking at the tables for S and $D = TST^{-1}$ in the form we have given them it certainly looks as though some of the pattern of S has survived into TST^{-1} but, at first sight, it is not clear quite what it is.

To find out what is happening let us consider the particular example we started with of the substitution code S given by the table (which we have had to split in two)

$$\text{ABCDEFGHIJKLMNOPQRSTUVWXYZ}$$
$$\text{UGCJWNKSLQFTOZPVDBYREXHMAI}$$

so that $D = TST^{-1}$ has the table

T(A) T(B) T(C) T(D) T(E) T(F) T(G) T(H) T(I) T(J) T(K) T(L) T(M)
T(U) T(G) T(C) T(J) T(W) T(N) T(K) T(S) T(L) T(Q) T(F) T(T) T(O)

T(N) T(O) T(P) T(Q) T(R) T(S) T(T) T(U) T(V) T(W) T(X) T(Y) T(Z)
T(Z) T(P) T(V) T(D) T(B) T(Y) T(R) T(E) T(X) T(H) T(M) T(A) T(I)

Let us try to work out a cycle form for D. We start by considering the effect of D on T(A). We see that D takes T(A) to T(U), T(U) to T(E), T(E) to

T(W), T(W) to T(H), T(H) to T(S), T(S) to T(Y) and T(Y) back to T(A). We thus have a cycle (T(A)T(U)T(E)T(W)T(H)T(S)T(Y)). In the same way, starting at B, we obtain the cycle (T(B)T(G)T(K)T(F)T(N)T(Z)T(I)T(L)T(T)T(R)), and so on. Thus the cycle form for D = TST^{-1} is

$$(T(A)T(U)T(E)T(W)T(H)T(S)T(Y))$$
$$(T(B)T(G)T(K)T(F)T(N)T(Z)T(I)T(L)T(T)T(R))$$
$$(T(C))(T(D)T(J)T(Q))(T(O)T(P)T(V)T(X)T(M))$$

Recalling that S has the cycle form

$$(AUEWHSY)(BGKFNZILTR)(C)(DJQ)(OPVXM),$$

we see that the cycle forms of S and of TST^{-1} look the same.

Exercise 14.1.3 *If a substitution code* S$'$ *has cycle form*

$$(AFYZ)(BLCQHI)(D)(ER)(G)(JK)(MOVXW)(NTS)$$

write down, without calculation, the cycle form of TS$'$T^{-1} *and then check your answer.*

What precisely do we mean by saying that the cycle forms of S and of TST^{-1} look the same? After a little thought, we see that it means that the two codes have the same number of cycles of each length. In the case given, S has one cycle of length 1, one cycle of length 3, one cycle of length 5, one cycle of length 7 and one cycle of length 10, and so does TST^{-1}. In Exercise 14.1.3, S$'$ has two cycles of length 1, two cycles of length 2, one cycle of length 3, one cycle of length 4, one cycle of length 5 and one cycle of length 6, and so does TS$'$T^{-1}. Group theorists call the list giving the number of cycles of each length the 'cycle type'. We thus have a theorem:

Theorem 14.1.4 *If* S *and* T *are two substitution codes then* S *and* TST^{-1} *have the same cycle type.*

In first year university algebra, the theorem is stated as 'cycle type is invariant under conjugation'. Few undergraduates would name this as the dullest theorem of the year, but most would consider it a contender. For the mathematicians who struggled with Enigma, it represented the first hint that the plugboard might not be as strong as it seemed.

Exercise 14.1.5 *(i) Describe how you can recognise a self-inverse substitution code by its cycle type. (If you are stuck, write down the table of some self-inverse substitution code and find its cycle form.) Show that a self-inverse substitution code must have one of 14 different cycle types.*

(ii) Using the fact that cycle type is preserved, show that if S *is a self-inverse substitution code, then so is* TST^{-1}.

(iii) Describe how you can recognise a substitution code which does not encipher letters as themselves by its cycle type.

(iv) Using the fact that cycle type is preserved, show that if S *is a substitution code which does not encipher letters as themselves, then so is* TST^{-1}.

(v) Show that every self-inverse substitution code which does not encipher letters as themselves has the same cycle type and that every code with that cycle type is self-inverse and does not encipher letters as themselves. Show that there are

$$\frac{26!}{2^{13}13!} \approx 7.9 \times 10^{12}$$

such codes.

Let us see how Theorem 14.1.4 might be used. Suppose that we know that S and T are substitution codes and that we know that $D = TST^{-1}$ enciphers GENERAL as LBXBYGA. Then we know that D takes G to L, L to A and A back to G. Thus $D = TST^{-1}$ has a cycle of length 3 and so S must have a cycle of length 3.

Exercise 14.1.6   *Find the possible cycle types of the 26 Caesar codes and show that the code* S *just discussed cannot be a Caesar code.*

Of course, we have dealt only with a single unchanging substitution code and the Commercial Enigma changes with each step. (Another depressing thought is that even if we could halt the Enigma mechanism so that it remained the same for each step we would learn nothing by studying its cycles since, by Exercise 14.1.5(v), every self-inverse substitution code which does not encipher letters as themselves has the same cycle type.) However, a theorem like Theorem 14.1.4 should not be treated as an isolated fact but as a signpost indicating a possible direction in which to proceed. I shall not discuss how Rejewski used this signpost in reconstructing the Enigma wiring but in the next section I shall show how using the nine-letter initial signal for the German Army's Enigma messages (the NDXERTFYU discussed at the end of the previous section), the Poles were able to reconstruct the daily setting of Enigma.

The reader who wishes to find out how the Enigma wiring was reconstructed will find Rejewski's own account in Appendix E of [132]. (An interesting point is that the interchanging of rotors which makes the solution of the daily keys harder, makes finding the wiring of the rotors easier since each rotor will, eventually, turn up as the outermost 'fast' rotor.) There is an instructive discussion of how to find rotor wirings for Primitive Enigmas in Kronheim's 1981 book [127]. It is a sign of the times that this last reference is, in effect, a textbook for those interested in using cryptography to maintain the security of computer systems.

14.2 Beautiful Polish females

As might be expected, the Third Reich devoted massive resources to intelligence and code-breaking. The Germans started with a core of able cryptanalysts, some with mathematical training. But, as Kahn remarks: 'though the cryptological agencies ... grew in size they did not necessarily grow in effectiveness.' There were some substantial successes, such as the joint Italian–German effort which enabled them to read the reports on the British forces in North Africa sent by the American Military Attaché, and major coups against the convoy codes used by the British. However, these successes were achieved against traditional codes and not against Enigma-type machines. Whereas the Poles and, later, the British recruited from young graduates, particularly in mathematics, the Germans used older, called-up reserve officers with a knowledge of foreign languages. The General Staff also tried recruiting archaeologists specialising in unknown ancient writing systems but the results were disappointing. The same went for mathematicians. Kozaczuk quotes the head of German Naval Intelligence to the effect that

> [a]s far as mathematical training goes, pure mathematicians were not well suited to the [job], since they tended to get lost in theoretical abstractions. Their speculative investigations would strike against an inpenetrable barrier when it was necessary to go beyond formulas to solve a problem that was insoluble from the standpoint of pure mathematics.

The Poles and the British were lucky to find pure mathematicians with the ability to 'go beyond formulas'.

As we have seen, a Military Enigma consists of a Commercial Enigma which acts as a substitution code E_1 for the first encipherment, a substitution code E_2 for the second encipherment and so on, together with a plugboard associated with a self-inverse substitution code P. The full machine acts as a substitution code

$$C_r = PE_rP$$

for the rth encipherment. In addition, we know that C_r is self-inverse and cannot encipher a letter as itself (we shall call such codes non-self-enciphering).

We saw in Section 13.3 that once the plugboard, the arrangement of rotors and counting mechanism had been set up according to the daily setting, the operator chose two three-letter combinations at random: the first the 'indicator setting' and the second the 'text setting'. He transmitted the indicator setting as it stood and then, using the indicator setting as a temporary text setting, he encoded the six-letter sequence consisting of the actual text setting repeated twice, obtaining a new six-letter sequence, and then transmitted the encoded sequence.

During the transmission of these six letters, the Commercial Enigma goes through six steps E_1, E_2, ..., E_6. In the argument that follows, we shall suppose that these are randomly and independently chosen self-inverse, non-self-enciphering substitution codes. As pure mathematicians, we know that this is not strictly true — the Enigma does not toss coins to establish how to move from E_1 to E_2 but moves in a mechanical and predictable manner from one state to the next. (Indeed, our ultimate purpose is to uncover the exact law governing the movement of Enigma.) On the other hand, a great deal of care has been lavished on Enigma to make it look random and there is no reason why we should not make use of this work, particularly when it is 'necessary to go beyond formulas to solve a problem that' appears 'insoluble from the standpoint of pure mathematics.'

The Poles noticed that among the initial nine-letter messages like FDW KRM GSA, RCX LAY JRC there were a certain number in which the 4th and 7th letters were the same, as in FMW THS TVU. When the 4th and 7th, 5th and 8th or 6th and 9th letters were repeated, they called them 'females'. How common are females? Let us start by looking at the 4th and 7th letters. In each case the same (unknown) letter is being enciphered first by $C_1 = PE_1P$ and then by $C_4 = PE_4P$. For the reasons given in the previous paragraph we may suppose C_1 and C_4 to be random and independent (self-inverse, non-self-enciphering) substitution ciphers. To fix ideas, suppose the unknown letter is A and $C_1A = B$. Then the 4th and 7th letters are females only if $C_4A = B$ also. But a random self-inverse, non-self-enciphering substitution code is equally likely to take A to any of the 25 remaining letters, so $\Pr(C_4A = B) = 1/25$. Since our choice of particular letters A and B is irrelevant to the computation,

$$\Pr(\text{the 4th and 7th letters are females}) = \frac{1}{25}.$$

Similarly

$$\Pr(\text{the } (3+i)\text{th and } (6+i)\text{th letters are females}) = \frac{1}{25}$$

for $i = 1, 2, 3$, and so

Pr(the message contains females)

$$= 1 - \Pr(\text{the message contains no females})$$

$$= 1 - \prod_{i=1}^{3} \Pr(\text{the } (3+i)\text{th and } (6+i)\text{th letters are not females})$$

$$= 1 - \left(1 - \frac{1}{25}\right)^3 \approx 0.115.$$

Thus, if 100 messages are sent each day, we may expect on average between 11 and 12 females (with, of course, more on some days and fewer on others).

The Poles realised that, although the plugboard P affects whether a female appears when one is possible for the Commercial Enigma, *the plugboard cannot make a female exist when no female is possible for the Commercial Enigma*. The plugboard does not hide all finger prints. Let us say that two substitution codes S_1 and S_2 'permit females' if there exists a letter x, say, such that $S_1 x = S_2 x$.

Theorem 14.2.1 *Let S_1, S_2 and T be substitution codes. Then TS_1T^{-1} and TS_2T^{-1} permit females if, and only if, S_1 and S_2 do.*

Proof If $S_1 x = S_2 x$, then

$$TS_1T^{-1}(Tx) = TS_1 x = TS_2 x = TS_2 T^{-1}(Tx).$$

Thus, if S_1 and S_2 permit females, so do TS_1T^{-1} and TS_2T^{-1}.

On the other hand, if $TS_1T^{-1}x = TS_2T^{-1}x$, then

$$S_1(T^{-1}x) = T^{-1}(TS_1T^{-1}x) = T^{-1}(TS_2T^{-1}x) = S_2(T^{-1}x).$$

Thus, if TS_1T^{-1} and TS_2T^{-1} permit females, so do S_1 and S_2. ∎

It follows, in particular, that $C_{i+3} = PE_{i+3}P$ and $C_{i+6} = PE_{i+6}P$ permit females if, and only if, E_{i+3} and E_{i+6} do. Thus if we observe females in the $(i+3)$th and $(i+6)$th place, we know (by direct observation) that C_{i+3} and C_{i+6} permit females and (by deduction) so must E_{i+3} and E_{i+6}. What proportion of possible daily settings are excluded by the existence of one pair of females? A good estimate would be $1 - \alpha$, where α is the probability that two randomly chosen self-inverse, non-self-enciphering, substitution codes E and E′ permit females. It is not obvious how to compute α, but it is not hard to show that $\alpha \leq 13/25$, and this will be our next task.

Lemma 14.2.2   *Two self-inverse, non-self-enciphering, substitution codes* E *and* E' *permit females if, and only if, when written in cycle form they contain (at least) one cycle of length 2 in common.*

Proof   Suppose that E and E' permit females. Then there is a letter X say, such that $EX = E'X = X'$ say. The cycle (XX') thus occurs in the cycle expansion of both E and E'.

Conversely suppose that the cycle (XX') thus occurs in the cycle expansion of both E and E'. Then, trivially, $EX = X' = E'X$ and E and E' permit females.   ∎

Lemma 14.2.3   *The probability* α *that two randomly chosen self-inverse, non-self-enciphering, substitution codes* E *and* E' *permit females satisfies the inequality* $\alpha \leq 13/25$.

Proof   Write E in cycle form as

$$(X_1X_2)(X_3X_4)\dots(X_{25}X_{26}).$$

By the previous lemma E and E' permit females if, and only if, $E'X_{2i-1} = X_{2i}$ for some $1 \leq i \leq 13$. Thus

$$\Pr(\text{E and E' permit females}) = \Pr(E'X_{2i-1} = X_{2i} \text{ for some } 1 \leq i \leq 13)$$

$$= \Pr\left(\bigcup_{1=1}^{13}\{E'X_{2i-1} = X_{2i}\}\right)$$

$$\leq \sum_{1=1}^{13}\Pr(E'X_{2i-1} = X_{2i})$$

$$= 13\Pr(E'X_1 = X_2) = 13/25,$$

as claimed.   ∎

Thus α is less than about $1/2$. We shall use this estimate in what follows but Exercise 14.2.4 at the end of the section shows that α can be computed exactly. If the reader uses the formula there, she will find that α is, in fact, quite close to 0.4 and so the method works rather better than our cruder estimates would suggest.

Remark The reader who has not been bludgeoned into acquiescence by the references to Group Theory, Probability and other Mysterious Capitalised Mathematical Theories may object that the probabilistic estimates of this section are unnecessary. If we want to know the average number of females among the initial nine-letter messages, all we have to do is take a month's worth of such messages and count them. In the same way we could obtain a good estimate of α by setting up a mock Enigma at random 1000 times and examining the substitution codes at the first and fourth encipherings. To this objection I would reply that the Polish mathematicians could make the probabilistic estimates much more quickly than they could gather the suggested statistics and then, having seen that these estimates were favourable, could then check them in the practical manner suggested. Before embarking on a project which will absorb most of our available resources (and, in particular, exclude other desirable projects) it is desirable to have both theoretical and experimental backing for it.

Do the females give us enough information to find the daily setting? Let us consider the simplest case in which the rotors of the Commercial Enigma and their order are known. There are then $26^3 = 17\,576$ possible daily settings. Each observed female allows about half of the possible daily settings. If we assume sufficient randomness, then the probability that a certain setting is permitted by n observed females should be about $(\frac{1}{2})^n$. Thus 10 females should reduce the number of possible settings by a factor of about $(\frac{1}{2})^{10} \approx 1/1000$ leaving about 17 possible settings. Of course, this probabilistic argument cannot be pushed too far since it appears to show that 15 females would reduce the possible settings to below 1. What it means, in practice, is that with more than 18 females we might expect only one possible consistent daily setting. Using this knowledge and our stock of initial messages (all of which consist of an unknown three-letter word enciphered *twice*), it should be easy to recover the plugboard P. If we have fewer females or are otherwise unlucky, we will obtain several possible daily settings and we will have to go through the procedure of trying to find P for each of them. In all but one case we will uncover an inconsistency or, if we do not, any attempt to read the main messages will produce rubbish. This procedure, which would take too long applied to all 26^3 daily settings, becomes perfectly feasible when we have reduced the problem to 20 or even 60 possibilities.

How might we proceed in practice? Suppose that we have the initial signal KNRTYSFYD or, in a clearer form, KNR TYS FYD. We note that there is a female in the second place of TYS and FYD, so we go to a room marked 2. In Room 2 (just as in Rooms 1 and 3) there are 26 bookshelves marked with the 26 letters of the alphabet. We go to the bookshelf marked K, corresponding to the first letter of the message. On this bookshelf (as on all the other bookshelves) there are 26 boxes marked with the 26 letters of the alphabet. We take down the box marked N, corresponding to the second letter of the message. In this box (as in all the other boxes) there are 26 perforated cards. We extract the card marked R corresponding to the third letter of the message. The card has a pattern of 26×26^2 squares on it, each square corresponding to one of 26^3 possible daily settings. The pattern is the same on all the cards, but our card (corresponding to the 'indicator setting' KNR with females in the $3+2=5$th and $6+2=8$th place) has holes punched in all those squares for which the daily setting permits females in the $3+2=5$th and $6+2=8$th place when the indicator setting KNR is used. When we have collected several such cards corresponding to different initial signals containing females, we place them in a pile so that the squares corresponding to the same daily settings are aligned and shine a light beneath the pile. Only those squares which let the light through will correspond to possible daily settings. (Welchman suggests that this accounts for the word 'female' by analogy between a punched hole, through which a light can shine, and a female socket into which a plug can be inserted to make an electrical connection.) By using a little extra ingenuity (see [253]) and slightly more complicated cards it is possible to reduce the number of cards needed from 26^3 to 26. The resulting cards were known as Zygalski sheets after their inventor.

Exercise 14.2.4 *This exercise is devoted to showing how α could be found. Since the journey is more interesting than the goal, we take the long way round. You will need enough knowledge of probability theory to interpret $\Pr(A \cap B)$ (the probability of events A and B both occurring) and similar expressions. We approach the matter via another problem.*

The wrong envelope problem considers a mathematician who has written letters to n different people and prepared n correctly addressed envelopes to them. Overcome by the effort involved, the mathematician now places the letters in the envelopes at random (but exactly one letter in each envelope). What is the probability that all the letters are now in the wrong envelopes? The reader should first attack this problem for herself and then start the problem sequence below.

(i) Let A and B be finite sets. We write $|A|$ for the number of elements of A and so on. Explain why

$$|A \cup B| = |A| + |B| - |A \cap B|$$

(In a particular instance, the number of people who speak at least one of the languages French and German equals the number of French speakers plus the number of German speakers minus the number of people who speak both languages.)

(ii) Show that if A, B and C are finite sets

$$|A \cup B \cup C| = |A| + |B| + |C| - |A \cap B| - |B \cap C| - |C \cap A| + |A \cap B \cap C|.$$

(iii) State and prove the corresponding result for four sets.

(iv) Satisfy yourself that you understand the following formula for n finite sets $A_1, A_2, \ldots, A_n$,

$$\left| \bigcup_{1 \le i \le n} A_i \right| = \sum_{i=1}^{n} |A_i| - \sum_{1 \le i < j \le n} |A_i \cap A_j| + \sum_{1 \le i < j < k \le n} |A_i \cap A_j \cap A_k| - \cdots.$$

(The notation is ghastly, but can you think of anything better?)

(v) Show that if A, B and C are events

$$\Pr(A \cup B \cup C) = \Pr(A) + \Pr(B) + \Pr(C)$$
$$- \Pr(A \cap B) - \Pr(B \cap C) - \Pr(C \cap A) + \Pr(A \cap B \cap C).$$

(vi) Satisfy yourself that you understand the following formula for n events $A_1, A_2, \ldots, A_n$,

$$\Pr\left(\bigcup_{1 \le i \le n} A_i \right) = \sum_{1 \le i \le n} \Pr(A_i) - \sum_{1 \le i < j \le n} \Pr(A_i \cap A_j)$$
$$+ \sum_{1 \le i < j < k \le n} \Pr(A_i \cap A_j \cap A_k) - \cdots.$$

(The results of parts (iv) and (vi) are known as 'inclusion–exclusion formulae'.)

(vii) Now let us look at the wrong envelope problem. If we take A_i to be the event that the ith letter finds itself in the right envelope, explain why

$$\Pr(\text{some letter in right envelope}) = \Pr\left(\bigcup_{1 \le i \le n} A_i \right)$$

and so

$$\Pr(\text{no letter in right envelope}) = 1 - \Pr\left(\bigcup_{1 \le i \le n} A_i \right).$$

Substitute in this last equation using the formula of part (vi).

(viii) Explain why if i, j, k, ... are distinct

$$\Pr(A_i) = \Pr(A_1) = \frac{1}{n}$$
$$\Pr(A_i \cap A_j) = \Pr(A_1 \cap A_2) = \frac{1}{n(n-1)}$$
$$\Pr(A_i \cap A_j \cap A_k) = \Pr(A_1 \cap A_2 \cap A_3) = \frac{1}{n(n-1)(n-2)}$$

and state the general formula.

If $N(1)$ *is the number of integers i with $1 \leq i \leq n$,*

$N(2)$ *is the number of pairs (i, j) with $1 \leq i < j \leq n$,*

$N(3)$ *is the number of triples (i, j, k) with $1 \leq i < j < k \leq n$,*

and so on, show that

$$N(1) = \frac{n}{1!}$$

$$N(2) = \frac{n(n-1)}{2!}$$

$$N(3) = \frac{n(n-1)(n-2)}{3!}$$

and state the general result.

Hence show that

$$\text{Pr(no letter in right envelope)} = 1 - N(1)\Pr(A_1) + N(2)\Pr(A_1 \cap A_2)$$
$$- N(3)\Pr(A_1 \cap A_2 \cap A_3) + \cdots$$
$$= 1 - \frac{1}{1!} + \frac{1}{2!} - \frac{1}{3!} + \cdots + (-1)^n \frac{1}{n!}.$$

(ix) Let p_n be the probability that every letter is in the wrong envelope when there are n envelopes. Use the formula at the end of part (viii) to compute p_1, p_2, p_3, $\dots$, p_{10}. Can you see a quick way of explaining why p_1 and p_2 have the values you have calculated?

(x) Many readers will know that, in fact,

$$1 - \frac{1}{1!} + \frac{1}{2!} - \frac{1}{3!} + \cdots + (-1)^n \frac{1}{n!} \approx e^{-1} \approx 0.36788$$

for large n. This part of the question is devoted to the problem of finding rapid estimates for sums of the form

$$s_n = u_0 - u_1 + u_2 - \cdots + (-1)^n u_n$$

where the u_j form a sequence of decreasing positive real numbers (in other words $u_j \geq u_{j+1} \geq 0$ for all j).

Show that

$$s_{2m} \geq s_{2m+2}, \quad s_{2m+3} \geq s_{2m+1} \quad \text{and} \quad s_{2m} \geq s_{2m+1},$$

for any $m \geq 0$. Deduce that

$$s_0 \geq s_2 \geq s_4 \geq \dots s_{2M} \geq s_{2M+1} \geq s_{2M-1} \geq s_{2M-3} \cdots \geq s_3 \geq s_1$$

and so, if $n \geq 2p$, then

$$s_{2p} \geq s_n \geq s_{2p+1}.$$

Conclude that

$$0 \leq s_{2p} - s_n \leq s_{2p} - s_{2p+1} = u_{2p+1}.$$

Show similarly that, if $n \geq 2p + 1$, then

$$0 \leq s_n - s_{2p+1} \leq s_{2p+2} - s_{2p+1} = u_{2p+2}.$$

Combining these two results, show that, if $0 \leq m \leq n - 1$, then

$$|s_n - s_m| \leq u_{m+1}.$$

This is sometimes stated as 'the error is less than the first term neglected'. Look at the values of p_j that you calculated at the end of part (ix) in the light of this result.

(xi) Now suppose that we have n pairs of cards each pair having the same label but the labels of each pair being different. (Thus we might have two aces, two kings, and so on.) We shuffle the pack and lay out the 2n cards in n pairs. What is the probability q_n that no pair consists of cards with the same label? Explain why the problem differs from the wrong envelope problem.

Use the same ideas as we used to solve the wrong envelope problem to show that

$$q_n = 1 - \binom{n}{1}\frac{1}{2n-1} + \binom{n}{2}\frac{1}{(2n-1)(2n-3)} - \cdots$$

$$+ (-1)^r \binom{n}{r}\frac{1}{(2n-1)(2n-3)\ldots(2n-r)} + \cdots$$

$$+ (-1)^n \frac{1}{(2n-1)(2n-3)\ldots 1}.$$

Explain why $\alpha = q_{13}$. Use the ideas of part (x) to show that $\alpha < 1/2$ and obtain better estimates.

Exercise 14.2.5 *If you enjoyed the solution of the wrong envelope problem and if you know the formula*

$$e^x \approx 1 + \frac{x}{1!} + \frac{x^2}{2!} + \cdots + \frac{x^N}{N!}$$

for large N, you may enjoy the following variation.

A large and cheerful crowd of N junior wizards leave their staffs in the porter's lodge on the way to a long night at the Mended Drum. On returning, each collects a staff at random from the pile, returns to his room and attempts to cast a spell against hangovers. If a junior wizard attempts this spell with his own staff, there is a probability p that he will turn into a bullfrog. If he uses someone else's staff he is certain to turn into a bullfrog. Show that the probability that in the morning the bedders will find N very surprised bullfrogs is approximately e^{p-1}.

14.3 Passing the torch

On 15 December, 1938 the German Army increased the number of rotors in use from three (giving six possible different configurations) to five (giving 60). Although the Poles reconstructed the two new rotors, the cost of finding the daily keys (measured in man-hours required to prepare suitable 'Zygalski sheets', the number of non-excluded daily settings to be examined, or in any other realistic terms) had now risen beyond any resources the Polish code-breakers could hope to command. On 30 March, 1939 the British and French Governments guaranteed Poland 'all support' in the event of an unprovoked German attack. On 27 April, Germany renounced her non-aggression pact with Poland. As tension rose, the Polish General Staff decided to share its cryptographic secrets with its allies.

The revelation of Polish success was performed in a suitably theatrical manner. The British and French delegations were ushered into a room containing tables on which stood several objects under covers. The covers were removed to show the Polish copies of the German Enigmas. When the British proposed bringing in their own draftsmen to copy the plans of the machines, the Poles announced that they had prepared two machines as gifts for their allies. The Poles also showed their *bomby* and Zygalski sheets and explained how they worked. One of the British representatives was 'Dilly' Knox, the doyen of British code-breakers. His attack on the Enigma had been foiled by his inability to work out the way the keyboard was connected to the input for the first disc. He knew that it was not the same as in the Commercial Enigma (where QWERTZU was connected up in the order 123456, following the layout of the German typewriter) and had given up hope of dealing with an unknown secret wiring. In answer to his question the Poles replied that the secret was simple — the military machine had ABCDEF connected up in the order 123456! (Had the Germans used a more complex system even Rejewski might have been foiled.) One of Knox's junior colleagues recalls the story that:

> after the meeting, Dilly returned to his hotel in a taxi with his French colleague, chanting, 'Nous avons le QWERTZU, nous marchons ensemble'†.

†'We have the QWERTZU, we march together.'

On 1 September, Germany and the USSR attacked Poland which was overwhelmed within three weeks. The nature of her conquerors is sufficiently illustrated by the chilling statistic that, during the Second World War, Poland was to lose 5 400 000 killed, all but 120 000 after her surrender. No hint of the pre-war success against Enigma reached German ears.

There is a story about a Cambridge don during the First World War who was asked why he was not at the front fighting for civilisation and replied 'Madam, I am the civilisation they are fighting to protect.' However, it is only a story, and a generation of Britain's most promising scientists had been included in the great slaughter. This time, the British were determined to make better use of their intellectual resources. By the time of the Polish meeting, the cryptological establishment had overcome its earlier prejudice against mathematicians (apparently based on the view that mathematicians were dreamy creatures liable to blurt out state secrets in a fit of absent-mindedness) and had started to recruit in earnest. (Those who still objected to the employment of mathematicians were told that everybody knew that mathematicians were good chess players and everybody knew that chess players made good codebreakers. It must also be said that the stresses of the time resulted in the recruitment of several experts on cryptogams†.) Since messages once deciphered have to be read and interpreted, linguists (including classicists, as the most likely to be able to learn Japanese), historians and others, thought to have special talents or just to be very bright, joined them. To these had to be added engineers, wireless experts and support staff. This ever-growing group worked in an ever-expanding collection of temporary huts and buildings in the grounds of Bletchley Park.

† Plants with no stamens or pistils, such as ferns and mosses. I have heard this story from two independent sources.

The superior British resources enabled them to cope with the tenfold increase in rotor arrangements. Production of the Zygalski sheets required examination of $60 \times 26^3 = 1\,054\,560$ rotor settings, and at the beginning of 1940 the British broke their first Enigma key. The best accounts of these events like [96] and [110] emphasise that the British code-breakers were involved in a race against the constant improvements that the German cryptographers made, both in the Enigma machines and in the way they were used. In a long-distance race any runner who falls behind the leaders will find it almost impossible to catch up. By giving their allies the wiring of the Enigma and the Zygalski sheets, the Poles had given the British the start they needed but the race was to be a close-run thing.

On 10 May, the Germans invaded France and, on the same day, in accordance with the best cryptographic principles, they changed their Enigma procedures in such a way that the 1560 Zygalski sheets, each with their carefully drilled 1000 or so holes, became just so much waste cardboard. The Zygalski method depended on the fact that the three-letter text setting was repeated twice giving rise to two different encipherments of the same three letters. The repetition guarded against the possibility of the text setting being garbled in transmission but, as the Germans now realised, went against the basic principle of cryptography that the same message should never be transmitted twice. From now on the text setting was enciphered only once.

This change should have blinded the British pursuers as effectively as a handful of pepper in the eyes but, even in its new form, the German Army's Enigma procedures had a basic flaw, though one which is more obvious to a generation brought up on computer passwords. It is very difficult for people to behave randomly. If asked to choose an integer between 1 and 10, more people will choose 7 than any other number. When asked to choose a computer password, people often used the names of boyfriends, girlfriends or well known dates. In [59], Feynmann describes how he made his reputation as a safe-cracker by exploiting these weaknesses. Not all German code clerks were paragons of patriotic efficiency, and few, even of the paragons, fully understood the limitations of the magic 'uncrackable' Enigma machines they were told to operate. In 1932, when the Poles started their operations, clerks would use triples like AAA. Orders were given forbidding the use of such triples so the clerks switched to ABC or running their fingers down the diagonals of the keyboards. Using their knowledge of the bad habits of individual code clerks and of code clerks in general, the British managed to overcome the crisis. Chapter 6 of Welchman's book gives a clear account of some of the methods used. Of course, if the Zygalski sheets had not been available in time, the detailed knowledge required to exploit the code clerks' weaknesses would not have existed; this was a race in which the code-breakers could not afford to fall behind.

The fruit of all this effort was distinctly bitter as Bletchley observed one of the greatest military disasters in British history.

> When any one of [the German] armoured units was held up by an Allied defensive position, we would probably decode an Enigma message from the unit's commander requesting air support. A little later we would hear that an attack by dive-bombing Stukas had been effective, and the armoured unit had resumed its advance. At intervals, all the major Panzer commanders would report their progress and their assessment of Allied capabilities.

The code-breakers consoled themselves that:

> Our decodes must have given early warning that the military situation was utterly hopeless. The mass of combat intelligence can hardly have failed to speed the extraordinary fleet of miscellaneous boats that brought so many back from the beaches of Dunkirk.

This was cold comfort indeed.

As the fall of France showed, even when Enigma decodes flowed freely, British commanders were only in the position of a poker player who catches an occasional glimpse of a card in his opponent's hand. Weak hands remained weak hands — the knowledge from breaking the German Railway's Enigma that a German invasion of Greece was imminent did not prevent the defeat of the Greek and British armies — and often, as with intercepts during the Battle of Britain, it is difficult to see how battles could have been fought differently, with or without the extra intelligence. What Enigma decodes did, time after time, was to shade the odds and, as it were, convert an unlucky card player into a lucky one.

A typical example, from a slightly later stage, occurred in 1941 when the German battleship *Bismarck*, having confirmed that she out-classed any single ship of the British or American navies by her destruction of the *Hood*, pride of the British fleet, vanished into the Atlantic. There followed many hours of 'desperate and sometimes irrational searching' by British ships and aircraft. After a day of wavering and contradictory advice and orders from the Admiralty, the Admiral in charge of the main British force decided to act on the assumption that the *Bismarck* was making for Brest. Although some Naval Enigma messages were being decoded, the time taken (three days or more) meant that no use could be made of this. On the other hand, the Air Force codes were being read quickly. The Luftwaffe chief of staff, General Hans Jeschonnek, was in Athens, directing his Air Force's part in the successful airborne invasion of Crete. His son was a junior officer aboard and the General abused the privilege of rank on 25 May to ask about the *Bismarck*'s progress and destination. Within less than an hour of the British Admiral's original decision, he was given confirmation of its correctness and for the next 15 hours the British battleships pounded towards Brest.

On the morning of 26 May, the *Bismarck* was finally sighted by the radar of a Coastal Command aircraft. An attack by torpedo bombers found one of the few weak points (or, for all anybody knows, the only weak point) of the *Bismarck* by hitting her rudder and putting her steering out of action. With only a few more hours of fuel left, the British battleships at last caught up with and destroyed their formidable opponent.

The sequel illustrated another problem with the use of Enigma intelligence. The German Navy had planned a three-month cruise for the *Bismarck* and its companion warship to disrupt the convoy system totally. Tankers and other ships had already been dispatched and, armed with Enigma decodes, the British set out to hunt them down.

Two ships which were to be spared in order to make the other losses appear accidental had the misfortune to encounter Royal Navy ships by chance and were also sunk. This massacre aroused German suspicions and a full-scale enquiry ensued. Although it concluded that Enigma remained unbroken, extra precautions were introduced which substantially increased the difficulty of reading the Submarine Enigma.

The code-breaking at Bletchley did something else as well. In a period when disaster followed disaster and the avoidance of defeat was hailed as victory, it gave Britain's leaders a glimmer of hope, however illusory that hope may have been†. It can be argued that, if the only result of the whole complex and expensive effort had been to keep Churchill's spirits up while Britain stood alone, it would still have been worth it.

Churchill's romantic soul loved the excitement and secrecy surrounding Bletchley. He relished the way that

> [t]he old procedures, like the setting up of agents, the suborning of informants, the sending of messages written in invisible ink, the masquerading, the dressing-up, the secret transmitters, and the examining of the contents of waste-paper baskets, all turned out to be largely cover for this other source, as one might keep some old-established business in rare books going in order to be able, under cover of it, to do a thriving trade in pornography and erotica.

Each morning, a summary of the previous day's decrypts together with the most important individual messages were brought to Churchill in a special buff-coloured dispatch box. He visited Bletchley to thank 'the geese who laid the golden eggs and never cackled'. Looking at the disparate, unkempt and definitely unmilitary crew formed by his top code-breakers, he is said to have added to his head of Intelligence 'I know I told you to leave no stone unturned to find the necessary staff, but I did not mean you to take me so literally!'

The early successes gained by the use of the Polish gift gave Bletchley the prestige and goodwill needed in the constant battles for scarce resources and personnel. Building on this base, it had been able to ride the blow dealt by new German procedures, but this success was based on the bad habits of a few operators (and the number of such mistakes was to decrease very rapidly after 1941). The German Navy used a different method to identify message settings which avoided the dangers inherent in the Army and Air Force systems, and no progress had been made against the Navy codes. As late as the summer of 1940, the administrative head of Bletchley told the head of the Naval Section, 'You know, the Germans don't mean you to read their stuff, and I don't

†An optimism shared at a lower level by the intercept operators whose tedious job it was to listen to faint morse code and take down gibberish with perfect accuracy. Welchman worried that the extra attention given to some messages might reveal they were being broken. 'I discovered, however, that the intercept operators believed that all their intercepts were decoded.'

suppose you ever will.' The main hope of any further success against Enigma rested with the new 'Turing bombes'.

The new bombes used the 'probable word method' discussed at the end of Section 13.2. The reader should reread that section and then ask herself how the simple method proposed there against the Commercial Enigma could possibly be adapted to work in the presence of a plugboard†.

†The Polish *bomby* depended on the fact that, initially, the Germans only 'steckered' 6 pairs of letters and so, from time to time, the short initial message would only involve 'self-steckered' letters (that is letters unaffected by the plugboard).

CHAPTER 15

Bletchley

15.1 The Turing bombes

Kahn writes that 'In Britain, Cambridge students and graduates were the cream of the nation and [Bletchley] took the cream of the cream.' Even among so many clever people, Turing was 'viewed with considerable awe because of his evident intellect and the great originality of his contributions. ... Many people found him incomprehensible, perhaps being intimidated by his reputation but more likely put off by his character and mannerisms. But all the Post Office engineers who worked with him ... found him very easy to understand. ... Their respect for him was immense,' though they also voiced the caveat at the end of Mitchie's recollection:

> He was intrigued by devices of every kind whether abstract or concrete — his friends thought it would be better if he kept to the abstract devices but that didn't deter him.

Before the war, Turing had himself designed an electrically operated enciphering machine and built part of it with his own hands. (He also made a start at building an ingenious analogue device for calculating the zeros of the Riemann zeta function. Just as importantly, some of his pre-war work was on what were then some of the deeper parts of probability theory.) It is interesting to note that Shannon, whose reputation also rests on rather deep and abstract results, is another great gadgeteer whose collected papers include one on the building of a calculator to work in Roman numerals (rather than binary).

Peter Hilton, an admiring younger colleague, recalled that

> [t]here was always a sense of immense power and ability to tackle every problem, and always from first principles. I mean, he not only, in our work during the war, did a lot of the theoretical work but he actually designed machines to help in the solution of problems — and with all the electrical circuitry that would be involved, as well. In all these ways he

always tackled the whole problem and never ran away from a calculation. If it was a question of wanting to know how something would in fact behave in practice, he would then do the calculations as well.

Hilton goes on to describe one of Turing's odder gadgets (a metal detector) adding 'But it worked — it worked. As with all things with Turing, it really did work.'

He also tells a typical Turing story.

Turing was a civilian, working in Intelligence, and he believed — again typical of Turing thinking in first principles — that the Germans might very well invade England and that he should be able to fire a rifle efficiently, and so he enrolled in what was called the Home Guard. The Home Guard was a civilian force, but which submitted to military training and in particular its members learnt how to fire a rifle. ... In order to enroll you had to complete a form, and one of the questions on this form was: 'Do you understand that by enrolling in the Home Guard you place yourself liable to military law?' Well, Turing, absolutely characteristically, said 'There can be no conceivable advantage in answering this question: "Yes"', and therefore he answered it "No". And of course he was duly enrolled, because people only look to see that things are signed at the bottom. And so he was enrolled, and he went through the training, and became a first class shot. Having become a first-class shot, he had no further use for the Home Guard, so he ceased to attend parades. And then in particular we were approaching a time when the danger of a German Invasion was receding and so Turing wanted to get on to other and better things. But, of course, the reports that he was missing on parade were constantly being relayed back to Headquarters and the officer commanding the Home Guard eventually summoned Turing to explain his repeated absence. It was a Colonel Fillingham, I remember him very well, because he became absolutely apoplectic in situations of this kind. This was perhaps the worst he had to deal with, because Turing went along and when asked why he had not been attending parades he explained it was because he was now an excellent shot and that was why he had joined. And Fillingham said: 'But it is not up to you whether you attend parades or not. When you are called on parade, it is your duty as a soldier to attend.' And Turing said: 'But I am not a soldier'. Fillingham: 'What do you mean, you are not a soldier! You are under military law!' And Turing: 'You know, I rather thought this situation could arise,' and to Fillingham he said: 'I don't know I am under military law.' And anyway, to cut a long story short, Turing said: 'If you look at my form you will see that I protected myself against this situation.' And so, of course, they got the form; they could not touch him; he had been improperly enrolled. So all they could do was to declare that he was not a member of the Home Guard. Of course that suited him perfectly.

It was quite characteristic of him. And it was not being clever. It was just taking this form, taking it at face value and deciding what was the optimal strategy if you had to complete a form of this kind. So much like the man all the way through.

He remembers that

[w]e were all very much inspired by him, [not only] his interest in the work but the simultaneous interest in almost everything else. As I say, it might be chess, it might be Go, it might be tennis and other things. And he was a delightful person to work with. He had great patience with those who were not as gifted as himself. I remember he always gave me enormous encouragement when I did anything at all noteworthy. And we were all very fond of him.

Turing was thus the natural person to develop machines to break codes produced by machines. It was an idea whose time had come. The code-breakers of all the major powers used the latest punch-card machinery. The code-breaking section of the German Foreign Office had many machines 'assembled out of standard parts for special purposes by Hans-Georg Krug, a former high school mathematics teacher who possessed a positive genius for this sort of thing'. We have already mentioned the Polish *bomby* which stepped through the 26^3 different positions of a given Enigma. But, in all these applications, the machines supplied the brute-force and human beings the subtlety. The Turing bombe supplied both.

The new bombes were the work of many people — one key idea came from Welchman and, no doubt, the fact that the Polish *bomby* had actually worked must have been a substantial encouragement — but it seems clear that the driving force and the source of many of the ideas was Turing. I shall content myself with trying to show why the Military Enigma was breakable in principle. The kind of electrical circuitry required is shown in the Appendix to Welchman's book. Let us suppose we guess a 12-letter stretch in some enciphered message. The method described at the end of Section 13.2 calls for us to set up 12 linked mock Enigmas (so that the outer rotor letter of each Enigma is one step further advanced than that of the previous one) and then drive the machine through its 26^3 possible states. As before, we shall assume that the inner rotors of the actual Enigma used to produce the message did not move during the encipherment of our 12 letters. The reasons for making this assumption are the same as those set out in Section 13.2.

We shall concentrate on the specific example given in Table 15.1. The second column shows the message, and the final column the enciphered message. The middle block shows the substitution alphabets supposed to be produced by the Commercial Enigma without plugboards. Thus looking at line 5 we see that, in the absence of a plugboard, the Enigma would encipher the 5th letter of the message according to the rule A goes to O, B goes to F and so on.

Let us write D ↔ J if D is steckered to J. Instead of considering all possible plugboard arrangements, let us ask simply which steckerings are possible for A. We begin by considering the possibility

(1) A ↔ A.

Looking at row 1, we see that the plugboard takes A to A, the Commercial Enigma takes A to J so, in order to arrive at the encipherment, the plugboard must take J to M. Thus J ↔ M. In the same way, looking at row 6, we see that V ↔ K. Thus

(2) A ↔ A, J ↔ M, V ↔ K.

Notice that at stage (2) we actually have five pieces of information since the symmetric plugboard arrangement also gives M ↔ J and K ↔ V. Notice also that any one of the three statements in (2) implies the two others.

Since M ↔ J, we can now use row 3 to show that O ↔ S. But we can do more. The reader may remember that in Section 13.3, using the advantages of 20/20 hindsight, I hinted that the self-inverse property of the Enigma was a cryptographic weakness. Since MLSYJKYHMLSH is encrypted as ADMIRALOFTHE, we may work from right to left as well as from left to right. Looking at row 5, we see that the plugboard takes J to M, the Commercial Enigma takes M to N so, in order to arrive at the encipherment, the plugboard must take N to R. Thus N ↔ R. In the same way, looking at line 9, we see that F ↔ Y. Thus

Table 15.1. *Successive Enigma encipherments.*

	In	ABCDEFGHIJKLMNOPQRSTUVWXYZ	Out
1	A	JFQXHBSEKAIYZTVUCWGNPORDLM	M
2	D	PNSKUZOWVLDJRBGATMCQEIHYXF	L
3	M	KDPBIQTMEOANHLJCFTGRXZYUWV	S
4	I	XODCHZLEPYUGQWBIMVTSKRNAJF	Y
5	R	OFYEDBZXLWQINMASKVPUTRJHCG	J
6	A	VWPTMXKUOLGJEZICRQYDKABFSN	K
7	L	DRTAGUEMZKJXIVYWSBQCFNPLOI	X
8	O	VZEJCQUNLDYIRHWXFNTSGAOPKB	H
9	F	HXIZGPEACYOVSTKFWUMNRLQBJD	M
10	T	HVXMZIKAFSGWDQURNPJYOBLCTE	L
11	H	OWMPYLTZKXIGCUADRQVFNSBJEH	S
12	E	SUZLJYITGEMDKXQROAPHBWVNFC	H

(3) $A \leftrightarrow A$, $J \leftrightarrow M$, $V \leftrightarrow K$, $O \leftrightarrow S$, $N \leftrightarrow R$, $F \leftrightarrow Y$,

and, as before, any one of the statements in (3) implies all the others.

Line 8 from left to right now gives $T \leftrightarrow H$, line 4 from right to left $Z \leftrightarrow I$ and line 11 from right to left $A \leftrightarrow H$ so

(4) $A \leftrightarrow A$, $J \leftrightarrow M$, $V \leftrightarrow K$, $O \leftrightarrow S$, $N \leftrightarrow R$, $F \leftrightarrow Y$, $T \leftrightarrow H$, $Z \leftrightarrow I$ $A \leftrightarrow H$

and, as before, any one of the statements in (4) implies all the others.

We have now obtained a contradiction since (4) contains both the statement $A \leftrightarrow A$ and the statement $A \leftrightarrow H$. The most obvious thing to do is to stop at this point and examine the remaining 25 possibilities one by one. Once we have shown that each of the 26 steckerings $A \leftrightarrow A$, $A \leftrightarrow B$, $A \leftrightarrow C$, ... $A \leftrightarrow Z$ are impossible, we know that the particular Enigma position cannot produce the correct encipherment *whatever the plugboard* and we can move on to the next one.

However, the most obvious thing to do is not necessarily the best one. Observe that we do not actually have to test for the possibility $A \leftrightarrow H$, since any one of the statements in (4) implies all the others so, in particular, $A \leftrightarrow H$ implies $A \leftrightarrow A$. This insight suggests that it might be worth while to press on to stage (5). Using line 10, together with the statement from (4) that $T \leftrightarrow H$, we obtain $A \leftrightarrow L$ so we will not have to test for this possibility. We observe also that we can reuse line 1 together with the statement $A \leftrightarrow H$ from (3) to get $E \leftrightarrow M$ and that line 6 can be reused in a similar manner.

With luck, as we move from stage (4) to (5), from (5) to (6) and onwards we will gather an avalanche of statements which will eventually include all of $A \leftrightarrow A$, $A \leftrightarrow B$, $A \leftrightarrow C$, ... $A \leftrightarrow Z$. All of these statements will be deducible from each other and all will thus give contradictions. (This would parallel the principle of formal logic which states *ex falso quodlibet* that is 'given any false statement we may deduce anything we want from it'†.) Thus working out all the consequences of the single statement $A \leftrightarrow A$ will (with luck) suffice *by itself* to exclude the possibility that the particular Enigma position could produce the correct encipherment with any plugboard whatsoever.

Moving a little closer (but not much) to the practical implementation of this scheme, I ask the reader to consider the 'skeletons' shown in Figure 15.1. Here we connect two letters if one is the encipherment of the other so that, for example, looking at row 1, A is joined to M, looking at row 2, D is joined to L. The letters then fall into a collection of connected components. In this case I have labelled them (a), (b), (c) and (d). It turns out that it is possible to set up electrical circuits in such a way as to perform all possible deductions (involving a specified letter, say A, and starting from a single specified steckering, say $A \leftrightarrow A$) essentially simultaneously. The only limitation is that we cannot use

†It is said that the philosopher McTaggart once challenged Bertrand Russell to use the statement '1=2' to prove that he was the Pope. This presented no problem. 'I and the Pope are two. But 1=2 so I and the Pope are one.'

more lines than there are mock Enigmas in a bombe. A standard bombe had 12 mock Enigmas, so we could use all of (a), (b), (c) and (d), but the first two prototypes only had 10 so we could only use (a) and (b) for them. We refer to the collection of components used (together with the list of the lines involved) as the menu.

This brings us to the second point. I have spoken of a possible 'avalanche of statements' but I have not given much evidence that it will occur. Actually, there are two separate points involved:

1   Will the avalanche start?

2   Once started, will it continue?

Both of these questions should more correctly be posed probabilistically as follows. Given a random choice of 12 (or whichever number of alphabets† is involved) is there a very high probability that:

†That is mock Enigmas, or lines involved by the menu.

1′   the avalanche will start, and

2′   once started will continue?

In Section 17.1 I will discuss the much easier problem of the propagation of surnames. To see a connection, consider a 'mother statement' say A ↔ A from which we deduce 'daughter statements' say J ↔ M, V ↔ K. Some statements will have no offspring, some will have one daughter, some two and so on. It is clear that in order to produce an avalanche, each mother must produce an average of more than one daughter (and, of course, the larger the family size the better). However, the problem is complicated by the fact that no 'generation' of daughters, granddaughters, great-granddaughters, or whatever, can exceed the number of lines involved by the menu.

It is clear on common sense grounds that (if avalanches occur at all) some menus must be more likely to provoke avalanches than others. The menu shown in Figure 15.1 is not a very good menu (for the

Figure 15.1: A possible menu.

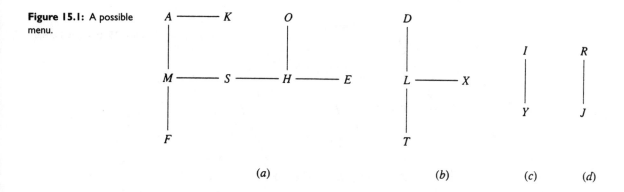

(a)   (b)   (c)   (d)

purposes of exposition, it helps if the number of deductions at each step is not too large) since it lacks closed circuits of the type shown in Figure 15.2. The reader should not find it hard to convince herself that the existence of closed circuits increases the probability that 'deductions get fed back into the system' and so increases the average family size.

Exercise 15.1.1   *Suppose that the encipherment of* THEGENERALHE *is*

ALAPHQNSNTTL

but the trial alphabets are as in Table 15.1. (For convenience we give the appropriate alphabets in Table 15.2.) Draw a menu and carry out a few stages of the appropriate deductions.

In Chapter 13 of [94], Derek Taunt discusses the work involved in setting up a menu. We can well believe that 'Much skill, ingenuity, and judgement could be expended on the composition of good menus from otherwise intractable material.' The construction of the prototype

Table 15.2. *A possible Enigma encipherment?*

	In	ABCDEFGHIJKLMNOPQRSTUVWXYZ	Out
1	T	JFQXHBSEKAIYZTVUCWGNPORDLM	A
2	H	PNSKUZOWVLDJRBGATMCQEIHYXF	L
3	E	KDPBIQTMEOANHLJCFTGRXZYUWV	A
4	G	XODCHZLEPYUGQWBIMVTSKRNAJF	P
5	E	OFYEDBZXLWQINMASKVPUTRJHCG	H
6	N	VWPTMXKUOLGJEZICRQYDKABFSN	Q
7	E	DRTAGUEMZKJXIVYWSBQCFNPLOI	N
8	R	VZEJCQUNLDYIRHWXFNTSGAOPKB	S
9	A	HXIZGPEACYOVSTKFWUMNRLQBJD	N
10	L	HVXMZIKAFSGWDQURNPJYOBLCTE	T
11	H	OWMPYLTZKXIGCUADRQVFNSBJEH	T
12	E	RUZLJYITFEMDKXQROAPHBWVNFC	L

Figure 15.2: A good menu.

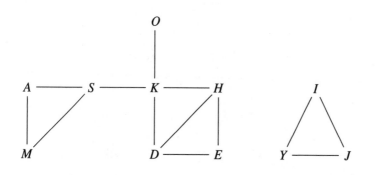

bombes must have been accompanied by many hand trials, much discussion of more or less plausible mathematical models and a great deal of nail-biting.

So far we have dealt with what will happen if our rotors and their positions do not correspond to those in the original Enigma. What will happen if they do? Table 15.3 shows an encipherment (for some unrevealed plugboard setting) corresponding to the 12 substitution codes shown.

Exercise 15.1.2 *Suppose we are in the situation given by Table 15.3. Draw the menu and carry out completely all the appropriate deductions which can be made starting from* E ↔ L.

I believe that the reader who carries out Exercise 15.1.2 will obtain the statements E ↔ L, N ↔ N, I ↔ I, Z ↔ B, D ↔ A, H ↔ Y, G ↔ C, X ↔ F, T ↔ Q and no more. This suggests that (as, in fact, is the case) we have the correct alphabets (and so the correct rotors in the correct positions) and the correct steckering E ↔ L. What will happen if we test another steckering, say E ↔ E?

Exercise 15.1.3 *Suppose we are in the situation given by Table 15.3 (and for which you drew the menu in Exercise 15.1.2). Carry out the first few stages of deductions starting from* E ↔ E.

Once again we expect an avalanche of statements to follow from the false assumption E ↔ E. However, *there is one statement about the steckering of* E *that cannot form part of the avalanche.* We have already seen that every statement in the avalanche is deducible from every other.

Table 15.3. *A correct Enigma encipherment.*

	In	ABCDEFGHIJKLMNOPQRSTUVWXYZ	Out
1	T	JFQXHBSEKAIYZTVUCWGNPORDLM	G
2	H	PNSKUZOWVLDJRBGATMCQEIHYXF	F
3	E	KDPBIQTMEOANHLJCFTGRXZYUWV	N
4	G	XODCHZLEPYUGQWBIMVTSKRNAJF	A
5	E	OFYEDBZXLWQINMASKVPUTRJHCG	I
6	N	VWPTMXKUOLGJEZICRQYDKABFSN	B
7	E	DRTAGUEMZKJXIVYWSBQCFNPLOI	F
8	R	VZEJCQUNLDYIRHWXFNTSGAOPKB	J
9	A	HXIZGPEACYOVSTKFWUMNRLQBJD	B
10	L	HVXMZIKAFSGWDQURNPJYOBLCTE	B
11	H	OWMPYLTZKXXIGCUADRQVFNSBJEH	L
12	E	RUZLJYITFEMDKXQROAPHBWVNFC	A

But in Exercise 15.1.2 we saw that the statement $E \leftrightarrow E$ was not among those deducible from $E \leftrightarrow L$ so it follows that the statement $E \leftrightarrow L$ will not be among those deducible from $E \leftrightarrow E$. On the other hand, there seems to be no reason to exclude any other statement concerning the steckering of E from the avalanche. Thus, with luck, statements deducible from $E \leftrightarrow E$ will include 25 of the 26 statements $E \leftrightarrow A$, $E \leftrightarrow B$, $E \leftrightarrow C$, ... , $E \leftrightarrow Z$ but cannot include the statement $E \leftrightarrow L$.

If we test a particular steckering, say $X \leftrightarrow X$ by running our mock Enigmas through their 26^3 positions, then at each step one of three desirable things may happen:

1   All 26 steckering possibilities for X are deducible. We then *know* that we have the wrong Enigma set up and the machine moves on.

2   No new steckering possibilities for X are deducible. There should be a good chance that we have the right Enigma rotors in the right positions (and, though this is far less important, that our test stecker for X is correct). We should investigate further 'by hand'.

3   All but one of the 26 steckering possibilities for X are deducible. There should be a good chance that we have the right Enigma rotors in the right positions (and, though this is far less important, that the excluded steckering possibility for X is correct). We should investigate further 'by hand'.

Of course, there remains a fourth, unpleasant, possibility:

4   The number of steckering possibilities for X which are deducible is k with $2 \le k \le 24$.

Provided this only occurs rarely, we can also investigate the associated Enigma setting by hand. If not, then (since the technology of 1940 will not permit a more complicated scheme) we will have to revert to our original naïve approach and examine each of the 26 steckerings $X \leftrightarrow A$, $X \leftrightarrow B$, $X \leftrightarrow C$, ... $X \leftrightarrow Z$ in turn. Such an approach will take 26 times as long.

Fortunately, as we now know, it was possible to find menus which provoked avalanches and the approach outlined above did actually work. The plugboard had been vanquished though the effectiveness of the approach remained crucially dependent on the number of bombes involved and the time available to use them. Examination of a 12-letter menu required 12 linked mock Enigmas (forming one bombe) which took perhaps 15 minutes to work through all the possibilities. Recall that the Army and Air Force Enigmas had 60 rotor orders and the Navy Enigma 336.

15.2 The bombes at work

There is a further problem. If the reader looks at the description at the end of Section 13.3 of how the German Army transmitted its 'text setting' (the stage of the Enigma cycle at which the message began), she will see that, once the daily setting has been discovered, it is as easy for the code-breakers to read the day's messages as it is for the intended recipient. The Navy provided no such obvious help. It thus appears that the deciphering of one message (the contents of which we may well already know, since our method depends on knowing or guessing several letters of the text) provides no help in deciphering another. A moment's reflection convinces us that this ought not to be so since our first successful decipherment revealed both the plugboard arrangement and the particular Enigma cycle (specified by the arrangement of the rotors and the state of the counting mechanism at one point in the cycle). We have thus reduced the problem of deciphering any given message from one involving astronomical numbers of possibilities to one of finding which particular stage, out of a possible 26^3 of the Enigma cycle, the machine was in at the start of the message.

In English, the most frequently occurring single letter is E, the most frequent sequence of two letters is TH and the most frequent sequence of three letters is THE. (Having made such a statement we must immediately qualify it. Presumably, different kinds of communication — military, scientific, and so on — have different frequencies for given sequences. Again different strategies for dealing with the spaces between words — replacing them by X, ignoring them, and so on — will also alter frequencies, possibly quite substantially.) Our attack will concentrate on what we expect to be a frequently occurring sequence of four letters, say THEX. By running an Enigma-like machine through the entire cycle we obtain the $26^3 = 17\,576$ different encipherings of THEX, corresponding to each starting point on the Enigma cycle, and record them on punch-cards. The kind of punch-card machinery available in 1940 was perfectly capable of scanning a text (once that, too, had been entered on punch-cards) and detecting coincidences. If a possible enciphering of THEX is detected we can set up an Enigma so that it is at the appropriate stage in its cycle and run it over the whole enciphered message and see if the trial decipherment makes sense.

Exercise 15.2.1 *(i) Show that the probability of a random sequence of four letters corresponding to one of our punch-cards is 1/26 and conclude that scanning a 200-letter message will produce, on average, about eight false alarms.*

(ii) What would happen if, instead of using a four-letter sequence, we used a three-letter sequence? What would happen if, instead of using a four-letter sequence, we used a five-letter sequence? Why do we use a four- rather than a three- or five-letter sequence?

Exercise 15.2.1 shows that the number of false alarms is not too great but, of course, the method is only useful if there is a reasonable probability of a positive outcome. Fortunately, the German language is rich in frequently occurring four-letter sequences ('tetragrams' in code-breakers jargon). According to [94] page 114 the tetragram EINS was used but it appears to me that the method would still have been feasible if the test had been run several times with different tetragrams.

Exercise 15.2.2 *Suppose that each of n tetragrams has probability p of occurring in a 200-letter message and that their occurrence is independent (this cannot be strictly true but will be close to the truth in practice). Roughly how many false alarms will occur if we test for each one and what is the probability that at least one of the tetragrams will appear? Look at the results you have obtained for various values of p (say p = 0.25, 0.1, 0.01) and n.*

The bombes were tended by Wrens, that is members of the Women's Reserve Naval Service. One of them, Diana Payne, recalls an interview at which she was asked if she could keep a secret, to which 'I answered that I really did not know as I had never tried.' A few weeks later she was ordered to Bletchley where in the morning

> [t]he conditions of the work were disclosed. It would involve shift work, very little hope of promotion, and complete secrecy. On this limited information we were given until lunchtime to decide whether we could face the ordeal.

She and the Wrens with her

> decided to face the challenge, and all signed the Official Secrets Act which committed us to the job for the duration of the war, and to keep the secret forever.

After suitable training the work began.

> The bombes were bronze-coloured cabinets about eight feet tall and seven feet wide. The front housed rows of coloured circular drums, each about five inches in diameter and three inches deep. Inside each was a mass of wire brushes, every one of which had to be meticulously adjusted with tweezers to ensure that the electrical circuits did not short. The letters of the alphabet were painted round the outside of each drum. The back

of the machine almost defies description — a mass of dangling plugs on rows of letters and numbers.

We were given a menu, which was a complicated drawing of numbers and letters from which we plugged up the back of the machine and set the drums on the front. ...

We only knew the subject of the key and never the contents of the messages. It was quite heavy work getting it all set up, and [we needed] good height and eyesight. All this work had to be done at top speed, and at the same time 100 per cent accuracy was essential. The bombes made a considerable noise as the drums revolved, each row at a different speed, so there was not much talking during the eight-hour spell. [From time to time] the bombe would suddenly stop, and we took a reading from the drums.

Since such a stop only indicated a possible setting and could have been due to what was called a 'legal contradiction', the bombe was restarted and the reading hurriedly phoned through to another room where operators using British Typex machines modified to model the German Enigmas would carry out the final decipherment. If they were successful

the good news would be a call back to say, 'Job up; strip machine'. It was a thrill when the winning stop came from one's own machine.

The Wrens worked on watches of four weeks duration: 8 a.m. to 4 p.m. the first seven days, 4 p.m. to midnight the second week, midnight to 8 a.m. the third and then a hectic three days of eight hours on and eight hours off, ending with a much needed four days leave. The bombes broke down frequently and could deliver unpleasant electric shocks. It was a monotonous 'life of secrecy and semi-imprisonment' overshadowed by the knowledge that 'any mistake or time wasted could mean lives lost.' The strain showed itself in nightmares and digestive troubles. Occasionally girls would simply collapse or go berserk on duty†.

More and more bombes were built and more and more Wrens were recruited to run them. Since they were in the Navy they worked under Naval discipline and the places they worked were treated as ships (some of them even had a quarter deck which all Wrens had to salute‡). And then after three and a half years

Helen Rance, one of the hundreds of ... Wrens of HMS *Pembroke* remembers the 'eerie silence' in the rooms of that stone frigate when the bombes were finally switched off. Several of the girls could not resist rolling the heavy drums across the floor — something they had never been allowed to do before; then ' ... we sat down and took all the drums to pieces with screwdrivers; everything was dismantled.'

†But, although at least one British dramatist has portrayed Bletchley as run along the lines of the society of H. G. Wells's *The Time Machine* with the Eloi and Morlocks of different sexes, the correct reference is surely to the same author's *The Land Ironclads.*

‡Naturally, the minibus in which they went on leave was known as 'The Liberty Boat'.

Exercise 15.2.3   *Welchman says that 'It was possible to choose a sequence of [rotor] orders that would not call for more than one drum
[corresponding to one rotor] to be changed between successive runs.' Set
up such a schedule for the 60 runs required by the five German Army
rotors.*

With the coming of the bombes, the breaking of German codes could
take place on an industrial scale. But, although many messages could be
read routinely, the vital Naval Enigma had a further layer of protection.
For the bombes to work they needed an initial crib, that is part of a
message known or guessed both in unenciphered and in ciphered form.
The knowledge, gained earlier, of the kinds of messages transmitted by
particular Army or Air Force sources gave such cribs, but the Naval
Enigma had not been read before, so no such cribs existed. Moreover,
the standard of cryptographic discipline of the German Navy was high.
Messages were kept short and often encoded by another method before
being enciphered by Enigma. The Navy maintained groups to monitor
its own communications and report on any lapses detected. Under these
conditions, where could the needed cribs be found?

For reasons ultimately, but subtly, connected with the earth's rotation,
weather conditions tend to move from West to East. The U-boats in the
Atlantic and the German armies in Europe needed weather forecasts
and the forecasters needed information from over the Atlantic. The
information was gathered in three ways: by reports from German
warships and merchant ships (and, after the first year, the surface
dominance of the British Navy meant that only U-boats were left in
the Atlantic), by aircraft flights and by weather ships. The British
managed to break the codes used in the retransmission of these reports
and the weather forecasts based on them. This had the immediate
advantage of giving the British meteorologists extra information on
which to base their own forecasts and raised the possibility of using
the decoded weather reports and forecasts as cribs for the U-boat
Enigma. It was not surprising that the British Navy reacted strongly
and unfavourably to an RAF proposal to shoot down German aircraft
engaged in meteorological flights.

However, the submarines did not transmit Enigma encipherings of
weather reports directly, but first encoded the reports in the form of
short signals of just a few letters long (we give an example in Section 16.2). This was done mainly to keep radio transmissions as short
as possible, but the extra level of security was an added bonus for
German cryptographers. The method for encoding the weather reports was given in a booklet which we shall call the Short Weather

†By a history student named Hinsley recruited straight from Cambridge. When I was an undergraduate, Hinsley and his colleagues were still around, some running the University, some trying to teach me algebra. They seemed to me pleasant, unadventurous but, above all, conventional people.

The oldest hath borne
   most; we that are young
Shall never see so much,
   nor live so long.

Cipher. A plan was conceived† to capture a German Weather ship and seize the month's Naval Enigma settings together with the precious Short Weather Cipher. How this was done, not once but twice, without arousing German suspicions forms the central episode of Kahn's exciting *Seizing The Enigma*.

Armed with these documents, Bletchley was at last able to read the U-boat Enigma and to read it fast enough to influence events. At a rough estimate, the evasive routing of convoys using the information thus yielded saved about 2 000 000 tons of shipping (or between 300 and 350 ships) during the last six months of 1941. The decrypts also enabled Naval Intelligence to build up a detailed picture of the U-boat organisation and tactics.

A naval officer posted to Bletchley vividly remembers

> the sense of shock produced on my first arrival at the Park by the grimness of its barbed-wire defences, by the cold and dinginess of its hutted accommodation, and by the clerk-work we were first set to do. But this was soon swept aside by the much greater shock of discovering the miracles that were being wrought at the Park. In Iceland I had been interrogating the survivors of the many merchant ships sunk in the, at first, highly successful offensives against the Atlantic convoys launched by the U-boats ... I had spent many hours trying to analyse their strength and tactics. I could have spared my pains. For I now discovered that all this, and everything else about U-boats, was known with precision by those privy to the Enigma decrypts‡. Leafing through the files of past messages ... I shivered at seeing the actual words of the signals passing between Admiral Dönitz and the boats under his command whose terrible work I had seen at first hand. ... No less shocking was the revelation of the bestiality that underlay this sophisticated form of warfare. This emerged vividly from Dönitz's exhortations to his captains — 'Kill, kill, kill!' and from the names given to the wolf-packs, such as *Gruppe Blutrausch* ('Blood Frenzy').

‡The information which Blackett attributes to 'the accounts of prisoners of war from sunken U-boats' presumably came, in great part, from Enigma decrypts still secret at the time he wrote.

It was well that Bletchley was so successful for those six months because the British advantage did not last much beyond them.

15.3 SHARK

From the first day of February 1942 darkness once more obscured Dönitz's communications from his opponents. The German Navy had introduced the four-rotor Enigma. A total redesign would have created many difficulties both in the manufacture and use of the new machine, so the new four-rotor Enigma was a clever modification of the old three-rotor one. By using a new thinner reflector it became possible to introduce a fourth 'thin rotor' which did not revolve but which could be set in any of 26 positions. In order to retain compatibility with the three-rotor Enigma used in the other armed services, the combination of the new reflector and a particular 'neutral' position of the new thin rotor replicated the effect of the old reflector.

Clues as to the nature of the new machine had accumulated at Bletchley during the previous six months, and an incident in which a message had been transmitted by mistake in the new code and then (by a much greater mistake) retransmitted in the old, enabled the British cryptologists to reconstruct the wiring of the fourth rotor and the new reflector before they entered service. Even with this knowledge, the introduction of the new machine remained a catastrophe for Bletchley. The fourth rotor multiplied the number of starting positions and thus the labour of the brute-force searching on which the decoding depended by a factor of 26. The new system was called SHARK by those who now sought to break it.

If bombes could have been brought forth in unlimited amounts, the situation would have been less serious, but at the end of 1941 there were only 12 bombes in action and by the end of 1942, only 49. On the 14 March 1942, a long message containing news of Dönitz's promotion to full Admiral was sent out in another code which Bletchley could read. On the assumption that a particularly long SHARK transmission was an encoding of the same message, the bombes were set to work and succeeded in recovering that day's keys. However, this required 6 bombes working for 17 days. It would have required 100 bombes working full time to reduce the decoding time to one day. (Nor was Dönitz promoted every day. The Germans had also changed their weather signals, so where were the cribs to come from to start the process?)

Plans were immediately made to build new four-rotor bombes working 26 times as fast, but the old three-rotor bombes had been built as close to the limits of existing technology as could be managed. It is not surprising that the task proved more difficult than expected and the first fast four-rotor bombes only entered service (and, according to Hinsley, experienced severe teething problems) on the British side in June 1943 and (slightly later but with fewer problems) on the American side in August 1943. By the end of 1943, a substantial number of four-rotor bombes would be in service but until then the old slow, three-rotor bombes would have to suffice.

In 1941, about 4 300 000 tons of shipping (representing about 1300 ships) had been lost world-wide, of which 2 400 000 tons (about 500 ships) were lost in the North Atlantic. German losses amounted to

35 U-boats. In 1942, the figures rose to 7 800 000 tons of shipping (representing about 1700 ships) lost world-wide, of which 5 500 000 tons (about 1000 ships) were lost in the North Atlantic. German losses amounted to 86 U-boats. It took the United States six months from its entry into the war, in December 1941, to organise proper coastal convoys and the resulting 'happy time' for the U-boats, whilst increasing the total tonnage sunk, took pressure off the Atlantic convoys for a time, but by the end of 1942 the main battle had returned to the Atlantic.

In retrospect, we may see some gleams of hope. Such was American industrial might that, by the end of 1942, new Allied ship building could more or less keep pace even with these dreadful losses. (It was not, however, clear that replacement crews could forever be found to man replacement ships.) This meant that, in the coldest strategic terms, the Battle of the Atlantic was not being lost, though neither was it being won. It was also true that, although the exchange rate of U-boats destroyed against ships sunk remained extremely unfavourable, the number of U-boats required to sink a given tonnage was steadily increasing. What had been courageous weighing of odds in the presence of newly-trained escorts became foolhardy risk-taking against more experienced and better trained opponents. One by one, the U-boat aces were killed or captured whilst the more cautious captains survived.

This, also, was little comfort to the convoys. As the American official historian wrote:

> By April 1943, the average kill per U-boat had sunk to 2000 tons [per month]. This might be interesting as a sort of sporting score, but the number of U-boats operating had so greatly increased that it was of little significance in solving the problem. When Daniel Boone, who shot fifty bears a year, was replaced by fifty hunters who averaged one each, the bears saw no occasion to celebrate the decline in human marksmanship.

Dönitz now had 200 submarines available. Storms in December 1942 and January 1943 greatly hampered the U-boats, though they sank seven out of nine tankers in one convoy heading for North Africa. In February, the great North Atlantic convoy battles resumed. A six-day battle in early February involved 21 submarines against 63 merchant ships with 12 escorts. Twelve merchant ships were sunk by torpedoes and another sank after collision, for the cost of three U-boats. The total of 380 000 tons for Allied losses in February was dwarfed by March's 590 000 tons. Two convoys, the one overtaking the other, were attacked day and night over a six-day period by 45 submarines. Twenty-one of the 92 merchant ships were sunk whilst the 18 escorts managed to sink only one submarine. Some 161 000 tons of cargo were lost including everything from steel and explosives to sugar, wheat and powdered milk.

The British official historian's summary runs as follows:

> The Admiralty ... recorded that 'the Germans never came so near to disrupting communications between the New World and the Old as in the first twenty days of March 1943'. Even at the present distance in time ... one [cannot] yet look back on that month without feeling something like horror over the losses we suffered. In the first ten days, in all waters, we lost forty-one ships; in the second ten days, fifty-six. More than half a million tons of shipping was lost in those twenty days: and what made the losses so much more serious than the bare figures can indicate, was that nearly two-thirds of the ships sunk during the month were sunk in convoy. 'It appeared possible,' wrote the Naval Staff after the crisis had passed, 'that we would not be able to continue [to regard] convoy as an effective system of defence.' It had, during three-and-a-half years of war, slowly become the lynchpin of our maritime strategy. Where could the Admiralty turn if the convoy system had lost its effectiveness? They did not know; but they must have felt, though no one admitted it, that defeat stared them in the face.

Just under half of the crews of merchant ships lost by enemy action could expect to survive. The official historian states that:

> It seems not unlikely that a quarter of the men who were in the Merchant Navy at the outbreak of war, and perhaps an even higher proportion, did not survive until the end, or if they survived, lived permanently damaged lives, still in the shadow of death.

Although certain élite corps (such as the bomber crews) suffered even higher casualty rates, proportionate losses among merchant seamen thus greatly exceeded those in the 'Fighting Services'. Traditionally, the occupation was poorly paid and insecure and though, mainly through the addition of 'war risk' (or 'danger') money, wages more than doubled during the first three years of war, the wage of an ordinary seaman in the middle of 1942 was less than £23 a month†.

†Agricultural labourers who were near the bottom of the wages ladder received about £10 a month. The poverty line for a family of five was considered to be about £18 a month. It must be remembered that wages were only paid for the voyage. The immemorial customs of the industry meant that survivors knew that their wages had ceased to be paid from the moment their ship sank.

> These, in the day when heaven was falling,
> The hour when earth's foundation fled,
> Followed their mercenary calling
> And took their wages and are dead.
>
> Their shoulders held the sky suspended;
> They stood, and earth's foundation stay;
> What God abandoned, these defended,
> And saved the sum of things for pay.

The effectiveness of the submarine campaign can be seen from the map in Figure 15.3 (copied from [111]). It shows how submarine activity had been pushed outward into the air-gap, but it also shows how Dönitz exploited every Allied weakness to sink yet more ships. This world-wide battle was directed by one man with a tiny staff, a Führer for submarine warfare moving his chess pieces in the service of a greater Führer.

But to control his chess pieces, the player needs to know where they are.

> Not only did returning U-boats always signal their expected time of arrival; every outward-bound U-boat reported on clearing Biscay or, if leaving from Norway or the Baltic, after crossing 60°N. Except when a U-boat sailed for a special task or a distant area, it received its destination point and its operational orders by [radio] after it had put to sea; these were matters which had to be settled in the light of the latest situation and of which all U-boats at sea had to be informed. No U-boat could deviate from its orders without requesting and receiving permission, and, without requesting and receiving permission, none could begin its return passage. In every signal it transmitted the U-boat was required to quote its present position; if it did not do so, or if it did not transmit for several days, it was ordered to report its position. To each of its signals, again, the U-boat was expected to append a statement of the amount of fuel on hand† ...

> The U-boat command ... ordered the formation and reformation of the patrol lines between specified geographical positions at regular intervals, addressing by name each U-boat commander who was to take place in a line and giving him his exact position in it. It informed each line of U-boats what to expect in the way of approaching convoys and, in addition, supplied for the attention of all U-boats at sea a steady stream of situation reports and general orders. When a convoy was sighted, the Command decided the time, the direction and the order of attack. The exercise of this degree of remote control ... required U-boats in contact with the target to transmit on high frequencies detailed descriptions of the situation.

†Hinsley gives the following signal as typical.

From: Schultze (U 432)
To: Admiral Commanding U-boats
In square 8852 have sunk one steamship (for certain) and one tanker probably. Set one tanker on fire on October 15th. My present position is square 8967. Have 69 cubic metres of fuel oil left. Have two (air) and one (electric) torpedoes left. Wind south-west, force 3 to 4. Pressure 996 millibars. Temperature 21 degrees centigrade.

The British had developed high frequency radio direction-finding equipment, 'Huff-duff', based on pre-war research on tracking thunderstorms. Under the impulsion of Blackett, eight land-based stations were set up from Land's End to the Shetlands. Exercise 3.3.2 explains why the line of stations was made as long as possible and also gives one of the many reasons why (as the Germans expected) they could not give accurate bearings from more than 500 kilometres away and were ineffective (at least for tactical purposes) in mid-Atlantic. However

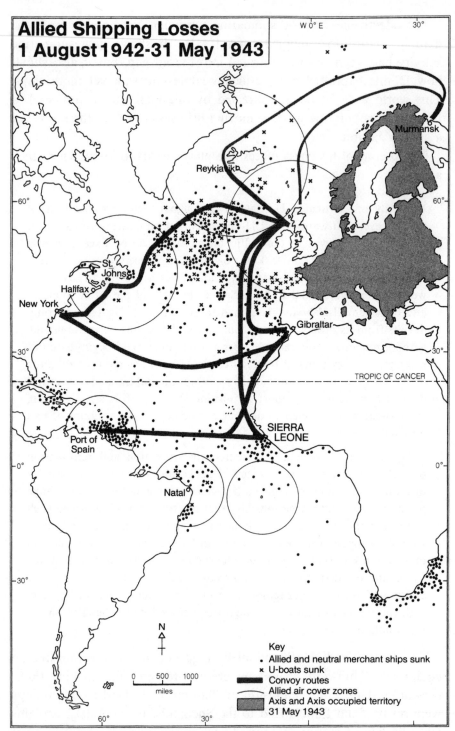

Allied Shipping Losses
1 August 1942-31 May 1943

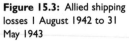
Figure 15.3: Allied shipping
losses 1 August 1942 to 31
May 1943

Key
• Allied and neutral merchant ships sunk
× U-boats sunk
— Convoy routes
— Allied air cover zones
▓ Axis and Axis occupied territory
31 May 1943

(as the Germans had not anticipated), Huff-duff sets were made small enough to go to sea with the escorts. This was a major tactical advance, but, to route convoys, only Enigma would do. It is no wonder that the Operational Intelligence Centre which sought to track the submarines urged the code-breakers to focus 'a little more attention' on SHARK and told them that the U-boat campaign was 'the one campaign which Bletchley Park are not at present influencing to any marked extent — and it is the only one in which the war can be lost unless BP *do* help. I do not think this is any exaggeration.'

In late October 1942, three British naval seamen boarded a sinking German submarine and rescued papers from it. First Lieutenant Fasson and Able Seaman Grazier went down with the submarine whilst seeking further papers. The third seaman, a cook's assistant called Brown, turned out to have lied about his age in order to join up. He was decorated for his bravery and sent home, only to die two years later trying to save his sister from a fire caused by a bombing raid.

Among the seized papers was the second edition of the Short Weather Cipher. Bletchley now had a supply of cribs to work on, but the fourth rotor still meant working through 26 times as many possibilities. On 13 December, they found a solution† and discovered, to their delight, that weather messages were transmitted with the fourth rotor in neutral, making the four-rotor machine in effect into a three-rotor machine. (This had been done so that the three-rotor machines of the weather service could continue to be used.) By itself, this should not have compromised SHARK, but it turned out that the same setting for the first three rotors was used in non-weather messages. Once the weather setting was known, the cryptanalysts had only to try the 26 possible settings of the thin rotor to solve the code completely. It is possible that the normally extremely competent German naval cryptologists made this mistake because of their obsessive fear of the enemy within. The fourth rotor was thus intended not to prevent the British breaking a system which was believed unbreakable, but to prevent unauthorised reading of submarine communications by other branches of the German military.

During the next few months, the British had increasing success in breaking keys, though there were crises when a new Short Weather Cipher went into effect and when a second thin rotor was introduced. SHARK keys were found for 90 of the 112 days between 10 March and the end of June. As we have seen, March witnessed the greatest German convoy victories of the war. Those victories were also the last. The figures of Table 15.4 tell their own story. At the end of May, Dönitz withdrew his forces from the North Atlantic to seek easier

†I cannot resist a little name-dropping here. The man in charge at the time that the break was made was Shaun Wylie, one of my predecessors at Trinity Hall.

targets elsewhere. In July, he tried the North Atlantic again when his submarines sunk 123 327 tons (18 ships) for a loss of 37 of their number but again withdrew, essentially for good. May marked the point when it became clear to Dönitz that 'We had lost the Battle of the Atlantic.'

Echoing Dönitz, Hitler attributed 'the temporary set-back to our U-boats ... to one single technical invention of our enemy,' that is the new 10 centimetre radar mounted in anti-submarine aircraft. To this, as Dönitz well realised, must be added the closing of the air-gap by very-long-range aircraft and improvised aircraft carriers. In theory, the submarine command was awake to the possibility that codes might be broken. In his memoirs, written before Allied code-breaking successes became known, Dönitz wrote:

> Our ciphers were checked and rechecked, to make sure they were un-breakable; and on each occasion the head of the Naval Intelligence Service at Naval High Command adhered to his opinion that it would be impossible for the enemy to break them.

In practice, to admit that the codes might be broken would destroy the entire Dönitz system and it was easier to blame radar, careless talk, spies† and bad luck. In the same way, although Huff-duff was more useful to escorts than 10 centimetre radar, all losses were blamed on radar.

†In 1945 an American Intelligence Officer noted that the bookcases of the German security services 'seemed to be lined with spy novels about the diabolically clever British.'

> Some U-boat commanders returning from patrol commented on the curious coincidence that a [wireless] transmission was so often followed by an attack, and suggested that there might be some connection. But [Naval Intelligence] could conceive of no such thing. Any U-boat commander who insisted was regarded as something of a crank.

One consequence of the lifting of the Enigma black-out was to administer a bitter though effective medicine to the British code-breakers who were also responsible for the security of British codes. The response of U-Boat Command to the re-routeing of convoys showed that the main British Naval cipher used for communication between the

Table 15.4. *Monthly losses in the North Atlantic, 1943.*

Month	Tonnage lost	Ships lost	U-boat losses
March	476 349	82	15
April	235 478	39	15
May	163 507	34	41

US and Royal Navies and convoy control had been comprehensively broken by German Naval Intelligence. The Royal Navy had the humiliating experience of being lectured to on security by its US counterpart. On 10 June, German Naval Intelligence noted that FRANKFURT had fallen silent. The new codes now introduced remained unbroken and Dönitz lost a source of information which, as he said, gave him half his intelligence.

The information provided by Enigma also enabled the Allies to hunt down nine out of the ten 'Milch Cows', large submarine tankers, on which Dönitz relied for the efficient conduct of operations in more distant waters. Although at one point the British dispatched a warning that some of the American successes were 'Too true to be good', the Germans seem to have taken the losses as just another disaster among many.

To this catalogue of German problems must be added new Allied weapons† and, thanks to operational research, more effective use of old ones. It is probable that the closing of the air-gap was the most important single factor in the victory but it was the combination of factors which made the victory overwhelming.

Dönitz continued his U-boat war until the bitter end. On the whole, naval historians hold this decision to be 'correct' in the sense that the U-boat threat continued to tie down large numbers of Allied naval ships and aircraft, although others point out that the U-boat campaign itself drew heavily on limited German resources. But, whether it was correct or not in a narrow military sense, in a broad strategic sense the decision was irrelevant. In the two remaining years of the war, only 337 Allied merchant ships were sunk at the cost of 534 U-boats. (Of the 45 U-boats which took part in the March convoy battles, only 2 survived the war and the rest were sunk, most without survivors.) Meanwhile a continuous stream of convoys crossed the Atlantic bearing the men, weapons and fuel required, first for the invasion of Italy and then for the Normandy landings and the subsequent battles.

We need not speculate about what would have happened if the Battle of the Atlantic had been lost. But if victory in that battle had required even a few more months, the Normandy landings would have been delayed for a year and we enter the province of 'alternative histories'. A more demonstrative age might have celebrated such a victory with a massive piece of statuary, perhaps showing a handsome but well-draped female representing Science with the armoured God of War prostrate under her foot, or perhaps the giant figures of Blackett and Turing gazing forever over the Western Approaches. In the absence of such a monument, Figure 15.4, taken from page 379 of Volume 2 of Roskill's splendid history, is a sufficient memorial.

†Including an air-dropped homing torpedo, 'Wandering Annie', developed by the Americans. Since it homed onto the cavitation produced by the submarine's propeller, the weapon could easily be countered by slowing the submarine. It was thus vital to keep the torpedo's existence secret. The first example was brought into Britain under tight security but the officer in charge relates that shortly afterwards 'I received a buff-coloured envelope with "OHMS" across the top, by ordinary post. Inside was a letter from His Majesty's Customs; they wanted to know why I had imported into the United Kingdom "packing cases containing what is believed to be some form of aerial homing torpedo for use against submarines." Why had I failed to declare them?'

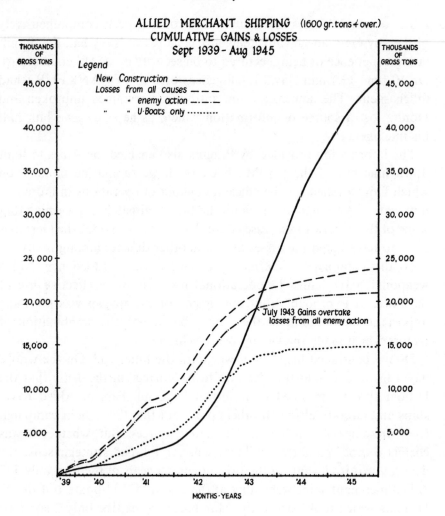

ALLIED MERCHANT SHIPPING   (1600 gr. tons + over.)
CUMULATIVE GAINS & LOSSES
Sept 1939 – Aug 1945

Legend
New Construction ————
Losses from all causes — — —
  "   " enemy action —..—..—
  "   " U-Boats only ....

July 1943 Gains overtake
losses from all enemy action

THOUSANDS
OF
GROSS TONS

MONTHS · YEARS

Figure 15.4: Cumulative gains and losses of shipping 1939–45.

CHAPTER 16

Echoes

16.1 Hard problems

At the end of the war, the British Government wished to hide two interlocked secrets — the fact that it and its American allies could read codes used by many other nations (there was a flourishing market in second-hand German Enigma machines) and the darker, greater secret, of how nearly the submarine war had ended in defeat†. Although thousands of people had worked in or with Bletchley, the secret was kept for 30 years. By the time the truth leaked out, the British had, finally, started to lose interest in their finest hour, and, in any case, there was hardly room for a homosexual pure mathematician in the pantheon of saviours of the nation‡.

However, fame, for mathematicians, consists in having their theorems remembered and their names mis-spelt. That Turing achieved this distinction in his own lifetime is revealed by the glossary of a 1953 book on computers.

> *Türing Machine.* In 1936 Dr Turing wrote a paper on the design and limitations of computing machines. For this reason they are sometimes known by his name. The umlaut is an un-earned and undesirable addition, due presumably to an impression that anything so incomprehensible must be Teutonic.

Turing also has the much rarer distinction, for a mathematician, of a first-class biography. Hodges's book [96] is a labour of love which had the unexpected, but fortunately temporary, side-effect of turning its hero into a cultural icon.

After the war, Turing worked on the development of the electronic computer foreshadowed in his Universal Machine and the later Bletchley machines. As might be expected, there was great debate at the time as to whether such machines might think, and, as might be expected from

†The nations of Western Europe still maintain extensive anti-submarine forces. In order to train these forces, they need substantial submarine fleets. In order to pay for these submarine fleets, they sell submarines and submarine technology to other nations. The possession of submarines by other nations increases the submarine threat against which it is necessary to maintain extensive forces.

‡Impressive and extensive vetting procedures now guard the security of Bletchley's successor establishment but, as I. J. Good who also worked on Naval Enigma observed, 'It was lucky that the security people didn't know about Turing's homosexuality early on, because if they had known, he might not have obtained his clearance and we might have lost the war.'

the writer of his 1936 paper, Turing held that:

1   It would eventually be possible to build machines whose outward
     'intellectual behaviour' would be indistinguishable from that of
     human beings.
2   Since the only way our inner 'thoughts' are displayed is by our
     outward 'intellectual behaviour', we would have to admit that such
     machines could think.

His views are set out in a beautifully clear paper entitled 'Computing
machinery and intelligence'. (It is much anthologized and appears, for
example, in Volume 4 of Newman's collection.)

A contrary view is set out by Roger Penrose in his book *The Emperor's
New Mind*. The reader who wishes to know more about Turing Machines
and related subjects can hardly do better than start with Penrose's book
(Hodges account [96] is also very lucid). In addition, Penrose provides
a penetrating overview of much of modern physics. I suspect that most
of my readers will incline to Turing's view but, even if correct, I do not
think that it is the last word.

One objection (with which he might well have agreed) is that an
aeroplane does not fly by imitating a bird, and a book records data
without imitating the brain. One of the aspects of human intelligence
that most interested Turing was chess playing and he outlined the
earliest chess program. Computers can now outplay 99.9% of the
human race, but they do it by using their 'inhuman' tactical strength
to overcome our 'human' strategic cunning. If machines cannot think,
it is futile to try and make them imitate human beings and, if they can
think, they can surely do better than copy a bunch of apes fresh out of
the trees.

The second objection is more fundamental. It says that, even if we
could construct a machine that behaved like a human, we would be
no closer to understanding the nature of thought. Three hundred years
ago, Leibnitz wrote

> ... supposing there were a machine, so constructed as to think, feel and
> have perception, it might be conceived as increased in size, while keeping
> the same proportion, so that one might go into it as in a mill. That being
> so, we should, on examining its interior, find only parts which work upon
> one another, and never anything to explain a perception.

Turing also thought deeply about morphogenesis, the process by
which an apparently undifferentiated, essentially symmetric, single cell
becomes a complex, highly differentiated fish or tiger or human being.
Some indication of the nature of Turing's problem may be found in the

sumptuous prose of D'Arcy Thompson, one of his predecessors in this field (though D'Arcy Thompson, writing early in this century, sought physical rather than biochemical answers to the question).

It is clear, I think, that we may account for many ordinary biological processes of development or transformation of form by the existence of trammels or lines of constraint, which limit and determine the action of the expansive forces of growth that would otherwise be uniform and symmetrical. The case has a close parallel with the operations of the glass-blower .... The glass blower starts his operations with a *tube*, which he first closes at one end so as to form a hollow vesicle, within which his blast of air exercises a uniform pressure on all sides: but the spherical transformation which this uniform expansive force would naturally tend to produce is modified into all kinds of forms by the trammels or resistances set up as the workman lets one part or another of his bubble be unequally heated or cooled. It was Oliver Wendell Holmes who first showed this curious parallel between the operations of the glass-blower and those of Nature, when she starts, as she so often does, with a simple tube. The alimentary canal, the arterial system including the heart, the central nervous system of the vertebrate, including the brain itself, all began as simple tubular structures. And with them Nature does just what the glass-blower does, and, we might say, no more than he. For she can expand the tube here and narrow it there; thicken the walls or thin them; blow off a lateral off-shoot or caecal diverticulum: bend the tube, or twist and coil it; and infold or crimp its walls as, so to speak, she pleases. Such a form as that of the human stomach is easily explained when it is regarded from this point of view; it is simply an ill blown bubble, a bubble that has been rendered lopsided by a trammel or restraint along one side, such as to prevent its symmetrical expansion — such a trammel as is produced if the glass blower lets one side of his bubble get cold, and such as is actually present in the stomach itself as a muscular band.

Just as Turing sought to reduce the most complex intellectual processes to a long series of simple steps, so he sought to find a few simple biochemical processes which would account for morphogenesis and allow the developing creature to be both the glass and the glass-blower. An indication of how this might be done is given in the second half of the discussion of Murray's law in Section 9.4.

There is a well known anecdote reported by de Morgan in his charming *Budget of Paradoxes*. (Laplace and Lagrange were the greatest mathematical physicists of Napoleonic times.)

Laplace once went in form to present [one of his books] to the Emperor. Napoleon, whom some wags had told that this book contained no mention of the name of God, and who was fond of putting embarrassing questions, received it with — 'Monsieur Laplace, they tell me you have written this large book on the system of the universe, and have never even mentioned its Creator.' Laplace, who, although the most supple of politicians, was as stiff as a martyr on every point of his philosophy or religion ... , drew himself up and answered bluntly, 'Je n'avais pas besoin de cette hypothèse-là' ['I had no need of that hypothesis']. Napoleon, greatly amused, told this reply to Lagrange, who exclaimed 'Ah! c'est une belle hypothèse; ça explique beaucoup de choses' ['Ah! it is beautiful hypothesis; it explains so many things'].

Turing was firmly on the side of Laplace.

After Turing's death, the unravelling of the structure and purpose of DNA showed that the structure of each living cell is, so far as we can now see, controlled by a tape in which four symbols are written out over and over again — striking, if not conclusive, evidence for Turing's view of the world. However, Turing's own work sought to deal not with the program but with the mechanism. He set up a mathematical model of a simple homogeneous mixture of chemicals and showed that small random disturbances (which must always occur) could lead to the growth of complex patterns. The work is suggestive and is still the subject of discussion and experiment (see, for example, [141] for an experiment which demonstrates the reality of such 'Turing Structures'), but though Turing showed how the Tiger might have got its stripes and the Leopard his spots, we still do not know if he was right†.

Morphogenesis is an example of an important but rather general question. (Incidentally, C. H. Waddington, whose book I used in the discussion of anti-submarine warfare, was a biologist deeply interested in this problem.) We close this section by discussing an important specific problem.

Recall that Turing's theorem tells us that there are problems for which there is no algorithm (more colloquially, there are problems with no solution). Developing this idea, people have shown that there are problems for which any algorithm must take a very large number of steps to find a solution (more colloquially, there are short problems with long solutions). Theorem 11.4.12 showed that, if $N \geq 4$, any method of sorting a set of 2^N unequal numbers $x_1, x_2, \ldots, x_{2^N}$ requires at least $N2^N/16$ comparisons. Mathematicians have constructed problems such that the number $f(n)$ of steps required to solve the problem with n pieces of data increases explosively fast.

Code-makers, like the manufacturers of the Enigma, seek to construct a problem (finding the code from the encoded message) which is hard to solve but easy to set (since it should be easy to encode the message). However, the code-breaking problem has a further property — once

†Stop press. The 31 August 1995 issue of *Nature* (Volume 376, Number 6543, pages 765–8) contains an article by S. Kondo and R. Asai which makes a strong case for the stripes of the angel fish as a Turing system.

we have found the solution it is easy to check that we have it. In the same way, although it requires at least $N2^N/16$, comparisons to sort 2^N unequal numbers, it only requires $2^N - 1$ comparisons to check that they are indeed sorted. We are thus led to ask whether there exist short problems with long solutions but short checks. Do there exist problems such that the number $f(n)$ of steps required to solve the problem with n pieces of data increases explosively fast but the number $g(n)$ of steps required to check whether a suggested solution is correct grows rather slowly? This is the so called P–NP problem and is one of the most important unsolved problems in mathematics today.

The reader may object that I have not made the problem precise enough. It is not hard to state the problem completely rigorously, but instead I shall give a precise statement of a specific problem which is believed to be just as hard.

The travelling salesman problem. Suppose that a salesman wishes to visit each of n towns once and only once, starting from a particular town. You are given the distance between each pair of towns (so there are $n(n-1)/2$ pieces of data). Estimate the number $F(n)$ of machine operations required by a program to find the shortest route.

Almost everybody believes that $F(n)$ grows faster than any power of n, that is

$$\frac{F(n)}{n^k} \to \infty$$

as $n \to \infty$ for all values of k. Anybody who can prove them right (or, who knows, wrong) will become a mathematical superstar overnight.

Exercise 16.1.1 *So far no one has succeeded in producing a ciphering system which has the advantages of Enigma but which is provably hard to break. One way of achieving this would be to produce a ciphering system such that if we had a general method of breaking it, we would also have a method of solving some mathematical problem which is known to be hard. In this exercise, we sketch a beautiful coding system due to Rivest, Shamir and Adleman which is believed to have the property that, if we had a general method of breaking it, we would also have a method of solving a mathematical problem which is believed to be hard. We use the mathematical shorthand*

$$a \equiv b \mod c$$

to mean that a and b leave the same remainder when divided by c (or, equivalently, $a - b$ is divisible by c). In theory this is all the reader needs *to know, but in practice the reader who has not met such expressions before may find what follows rather heavy going.*

(i) (This will hardly be new for most readers.) B *is the second letter of the alphabet,* A *is the first and* D *is the fourth. If we associate* BAD *with the integer*

$$(2-1) + (1-1)26^1 + (4-1)26^2,$$

describe an appropriate rule which associates every sequence π of n letters with a unique integer N_π such that $0 \le N_\pi \le 26^n - 1$. Explain why this means that instead of encoding letter sequences we can encode integers.

(ii) Suppose that $n \ge 1$. We say that m and n are coprime if m and n have no common factors. Use Bezout's theorem (the consequence of Euclid's Algorithm obtained in Exercise 10.3.8) to show that there exist integers r and s such that

$$rm + sn = 1$$

and so

$$rm \equiv 1 \mod n.$$

If $ml \equiv 0 \mod n$ show that $l \equiv 0 \mod n$.
If $ml \equiv mk \mod n$ show that $l \equiv k \mod n$.

(iii) If p and q are distinct primes, show by considering the sets

$$A = \{pq - 1 \ge r \ge 0 : r \equiv 0 \mod p\}$$
$$B = \{pq - 1 \ge r \ge 0 : r \equiv 0 \mod q\}$$
$$C = \{pq - 1 \ge r \ge 0 : r \equiv 0 \mod pq\}$$

that there are exactly $(p-1)(q-1)$ integers between 0 and pq which are coprime to pq.

(iv) In (v) we shall prove that if p and q are distinct primes and k is coprime to pq, then

$$k^{(p-1)(q-1)} \equiv 1 \mod pq.$$

Check the following verification for $p = 7$, $q = 3$, $k = 2$

$$k^{(p-1)(q-1)} \equiv 2^{12} \equiv (2^4)^3 \equiv 16^3 \equiv (-5)^3 \equiv -125 \equiv -(-1) \equiv 1 \mod 21.$$

Continue, using other values of p, q and k and verifying the result, until you are convinced that it is likely to be true.

(v)(a) Let $a_1, a_2, \ldots, a_{(p-1)(q-1)}$ be the $(p-1)(q-1)$ integers between 0 and pq which are coprime to pq. If k is coprime to pq explain why ka_r is coprime to pq and there is exactly one integer b_r between 0 and pq such that $b_r \equiv ka_r \mod pq$. Explain why b_r is coprime to pq.

(b) Show that if $1 \le r \le (p-1)(q-1)$, then $b_r = b_s$ only if $r = s$. Conclude that the sequence $b_1, b_2, \ldots, b_{(p-1)(q-1)}$ is just the sequence $a_1, a_2, \ldots, a_{(p-1)(q-1)}$ rearranged in some order.

(c) Deduce that

$$b_1 b_2 \ldots b_{(p-1)(q-1)} = a_1 a_2 \ldots a_{(p-1)(q-1)}$$

and so

$$ka_1 ka_2 \ldots ka_{(p-1)(q-1)} \equiv a_1 a_2 \ldots a_{(p-1)(q-1)} \quad \bmod pq$$

or, rearranging,

$$k^{(p-1)(q-1)} a_1 a_2 \ldots a_{(p-1)(q-1)} \equiv a_1 a_2 \ldots a_{(p-1)(q-1)} \quad \bmod pq.$$

Conclude, using (ii), that

$$k^{(p-1)(q-1)} \equiv 1 \quad \bmod pq$$

whenever k is coprime to pq. (This is a special case of the Fermat–Euler little theorem, see Chapter 11 of [32] or Chapters 6 and 7 of [85].)

(vi) Suppose that p and q are distinct primes and write n = pq. Let a be a number coprime to $(p-1)(q-1)$. By the method of Exercise 10.3.8 we can find b' and b'' such that

$$b'a + b''(p-1)(q-1) = 1,$$

and so

$$b'a \equiv 1 \quad \bmod (p-1)(q-1).$$

It is convenient to choose $1 \le a \le (p-1)(q-1)$. Explain how to find $1 \le b \le (p-1)(q-1)$ such that

$$ba \equiv 1 \quad \bmod (p-1)(q-1).$$

We call the pair (n,a) the enciphering key *and the pair (n,b) the deciphering key.*

Suppose that k (our secret number) is such that $0 \le k \le n-1$. To encipher k using the enciphering key (n,a), we compute $l \equiv k^a \bmod n$ with $0 \le l \le n-1$ and let $l = f(k)$ be our transmitted message.

To decipher a message l using the deciphering key (n,b), we compute $k' \equiv l^b \bmod n$ with $0 \le k' \le n-1$ and let $k' = g(l)$ be our transmitted message.

Explain why $ab = 1 + r(p-1)(q-1)$ for some integer r and verify that if $0 \le k \le n-1$,

$$g(f(k)) \equiv (k^a)^b \equiv k^{ab} \equiv k^{1+r(p-1)(q-1)} \equiv k(k^{(p-1)(q-1)})^r \equiv k$$

justifying each step. If $0 \le l \le n-1$, what is the value of $f(g(l))$?

(vii) Suppose that you have set up such a system with $p = 131$, $q = 151$ and $e = 11\,143$ (the numbers have been chosen for ease of calculation rather than realism). Which positive integer less than $pq = 19\,781$ is encoded by the number 141? [Note that a relatively quick method of computing a^{26} is to write $a^{26} = ((((a^2)^2)^2)^2((a^2)^2)^2 a^2.]$

There is good reason to believe that, provided suitable precautions are taken with the choice of p, q and a,

(A) any method of decoding the enciphered messages will give the deciphering key (n, b) and

(B) using the deciphering key it will be easy to factor n.

Thus we believe that any method of decoding the enciphered messages will give a way of factoring n. However, although mathematicians have thought about the problem for many years, no one has found an easy way of factoring large numbers. Thus there is good reason to suppose that cracking the Rivest–Shamir–Adleman system is hard. However, nobody is quite sure and research continues. (For more details see Chapter 15 of Childs' A Concrete Introduction To Higher Algebra. *Although Koblitz's book* A Course In Number Theory And Cryptography *contains some advanced mathematics, Chapter I and Sections 1 and 2 of Chapter IV are very readable and give a deeper view of what is involved.)*

16.2 Shannon's theorem

Historic meetings like the 1941 conference of Churchill and Roosevelt on a battleship off the American coast carry a powerful symbolic weight. We see reflected in them the rise and fall of nations and the combat of ideas. For mathematicians, the 1943 Bell Labs tea-room meetings of Turing and Shannon — the one the patron saint of modern computing, the other that of modern communication systems — have similar resonances. Of course, the mathematicians were half the age of the politicians. Turing, though the centre of the most secret branch of his country's war effort, was only just past 30 and Shannon was even younger. The statesmen discussed the future of the world, the mathematicians discussed how to make a machine think. (Shannon thought they should play music to it as well as trying to stuff it with data.)

Just as Turing's main interest was not in code-breaking but in computing, so Shannon's main interest was not in hiding information but in transmitting it clearly. But just as Turing's main and secondary interests illuminated each other, so Shannon's two interests turn out to be closely connected.

Consider mankind's second most important method of communication — writing. Written language is highly redundant. If I replace every sixth letter in what I write by x then ix is stixl easy xo work xut whax I mean. Xhen yox solvex the sixple repxacemext ciphxrs in Sxction 13.1 xou madx great xse of txis facx. By omitxing evxry sixxh lettxr I couxd prodxce a boxk whicx was shxrter axd so chxaper bxt stilx made sxnse. Shrthnd systms mk use of rdndncy in a mr snsbl way†.

The weather observations from U-boats were condensed

... into single letters using tables in the Short Weather Cipher. Thus, in Table 3 of the edition in use in 1941, atmospheric pressure of 971.1–973.0 millibars was represented by N, 973.1–975.0 by M. In Table 6, cirrus cloud cover of 1/10–5/10 became E. Using the Short Weather Cipher, the observer aboard ship converted his measurements into letters in a prescribed order. For example, a surface observation from 68° north latitude, 20° west longitude (northwest of Iceland) reporting atmospheric pressure of 972 millibars, temperature of minus 5° Celsius, wind from the northwest with force 6 on the Beaufort scale (a strong breeze of 25 to 31 miles per hour), 3/10 cirrus cloud cover, and visibility up to 5 nautical miles, would become MZNFPED.

It is an impressive feat to reduce an entire weather report to a seven-letter word. The object of this brevity was to reduce transmission time and so thwart the effort of enemy radio location. A by-product of the reduction in redundancy was to make the weather code hard to break and we have seen that much of the British success against the submarine Enigma depended on the capture of various editions of the Short Weather Cipher. Another, less welcome result was that any mistakes in transmitting or receiving the message could not be corrected.

The same unpleasant effect of low redundancy can be seen in mathematical formulae. Five hundred years ago a mathematician might write: 'If we add three to twice the unknown quantity we get fifteen. What is the unknown quantity?' Nowadays we would simply say

$$\text{'Solve } 3 + 2x = 15.\text{'}$$

The modern version has many advantages but a single misprint in the formula would be disastrous‡.

Shannon gives a further, relatively frivolous example of the use of redundancy.

The redundancy of language is related to the existence of crossword puzzles. If the redundancy is zero, any sequence of letters is a reasonable text in the language and any two dimensional array of letters forms a crossword puzzle. If the redundancy is too high, then language imposes too many constraints for large crossword puzzles to be possible.

Language has many purposes besides communication. Penelope Leach, in her splendidly reassuring book of advice for parents, points out that:

Normal infants begin to acquire real speech before they actually need it. They begin to use words at an age and stage when crying, sound-making and gesture are still sufficient to ensure the meeting of their needs. Furthermore, those first words ... are very seldom ones which have anything to do with the infant's needs in a physical sense. ... Infants very seldom begin their speech with need fulfilling words such as 'milk' or 'up' or 'sleep' or 'want'. The first words are almost always the names of deeply loved people or animals or of highly significant objects which give them pleasure. Such early words are almost always used in the context of calling an adult's attention to something; inviting her to share the experience.

We use language for companionship, for amusement, to establish solidarity and, sometimes, just to drive away silence. The language we use is not the product of design but of historical accident. The Germanic 'swine', 'cow' and 'sheep' become the Romance 'pork', 'veal' and 'mutton' when dead because, at one time, an Anglo-Saxon peasantry was ruled over by a Norman, French-speaking aristocracy†.

Having issued all these caveats, it is still interesting to look at language as a device for efficient communication‡. We notice that words which are often used like 'yes', 'no', 'is' and 'me' tend to be short. Less frequently employed elements of our vocabulary are usually longer. This is precisely what we might expect in an efficient system. Sometimes, as when 'motor car' becomes 'car' or 'aeroplane' becomes 'plane', we can actually see words becoming shorter when they need to be used more often. In spite of this, our language is, as we have seen, still highly redundant. Why is this? One possibility is that redundancy allows us to mis-hear parts of a discourse without losing the sense.

All this is just interesting speculation. Shannon went much deeper§. In order to understand thought, Turing considered not actual people but simple machines. In the same way Shannon ignored the richness of human language and considered a simple 'mechanical' model. In Shannon's model, we are allowed to transmit a series of zeros and ones so that a message is something like

11100010001001001.

What 'meaning' we assign to such a string is up to us. In the Short Signal Cipher used by the German U-Boats in 1941

aaaa

meant 'Intend to attack reported enemy forces.' Each digit that we transmit costs us a certain fixed amount of money (one unit, say) so that we wish to be brief. If we could be sure that our message was transmitted correctly then

†The English, it is said, have a language which is the union of two great tongues, but unfortunately they speak the intersection.

‡But, before plunging into this topic, let me recommend Pinker's *The Language Instinct* which offers a fascinating account of how thought becomes word. The author's clear style is in itself an argument for 'universal grammar'.

§The link in Shannon's ideas between the three subjects of secret codes, efficient communication through error-prone systems and natural language is shown in the titles of the two major papers which Shannon published in 1948 and 1949 summarising his earlier researches, and a later one published in 1951: 'A mathematical theory of communication', 'Communication theory of secrecy systems', and 'Prediction and entropy of printed English'. All the papers appeared in the *Bell System Technical Journal* and are reprinted in his Collected Works. The first, key, paper is reprinted in the book *The Mathematical Theory of Communication* together with a non-mathematical account by Weaver. With a little judicious skipping, the first few pages of Shannon's paper would be rewarding reading for most readers of this book. Weaver's article, though less incisive, is also worth reading.

(1) for 1 unit we could transmit 2 possible messages 0 and 1

(2) for 2 units we could transmit 4 possible messages 00, 01, 10, and 11,

(3) for 3 units we could transmit 8 possible messages 000, 001, 010, 011, 000, 101, 110, and 111.

$$\vdots$$

(n) for n units we could transmit 2^n possible messages.

Unfortunately there is a probability p, say, that a particular digit is wrongly transmitted (so that 0 becomes 1 and 1 becomes 0). (For simplicity we shall assume that errors are independent of each other.) What can we do now?

If we have only two messages to transmit ('war' or 'peace', say) and n units to spend, then we send n zeros or n ones. If $p < 1/2$, the probability that a message of n ones is received with more zeros than ones can be made arbitrarily small by taking n large enough. However, we wish to transmit many possible messages using our n units. Further, although we cannot rule out the possibility of error, we wish to reduce the probability of error to some small value ϵ. (In practice $\epsilon = 10^{-8}$ would surely be satisfactory.)

A way ahead is provided by the observation that if we toss a fair coin many times, then it is very unlikely that the proportion of heads will differ very greatly from 1/2. More generally the following 'law of large numbers' holds†.

†In Solzhenitsyn's *The First Circle* the Gulag engineers are set to build a telephone scrambler – a form of device which interested both Shannon and Turing. The English translator is not entirely at home with some of the details and renders the 'law of large numbers' as the 'law of great numbers'.

Theorem 16.2.1 *Let $\epsilon, \eta > 0$. Suppose we make N tosses of a coin which has probability p of coming down heads. Provided that N is sufficiently large, the probability that the number of heads M fails to satisfy the inequality*

$$\left| p - \frac{M}{N} \right| < \eta$$

is less than ϵ.

Most of my readers will accept the truth of this theorem without argument. (What is the theorem after all but a correctly stated form of the 'law of averages'?) A proof is given in every first course on probability. We need the following variation.

Theorem 16.2.2 *Let $\epsilon > 0$ and $0 \le p < q < 1/4$. Suppose we transmit a string of N zeros and ones using a system which has a probability p of wrongly transmitting any particular digit (independently of what it does* *to any other). Provided that N is sufficiently large, the probability that the number of errors M is greater than or equal to qN is less than ϵ.*

Theorem 16.2.2 follows from Theorem 16.2.1 on replacing the words 'wrongly transmitting a digit' by the words 'coming down heads' and setting $\eta = q - p$.   ■

This suggests that, if we receive a string of N zeros and ones through the system of Theorem 16.2.2 which differs in less than qN places from a possible message, we should assume that it was that message which was sent. (In the same way, if you receive a message 'Hoppy barthday' on your birthday, you can easily work out what was meant.) The only problem occurs if the string received differs in less than qN places from more than one possible message† and this cannot occur if all the possible messages differ in at least $2qN+1$ places from each other. This gives us the following theorem.

†Recall Churchill's remark on being told of an MP called Bossom: 'Curious name that, neither one thing nor the other.'

Theorem 16.2.3   *Let $\epsilon > 0$ and $0 \le p < q < 1/4$. Suppose we transmit a string of N zeros and ones using a system which has a probability p of wrongly transmitting any particular digit (independently of what it does to any other). Suppose further that the possible transmitted messages all differ in at least $2qN+1$ places from each other and that we use the rule:* If we receive a string which differs in less than qN places from a possible message we should assume that it was that message. If there is no possible message with this property, we give up. *Provided that N is sufficiently large, the probability that we give up or that our choice of message is mistaken is less than ϵ.*

Theorem 16.2.3 brings us to the central question.

Central question   *How large a collection of strings of zeros and ones of length N can we find which has the property that each string differs in at least $2qN + 1$ places from every other string in the collection?*

The next lemma brings us close to a satisfactory answer.

Lemma 16.2.4   *(i) If $0 \le u \le 1/2$ and uN is an integer, then*

$$\sum_{r=0}^{uN} \binom{N}{r} \le u^{-uN}(1-u)^{-(1-u)N}.$$

(ii) If $0 \le u \le 1/2$ and we write $[uN]$ for the greatest integer less than uN, then

$$\sum_{r=0}^{[uN]} \binom{N}{r} \le u^{-uN}(1-u)^{-(1-u)N}.$$

(iii) Suppose that $0 \le q \le 1/2$. If x is a string of zeros and ones of length N, then the number of strings of length N which differ from x in less than $2qN + 1$ places is at most $(2q)^{-2qN}(1 - 2q)^{-(1-2q)N}$.

Proof *(i)* By the binomial theorem

$$1 = (u + (1 - u))^n$$

$$= \sum_{r=0}^{N} \binom{N}{r} u^r (1 - u)^{N-r}$$

$$\ge \sum_{r=0}^{uN} \binom{N}{r} u^r (1 - u)^{N-r}$$

$$\ge \sum_{r=0}^{uN} \binom{N}{r} u^{uN} (1 - u)^{1-uN}$$

$$= u^{uN} (1 - u)^{1-uN} \sum_{r=0}^{uN} \binom{N}{r}.$$

Multiplying both sides of the inequality by $u^{-uN}(1 - u)^{-(1-uN)}$ we have the result.

(ii) This is just a simple modification of the proof of (i). Any reader interested enough to want the proof in detail should be interested enough to write it out for herself.

(iii) The number of strings which differ from x in exactly r places is the number of ways of choosing r places from N possible places, that is the number $\binom{N}{r}$ of ways of choosing r objects from N. Thus, using part (ii),

number of strings which differ from x in less than $2qN + 1$ places

$$\le \sum_{r=0}^{[2qN]} \text{number of strings which differ from x in exactly } r \text{ places}$$

$$= \sum_{r=0}^{[2qN]} \binom{N}{r}$$

$$\le (2q)^{-2qN}(1 - 2q)^{-(1-2q)N},$$

which is the result we wanted. ∎

Our results become easier to interpret if we observe that

$$(2q)^{-2qN}(1 - 2q)^{-(1-2q)N} = ((2q)^{-2q}(1 - 2q)^{-(1-2q)})^N$$

and so, choosing η such that $2^\eta = (2q)^{-2q}(1-2q)^{-(1-2q)}$, we have

$$(2q)^{-2qN}(1-2q)^{-(1-2q)N} = 2^{\eta N}.$$

Lemma 16.2.5   (i) *If* $0 < p < 1/2$, *then*

$$1 \geq p^p(1-p)^{1-p} > 1/2.$$

(ii) *If* $0 \leq q < 1/4$ *and* $2^\eta = (2q)^{-2q}(1-2q)^{-(1-2q)}$, *then* $1 > \eta \geq 0$.

Proof   The following proof requires a second year of calculus. Let

$$g(p) = \log(p^p(1-p)^{1-p}) = p \log p + (1-p) \log(1-p).$$

Differentiating we have

$$g'(p) = \log p - \log(1-p),$$

so $g'(p) < 0$ for $1/2 > p > 0$. Thus g is strictly decreasing as p runs from 0 to $1/2$ and so $g(p) > g(1/2)$ and

$$\log(p^p(1-p)^{1-p}) > \log((1/2)^{1/2}(1-1/2)^{1-1/2}) = \log 1/2$$

for all $1/2 > p > 0$. Taking exponentials, we have

$$p^p(1-p)^{1-p} > 1/2$$

for all $1/2 > p > 0$. (To see that $1 > p^p(1-p)^{1-p}$, just observe that $x^\alpha < 1$ whenever $0 \leq x \leq 1$ and $\alpha \geq 0$.)

If you do not have sufficient calculus, you can convince yourself of the truth of the result by graphing $p^p(1-p)^{1-p}$ using a pocket calculator.
   (ii) Set $p = 2q$.   ∎

Using Lemma 16.2.4 we can now obtain a satisfactory answer to our 'central question'.

Lemma 16.2.6   *Suppose that*

$$0 \leq q < 1/4 \text{ and } 2^\eta = (2q)^{-2q}(1-2q)^{-(1-2q)}.$$

(i) *Suppose that* m *is a positive integer with*

$$(m-1)2^{\eta N} < 2^N.$$

If $x_1, x_2, \ldots, x_{m-1}$ *are any strings of zeros and ones of length* N, *then there is a string* x_m *which differs from each of* $x_1, x_2, \ldots, x_{m-1}$ *in at least* $2qN + 1$ *places.*

(ii) *Suppose that* M *is a positive integer with*

$$M2^{\eta N} \leq 2^N.$$

Then there exist M *strings of zeros and ones of length* N *such that each string differs in at least* $2qN + 1$ *places from every other string in the collection.*

Proof (i) We use a counting argument. Let A_j be the set of strings of length N which differ from x_j in less than $2qN + 1$ places. By Lemma 16.2.4(iii), each A_j contains at most $2^{\eta N}$ strings. If we let

$$A = A_1 \cup A_2 \cup A_3 \cup \cdots \cup A_{m-1}$$

(so that A is the set of strings of length N which differ from at least x_j in less than $2qN + 1$ places), then A can contain at most $(m - 1)2^{\eta N}$ strings. But there are 2^N possible strings of zeros and ones of length N so, since $2^N > (m - 1)2^{\eta N}$, there must be a string x_m which does not belong to A, that is which differs in at least $2qN + 1$ places from each of $x_1, x_2, \ldots, x_{m-1}$.

(ii) Start with any string x_1. By part (i) we can find a string x_2 which differs in at least $2qN + 1$ places from x_1. Applying part (i) again we can find a string x_3 which differs in at least $2qN + 1$ places from each of x_1 and x_2. We now find a string x_4 which differs in at least $2qN + 1$ places from each of x_1, x_2 and x_3 and so on. Since $M2^{\eta N} \le 2^N$, we can continue in this way until we have M strings $x_1, x_2, \ldots, x_M$ each of which differs in at least $2qN + 1$ places from every other string in the collection. ∎

Exercise 16.2.7 *(i) By taking $q = 1/10$ and using Lemma 16.2.6 show that there exist at least 6 strings of length 10, each of which differs in at least three places from any other.*

(ii) By explicitly calculating $\sum_{r=0}^{2} \binom{10}{r}$ and then using the arguments of Lemma 16.2.4 (iii) and Lemma 16.2.6 (ii) show that there exist at least 19 strings of length 10, each of which differs in at least three places from any other.

(iii) Can you find 19 strings of length 10, each of which differs in at least three places from any other? If you can, I take my hat off to you and ask you to rework this exercise with $q = 1/10$ and strings of length 20. If you cannot, you will have learnt that it is one thing to show that the strings of Lemmas 16.2.4 and 16.2.6 exist and quite another thing to find them.

We have thus answered our 'central question' and, by combining Theorem 16.2.3 with Lemma 16.2.6, we obtain a simple version of Shannon's remarkable theorem.

Theorem 16.2.8 *Let $\epsilon > 0$ and $0 \le p < q < 1/4$ and*

$$2^\eta = (2q)^{-2q}(1 - 2q)^{-(1-2q)}.$$

Suppose we transmit a string of N zeros and ones using a system which has a probability p of wrongly transmitting any particular digit (independently *of what it does to any other). If N is sufficiently large and M is a positive integer with*

$$M2^{\eta N} \leq 2^N,$$

then we can find M strings of length N such that if we use the rule:

If we receive a string which differs in less than qN places from a possible message we should assume that it was that message. If there is no possible message with this property we give up.

then the probability that we give up or that our choice of message is mistaken is less than ϵ.

We may restate the theorem in simple terms.

Theorem 16.2.9   *Suppose we have a system which has a probability p of wrongly transmitting any particular digit (independently of what it does to any other). Then, provided that $p < 1/4$, there exists an η depending on p such that (provided M is large) we can transmit 2^M different messages with arbitrarily small probability of error, provided only that we use strings of length $\eta^{-1}M$.*

If each digit takes a fixed time to transmit, this tells us that we can make our error-prone system as good as an error-free system at the cost of slowing it down by a factor η. In money terms, if each digit costs a fixed sum to transmit, we can make our error-prone system as good as an error-free system but it will cost us η^{-1} as much.

In practice, we would not expect to use communication systems with error probability p anywhere near as large as $1/4$, but mathematicians like Shannon always want the 'best possible' version of a theorem. By introducing an extra twist into the argument, Shannon was able to prove the best possible result.

Theorem 16.2.10   *If $0 \leq p < 1/2$, $2p^p(1-p)^{1-p} > \eta$ and $\epsilon > 0$, then, provided only that N is large enough, we can find 2^N different messages of length $N\eta^{-1}$ such that the probability that a transmitted message is wrongly received is less than ϵ.*

Exercise 16.2.11   *On the principle that someone who is known always to give the wrong advice is as useful as someone who is known always to give the right advice, explain why we could also take $p > 1/2$ in Theorem 16.2.10. Explain why a transmission system with $p = 1/2$ is totally useless.*

The communication system discussed in Shannon's theorem seems rather special. What about communication systems involving continuous (analogue) signals rather than (discrete or digital) strings of zeros and ones? Just as Turing showed that his simple computer was a universal model for all computers, so Shannon showed that his simple system was a model for all communications systems. The steady advance of digital transmission and recording of analogue signals bears witness to the truth of his insight.

Shannon's theorem is an 'existence theorem'. It tells us that, in theory, we can use an error-prone communication channel almost as efficiently as an error-free one, but it does not tell us how to do this in practice. The first practical scheme for correcting errors in a noisy communication system was invented by Hamming. It is said that the moment his employers realised what he had done, they swept the contents of his desk into sacks and dumped them in the nearest patent office! Since then many 'error-correcting codes' have been developed. To be practical, an error-correcting code must be easy to specify and use, and no known practical code approaches the efficiency of Shannon's scheme. However, most communication channels have very low error rates to start with, so we do not need Shannon efficiency. Error-correcting codes are now used in everything from computer communication to compact discs.

Exercise 16.2.12 *On the inner title page of most recent books the reader will find publication details including their International Standard Book Numbers (ISBN). (For this book the hardback has ISBN 0 521 56087 X and the paperback 0 521 56823 4.) The ISBN uses single digits selected from 0, 1, 2, ... , 8, 9 and X representing 10. Each ISBN consists of nine such digits $a_1, a_2, \ldots, a_9$ followed by a single check digit a_{10} such that*

$$10a_1 + 9a_2 + 8a_3 + \cdots + 3a_8 + 2a_9 \equiv a_{10} \mod 11 \qquad (*)$$

(in simpler terms, a_{10} is the remainder when we divide $10a_1 + 9a_2 + 8a_3 + \cdots + 3a_8 + 2a_9$ by 11).

(i) Check that (*) holds for the ISBNs of the present book and for a couple of other books.

(ii) Show that, if you write down the ISBN of a book, but make a mistake in one digit, then relation (*) will not work for the new number.

(iii) Show that, if you write down the ISBN of a book but interchange two adjacent digits, then relation (*) will not work for the new number. (Mistakes of the type given in (ii) and (iii) are the most common in typing.)

(iv) Show, however, that it is possible to interchange two adjacent digits and make a mistake in another in such a way that the relation (*) still holds for the new number.

(v) The ISBN is designed to detect errors but not to correct them. Show that even if we make a mistake in only one digit, there is no way of telling from the new number what the correct number was.

Exercise 16.2.13 (A Hamming code)   *(This exercise describes one of the error-correcting codes invented by Hamming. It uses more abstract algebra than the rest of the book. In particular the reader will need to know about matrices and vectors.) In this exercise we work with arithmetic modulo 2. Thus*

$$0 + 0 = 0, \ 1 + 0 = 1, \ 0 + 1 = 1, \ 1 + 1 = 0$$

and

$$0 \times 0 = 0, \ 1 \times 0 = 0, \ 0 \times 1 = 0, \ 1 \times 1 = 1.$$

We consider the Hamming matrix given by

$$H = \begin{pmatrix} 1 & 0 & 1 & 0 & 1 & 0 & 1 \\ 0 & 1 & 1 & 0 & 0 & 1 & 1 \\ 0 & 0 & 0 & 1 & 1 & 1 & 1 \end{pmatrix},$$

and write e_j for the column vector corresponding to the jth column of H. Thus

$$e_1 = \begin{pmatrix} 1 \\ 0 \\ 0 \end{pmatrix}, \ e_2 = \begin{pmatrix} 0 \\ 1 \\ 0 \end{pmatrix}, \ e_3 = \begin{pmatrix} 1 \\ 1 \\ 0 \end{pmatrix},$$

and so on. We also write f_j for the column vector of length 7 with a 1 in the jth place and a zero in all the others. Thus

$$f_1 = \begin{pmatrix} 1 \\ 0 \\ 0 \\ 0 \\ 0 \\ 0 \\ 0 \end{pmatrix}, \ f_2 = \begin{pmatrix} 0 \\ 1 \\ 0 \\ 0 \\ 0 \\ 0 \\ 0 \end{pmatrix}, \ f_3 = \begin{pmatrix} 0 \\ 0 \\ 1 \\ 0 \\ 0 \\ 0 \\ 0 \end{pmatrix},$$

and so on.

(i) Show that $Hf_j = e_j$.

(ii) Suppose that we wish to send a message a, b, c, d of four zeros and

ones. If we set

$$x = a + b + c$$
$$y = a + c + d$$
$$z = b + c + d$$

show that

$$x + a + b + c = 0$$
$$y + a + c + d = 0$$
$$z + b + c + d = 0.$$

- *If we write*

$$t = \begin{pmatrix} x \\ y \\ a \\ z \\ b \\ c \\ d \end{pmatrix},$$

show that $Ht = 0$.

(iii) Suppose we transmit the sequence a, b, c, d, x, y, z *of seven zeros and ones (consisting of the message* a, b, c, d *we wish to send followed by the 'check digits'* x, y, z) *and it is received as the sequence* $a', b', c', d', x', y', z'$. *Set*

$$t' = \begin{pmatrix} x' \\ y' \\ a' \\ z' \\ b' \\ c' \\ d' \end{pmatrix}.$$

Explain why, if the sequence is transmitted correctly, then $t' = t$ *and*

$$Ht' = 0.$$

Explain why, if the sequence is transmitted correctly except for an error in the jth place, then $t' = t + f_j$ *and*

$$Ht' = e_j.$$

(iv) Explain how, if we assume that at most one error has occurred in the transmission of a, b, c, d, x, y, z, *we can recover the original sequence from the received sequence* $a', b', c', d', x', y', z'$.

(v) Suppose that the probability of an error in the transmission of a digit is p independent of what happens to the other digits. Show that the probability of more than one error occurring in the transmission of the sequence a, b, c, d, x, y, z is

$$q = 1 - (1 - p)^7 - 7(1 - p)^6 p.$$

Compute q for p = 0.1 and p = 0.01. If p is small show, by using the binomial theorem, or otherwise, that

$$q \approx 21p^2.$$

Thus if we wish to send a message of four digits by a system which costs 10^{-6} cents a digit and has a probability 10^{-6} of garbling a given digit, we can spend 4×10^{-6} cents and send the message as it stands with a probability of approximately 4×10^{-6} of having it garbled, or use Hamming's code, spending 7×10^{-6} cents, but, in return, reducing the probability of garbling to approximately 21×10^{-12}.

In his book *From Error-Correcting Codes through Sphere Packings to Simple Groups* T. M. Thompson traces the path which led from Hamming's original practical insight to remarkable discoveries in the abstract field of group theory. Thompson's unique glimpse into the way mathematics is made should be on the bookshelf of anyone interested in the process of mathematical discovery. As a taster, here is Conway's account of how, after hours of intensive work trying to construct a new simple group,

> [a]t quarter past midnight, I telephoned [J. G. T.] Thompson [the world expert on such matters] again, saying it was all done. The group is there. It was absolutely fantastic — twelve hours had changed my life. Especially since I had envisioned going on for months — every three days spending six to twelve hours on the damned thing.

PART V

The pleasures of thought

CHAPTER 17

Time and chance

17.1 Why are we not all called Smith?

In Section 15.1 we saw that the question of whether the Turing bombes would work depended on whether an effectively random process in which 'mother statements' give rise to a certain random number of 'daughter statements' which in turn give rise to 'granddaughter statements' and so on would die away quickly or give rise to a family containing all (or almost all) possible statements. The bombe problem is complicated by several special features and, though I hope that the discussion of the next two sections will shed some light on it, I shall concentrate on simpler but related problems.

Sir Thomas Browne wrote 'Generations pass while some trees stand and old families last not three oaks'. (The quotation is taken from Kendall's fascinating papers [114] and [115].) Less concisely, Galton wrote:

> The decay of the families of men who occupied conspicuous positions in past times has been a subject of frequent remark, and has given rise to various conjectures. It is not only the families of men of genius or those of the aristocracy who tend to perish, but it is those of all whom history deals, in any way, even such men as the burgesses of towns, ... The instances are very numerous in which surnames that were once common have since become scarce or have wholly disappeared.

†Social Darwinism applies the Darwinian doctrine of the survival of the fittest to human society. Rich social Darwinists take wealth as the best indication of fitness to survive, academic social Darwinists take intellectual achievements as the best indication and so on. They are often haunted by the fear that the unfit do not understand this and may outbreed the fit.

As a social Darwinist†, Galton was unwilling to accept the explanation 'that a rise in physical comfort and intellectual capacity is naturally accompanied by a diminution in fertility' and seized on the suggestion that the disappearance of surnames occurred 'by the ordinary laws of chances'. He appealed to a mathematically inclined friend, the Rev. H. W. Watson, who produced a discussion along the following lines.

As a simple model, we may suppose that each male has a probability p_r of producing r male sons independently of what happens for other

413

males. The average number of sons is thus

$$\kappa = \sum_{r=1}^{N} r p_r,$$

where N is the maximum number of sons possible. It is intuitively clear and in fact correct that the average number of sons of sons will be

average number of sons × average number of sons per son = κ^2,

and more generally

average number in nth generation directly in male line = κ^n.

It is thus pretty obvious that, if $\kappa < 1$, the male line will eventually die out. A simple example shows that matters are less clear if $\kappa \geq 1$.

Exercise 17.1.1 (i) If $p_0 = 1/2$, $p_3 = 1/2$, $p_r = 0$ otherwise, show that $\kappa = 3/2$ but there is a probability of at least $1/2$ that a surname initially borne by a single male will die out.

(ii) If $p_1 = 1/2$, $p_2 = 1/2$, $p_r = 0$ otherwise, show that $\kappa = 3/2$, but there is a probability 0 that a surname initially borne by a single male will die out.

It turns out that a key role in our discussion will be played by the polynomial

$$\phi(t) = \sum_{r=0}^{N} p_r t^r.$$

Let q be the probability of extinction for the male line starting with one man. If we start with r men in the same generation, the probability that all their male lines become extinct is then q^r. It follows, in particular, that if a man has m sons, the probability of extinction of his male line is now q^m. Using this observation we see that

$$q = \text{Pr(male line of one man becomes extinct)}$$

$$= \sum_{r=0}^{N} \text{Pr(one man has } r \text{ sons)} q^r$$

$$= \phi(q),$$

so the probability of extinction q is a root of $x = \phi(x)$ in the region $0 \leq x \leq 1$.

Thus to study q we must study ϕ. We know that

$$\phi(1) = \sum_{r=0}^{N} p_r = 1$$

(since probabilities add up to 1). We also know that

$$\phi(0) = p_0.$$

The case when the probability that a man has no sons is zero is rather uninteresting, since extinction is impossible, so we shall assume that this is not the case and so

$$\phi(0) > 0.$$

Since $\phi(t) = \sum_{r=0}^{N} p_r t^r$ and p_r is a probability and so positive, ϕ is a polynomial curving upwards as shown in Figure 17.1. (If the reader thinks the words 'curving upwards' are non-mathematical, she will find a more mathematical formulation in Exercise 17.1.4.)

If we also draw in the straight line $y = x$, we see that there are three possibilities which we show in Figure 17.2. If the slope of ϕ is greater than 1 at $x = 1$ ($\phi'(1) > 1$ in the language of the calculus), then the graph $y = \phi(x)$ cuts the line $y = x$ for some $x = \alpha$ between 0 and 1. We will have $\phi(\alpha) = \alpha$ and the graph of ϕ will look as in Figure 17.2(a). If the slope of ϕ is less than 1 at $x = 1$ ($\phi'(1) < 1$), then the graph $y = \phi(x)$ will not cut the line $y = x$ for any x between 0 and 1 and the graph of ϕ will look as in Figure 17.2 (c). If the slope of ϕ is exactly 1 at $x = 1$ ($\phi'(1) = 1$), then the line $y = x$ is a tangent to the graph of ϕ at the point $(1, 1)$ and, once again the graph $y = \phi(x)$ will not cut the line $y = x$ for any x between 0 and 1. The graph of ϕ will look as in Figure 17.2(b).

Figure 17.1: Watson's function.

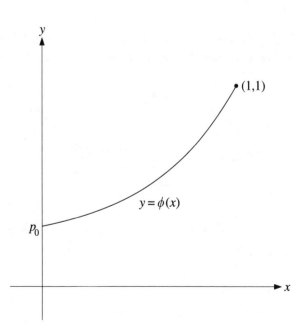

The difference between the various cases depends on the slope of the graph of ϕ at 1. To find this slope we need calculus. Let X be the number of sons from some given father so that, as before, $p_r = \Pr(X = r)$. We note that

$$\phi'(t) = \sum_{r=0}^{N} r p_r t^r$$

and so the slope of the graph of ϕ at 1 is

$$\phi'(1) = \sum_{r=0}^{N} r p_r = \sum_{r=0}^{N} r \Pr(X = r) = \mathbb{E}X = \kappa,$$

where $\mathbb{E}X$ is the expected (that is the average) value of X. Thus $\phi'(1)$ is the expected (average) number κ of sons per father. Figure 17.2(a) corresponds to the case $\kappa > 1$ and Figures 17.2(b) and (c) correspond to the cases $\kappa = 1$ and $\kappa < 1$. If $\kappa = 1$ or $\kappa < 1$, we know that $\phi(x) = x$ has exactly one solution with $0 \leq x \leq 1$, to wit $x = 1$ so $q = 1$. If $\kappa > 1$, then $\phi(x) = x$ has two solutions with $0 \leq x \leq 1$, to wit $x = 1$ and $x = \alpha$. Now, as we noted near the beginning of the discussion of Galton's problem, the average number of members of the male line at the nth generation will be κ^n so, if $\kappa > 1$, this number increases exponentially. We are thus led to reject the possibility $q = 1$ of extinction with probability 1 and conclude that $q = \alpha$. (We return to this point in the appendix to this section.)

Theorem 17.1.2 *Suppose $p_0 \neq 0$. If the average (mathematically, the expected) number κ of sons is greater than 1, the male line will become*

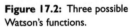

Figure 17.2: Three possible Watson's functions.

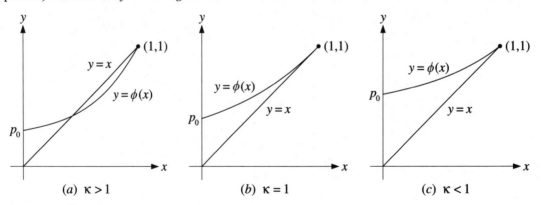

(a) $\kappa > 1$ (b) $\kappa = 1$ (c) $\kappa < 1$

extinct with probability α, *where* α *is the unique root of*

$$\sum_{r=0}^{N} p_r t^r = t$$

with $0 < t < 1$.

If the average number of sons κ *is 1 or less, then the male line will become extinct with probability 1.*

Exercise 17.1.3 *In our discussions above we have excluded the case* $p_0 = 0$. *What will happen if* $p_0 = 0$? *You should consider the two cases:*

(a) $p_1 = 1$,

(b) $p_1 < 1$.

Some of the most interesting ideas are connected with the 'transitional case' in which the average number κ of sons is exactly 1. The theorem tells us that any male line will become extinct with probability 1. On the other hand, the average number in the nth generation will be $\kappa^n = 1$. Thus the average number of members of the male line of all generations will be infinite in spite of the fact that the line is doomed to extinction! I shall refer to this as Watson's paradox. Let p_n be the probability that the male line lasts until at least the nth generation. Since the average size of the nth generation is 1

$p_n \times$ (average size of nth generation for a male line

which lasts at least n generations) $= 1$.

Thus the average size of the nth generation among those which have lasted that far is p_n^{-1}. Since p_n becomes small as n becomes large, it follows that p_n^{-1} becomes large. In other words, few male lines will last to the nth generation but those that do will have (on average) many members in the nth generation.

We shall see that Theorem 17.1.2 occurs not merely in the study of surnames but also in the study of epidemics and evolution. However, Nature is subtle and our rather simple approach is unlikely to give anything more than a glimpse of what might be the case. The reader should be aware that much of the rest of this chapter is more or less speculative and should pause from time to time to ask, 'How can we know that?'

> The Microbe is so very small
> You cannot make him out at all,
> But many sanguine people hope
> To see him through a microscope.
> His jointed tongue that lies beneath

A hundred curious rows of teeth;
His seven tufted tails with lots
Of lovely pink and purple spots,
On each of which a pattern stands,
Composed of forty separate bands;
His eyebrows of a tender green;
All these have never yet been seen —
But Scientists, who ought to know,
Assure us that they must be so ...
Oh! let us never, never doubt
What nobody is sure about!

At first sight, the answer to the question 'Why are we not all called Smith?' is the population explosion. During most of the period that surnames have existed in Europe, the population has been increasing, so κ, the average number of male offspring from a given man, has been greater than 1. The probability α of a given surname becoming extinct was thus strictly less than 1 and, although many surnames have perished, many have survived.

However, endless population expansion has not been possible for animal species in the past, and only the kind of wild-eyed enthusiast who frequents science fiction conferences and a certain kind of economics department believe it will be possible for the human race to expand endlessly. What happens if $\kappa = 1$? Our discussion of Watson's paradox told us that (for large n) few surnames will last to the nth generation, but those that do will have (on average) many bearers in the nth generation. In isolated mountain villages where the population size has remained much the same for many generations, we do, indeed, observe many bearers of few surnames.

We all possess a more fundamental surname: one that passes not through the male but through the female line. Mitochondrial DNA is inherited from the mother. According to some geneticists, we all possess, in effect, a particular type of mitochondrial DNA. If these geneticists are correct, about 10 000 generations ago, the whole human race consisted of a few thousand individuals who lived through many generations in the same location without much change in the community size. Eventually, through the operation of Watson's paradox, that community had only one kind of mitochondrial DNA and it is this mitochondrial DNA which we all inherit (through the female line) and so, in some sense, we are all called Smith. (Unfortunately, there are problems with the account given in this paragraph, but those of us who like striking stories still hope that it will turn out to be

†Stop press. The journal *Science* of 26 May 1995 (pages 1183–5 and 1141–2) contains a study of a DNA 'male surname' (more specifically 'a 729-base-pair intron located between the third exon and the zinc-finger-encoding fourth exon of the *ZFY* locus') which avoids some of the problems associated with the research reported above and confirms, via an 'African Adam', its general outline. However, the matter is not completely settled.

correct†. Those who wish to know more can consult J. Reader's *Missing Links*.) We shall give a slightly more sophisticated version of the operation of Watson's paradox in evolution later in this chapter.

Conscience-stricken appendix

Unfortunately the discussion above contains one point at which I exercised some cunning sleight of hand. Since students believe that rigour is something to do with (a) not using diagrams and (b) proving the obvious, you may suspect that this occurred when I used Figure 17.2 to study the roots of $x = \phi(x)$. This is not the case, as those readers who doubt my word may check by doing the following exercise (which requires a year or so of calculus).

Exercise 17.1.4 *(If you have not done a year or so of calculus or if you find the diagrammatic treatment above sufficiently convincing you should not even bother to read this exercise.) We will make repeated use of the fact that, if $f'(t) > 0$ for all $a < t < b$, then f is strictly increasing between a and b. As usual, we shall suppose $p_0 \neq 0$.*

(i) Show that $\phi'(t) > 0$ for all $0 < t < 1$ and deduce that $\phi(t)$ is strictly increasing in the region $0 < t < 1$. Show that $\phi''(t) > 0$ for all $0 < t < 1$.

(ii) Set $\chi(t) = \phi(t) - t$. Find $\chi(1)$ and show that $\chi(0) < 0$. Show that $\chi''(t) > 0$ for all $0 < t < 1$. What does this tell us about the behaviour of $\chi'(t)$ for t between 0 and 1? Show that $\chi'(0) < 0$. Sketch some possible graphs for χ in the manner of Figure 17.2.

(iii) If $\chi'(1) \leq 0$ show that $\chi'(t) < 0$ for all $0 < t < 1$. Deduce that $\chi(t) < 0$ for all $0 \leq t < 1$.

(iv) If $\chi'(1) > 0$, show that there exists a β with $0 < \beta < 1$ such that $\chi'(t) < 0$ for $0 \leq t < \beta$, $\chi'(\beta) = 0$ and $\chi'(t) > 0$ for $\beta < t \leq 1$. Deduce that $\chi(t) < 0$ for $\beta \leq t < 1$ (so that, in particular, $\chi(\beta) < 0$). Now show that there exists a unique α with $0 < \alpha < \beta$ such that $\chi(\alpha) = 0$. Explain why α is the only root of $\chi(t) = 0$ in the region $0 < t < 1$.

(v) Conclude that the equation $\phi(t) = t$ has exactly one root (which we continue to call α) in the region $0 < t < 1$ if $\phi'(1) > 1$ and none if $\phi'(1) \leq 1$.

In fact the problem occurs in the apparently innocuous argument '... *the average number of members of the male line at the nth generation will be κ^n so, if $\kappa > 1$, this number increases exponentially. We are thus led to reject the possibility $q = 1$ of extinction with probability 1 and conclude that $q = \alpha$.*' Plausible as this argument may seem at first sight, it ceases to carry any conviction after we have understood Watson's paradox where $\kappa = 1$ and we know that, although the average number of members in a male line is infinite, yet any male line will become extinct with probability 1. Indeed, the great majority of those who thought about the matter before 1930 thought that, even when $\kappa > 1$, extinction would occur with probability 1. Their number included Watson who, although he produced the arguments on which this appendix is based, was so convinced that extinction would occur that he unconsciously fudged the crucial point to obtain the result he expected! Although it seems likely that several mathematicians must have known the correct result, the first correct published account is due to Steffensen.

How can we prove Watson's theorem without making use of an unsatisfactory argument? The answer lies in the closer study of

$$q_n = \Pr\{\text{male line extinct by the } n\text{th generation}\}.$$

Observe that if the initial single male, call him Adam, has r sons the probability that Adam's line will be extinct after $n + 1$ generations is the probability q_n^r that each of the r male lines of his r sons will be extinct after n generations. Thus

$$q_{n+1} = \Pr\{\text{Adam's line extinct by the } n\text{th generation}\}$$

$$= \sum_r \Pr\{\text{Adam has } r \text{ sons}\}$$

$$\times \Pr\{\text{Adam's line extinct by the } n\text{th generation if he has } r \text{ sons}\}$$

$$= \sum_r p_r q_n^r = \phi(q_n)$$

The sequence of q_n is thus given by the two rules: $q_0 = 0$ and

$$q_{n+1} = \phi(q_n).$$

Let us consider the most interesting case when $\kappa > 1$, corresponding to Figure 17.2 (a). If we plot successively $(q_0, q_0) = (0,0)$, $(q_0, q_1) = (q_0, \phi(q_0))$, (q_1, q_1), $(q_1, q_2) = (q_1, \phi(q_1))$, (q_2, q_2), $(q_2, q_3) = (q_2, \phi(q_2))$ and so on, we see a typical 'staircase' pattern emerge in Figure 17.3. It is, I think, clear from the diagram that q_n increases towards α. In other words, the probability of extinction after at most n generations increases towards α and so the probability of eventual extinction is α.

If we draw similar staircases for the cases corresponding to Figures 17.2(b) and 17.2(c) we obtain the diagrams shown in Figure 17.4 where q_n increases towards 1. In other words, the probability of extinction after at most n generations increases towards 1 and so the probability of eventual extinction is 1. This completes our proof of Theorem 17.1.2.

In some vague sense, the 'transitional case' when $\kappa = 1$ represents 'unwilling extinction', and this observation is borne out by a closer inspection of Figure 17.4. In case (c), the staircase moves towards the point $(1,1)$ quite rapidly, but in the 'transitional case' (b) the staircase

Figure 17.3: The staircase to (α, α). (Case (a) of Figure 17.2)

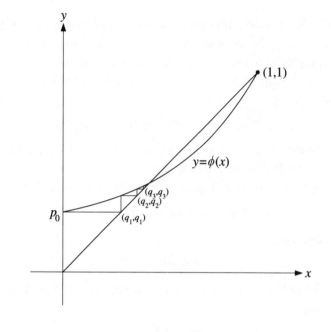

Figure 17.4: Staircases to extinction. (Cases (b) and (c) of Figure 17.2)

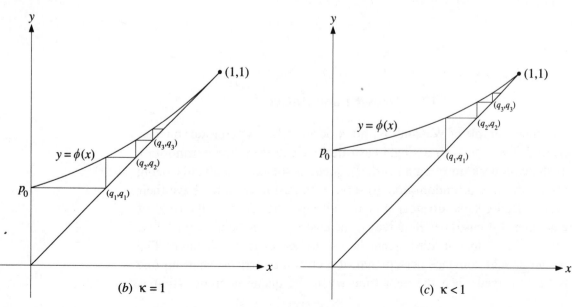

(b) $\kappa = 1$ (c) $\kappa < 1$

is squeezed between the curve $x = \phi(y)$ and its tangent $x = y$ at $(1, 1)$ so that the steps become very small quite quickly and $r_n = 1 - q_n$ approaches 0 rather slowly.

The reader who prefers not to rely on diagrams and is familiar with the idea of limits may like to do the final exercise but, once again, it is not an essential part of the argument.

Exercise 17.1.5 *(i) We start with the more interesting case $\phi'(1) > 1$. Let the sequence q_n be defined, as before, by the rules $q_0 = 0$, $q_{n+1} = \phi(q_n)$. By applying ϕ to each of the three elements in the inequality, show that if $q_n < q_{n+1} < \alpha$, then $q_{n+1} < q_{n+2} < \alpha$. Show that $q_0 < q_1 < \alpha$ and deduce that*

$$q_0 < q_1 < q_2 < q_3 < \ldots < q_{n-1} < q_n < \alpha$$

for all n. Thus q_n is an increasing sequence bounded by α and so must converge to some limit γ with $0 < \gamma \le \alpha$.

Since $q_n \to \gamma$ as $n \to \infty$ and ϕ is continuous, it follows that

$$q_{n+1} = \phi(q_n) \to \phi(\gamma)$$

as $n \to \infty$. But $q_{n+1} \to \gamma$ as $n \to \infty$, so

$$\gamma = \phi(\gamma).$$

Since α is the only root of $\phi(t) = t$ in the region $0 < t < 1$, it follows that $\gamma = \alpha$ and

$$q_n \to \alpha$$

as we asserted from inspection of Figure 17.3.

(ii) Show similarly that, if $\phi'(1) \le 1$, then

$$q_n \to 1$$

as $n \to \infty$.

17.2 Growth and decay

Another example of Watson's paradox occurs when we consider queues. Consider a shop with a single cash desk. Customers arrive randomly at the cash desk, forming an orderly queue if someone is already being served. It takes a random time to serve each customer (some have their money ready, some do not). Let us call customer *B* the 'offspring' of customer *A* if customer *B* arrives while customer *A* is being served, and say that customer *A* 'dies' when he or she leaves the cash desk. The reader should convince herself that, if a customer arrives when no one is being served, then the next time when the queue is empty (that is nobody is being served) will occur when the tribe of offspring, offspring of offspring and so on through the generations finally becomes extinct. Theorem 17.1.2 now tells us that, as we would expect, the behaviour of the queue is governed by κ, the average number of customers who arrive whilst another customer is being served. Consider a particular customer who arrives when the queue is empty. If $\kappa < 1$, the queue which now forms will eventually become empty again with probability 1. If $\kappa > 1$, there is a probability $1 - \alpha > 0$ that the queue will never be empty again. Viewed from the cash desk this means that, if $\kappa < 1$, the cashier will have a rest from time to time but, if $\kappa > 1$, the cashier may have a few initial rest periods (when the cashier is lucky and the queue eventually empties) but eventually the cashier will be unlucky, the queue will never empty again and the cashier will work continuously.

What happens if $\kappa = 1$? Watson's paradox tells us that, if n is large, very few queues will produce an nth generation; but that those that do will have, on average, very large nth generations. Thus the cashier will have rest periods but, from time to time, very large queues will build up. Of course, it is not likely that, in practice, κ will be exactly 1, but if κ is less than but close to 1, we will still expect large queues to build up from time to time and if κ is greater than but close to 1, we expect $1 - \alpha$ to be small and so (on average) there will be many rest periods before the queue never empties again. In practice, cash desks shut from time to time, people go elsewhere if they see a very long queue and, in any case, κ is likely to vary with time, so it will not be possible to tell the difference between the various cases when κ is close to 1. Stepping back a little, we see that when the cash desk is 'operating near capacity' (that is κ is close to 1), the behaviour of the queue will be extremely erratic with some periods when it clears completely on a fairly regular basis and others when the queue builds up to a very long line. This phenomenon reoccurs for more complicated queueing schemes (for example, several cashiers in a supermarket) and it may help to while away the time spent in doctors' waiting-rooms to meditate on this. In general it shows that queuing systems should be run under their 'theoretical capacity'†.

†One of the readers of my manuscript added the heartfelt marginal note 'Try convincing a Government Minister of this!'

The most typical use of Theorem 17.1.2 is illustrated by a disease carrier arriving at a previously healthy town. The number of people he infects depends on chance accidents — he may infect none or he may infect many. Those whom he infects will each infect a random number of people and so on. Theorem 17.1.2 tells us that if κ the average number of people infected by each disease carrier is less than 1, we can expect the epidemic will fade out but that if $\kappa > 1$, there is a probability $p > 0$ that it will spread. If, at some stage, there are n infectious people (or if n infectious people enter the town), then the epidemic will die out only if the n separate epidemics traceable to the n separate sources all die out and this will occur with probability $(1 - p)^n$. Thus, even if p is small, once many people are infected the epidemic is pretty certain to spread. However, the theorem by itself tells us very little about the course of the epidemic once it has started.

The study of the progress of epidemics is complicated by the fact that people cease to be infectious after a time. We therefore start with the simpler case of the progress of a rumour. Here, everybody who has heard a rumour remembers it for ever and tells, on average, λ people about it each hour. Of course the number of people to whom a rumour carrier communicates it in a given hour will be random — he may infect none or he may infect many. Thus if, initially, only one or two people have heard the rumour the number of people who have heard it after a few hours will itself be random. Eventually, however, a relatively large number of people will have heard the rumour and its progress now becomes more predictable. The reader will be familiar with 'the law of averages' which says that the aggregate of a large number of similar random events will (usually) be predictable. If x people have heard the rumour at some time then they will tell about λx people in the next hour, and so about $(1 + \lambda)x$ people will have heard the rumour after another hour and the number $x(t)$ who have heard the rumour after t will increase exponentially as shown in Figure 17.5.

Exercise 17.2.1 *(A very simple differential equation) Explain why the spread of a rumour (once established) is given, according to the account above, by*

$$x'(t) = \lambda x(t).$$

Deduce that $x(t) = Ae^{\lambda t}$ for large t. Explain why A is a random number.

However, as the reader has been muttering to herself for the last two paragraphs, things are not that simple. To start with, trees do not grow up to reach the sky. An epidemic cannot infect, and a rumour cannot have been heard by, more people than there are in a town. Again, initially, almost everybody an infected person comes into contact with will be uninfected, but as an epidemic spreads, the average number of uninfected people that an infected person meets will fall and a similar thing will happen for rumours. Once again, the fact that infected persons will cease to be infectious makes the treatment of epidemics harder than that of rumours, so we stick to rumours.

For rumours we expect the rate of growth to slow down as more and more people have heard the rumour and drop towards zero as the total number of rumour bearers approaches the total population P. When there are only a few people who do not know the rumour, the question of when they hear it ceases to be governed by the law of averages and becomes a random matter again. We expect $x(t)$ the number of rumour bearers to have a graph something like that shown in Figure 17.6. (Note that we have chosen the time origin so that $x(0) = P/2$.)

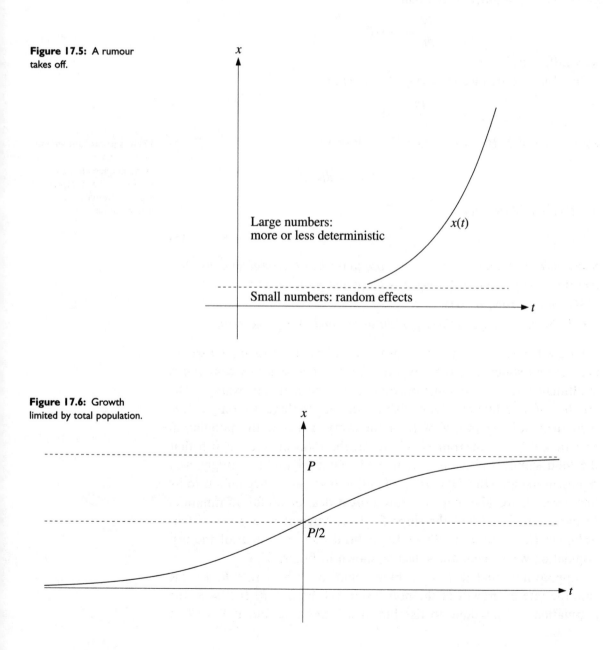

Figure 17.5: A rumour takes off.

Large numbers: more or less deterministic

$x(t)$

Small numbers: random effects

Figure 17.6: Growth limited by total population.

P

$P/2$

Exercise 17.2.2 *(This involves the solution of first-order separable differential equations.) (i) At time t there will be x(t) people who know and (P−x(t)) who do not know the rumour. Thus on average (P−x(t))/(P−1) of the people to whom a rumour bearer speaks will not have heard the rumour. (We have P − 1 rather than P because rumour bearers do not speak to themselves. However, we shall assume that P is large and the difference between P and P −1 is negligible.) Explain why, if P is large, it would be reasonable to represent the kind of rumour spreading just described by the differential equation*

$$\frac{dx}{dt} = \frac{\lambda}{P}x(P - x)$$

with x(0) = P/2.

(ii) Let us start by considering the equation

$$\frac{dx}{dt} = x(1 - x),$$

with x(0) = 1/2. By writing down something like†

$$\left(\frac{1}{x} + \frac{1}{1 - x}\right) dx = dt,$$

or otherwise, show that

$$\frac{x}{1 - x} = e^t. \qquad (*)$$

Show, directly from (*) that x(t) is close to 0 when t is large and negative and that x(t) is close to 1 when t is large and positive.

Sketch the graph of x(t).

(iii) Solve the differential equation of (i) and sketch the result.

†This is a standard step in solving such equations, but many mathematicians would say that it represents a purely formal manipulation.

Figure 17.6 seems so natural that we would expect it to appear whenever growth which would be exponential in the absence of constraints is ultimately limited to some maximum P. Consider, for example, the number of a certain species of fish in the sea. If there are only a few, then there will be plenty of food for every fish and the population growth will be unconstrained. However, the total number of fish that the food supply will support is limited. Thus Figure 17.6 might well represent the growth of the fish population from small beginnings to its final size. Figure 17.6 then contains a great deal of useful information. In particular, if we find the slope S of the tangent at a point C (coordinates (X, T) say), it tells us the natural rate of growth of the fish population when there are X fish as shown in Figure 17.7.

Suppose now that we start fishing, removing fish at rate R. In this case the rate of growth of the population will be $S − R$. If $S > R$, the population will continue to rise but at a lower rate, but if $S < R$, it will fall. In Figure 17.8 we have marked two points A and B (with x coordinates x_A and x_B, say) on the curve such that the slope S is smaller than the rate of fishing R to the left of A and to the right of B but greater between A and B. Thus if we start fishing with a population x greater than x_B, the population will fall towards x_B, if we start fishing with a population x between x_A and x_B the population will rise towards x_B, and if we start fishing with a population x less than x_A, the population will fall towards 0 (that is towards extinction). In other words, if we fish at rate R, and our initial population is greater than x_A, the population will eventually stabilise at x_B but, if the initial population is smaller than x_A, we will drive the fish to extinction. Of course, if R is too big, there will be no points A and B because $R > S$

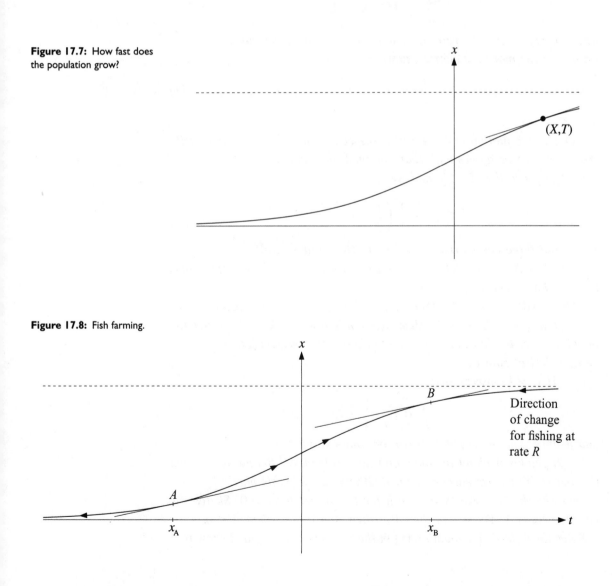

Figure 17.7: How fast does the population grow?

Figure 17.8: Fish farming.

Direction of change for fishing at rate R

at each point on the curve and, whatever the initial conditions, the population will fall towards 0.

Suppose we wish to 'harvest' the fish on a regular basis and not to drive them to extinction. As we increase R, the points A and B will move closer together so the difference between the 'stabilised population' x_B and the 'danger of extinction point' x_A will get smaller. The cost of an increased harvest is a reduction in the margin of safety.

Exercise 17.2.3 (*This involves the solution of first-order separable differential equations.*)

(i) *Let us suppose that a population of fish is governed by the differential equation*

$$\frac{dx}{dt} = x(1-x) - L$$

where L represents the rate of fishing. Show that, if we put $y = x - \frac{1}{2}$, we obtain the more symmetric equation

$$\frac{dy}{dt} = -y^2 + K \qquad\qquad (*)$$

with $K = L - \frac{1}{4}$.

(ii) *Solve Equation (*) directly in the case when $K = -\frac{1}{4}$ and verify that your answer agrees with that obtained in Exercise 17.2.2.*

(iii) *Suppose that $L < \frac{1}{4}$, and we set*

$$C(L) = \frac{1}{2}\left(\frac{1}{4} - L\right)^{1/2}.$$

Show that three cases arise according to the value of $x(0)$.

(a) *If $x(0) < \frac{1}{2} - C(L)$, then $x(t)$ eventually becomes negative (thus the population becomes extinct).*

(b) *If $x(0) = \frac{1}{2} - C(L)$, then $x(t)$ remains fixed at this constant value.*

(c) *If $\frac{1}{2} - C(L) < x(0)$, then $x(t)$ tends towards $\frac{1}{2} + C(L)$ and the population settles down at a new equilibrium with a population of $\frac{1}{2} + C(L)$ being fished at rate L.*

(iv) *If $L > \frac{1}{4}$ observe that*

$$\frac{dy}{dt} \leq L - \frac{1}{4} < 0$$

and deduce that the population will become extinct.

(v) *If you know about the function $\tan^{-1}$, solve the differential equation (*) and verify the conclusion of (iv). Otherwise omit this part.*

(vi) *Sketch the curve $tx = 1$ both for $t < 0$ and for $t > 0$. Sketch the entire curve $(t + A)x = 1$ in the case when $A > 0$ and when $A < 0$. Solve the differential equation (*) in the case when $L = \frac{1}{4}$ and show that* *if* $x(0) \geq \frac{1}{2}$, *then, if the model is exact, the population will settle down to the value* $\frac{1}{2}$ *but if* $x(0) < \frac{1}{2}$, *then the population will become extinct. By considering the effect of small random fluctuations, explain why, in practice, we would expect the population to become extinct whatever the initial value* $x(0)$.

(vii) For the rest of this exercise we shall consider the more general differential equation

$$\frac{dx}{dt} = kx(P - x) - l \qquad (**)$$

describing the behaviour of a population of fish with fishing at rate l. *Explain to a non-mathematician the meaning of* P *and* k *and why we take* $P, k > 0$ *and* $l \geq 0$.

(viii) Show that making the substitution $X = ux$, $T = vt$ *with suitable values of* u *and* v *converts Equation* (**) *to the form*

$$\frac{dX}{dT} = X(1 - X) - L.$$

What is the value of L?

(ix) Show that if we choose l *so as to stabilise the population at* $(\frac{1}{2} + \delta)P$ *(starting from* $x(0) = P$, *say) with* $\frac{1}{2} \geq \delta \geq 0$, *then, if some accident reduces the population to some value greater than* $(\frac{1}{2} - \delta)P$, *it will recover but if the value is reduced below* $(\frac{1}{2} - \delta)P$, *then, in the absence of any change in fishing rates, the population will become extinct. Thus we have a margin of safety of* $2\delta P$.

Let $l(\delta)$ *be the value of* l *required to carry out the policy in the previous paragraph. Show that*

$$l(\delta) = k \left(\frac{1}{4} - \delta^2 \right).$$

Explain why you would expect $l(\delta)$ *to depend on* k *as shown. In some sense,* $l(0)$ *represents the maximum sustainable rate of fishing and* $l(\delta)/l(0)$ *the efficiency of our chosen policy. Show that*

$$\frac{l(\delta)}{l(0)} = (1 - 4\delta^2).$$

In choosing δ, *we thus balance efficiency against a suitable margin of safety. Show that (if we are not too greedy) we can have quite a wide margin of safety without greatly reducing efficiency.*

(x) 'Medium-size populations give the most fish.' Comment.

In 1945 Steinbeck wrote

> Cannery Row in Monterey in California is a poem, a stink, a grating noise, a quality of light, a tone, a habit, a nostalgia, a dream. Cannery Row is the gathered and scattered, tin and iron and rust and splintered wood,

chipped pavement and weedy lots and junk heaps, sardine canneries of corrugated iron, honky-tonks, restaurants and whore-houses, and little crowded groceries and laboratories and flop houses. (*Cannery Row*, Steinbeck)

Cannery Row lived by the Pacific sardine. Table 17.1, gives the annual catch from 1922 to 1945. In 1931, the California Fish and Game Commission recommended limiting the catch to 200 thousand tons a year, and this figure and the slightly higher one of 250 thousand tons recur in expert advice from then on. The fishing industry responded by setting up factory ships outside state waters. Whenever it looked as though legislation limiting the sardine fisheries was possible, the industry

> ... resorted to a plan (used before and since) by which ... legislation could be postponed by asking for a special study of the abundance and thus disregard the work of the State Fisheries Laboratory or at least throw doubts upon its findings.

It was always possible to find scientists who believed that it was virtually impossible to overfish a species like the sardine.

In the years after 1945 the catch began to fall in a marked but erratic manner. The industry pressed for more research, and in 1947 a Marine Research Committee was set up to 'to seek out the underlying principles which govern the Pacific sardine's behaviour, availability and total abundance.' The research was paid for out of a tax on fish landings and the fishing industry had the majority on the committee. Table 17.2 gives the annual catch from 1939 to 1962. In 1967, the California Legislature imposed a moratorium on further sardine fishing.

Cannery Row now caters for the tourists with a fine range of medium-price fish restaurants and souvenir shops. There is also a splendid aquarium with a main tank three stories high containing sharks, rays and other denizens of the deep. If you wait a little, you may see

Table 17.1. *Annual catch of sardines along the Pacific coast in thousands of tons 1922–45. Each season runs from June in the stated year to the next May.*

Year	22	23	24	25	26	27	28	29	30	31	32	33
Catch	66	85	174	153	201	256	335	412	260	238	295	387

Year	34	35	36	37	38	39	40	41	42	43	44	45
Catch	638	632	791	498	671	583	493	680	573	579	614	440

a small shoal of handsome fish sweep by, and a glance at one of the helpful identification plaques tells you that you have just seen the Pacific sardine.

The story is not, of course, as simple as this. It seems likely that the collapse of the sardine population may have been precipitated by changes in the Pacific ocean currents. Other changes in these currents may have triggered the more recent but equally spectacular collapse of the Peruvian anchovy fisheries (where the catch increased from near zero to well over 10 million tons a year between 1957 and 1971, and then dropped to less than 5 million in the next year, continuing to fall to an average of less than 2 million tons a year at the end of the 1970s). But, in both cases, it was overfishing which turned a natural population change into a population collapse.

It is also true that Cannery Row reads better than it lived. When we regret the collapse of the California fisheries, are we not just being sentimental?

Exercise 17.2.4 *You are in charge of the 'Fund For Extremely Deserving Widows And Orphans'. The fund has just been left a large lake with the instructions that it is to be managed so as to produce the greatest possible income for the widows and orphans under the fund's devoted care. Up to now, the lake has been 'farmed', in that one fishing expedition each year removes kP fish and the fish population recovers its former size P by the time of the next year. The cost of a fishing expedition is independent of the number of fish that are caught and this cost can be recovered by selling l fish; the sale of the rest of the fish is pure profit. The interest on money deposited in a bank is η units of currency per unit of currency deposited per year $[P > l > 0, \eta > 0]$. (You should assume zero inflation.) You have the choice between*

(a) continuing the previous policy and giving the profits to the widows and orphans, or

(b) organising a single fishing expedition, catching all the fish in the

Table 17.2. *Annual catch of sardines along the Pacific coast in thousands of tons 1940–62.*

Year	39	40	41	42	43	44	45	46	47	48	49	50
Catch	583	493	680	573	579	614	440	248	130	189	339	353

Year	51	52	53	54	55	56	57	58	59	60	61	62
Catch	145	15	19	81	79	47	32	126	59	49	47	19

lake, banking the proceeds and giving the interest to the widows and or-phans.

Show that you should do (b) if

$$\eta > \frac{kP - l}{P - l},$$

and (a) if the inequality is reversed.

Show, in particular, that if $\eta > k$ then you should do (b). Why should we have foreseen this?

Using the model above, explain why we may expect fishing to extinction may be rational, in a simple economic sense, for species which reproduce slowly†. What will be the effect of high interest rates on such decisions? (Compare the discussion at the end of Section 4.5.)

Since the destruction of a major fishery involves much immediate hardship and reduces the world's food supply either permanently or for some years, it does seem, none the less, that the general good is not served by overfishing‡. But here we come to a problem which flickers in and out of this book just as it flickers in and out of real life — the good of the community is not identical with the good of each separate individual making it up. If a large number of fishing boats fish from some large fish population, then, if a single boat doubles its catch, it will make no difference to the sustainability of the population. If the fish are being exploited with a proper margin of safety, then the extra catch will not matter. If there is no proper margin of safety, then the fishery is doomed to collapse anyway and it is all the more important to make what profit we can while we can. Since what applies to one boat applies to all, all boats will catch as much as they can, *even if each skipper knows that everybody would ultimately be better off if everybody caught less.*

It is only the effects of custom which make us believe such situations are rare. Much of the legislation with which we surround ourselves consists of rules which we are only too pleased to obey *provided we can be assured that everybody else will obey them too.* The problem with fisheries is that suitable laws are hard to make and to enforce. In addition, the life cycle of a fishery is not such as to encourage prudence. Initially, catches will be large since there will be few boats and the fish population will be at its high 'natural' level. Encouraged by this, more people will borrow to buy boats to take part. At the same time technological improvements will increase the catch per boat (and increase the indebtedness of the fishermen who need to buy the equipment involved). By the time the first warnings of overfishing appear, the capacity of the fishing fleet will far exceed the sustainable

†This will disturb only those who believe *both* in saving the whale and the substitution of economic for moral choices.

‡Though, no doubt, the vast output of Panglossian economic literature contains arguments to the contrary.

fishing level. If each boat reduces its catch to the 'correct' level, the owners will be unable to repay their borrowings. Faced with the choice of certain bankruptcy this year or possible bankruptcy in three years (after all, the experts have been wrong before), the owners will make the rational choice and continue to take as much fish as they can.

> Between 1964 and 1968, Norwegians invested almost a hundred million dollars building up their herring purse seine fleet. Unfortunately, the herring stocks were unable to increase their productivity to match the increase in harvesting capacity. In 1967, Norway took 1.2 mmt [million metric tons] of herring; in 1968, 0.70; and in 1969, in spite of the expanded fleet, the catch dropped to 0.10 mmt. Norway now [owned] a huge surplus of herring purse seiners with nothing to catch. The herring stocks [had] dropped to less than 1/15 of their abundance in 1950.

Between 1948 and 1989, the tonnage of fish caught from the world's oceans increased fivefold, but this increase is levelling off. We may hope that this levelling-off represents the worldwide adoption of rational fishing management so that the world's stock of fish is being exploited at a sustainable yield. Or we may look at Table 17.1 and wonder.

17.3 Species and speculation

Few mathematical ideas have excited such general interest as the 'butterfly effect'. Lorenz's question, 'Can the flap of a butterfly's wings in Brazil set off a tornado in Texas?' and his affirmative answer have become part of popular culture. (In less vivid terminology, Richardsonian numerical weather forecasting is possible over 24 hours since we only need to know the present weather to a reasonable degree of accuracy to predict the weather tomorrow to a fair degree of accuracy. To predict the weather in a month's time to a fair degree of accuracy we would need to know the weather now to an impossible degree of accuracy†.) But the answer gives rise to another question, 'Why does every flap of a butterfly's wing not set off a tornado in Texas?' As I look out over the dripping leaves and grey skies which announce a Cambridge autumn, I know that it is impossible in practice and, almost certainly, in principle to predict the weather in six months' time, but I also know that by that time spring will have arrived. I know there will have been storms in the meantime but I also expect that, though they may be powerful enough to bring down trees, they will not reduce the town centre to a heap of rubble. The problem of meteorology is to account for the coexistence of change and permanence, of instability and stability within the same system.

†As might be expected the moment that Lorenz's butterfly made its appearance in the media, a queue of military men formed outside forecasting offices round the world enquiring about the use of controlled explosions against other peoples weather systems. Can the reader see why they were doomed to disappointment?

In the same way, any theory of evolution has to account not merely for the disappearance of old species and the appearance of new species but for the persistence of certain species, like sharks, essentially unchanged for hundreds of millions of years. Following Darwin, we see the key as lying in the concept of selective advantage. If two groups of animals compete for the same set of resources (so that total numbers are limited), one group will, on average, reproduce above its replacement rate (so that if the population consists of males and females, each male will, on average, have more than one son) and the other will, on average, reproduce below its replacement rate. Given our tendency to cheer the victor, it is natural to think of the group which reproduces above its replacement rate as 'better adapted' or 'fitter', but we must remember that the same ideas must apply to tapeworms as to gazelles.

To see why sharks, say, might remain unchanged for millions of years, consider the effect of some genetic change. If the change is large, it is almost certain to be unfavourable, and, if it is small, it may be favourable or unfavourable but, by definition, will not have much selective advantage (or disadvantage). Our standard diagram, drawn as Figure 17.9, shows how a genetic change with some selective advantage may propagate through a population. Initially we have a Galton–Watson problem: the favourable genetic change may be considered as a surname and the probability that it will establish itself

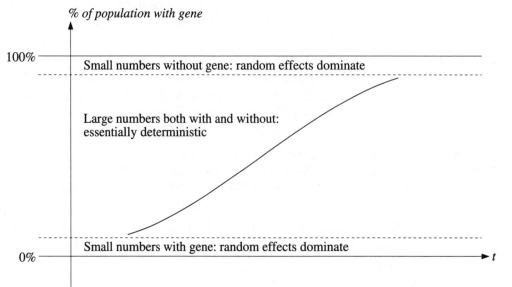

Figure 17.9: Propagation of a favourable genetic change.

amongst a reasonably large group is roughly equal to the probability of non-extinction of the surname. Once established, it will presumably propagate along some sort of a growth curve *AB* and, once almost all of a certain generation of the population have the gene†, the problem of how long it takes for all to have the gene becomes probabilistic again.

In the unusual event of a genetic change which is large and favourable, the extinction probability in the Galton–Watson problem will be substantially less than 1 and the growth curve will be steep, allowing the gene to spread over most of the population in relatively few generations. When the factories of the British Midlands covered every surface with black soot, several species of moth which had previously possessed light coloured wings very rapidly changed to dark wings. When the pollution ceased the change reversed itself. Similar considerations apply, presumably, to drug resistance in bacteria.

However, if the selective advantage of some change is small, we are back with a Galton–Watson problem with κ close to 1 and we know that the probability of extinction is very close to 1. Furthermore, even if the gene does become established, it will take a very long time indeed to spread through a large population.

Exercise 17.3.1 *If $x_{n+1} = 1.01x_n$ and $x_0 = 1$, find x_{10}, x_{100} and x_{1000}.*

If large changes are very likely to be unfavourable and small changes are unlikely to establish themselves, we have excellent reasons to expect species to be stable over long periods of time. But we appear to have over-egged the cake since it is now hard to see how any changes can arise.

However, the consequences of Watson's paradox allow us to deal not only with the extinction of old species but with the creation of new ones! To see how this could be so we need the concept of geographical isolation by which a small group of our species can be cut off from the rest. As a simple example, we might suppose that a flock of small birds might be blown off the mainland and onto an island so distant that they cannot return.

However, there are many more ways in which isolation can occur. One of the high points of Darwin's *On the Origin of Species* is his explanation of

> [t]he identity of many plants and animals, on mountain-summits, separated from each other by hundreds of miles of lowlands, where the Alpine species could not possibly exist.

† For the purpose of this discussion the gene is simply a shorthand way of referring to the 'surname associated with the genetic change'.

In a passage which is, unfortunately, to long to quote in full† he first describes the evidence for recent ice ages.

> The ruins of a house burnt by fire do not tell their tale more plainly, than do the mountains of Scotland and Wales, with their scored flanks, polished surfaces and perched boulders, of the icy streams with which their valleys were lately filled.

Darwin goes on to describe how, as the climate grew colder, arctic species moved south so that

> [b]y the time the cold had reached its maximum, we should have a uniform arctic fauna and flora, covering the central parts of Europe, as far south as the the Alps and Pyrenees, and even stretching into Spain. The now temperate regions of the United States would likewise be covered by arctic plants and animals, and these would be nearly the same as Europe; for the present circumpolar inhabitants which we suppose to have everywhere travelled southward, are remarkably uniform round the world. ...
>
> As the warmth returned, the arctic forms would retreat northward, closely followed up in their retreat by the productions of the more temperate regions. And as the snow melted from the bases of the mountains, the arctic forms would seize on the cleared and thawed ground, always ascending higher and higher, as the warmth increased, whilst their brethren were pursuing their northern journey. Hence, when the warmth had fully returned, the same arctic species, which had lately lived in a body together on the lowlands of the Old and New Worlds, would be left isolated on distant mountain-summits (having been exterminated on all lesser heights) and in the arctic regions of both hemispheres.

Returning to our main argument, suppose that we have a geographically isolated group which forms a breeding community of a few hundred individuals or less over many generations. If we consider small genetic changes with little selective advantage (or disadvantage), Figure 17.9 is then compressed so that the 'deterministic curve AB' vanishes and the 'random margins' now occupy the entire width of the figure shown as Figure 17.10.

If we consider a neutral genetic change (corresponding to $\kappa = 1$ in the Galton–Watson problem), then the Galton–Watson paradox tells us that although the associated gene will usually become extinct, occasionally it will spread through the entire population. But, once every member of the population has that gene, it cannot be extinguished. What is true for $\kappa = 1$ will be true for κ close to 1, so *in a small population* genes which have a slight selective advantage, or are neutral, or even genes which have a slight selective disadvantage can start with one individual and end up being possessed by every member of the population several generations later.

Over many generations the accumulation of many such small neutral or near neutral changes will change the nature of the group and we get a new species. This new species is not 'better' or 'worse' than the old, it is simply, and merely by the operation of a long sequence of chance events, *different*. If the causes of the isolation of our breeding group are removed, then the old species *U* and the new species *V* will come into contact. If they are now so different that they do not compete for the same resources, the two species will coexist. If they do compete for the same resources, the one with the reproductive advantage will displace the other.

Most biologists believe that the kind of mechanism I have outlined provides a route for the origin of new species. It is not the only route and there is much room for disagreement on the relative importance of the various possible routes. (One of those who believe in the importance of the route described above is S. J. Gould whose many excellent collections of popular essays contain spirited defences of this thesis.) However, we shall now leave evolutionary speculations to see how the ideas surrounding the Galton–Watson problem apply to the rather more immediate concern of tracing the course of epidemic illnesses.

It is easy to see how epidemics can start; it is slightly less easy to see how they stop and how, once they stop, the disease can survive. One important reason is that for many infectious diseases, sufferers, if they survive, have developed defences against that disease and are immune

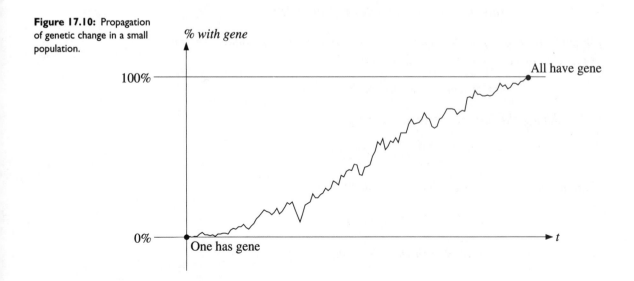

Figure 17.10: Propagation of genetic change in a small population.

from further attacks. Thus at any point the population will contain a proportion μ of immune members. If each sufferer would (in the absence of such immunity) transmit the disease to κ others on average, we would now expect them to transmit the disease to $(1 - \mu)\kappa$ others on average. If $(1 - \mu)\kappa > 1$ the disease will tend to spread, but this in turn will increase μ and so decrease $(1 - \mu)\kappa$. On the other hand if $(1 - \mu)\kappa < 1$ the incidence of the disease will diminish. Births (and possibly immigration) will decrease μ and so increase $(1 - \mu)\kappa$. This suggests that if the population is sufficiently large, the disease will not die out and the disease level will be found close to the point where

$$(1 - \mu)\kappa = 1.$$

If the disease level is well below this 'natural level' (for example, if the population is free from the disease and it is introduced from the outside), it will rise rapidly, overshooting the 'natural level', possibly catastrophically. Once established, the disease level will presumably oscillate about its natural level and become 'endemic'.

Exercise 17.3.2   *(This requires the solution of a second-order linear differential equation. It is not central to the argument.) (i) Consider a population which at time t contains, in the jargon of the trade, S(t) susceptibles (that is people who have not yet had the disease), I(t) infectives (people who have the disease and can pass it on) and R(t) 'removed' (those people who are isolated, dead or immune so they cannot pass on the disease). It is natural to suppose that the rate at which susceptibles catch the disease is proportional both to S(t) the number available to catch the disease and to I(t) the number of infectives. Thus*

rate at which susceptibles fall ill $= \beta SI$

for some $\beta > 0$. If we assume that new susceptibles are added to the population at constant rate $\sigma > 0$ we have

rate at which new susceptibles arrive $= \sigma$,

so, combining the last two statements, we get

$$\frac{dS}{dt} = \sigma - \beta SI.$$

Explain, in a similar way, why it is plausible that

$$\frac{dI}{dt} = \beta SI - \gamma I.$$

We are not interested in R(t) in this question.

(ii) We now have two equations

$$\frac{dS}{dt} = \sigma - \beta SI$$

$$\frac{dI}{dt} = \beta SI - \gamma I$$

(with $\sigma, \beta, \gamma > 0$), but we cannot solve them. Nonetheless, it is possible to extract a fair amount of information. Let us ask first whether it is possible for the system to be in a steady state with $S(t) = S_0$ and $I(t) = I_0$ constant. Show that such a steady state is only possible with $I_0 = \sigma/\gamma$ and $S_0 = \gamma/\beta$.

(iii) Now let us ask how $I(t)$ and $S(t)$ behave near the steady state. The natural approach is to write $I(t) = I_0 + i(t)$, and $S(t) = S_0 + s(t)$ so that $s(t)$ and $i(t)$ are small. Show that, if we neglect the very small quantity $i(t)s(t)$, our equation becomes

$$\frac{ds}{dt} = -\gamma i(t) - \frac{\beta\sigma}{\gamma}s(t)$$

$$\frac{di}{dt} = \frac{\beta\sigma}{\gamma}s(t).$$

By differentiating the first equation and substituting from the second, show that

$$\frac{d^2s}{dt^2} + \frac{\beta\sigma}{\gamma}\frac{ds}{dt} + \beta\sigma s = 0.$$

Hence show that

$$s(t) = Ae^{-vt}\cos(\omega t + \theta),$$

where v and ω are to be found and A and θ depend on the initial conditions.

At first sight this seems very satisfactory, since it predicts epidemic waves, but our feeling of satisfaction is much diminished when we note that (a) the period predicted does not seem to fit well with the observation that many epidemics have a seasonal character and (b) the model predicts that the cycles will die out. Various modifications to the model have been proposed (note, for example, that the beginning of the school year can be characterised as a time that children meet to exchange diseases, and this will tend to strengthen a tendency for periods to be a multiple of year), but we shall not take the matter further. (The reader will find much of interest in Chapter 6 of [4], where it is shown that our primitive model is rather more successful than at first appears. Further mathematical progress requires some reflection on the theory of non-linear differential equations.)

A more satisfying application of mathematical theory to epidemics will be found in Exercise 17.3.3 which follows.

Exercise 17.3.3   *This long exercise requires a substantial amount of calculus and is not essential to the main argument.*

(i) Consider a population which at time t contains S(t) susceptibles (that is people who have not yet had the disease), I(t) infectives (people who have the disease and can pass it on) and R(t) 'removed' (that people who are isolated, dead or immune). In this question we are interested in the short term behaviour of the disease, so we ignore births and assume that the total population N does not change for other reasons. Explain why the following equations might provide a good model for what happens.

$$S + I + R = N,$$
$$\frac{dS}{dt} = -\beta SI,$$
$$\frac{dI}{dt} = \beta SI - \gamma I,$$
$$\frac{dR}{dt} = \gamma I$$

with $\beta, \gamma > 0$.

(ii) As we said in the previous exercise, it is one thing to set up a model and quite another thing to solve it. However, we can make a start by finding the I in terms of S. Observe that, using the second and third equations of (i), we have

$$\frac{dI}{dS} = \frac{dI}{dt}\frac{dt}{dS} = \frac{\beta SI - \gamma I}{\beta SI} = 1 - \frac{\gamma}{\beta S}.$$

Now solve the differential equation to obtain

$$I = I_0 + S_0 - S + \frac{\gamma}{\beta} \log \frac{S}{S_0},$$

where I_0 and S_0 are the values of I and S when $t = 0$ and $\log$ is the function which the reader may know as $\log_e$ or $\ln$.

(iii) Let us write $\rho = \gamma/\beta$. Show that I (considered as a function of S) has a unique maximum when $S = \rho$. Check that the curves of I against S look as in Figure 17.11(a).

(iv) At first glance, Figure 17.11(a) merely tells us the possible values of $(S(t), I(t))$ but does not tell us whether they occur. By referring back to the equations of (i), explain why S(t) must decrease as t decreases and so the $(S(t), I(t))$ must move along the curves in the direction marked by the arrows of Figure 17.11(b).

Referring back to the equations of (i), explain why it is not possible that $S(t), I(t) \geq \delta$ for some fixed $\delta > 0$ and all $t > 0$. Conclude that $(I(t), S(t))$ must eventually traverse the whole of the curves shown in Fig-

ure 17.11(b) and so $S(t) \to S_\infty$ as $t \to \infty$ where S_∞ is the smaller root of

$$0 = I_0 + S_0 - S_\infty + \rho \log \frac{S_\infty}{S_0}.$$

(v) By combining the results so far, show that if $S_0 \leq \rho$, then $I(t)$ decreases to zero and $S(t)$ decreases to S_∞, but that if $S_0 > \rho$, then $S(t)$ will still decrease from S_0 to S_∞ but now $I(t)$ will increase to a maximum when $S(t) = \rho$.

Explain how these results lead to the conclusions:

(A) An epidemic will occur if the disease is introduced only if the number of susceptibles S_0 in the population exceeds a threshold level $\rho = \beta/\gamma$.

(B) The spread of the disease stops not because there are no more susceptibles, but because there are no more infectives. In particular, some individuals will escape the disease.

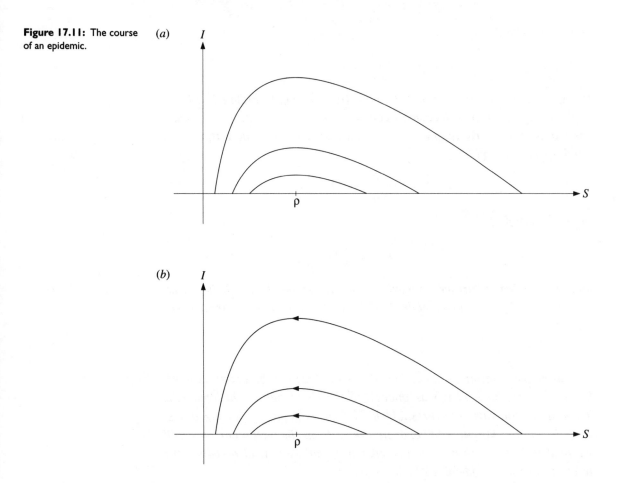

Figure 17.11: The course of an epidemic.

This exercise deals with the short term. How is ρ related to the μ and κ of the previous discussion of long-term behaviour?

(vi) Assume that $S_0 > \rho$. We are interested in the number J of people who will catch the disease during the epidemic. Explain why $J = S_0 - S_\infty$. If, as will normally be the case with an epidemic, the initial number of infectives I_0 is very small show, using the formula of (iv), that, to a good approximation,

$$0 = J + \rho \log \left(1 - \frac{J}{S_0} \right). \tag{*}$$

We can always use a standard root-finding method to obtain as good an approximation as we wish to J for given ρ and S_0. However, if $v = \rho - S_0$ is small compared with ρ (so we are not far from the threshold level), we can proceed directly.

It is known that if $|x| < 1$, then

$$\log(1 + x) = x - \frac{x^2}{2} + \frac{x^3}{3} - \frac{x^4}{4} + \dots$$

and so, in particular,

$$\log(1 + x) = x - \frac{x^2}{2}$$

to a good approximation when x is small. By referring to Figure 17.11(b), explain why if $v = \rho - S_0$ is small compared with ρ, we will have J small compared with ρ. By taking $x = -J/S_0$ and using (*) show that, to a good approximation,

$$J = S_0 \left(\frac{S_0}{\rho} - 1 \right).$$

Show, by simple algebra, that

$$S_0 \left(\frac{S_0}{\rho} - 1 \right) = 2v \left(1 + \frac{v}{\rho} \right)$$

and, by making a further approximation (to be justified), deduce that if $v = \rho - S_0$ is small compared with ρ, then, to a good approximation,

$$J = 2v,$$

and, after the epidemic, the number of susceptibles will be about as much below the threshold as it was above the threshold before the epidemic. Thus, if we can raise the threshold level (for example, by having a system of isolation hospitals to separate the ill from the well), not only will we tend to reduce the number of epidemics, but we will also reduce the intensity of those epidemics which do occur.

These results are due to Kermack and McKendrick. By working rather harder than we have done, they were able to find good approximate solutions for R(t) and R'(t) in the case of small epidemics (when v = ρ − S₀ is small compared with ρ) and show that their predictions coincided very closely with the figures for an outbreak of plague in Bombay in 1907. (References to the original paper together with a much broader, but highly mathematical, view of the topic will be found in Bailey's The Mathematical Theory of Epidemics. *Most readers will prefer the treatment in Chapter 19 of Murray's* Mathematical Biology.*)*

Claude Cockburn used to say that the only question journalists need ask themselves in the presence of a politician was, 'Why is this bastard lying to me?' My readers, lulled by the plausible analogies, righteous indignation and authoritative-looking differential equations of the last two sections, ought to ask themselves the same question. The same diagram (Figure 17.9 and elsewhere) has been recycled over and over again as a typical growth curve, but is there any evidence that it applies in the situations to which we have applied it?

In the case of the growth of fish stocks, the answer must be — not much, and if someone did produce figures which fitted the curve neatly, I would suspect fraud. Even without the intervention of man, we would expect a fish population to vary considerably from year to year in responses to changes in the weather and to changes in other populations (both those fed on by and those feeding on the population studied). Thus P, the 'undisturbed maximum population', is not constant but subject to substantial annual fluctuations from year to year. Drawing a smooth curve with asymptotes 0 and P becomes a much less attractive process if we do not believe in a constant P.

However, Cockburn's dictum has its limitations. If we assume that all politicians are crooks, then there will be no advantage to a politician in being honest and all politicians will become crooks. 'So voter, give an honest curse and choose the bad against the worse.' By examining our simple model, we have gained a clearer insight into how we might harvest fish on a regular basis. We have also seen that there are two key variables — population size and the rate of population growth. It might be argued that such conclusions could have been obtained by purely verbal arguments, but the mathematical model makes the underlying assumptions explicit and makes it much easier to see how those assumptions might be modified. Provided we do not treat it as anything but a preliminary attack on a difficult problem, the model yields a fair amount of insight.

In the case of epidemics, it is clear that our treatment considers their development in time but not in space. A discussion which compares various models which attempt to take this into account with actual data will be found in Chapters 12–15 of [36]. However, I hope to show in the next section that even simple models add to our understanding of important issues.

17.4 Of microorganisms and men

For an epidemic to spread, the average number κ of further people infected by a particular individual with a disease must be greater than 1 and preferably (so far as the disease is concerned) substantially greater than 1. But κ depends not only on the disease but on the mode of life of the society in which it finds itself. Hunter-gatherers move in small groups and are sparsely scattered across the land. Farming enables and usually requires larger concentrations of people to gather together and yields a higher population density. In towns, many people live together at close quarters. Thus κ will be small for hunter-gatherers, larger for farmers and much larger still for town dwellers. Cities like London were proverbially unhealthy and it seems likely that, for most of history, cities required a constant influx of newcomers from the country simply to maintain their population at the same level.

We believe that North and South America were first colonised by hunter-gatherers entering through Siberia and Alaska, and then spreading southwards. Even if some of the early arrivals had brought smallpox and measles with them, the small κ associated with their mode of life would have extinguished these and similar crowd diseases. (In modern times measles dies out even on quite large inhabited islands and a mixture of empirical evidence and theoretical argument suggests that a 'herd' of humans of about 250 000 is required to support the disease.) By the time the Spaniards arrived in the New World, the successors of the small bands had formed civilised and complex states with large populations and large towns. The diseases of the Old World, led by smallpox, measles and mumps, erupted into these populations with no natural immunity, and death rates may have been as high as 50%. The Spaniards who had survived childhood attacks of all these diseases were nearly immune — a circumstance which seems to have been attributed to divine favour by both sides, and the chaos consequent on the epidemics was one of the main factors that enabled a handful of adventurers to overthrow entire societies.

The phenomenon repeated itself throughout the Americas. In 80 years, the population of Lower California fell by 90%. In the words of a German missionary in 1699, 'The Indians die so easily that the bare look and smell of a Spaniard causes them to give up the ghost.'

Such figures must be estimates, but the military doctors of the American Civil War, although not able to do much for their patients, were well equipped to count them. Two Union soldiers died of disease for every one killed in combat (but in the Crimean War, fought a little earlier, four British soldiers died of disease for every one killed in battle). The new recruits from rural areas promptly went down with measles and mumps (both of which mainly killed indirectly by leaving the weakened victim prey to other illnesses) followed by smallpox and erysipelas (which killed directly). The death rate from disease was 43% higher among 'hardy pioneers' (that is Union soldiers from the west of the Appalachians) than among the 'pasty-faced city dwellers' from the East. Disease halved the size of most regiments from their initial complement before they even went into battle†.

†The worst case was presented by the 65th Regiment which saw no action during its three-year history but lost 772 of the 1769 men who served in it to disease and camp injuries.

Modern European cities are not the unhealthy places that they were in past times. Public health measures to ensure pure food and water, proper sewage disposal and so on, together with private affluence enabling a high degree of hygiene, have reduced κ for infectious diseases in general. The historical examples of the previous paragraphs suggest that under certain circumstances this may not be an unmixed blessing (consider the problems of a tourist from such a place, or the problems facing the whole city if hygiene breaks down following a disaster). A good example is given by polio which confounded expectations by apparently increasing in incidence and severity with increasing standards of living. The reason is now thought to be as follows. The lower the prevalence of an infectious disease, the higher the average age at which someone can expect to catch it. Many diseases are less serious for adults than young children, but polio reverses this. In the kind of city which has prevailed through most of history, most children will have had an attack and so acquired immunity by the age of three. Only with increased cleanliness and reduced crowding can polio become a disease of adults.

Artificial immunization by vaccination provides a way round the problem since, when possible, it enables us to increase the proportion μ of immune members of a society. The discussion of the previous section shows that vaccination not only protects the individual but, by reducing $(1-\mu)\kappa$, the average number of individuals infected by a given bearer of the disease, decreases the probability of epidemics and the severity of epidemics when they do occur.

Indeed, vaccination raises the exciting possibility of eliminating some diseases entirely. If, by vaccinating new arrivals (and revaccinating others if the immunity conferred does not last a lifetime), we can keep $(1 - \mu)\kappa < 1$ then the disease should tend to die out. Partly because of the random element in the behaviour of the disease when only small numbers are infected, partly because vaccination levels will vary over the community leaving pockets with $(1 - \mu)\kappa > 1$ and partly because Nature is not as simple as our equations make out, there will be isolated outbreaks; but a policy of isolation and mass vaccination may enable us to contain them and, if so, such outbreaks will become rarer and rarer and eventually cease.

Such campaigns have cleared whole regions in the past, but, because of the risk of infection from outside, it requires either efficient quarantine (virtually impossible in the age of the jumbo jet), some mechanism for dealing swiftly with any outbreaks (and the mechanism must work perfectly every time), or the continued maintenance of high levels of vaccination (difficult when people no longer feel threatened), or some combination of all three to keep the region clear. Might it not be possible to eradicate a disease from the entire world?

This has been done once† in the case of smallpox. The total cost over ten years for the entire World Health Organisation campaign to eliminate smallpox world-wide was $313 million. At the same time, the United States was spending $150 million per year on vaccination and quarantine measures intended merely to protect its own population, and similar expenditure world-wide may have reached $1000 million a year. The reason why countries were prepared to spend so much money on their own defence was the nature of the disease which had no known cure, killed one in five of its victims (and could kill two in five in bad outbreaks) and left many of those who recovered horribly scarred and sometimes blind. The reasons why only relatively small sums were allocated to the eradication program include a general scepticism about the possibility of success (particularly after the expensive failure of an attempt to eradicate malaria world-wide; this failure still casts a long shadow). The reasons why the developed countries did not shower WHO with gold when the campaign was concluded reflect those of the burghers of Hamelin:

†Twice, if you count vesicular exathema of swine, whatever that may be.

> Our business was done at the river's brink;
> We saw with our eyes the vermin sink,
> And what's dead can't come to life, I think.
> So friend, we're not the folks to shrink
> From the duty of giving you something for drink,
> And a matter of money to put in your poke;
> But as for the guilders, what we spoke

Of them, as you very well know, was in joke.
Besides, our losses have made us thrifty.
A thousand guilders! Come, take fifty!'

Why has this success not been repeated with other diseases? The cynic might suggest that none of the remaining old infectious diseases carry much terror to the inhabitants of wealthy countries. In England, measles is a disease which causes a child to miss a week of school, yet measles kills over a million children a year in developing countries. There is an old joke about the East German leader Ulbricht, inspecting his country's social services. He is first shown a lunatic asylum and, visibly impressed by the excellence of the care, tells his aide to give it an extra million marks in the next budget. Next, he is shown a model prison and the same sequence of events occurs. Finally, he visits a nursery school and again admires all he sees, but at the end tells his aide to give it a mere thousand marks. 'But, Herr Ulbricht, you gave the others much more.' 'I'm hardly likely to end up in a nursery school, am I?'

However, the reflections of the previous paragraph are not entirely justified and arise from the mathematician's habit of looking at the general properties which unite rather than the specific differences which divide. If we look more carefully we discover that the eradication of smallpox was only possible because it possessed a cluster of characteristics.

1 Smallpox cases only became infectious close to the point when it became clear that the sufferers had the disease and ceased to be infectious when they recovered. Thus, isolating cases reduced κ drastically. This cannot be done with a disease like diphtheria that is infectious before clear symptoms show (indeed many children carry the disease without themselves ever showing signs of ill-health). It is notorious that some of those who recover from typhoid continued to carry the disease.

2 There were no animal reservoirs of the disease. Since yellow fever is carried by monkeys as well as men, no amount of vaccination of the human population will eradicate the disease.

3 Modern vaccination against smallpox gave certain protection for at least three years. There exist many diseases (for example, influenza) where vaccination is not a certain guard against the disease.

4 There was only one kind of smallpox. Influenza is forever changing its form and a vaccine against this year's flu will be ineffective against that of the year after next.

Finally, although smallpox's terrible reputation included memories of how it 'spread like wildfire', it was actually much less infective (had a smaller κ) than most of the other major infectious diseases and was almost always transmitted by direct contact (making isolation easier and more effective). This means that the μ required to produce a sufficiently small $(1 - \mu)\kappa$ does not have to be unrealistically close to 1.

The only reasonable candidate for eradication appears to be measles for which 2, 3 and 4 hold, though 1 does not (carriers are infectious before the disease declares itself though they cease to be infectious when they recover). Unfortunately, measles is very infectious and though rich countries like the United States can maintain μ sufficiently close to 1 to keep themselves free of the disease, it is not clear that, even granted the money and the will, such levels are practicable world-wide†.

We are prone to many diseases caused by microorganisms of many different sizes and kinds. Unfortunately, such microorganisms leave few lasting traces and even over the short span of written history it is hard to be sure when a disease is really new to a certain society. Nonetheless we can make some plausible speculations.

A large animal like man represents a potentially almost endless source of dinners to a very small one. In order to reduce this potential, large animals possess a collection of very powerful defences. It has been suggested that, in the absence of the threat represented by microorganisms, asexual reproduction, where an individual makes essentially perfect copies of itself, would be the norm. Sexual reproduction requires a greater investment (because, for example, of the need to develop specialised organs, the requirement that a mate be found and so on) to produce new organisms which only partly resemble each of the two parents. However, this general shuffling of characteristics involved in sexual reproduction means that each member of a species represents a different problem for attacking microorganisms‡.

From time to time, a microorganism from some other habitat finds a large animal without the appropriate defences. Under these circumstances we would expect the microorganism to flourish and breed explosively — probably killing the large animal in the process. With luck (for the line of microorganisms) some of the microorganisms will have transferred to other large animals of the same species and the cycle will repeat itself. We might expect one of three outcomes:

1   The microorganisms fail to establish themselves as permanent predators on the large animals. For example, they may not be

†Anderson and May suggest that whilst vaccination levels of 70–80% in the neighbourhood of known infections might be sufficient to eradicate smallpox, vaccination levels of 90–95% would be required for measles ([4] page 88).

‡In early nineteenth century Ireland the peasantry depended on the potato. In many of the poorer parts cultivation centred round one variety, the 'Connaught lumper' which, like all potato varieties, was propagated asexually. Thus large numbers of potato plants were, effectively, identical and once the potato blight (a kind of fungus) took hold it destroyed the entire crop. One million Irish died and at least a million emigrated. As to the 'question of how the Irish alone, of all the populations of Europe arrived at [this] abject wretchedness' it is hard to disagree with the general view that the answer lay in a system of 'unwise laws and unrestrained extortion' [19] which forced the peasants into total dependence on one crop. Once the famine had taken hold, the prevailing economic theories encouraged the British Government to respond in a way which 'was inadequate in terms of humanitarian criteria and increasingly ... systematically and deliberately so' [117].

able to establish a sufficiently reliable route from one animal to another. Every year we read reports of such frightening but rare 'exotic' illnesses.

2   The microorganisms kill off the entire species.

3   The microorganisms establish themselves as a permanent disease but do not kill off the entire species.

What will happen in case 3? Clearly, those animals which can cope best with the disease are likely to have more offspring on average so the species will evolve in a direction which will reduce the effects of the disease. What of the disease? It, too, will evolve in such a way as to adapt to its new habitat, perhaps developing better methods of out-flanking the animal's defences, or improving its mode of transmission. However, there is another, more subtle direction it might take. The long-term survival of a particular line of the disease depends crucially on κ, the average number of other animals that an infected animal will pass on the disease to before it dies. If the disease kills quickly, then there will not be much time for transmission and κ is likely to be small. It is therefore 'in the disease's own interest' not to kill its host quickly or even not to kill it at all. We may therefore expect a disease to evolve in the direction of mildness†.

†But other strategies are possible; the life cycle of anthrax requires the death of the host.

Some evidence for this theory is found in the multitude of minor diseases which circulate. Children, as one might expect, are particularly prone to 'something going the rounds' which makes them 'off-colour' for a few days.

At the other extreme we have the Black Death which, starting in the Orient, swept through Europe around 1350. An Irish chronicler tells how the plague first arrived in Ireland 'near Dalkey and Drogheda, and almost destroyed and denuded of human inhabitants the towns of Dublin and Drogheda themselves.' He continues:

> In scarcely any house did only one die but all together man and wife with their children and household, traversed the same road, the road of death. Now I, brother John Clyn, of the Order of Minor Friars and the community of Kilkenny, have written in this book these notable events that have occurred in my own time, which I have learned from the evidence of my eyes or upon reliable report. And lest [these] notable events should perish with time and fade from the memory of future generations ... while waiting among the dead for the coming of death, I have set them down in writing just as I have truthfully heard and examined them. And lest the writing should perish with the writer and the work with the workman, I leave the parchment for the work to be continued in case in the future any human survivor should remain or

someone of the race of Adam should be able to escape the plague and
continue what I have begun†.

Perhaps one third of the population of Europe died and, under repeated
attack by the plague, the population continued to decline for another
100 years. Yet plague is normally

> not a disease of human beings at all. The great plagues of history were
> biologically unimportant accidents, the result of human entanglement
> with a self-contained triangular interaction of rodent, flea and plague
> bacillus.

It must certainly be the case that many human diseases cannot have
existed in their present form for very long. As we have seen, measles
and smallpox are 'crowd diseases' which cannot survive away from
dense human settlement. They cannot be older than the invention
of agriculture (about 10 000 BC) and are probably not older than
the first large states (about 3000 BC). Thus the diseases are very
new on a human time-scale, though less so on the time-scale of the
microorganism‡. The measles virus has molecular similarities with that
responsible for distemper in dogs and rinderpest in cattle. It is possible
that we have acquired measles and many other of our diseases from
our domestic animals (though, of course, the traffic will also have gone
the other way). It has been suggested that we have leprosy from our
water buffalo, diphtheria from our cattle and some of the collection of
rhinoviruses which cause the common cold from our horses.

Nature is never as simple as we try to make her. Even if the
speculations above are partly correct, they are unlikely to give the
whole truth. None the less, if they are even partly correct, they suggest
that new 'human infectious diseases' may still arise and that, initially,
they may be exceptionally virulent.

Whilst admitting this possibility, Burnet and Wright wrote in 1970:

> The most likely forecast about human infectious disease is that it will be
> very dull. There may be some wholly unexpected emergence of a new
> and dangerous infectious disease, but nothing of the sort has marked the
> last fifty years.

AIDS has now been with us for 25 years. Within 40 years of the arrival
of cholera in Europe, our Victorian predecessors were already taking in
hand the measures which were to conquer it. It seems unlikely that we
shall do so well against AIDS.

Moreover, most of the old infectious diseases still afflict hundreds of
millions in the third world.

†There follow two words
in the same hand, *magna
karistia* – 'great dearth',
and then, in another hand,
'Here it seems that the
author died.'

‡If measles and humans
have been in association
for 5000 years this
represents some 200 human
generations and some
130 000 virus generations.

During the last two decades when biotechnology has made so many stunning advances, the health of tropical people has worsened. Eradication and control schemes have collapsed. Old, proven therapies have become impotent in the face of drug-resistant microorganisms. New, affordable, non-toxic chemotherapeutics have not been developed; a drugs-for-profit pharmaceutical industry gives low priority to the diseases of poor people. So, too, with insecticides.

At the end of *La Peste*, which describes a visitation of the plague to a modern city, Camus describes how, during the celebrations which greet its inexplicable end, his main protagonist, Doctor Rieux, determines to record the history of the event.

He knew the tale he had to tell could not be one of final victory. It could only be a record of what had had to be done, and what would surely have to be done again in the never-ending fight against terror and its relentless onslaughts, despite their personal afflictions, by all those who, while unable to be saints but refusing to bow down to pestilences, strive their utmost to be healers.

And indeed, as he listened to the cries of joy rising from the town, Rieux remembered that such joy is always imperilled. He knew what those jubilant crowds did not know but could have learnt from books: that the plague bacillus never dies or disappears for good; that it can lie dormant for years and years in furniture and linen-chests; that it bides its time in bedrooms, cellars, trunks and bookshelves; and that perhaps the day would come when, for the bane and the enlightening of men, it roused up its rats again and sent them forth to die in a happy city.

CHAPTER 18

Two mathematics lessons

18.1 A Greek mathematics lesson

[We know a great deal about ancient Greek mathematics but very little about the Greek mathematicians who produced it. In his text, the *Elements*, Euclid organises an extraordinary amount of beautiful mathematics into a deductive whole†, but we can only guess how he expected students to learn from it.

†The reader who is interested should probably start by looking at a modern summary like Chapter 4 of Kline's *Mathematical Thought from Ancient to Modern Times*.

We do have one passage from Plato, the philosopher who founded the intellectual school to which Euclid belonged. It takes the form of an imaginary dialogue between Socrates (Plato's teacher), Meno (a young aristocrat) and a slave. (I shall use the translation [180] by W. K. Guthrie in the splendid Penguin Classics series.) The reader may sometimes wonder how we can ever understand anything new in mathematics. For if we can understand it, then it cannot be essentially new, and if it is essentially new, we have no way of understanding it. Meno repeats this argument in a different context to show that we cannot learn the nature of virtue.]

MENO But how will you look for something when you don't in the least know what it is? How on earth are you going to set up something you don't know as the object of your search? To put it another way, even if you come right up against it, how will you know that what you have found is the thing you didn't know?

SOCRATES I know what you mean. Do you realise that what you are bringing up is the trick argument that a man cannot try to discover either what he knows or what he does not know? He would not seek what he knows, for since he knows it there is no need of the enquiry, nor what he does not know, for in that case he does not even know what he is to look for.

MENO Well, do you not think it a good argument?

SOCRATES No.

MENO Can you explain how it fails?

[Socrates argues that the soul is immortal and that it has already acquired all knowledge. What we call learning is merely the recollection of what, in fact, we already know but have forgotten. Socrates continues:]

... We ought not then to be led astray by the contentious argument you quoted. It would make us lazy, and is music in the ears of weaklings. The other doctrine produces energetic seekers after knowledge; and being convinced of its truth, I am ready, with your help, to enquire into the nature of virtue.

MENO I see, Socrates. But what do you mean when you say that we don't learn anything, but that which we call learning is recollection? Can you teach me that it is so?

SOCRATES I have just said that you're a rascal, and now you ask me if I can teach you, when I say there is no such thing as teaching, only recollection. Evidently you want to catch me contradicting myself straight away.

MENO No, honestly Socrates, I wasn't thinking of that. It was just habit. If you can in any way make clear to me that what you say is true, please do.

SOCRATES It isn't an easy thing, but still I would like to do what I can since you ask me. I see you have a large number of retainers here. Call one of them, anyone you like, and I will use him to demonstrate it to you.

MENO Certainly. *(To a slave-boy.)* Come here.

SOCRATES Listen carefully then, and see whether it seems to you that he is learning from me or simply being reminded.

MENO I will.

SOCRATES Now boy, you know that a square is a figure like this? *(Socrates begins to draw figures in the sand at his feet. He points to the square ABCD (Figure 18.1).)*

BOY Yes.

SOCRATES It has all these four sides equal?

BOY Yes.

SOCRATES And these lines which go through the middle of it are also equal? (The lines *EF, GH*.)

BOY Yes.

SOCRATES Such a figure could be either larger or smaller, could it not?

BOY Yes.

SOCRATES Now if this side is two feet long, and this side the same, how many feet will the whole be? Put it this way. If it were two feet in this direction and only one in that, must not the area be two feet taken once?

BOY Yes.

SOCRATES But since it is two feet this way also, does it not become twice two feet?

BOY Yes.

SOCRATES And how many feet is twice two? Work it out and tell me.

BOY Four.

SOCRATES Now could one draw another figure double the size of this, but similar that is with all its sides equal like this one?

BOY Yes.

SOCRATES How many feet will its area be?

BOY Eight.

SOCRATES Now then, try to tell me how long each of its sides will be. The present figure has a side of two feet. What will be the side of the double sided one?

BOY It will be double, Socrates, obviously.

SOCRATES You see, Meno, that I am not teaching him anything, only asking. Now he thinks he knows the length of the side of the eight-feet square.

MENO Yes.

SOCRATES But does he?

MENO Certainly not.

SOCRATES He thinks it is twice the length of the other.

MENO Yes.

SOCRATES Now watch how he recollects things in order — the proper way to recollect.

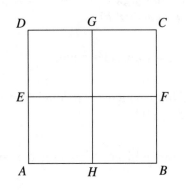

Figure 18.1: A four-foot square.

You say that the side of double length produces the double-sized figure? Like this I mean, not long this way and short that. It must be equal on all sides like the first figure, only twice its size, that is eight feet. Think a moment whether you still expect to get it from doubling the side.

BOY Yes, I do.

SOCRATES Well now, shall we have a line double the length of this (*AB*) if we add another the same length at this end (*BJ*)?

BOY Yes.

SOCRATES It is on this line then, according to you, that we shall make the eight-feet square, by taking four of the same length?

BOY Yes.

SOCRATES Let us draw in four equal lines (*i.e. counting AJ, and adding JK, KL, and LA made complete by drawing in its second half LD*), using the first as a base. Does this not give us what you call the eight-feet figure?

BOY Certainly.

SOCRATES But does it contain these four squares, each equal to the original four-feet one.

(Socrates has drawn in the lines CM, CN to complete the squares that he wishes to point out (Figure 18.2).)

BOY Yes.

SOCRATES How big is it then? Won't it be four times as big?

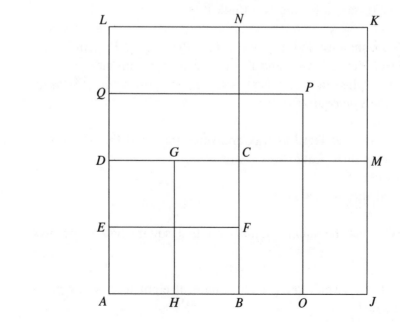

Figure 18.2: Socrates's figure.

BOY Of course.

SOCRATES And is four times the same as twice?

BOY Of course not.

SOCRATES So doubling the side has given us not a double but a fourfold figure.

BOY True.

SOCRATES And four times four are sixteen, are they not?

BOY Yes.

SOCRATES Then how big is the side of the eight-feet figure? This one has given us four times the original area, hasn't it?

BOY Yes.

SOCRATES And a side half the length gave us a square of four feet?

BOY Yes.

SOCRATES Good. And isn't a square of eight feet double this one and half that?

BOY Yes.

SOCRATES Will it not have a side greater than this one but less than that?

BOY I think it will.

SOCRATES Right. Always answer what you think. Now tell me: was not this side two feet long, and this one four?

BOY Yes.

SOCRATES Then the side of the eight-feet figure must be longer than two feet but shorter than four?

BOY It must.

SOCRATES Try to say how long you think it is.

BOY Three feet.

SOCRATES If so, shall we add half of this bit *(BO, half of BJ)* and make it three feet? Here are two, and this is one, and on this side similarly we have two plus one; and here is the figure you want. *(Socrates completes the square AOPQ.)*

BOY Yes.

SOCRATES If it is three feet this way and three that, will the whole area be three times three feet?

BOY It looks like it.

SOCRATES And that is how many?

BOY Nine.

SOCRATES Whereas the square double our first square had to be how many?

BOY Eight.

SOCRATES But we haven't yet got the square of eight feet even from a three-feet side?

BOY No.

SOCRATES Then what length will give it? Try to tell us exactly. If you don't want to count it up, just show us on the diagram.

BOY It's no use, Socrates, I just don't know.

SOCRATES Observe, Meno, the stage he has reached on the path of recollection. At the beginning he did not know the side of the square of eight feet. Nor indeed does he know it now, but then he thought he knew it and answered boldly, as was appropriate — he felt no perplexity. Now, however, he does feel perplexed. Not only does he not know the answer; he does not even think he knows.

MENO Quite true.

SOCRATES Isn't he in a better position now in relation to what he didn't know?

MENO I admit that too.

SOCRATES So in perplexing him and numbing him like the sting-ray, have we done him any harm?

MENO I think not.

SOCRATES In fact we have helped him to some extent towards finding out the right answer, for now not only is he ignorant of it, but he will be quite glad to look for it. Up to now, he thought he could speak well and fluently, on many occasions and before large audiences, on the subject of a square double the size of a given square maintaining that it must have a side of double the length.

MENO No doubt.

SOCRATES Do you suppose then that he would have attempted to look for, or learn, what he thought he knew (though he did not), before he was thrown into perplexity, became aware of his ignorance?

MENO No.

SOCRATES So the numbing process was good for him.

MENO I agree.

SOCRATES Now notice what, starting from this state of perplexity, he will discover by seeking the truth in company with me, though I simply ask him questions without teaching him. Be ready to catch me if I give him any instruction or explanation instead of simply interrogating him on his own opinions.

(Socrates here rubs out the previous figures and starts again (Figure 18.3)).

Tell me boy, is this not our square of four feet? (*ABCD.*) You understand?

BOY Yes.

SOCRATES Now we can add another equal to it like this? (*BCEF.*)

BOY Yes.

SOCRATES And a third here, equal to each of the others? (*CEGH.*)

BOY Yes.

SOCRATES And then we can fill in this one in the corner? (*DCHJ.*)

BOY Yes.

SOCRATES Then here we have four equal squares?

BOY Yes.

SOCRATES And how many times the size of the first square is the whole.

BOY Four times.

SOCRATES And we want one double the size. You remember?

BOY Yes.

SOCRATES Now does this line going from corner to corner cut each of the squares in half?

BOY Yes.

SOCRATES And these are four equal lines enclosing this area.(*BEHD*)

BOY They are.

SOCRATES Now think. How big is this area?

BOY I don't understand.

SOCRATES Here are four squares. Has not each line cut off the inner half of each of them?

BOY Yes.

SOCRATES And how many such halves are there in this figure? (*BEHD.*)

BOY Four.

SOCRATES And how many in this one? (*ABCD.*)

BOY Two.

SOCRATES And what is the relation of four to two?

BOY Double.

SOCRATES How big is this figure then?

BOY Eight feet.

SOCRATES On what base?

BOY This one.

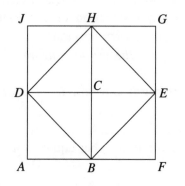

Figure 18.3: The dissected square.

SOCRATES The line which goes from corner to corner of the square of four feet?

BOY Yes.

SOCRATES The technical name for it is 'diagonal'; so if we use that name, it is your personal opinion that the square on the diagonal of the original square is double its area.

BOY That is so, Socrates.

SOCRATES What do you think, Meno? Has he answered with any opinions that were not his own?

MENO No, they were all his.

SOCRATES Yet he did not know, as we agreed a few minutes ago.

MENO True.

[Socrates concludes that, since the slave boy has discovered the result for himself, the knowledge must have been somewhere within him to start with.]

SOCRATES And if the truth about reality is always in our soul, the soul must be immortal, and one must take courage and try to discover — that is to recollect — what one doesn't happen to know, or (more correctly) remember, at the moment.

MENO Somehow or other I think you are right.

SOCRATES I think I am. I shouldn't like to take my oath on the whole story, but one thing I am ready to fight for as long as I can, in word and act: that is that we shall be better, braver, and more active men if we believe it right to look for what we don't know than if we believe there is no point in looking because what we don't know we can never discover.

MENO There too I am sure you are right.

[Thus, Socrates says, even if we do not accept that the soul is immortal and already contains all knowledge in a latent form, he has demonstrated that it is possible to acquire new knowledge.]

18.2 A modern mathematics lesson I

[We imagine a university teacher in her office with two bright students, Eleanor and Stuart. The students have just read through a photocopy of the preceding section.]

TEACHER Well, what did you think of it?

STUART I wasn't convinced by the solution.

ELEANOR And I wasn't convinced by the problem.

TEACHER What do you mean?

ELEANOR I don't think you have to know what you're looking for in order to find it. After all, if you step into a new room you see things you didn't know were there.

TEACHER But they are not really new to you; they are just variations on familiar objects. Think of a baby. How does it learn to make sense of the world?

STUART With difficulty!

TEACHER But it does! And if you put a CRAY computer on wheels and hooked it up to a TV camera it wouldn't.

STUART But if it was properly programmed it would. A baby is programmed — hard-wired, in fact — to learn.

TEACHER So Plato thinks we can understand the world because we have immortal souls, but you think it is because we have been hard-wired. To each generation its own myth.

STUART 'Infinite are the arguments of mages.'

TEACHER But Plato is doing more than just arguing about the possibility of new knowledge. Many of his dialogues involve attempts to discover the nature of correct arguments. His pupil, Aristotle, was the first to codify a system to determine the validity of arguments. Thus

1   Socrates is a man.
2   All men are mortal.
3   Therefore Socrates is mortal.

is a valid argument; but

1   All cats are mortal.
2   Socrates is mortal.
3   Therefore Socrates is a cat.

is not.

STUART I have a cat called Socrates.

ELEANOR So an invalid argument may have valid premises and a valid conclusion.

TEACHER Leaving Stuart's cat out of it, Plato's dialogue also deals with two interlinked pieces of mathematics. The first is Pythagoras's theorem.

STUART The square on the hypotenuse of a right-angled triangle equals the sum of the squares on the other two sides.

TEACHER Exactly. Had the slave not been allowed a well earned rest, Socrates might well have persuaded him to look at the following two geometric figures.

(She draws Figure 18.4)

In both figures $EFGH$ is a square of side $a + b$. In the first figure N has been placed on straight line HG in such a way that NG has length a and HN has length b. Similarly FM, GP, LH have length a and ME, PF, EL have length b. The point Q is the intersection of the straight lines LP and MN. In the second figure, N and L occupy the same positions as before and R and S have been placed on FG and EF in such a way that GR and FS have length b and RF and SE have length a.

ELEANOR I can see what the slave would say!

TEACHER OK, what would he say?

ELEANOR O Socrates, I see it all now! The eight triangles HLN, QNL, FMP, QPM, GNR, FRS, ESL, LHN are all right-angled triangles with their shorter (non-hypotenuse) sides of length a and b. They are thus all exact copies of each other and so have the same area, let's call it Δ. Now in the first diagram we see that the square $EFGH$ is decomposed (like a jigsaw) into a square $QPGN$ of side length a and so of area a^2, a square $EMQL$ of side length b and so of area b^2, and four triangles HLN, QNL, FPM, QPM each of area Δ. Thus *(She writes.)*

$$\text{Area}(EFGH) = a^2 + b^2 + 4\Delta.$$

In the same way, Socrates, looking at the second diagram, we see that *(She writes.)*

Figure 18.4: A proof of Pythagoras's theorem.

$$\text{Area}(EFGH) = \text{Area}(RSLN) + 4\Delta,$$

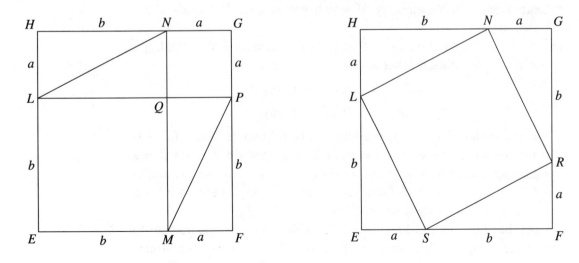

and so, comparing the two equations,

$$\text{Area}(RSLN) = a^2 + b^2.$$

But *RSLN* is a square each of whose sides is a hypotenuse (of length *c*, say) of one of our standard small triangles. Thus

$$c^2 = \text{Area}(RSLN) = a^2 + b^2,$$

and the square on the hypotenuse of our right-angled triangles equals the sum of the squares on the other two sides.

STUART So you see, Meno, that even an uneducated slave can sometimes do something well. But slave, do you not remember the puzzle in which four pieces making an eight by eight square are reassembled to form a five by thirteen rectangle†?   †See Figure 10.1.

ELEANOR That, O Socrates, was a trick; this is for real.

STUART Says you, O slave!

TEACHER No, I think Socrates must either find a flaw in the reasoning or shut up.

STUART *(After some thought.)* OK. Tell me slave, why do you say that *RSLN* is a square?

ELEANOR *(Also after some thought.)* Good point. It certainly looks like a square. ... You agree that all the sides *RS*, *SL*, *LN*, *NR* are equal.

STUART Yes, but *RSLN* might still only be a rhombus. You must show me that *RSL* forms a right-angle.

ELEANOR But, by symmetry all the four angles $\angle RNL$, $\angle NLS$, $\angle LSR$ and $\angle SRN$ are equal and the angles of a quadrilateral add up to four right-angles so each of the angles $\angle RNL$, $\angle NLS$, $\angle LSR$ and $\angle SRN$ is a right-angle.

STUART What is this symmetry of which you speak? It sounds like a fudge to me.

ELEANOR No it isn't. You agree that the two triangles *RNG*, *NHL* are perfect copies of each other and so *(she writes)*

$$\angle RNG = \angle NLH = \alpha, \text{ say}$$
$$\angle GRN = \angle HNL = \beta, \text{ say}.$$

But the angles of a triangle add up to two right-angles and $\angle LHN$ is a right-angle, so looking at the triangle *NHL* we see that $\alpha + \beta$ is a right-angle. On the other hand *HNG* is a straight line so the angles $\angle LNH$, $\angle GNR$ and $\angle RNL$ add up to two right-angles and so $\angle RNL$ by itself must be a right angle.

STUART But how do you know that the angles of a triangle add up to two right-angles? Or, come to that, how do you know that right-angled triangles with their shorter sides of length *a* and *b* are all exact copies of each other (whatever 'exact copy' may mean)?

ELEANOR That isn't fair. You ask 'Why A?', I reply 'A follows from B', so you say 'Why B?' and I say, 'Because C', so you say, 'Why C?' That way I can never win.

TEACHER And that is precisely why the Greeks invented the axiomatic method. You and Stuart have to agree in advance on a number of statements that you both accept.

STUART For the sake of argument and for the time being!

TEACHER Precisely: for the sake of argument. Then, if Eleanor succeeds in showing that her assertion follows from the previously agreed assertions or, as we shall call them, axioms, she wins and Stuart has to shut up.

ELEANOR Supposing that to be possible. But I thought that the Greeks believed their axioms to be self-evident statements.

TEACHER So people say. But chaps like Archimedes were clearly so clever, and so aware of the subtlety of mathematical thought, that I would hesitate to say what exactly they believed. In any case, now, after a hundred years of wrangling over the foundations of mathematics, it is clear that there is no statement whatsoever which cannot be doubted by someone sufficiently clever. For us, axioms are just the rules of the game. Different axioms give different games, that's all.

STUART So all I have to do is to collect a job lot of axioms, call them Stuartian geometry, and I'm in business.

TEACHER Not quite. It's easy to find someone else to play chess with, but not so easy to find someone to play Stuartian chess (governed as it is by a job lot of rules). In order that a set of axioms survive as an object of study, it must give rise to interesting mathematics.

STUART And who defines interesting?

TEACHER Dieudonné says that good mathematicians are people who do good mathematics — and good mathematics is what good mathematicians do! I think that, whilst the logical correctness of a piece of mathematics is an intrinsic property, the decision as to its interest is a social one. If sufficiently many mathematicians find it interesting, it is interesting. If not, not. There is another interesting thing about the example chosen by Plato. You remember that he asks the slave for the length of the side of 'a square of eight feet' (or, as we would say, a square of area eight square feet) and the slave first guesses four feet and then three feet?

ELEANOR Yes.

TEACHER Well, what is it, in fact?

ELEANOR Obviously, the square root of eight feet.

TEACHER Obviously, to someone well trained, perhaps. And what's so special about the square root of eight?

ELEANOR It's irrational.

TEACHER Yes. It is not the ratio of two integers — not a fraction. The Greeks had only just discovered the existence of lengths which weren't rational multiples of each other and were very excited by it. Can you prove that the square root of eight is irrational?

STUART Yes: the square root of eight is two times the square root of two and we proved in school that the square root of two is irrational.

ELEANOR So did we.

TEACHER You obviously had good teachers. Let Stuart do it.

(She hands a pad to Stuart who writes as he speaks.)

STUART Suppose $\sqrt{2}$ is rational with value n/m where n and m are integers. By dividing out any common factors we can write $\sqrt{2} = n/m = p/q$ where p and q are integers greater than zero with no common factor. Now observe that

$$2 = (\sqrt{2})^2 = \frac{p^2}{q^2}$$

and so $2q^2 = p^2$.

Thus p^2 is even, and, since the square of an odd number is odd, p must be even and $p = 2r$ for some positive integer r. But now we have

$$2 = (\sqrt{2})^2 = \frac{p^2}{q^2} = \frac{4r^2}{q^2}$$

and so $q^2 = 2r^2$. Precisely the same argument as before now shows that q is even. Thus both p and q are even, contradicting our statement that they have no common factor. Since the assumption that $\sqrt{2}$ is rational leads to a contradiction, $\sqrt{2}$ must be irrational.

ELEANOR General applause! But how do you know that you can always divide out all the common factors, or that the square of an odd number can't be even?

STUART This looks like a job for Axiom Man.

TEACHER I agree that this is a good test-bed for the axiomatic method, but before going any further I need a coffee. Do you want some?

STUART Yes, with a little milk, please.

ELEANOR The same for me, please.

18.3 A modern mathematics lesson II

TEACHER Right, enough coffee time! Let me show you some axiomatics in action. I will try to remember the axioms for the integers and then we will try to prove the irrationality of the square root of 2 starting from those axioms.

STUART Are there going to be a lot of axioms?

TEACHER I'm afraid so. To make things easier I'll take them in three groups. The first group form the basic rules of arithmetic for integers. *(She writes them down as she speaks)*

Laws of addition:

(A1) $a + b = b + a$.   (commutative law)

(A2) $a + (b + c) = (a + b) + c$.   (associative law)

(A3) There is an integer called 0 which has the property that $a + 0 = a$ for all a.   (zero element)

(A4) If a is an integer then there is an associated integer $(-a)$ with the property that $a + (-a) = 0$.   (additive inverse)

Laws of multiplication:

(M1) $ab = ba$.   (commutative law)

(M2) $a(bc) = (ab)c$.   (associative law)

(M3) There is an integer called 1 which does not equal 0 and which has the property that $a1 = a$ for all a.   (unit element)

(M4) If $c \neq 0$ and $ac = bc$, then $a = b$.   (cancellation law)

Joint law:

(D) $a(b + c) = ab + ac$.   (distributive law)

(Stops writing.)

And that's the end of the first group.

STUART Ah! Why, ye Gods, should two and two make four?

TEACHER If you tell me what two is, and what four is, then I think I can answer your question. The rules above only mention one and zero.

STUART Well, I define two by $2 = 1 + 1$.

TEACHER And three?

STUART $3 = 2 + 1$ and $4 = 3 + 1$.

TEACHER *(Writing)* So

$$
\begin{aligned}
2 + 2 &= 2 + (1 + 1) && \text{(By definition)} \\
&= (2 + 1) + 1 && \text{(By the associative law (A2))} \\
&= 3 + 1 && \text{(By definition)} \\
&= 4 && \text{(By definition)}
\end{aligned}
$$

and we are done.

STUART Isn't that rather a fuss about nothing?

TEACHER Not compared with the fuss you would make if I couldn't prove it from the axioms. In the same way, we can build up some standard rules from the axioms.

(She starts writing.)

(A3′) $0 + a = a$.

(A4′) $(-a) + a = 0$.

(M3′) $1a = a$ for all a.

(D′) $(b + c)a = ba + ca$.

ELEANOR Those are easy.

TEACHER They are all easy. What about the cancellation laws for addition?*(Writes.)*

(A4″) If $a + c = b + c$ then $a = b$.

(A4‴) If $c + a = c + b$ then $a = b$.

ELEANOR *(Writes.)* If $a + c = b + c$ then

$$a = 0 + a = ((-c) + c) + a = (-c) + (c + a) = (-c) + (c + b)$$

using (A3′), (A4′), the associative law (A2) and the hypothesis. But the same arguments give $b = (-c) + (c + b)$ so $b = a$. Thus (A4″) holds. We can prove (A4‴) in much the same way, or we can prove it from (A4″) by using the commutative law (A1).

TEACHER And what we might call 'the uniqueness of the zero'. *(Writes)*

(UA3) If $a + z = a$ then $z = 0$.

ELEANOR *(Writes.)* If $a + z = a$ then, by (A3), $a + z = a + 0$ and so by the cancellation law (A4″), $z = 0$. Similarly, or by using the commutative law (A1),

(UA3′) If $z + a = a$ then $z = 0$.

TEACHER And, for Stuart, the proof that $(-(-a)) = a$

STUART *(Writes.)* By (A4) applied to $(-a)$ we have $(-a) + (-(-a)) = 0$ and by (A4′) we have $(-a) + a = 0$. Thus

$$(-a) + (-(-a)) = 0 = (-a) + a,$$

so, by the cancellation law (A4″), $(-(-a)) = a$.

(He stops writing.) It seems to me that this kind of thing could easily pall.

TEACHER But our object is to deduce everything from a limited set of rules, not to have fun. It's a bit like doing an instrument check in an aircraft before take-off. The fun is in the flying, but the instrument check makes the flying safer. However, if you will accept that, in principle, we could deduce all the standard properties of addition and multiplication from the axioms above I will move on to the next set of axioms.

STUART OK, but how do we know that we need more axioms?

TEACHER Well, for one thing, there are lots of systems which obey the axioms above and which obviously aren't the integers.

STUART Such as?

TEACHER Consider $\mathbb{Z}_2$ which just has two distinct elements 0 and 1 with the following rules for addition and multiplication *(writes)*

$$0 + 0 = 1 + 1 = 0, \text{ and } 0 + 1 = 1 + 0 = 1,$$
$$0 \times 0 = 1 \times 0 = 0 \times 1 = 0, \text{ and } 1 \times 1 = 1.$$

ELEANOR I recognise that! It's arithmetic mod 2. But how would I check the axioms if I didn't know that?

TEACHER We could just check them case by case. Take the distributive law (D) for example. If $a = 0$, then

$$a(b + c) = 0(b + c) = 0 = 0 + 0 = 0b + 0c = ab + ac,$$

and if $a = 1$, then

$$a(b + c) = 1(b + c) = b + c = 1b + 1c = ab + ac,$$

so in either case $a(b + c) = ab + ac$ and the distributive law holds.

STUART I agree it's easy to check. But what happened to our proof that $2 + 2 = 4$?

TEACHER It's still valid, but you defined $2 = 1 + 1$, $3 = 2 + 1$ and $4 = 3 + 1$.

STUART I see, so, with my definitions, $2 = 0$, $3 = 1$ and $4 = 0$ and we have proved $0 = 0 + 0$.

ELEANOR Which we call Stuart's theorem in honour of its discoverer.

TEACHER The next set of axioms is concerned with the notion of greater than (what we call order relations). *(Writes.)*

Laws of order:

(*O*1) Given any a and b *exactly one* of the following possibilities holds: $a > b$, $b > a$ or $a = b$.

(*O*2) If $a > b$ and $c > d$ or $c = d$ then $a + c > b + d$.

(*O*3) If $a > b$ and $c > 0$ then $ac > bc$.

(Stops writing).

And that's the end of the second group. (Incidentally the law (*O*1) is said to be a trichotomy.) We say that $a \geq b$ if $a > b$ or $a = b$, we say that a is positive if $a \geq 0$ and so on and so forth†. Now what would you like to prove?

ELEANOR How about the product of two negative numbers being positive?

TEACHER OK. Any ideas?

STUART Well, what we would like to say, I suppose, is that if a and b are negative then $-a$ and $-b$ are positive and so $ab = (-a)(-b)$ is positive.

TEACHER And how are you going to show that $ab = (-a)(-b)$?

ELEANOR Clearly that does not involve the order axioms.

†Mathematicians say that a is positive if $a \geq 0$ and a is *strictly positive* if $a > 0$. Similarly they say that a is negative if $a \leq 0$ and strictly negative if $a < 0$.

TEACHER True. So both of you try and prove it from the earlier axioms.
  (She takes out her post and reads it while the students write.)

Exercise 18.3.1   *Try to do this yourself.*

TEACHER (*Looking at their proofs.*) Very good. And just as there's
  more than one way of skinning a cat, so you have chosen two
  different ways forward. Eleanor first proves that $0a = 0$, by using the
  relation $0a + 0a = (0 + 0)a = 0a$. She then shows that $(-a)b = -(ab)$
  and goes on from there, whilst Stuart has started by showing that
  $(-a) = (-1)a$.
STUART Which method is better?
TEACHER Who cares? The distance between no proof and some proof is
  much greater than the distance between some proof and some other
  proof. In this case, Stuart needs to go through much the same steps
  to show that $(-1)(-1) = 1$ as Eleanor uses to show directly that
  $ab = (-a)(-b)$ so Eleanor's proof is neater. On the other hand, we
  need to know that $(-a) = (-1)a$ so the extra work in Stuart's proof is
  not all wasted. Now, how are you going to prove that if a is negative
  then $-a$ is positive?
STUART Easy. If a is negative then either $a = 0$ or $a < 0$. If $a = 0$
  then $-a = (-1)a = (-1)0 = 0$ so $-a$ is positive. If $a < 0$ and
  $-a$ is not positive then $0 > a$ and $0 > (-a)$ so, by Axiom $(O2)$,
  $0 = 0 + 0 > a + (-a) = 0$ contradicting Axiom $(O1)$. Thus if $a < 0$
  then $-a$ is positive and we are done!
TEACHER Fine. Now can you show that if $a > b$ and $b > c$ then $a > c$?
STUART Doesn't that follow from the definition?
TEACHER No, because we didn't define $>$, we just gave the laws $(O1)$,
  $(O2)$ and $(O3)$ it obeys.
STUART Well, why don't we just add a fourth rule?
TEACHER Bertrand Russell says something to the effect that 'The method
  of "postulating" what we want has many advantages; they are the
  same as the advantages of theft over honest toil.'
ELEANOR *(Who has been writing whilst the other two have been talking.)*
  And I think I can see a way of getting round the problem by honest
  toil. All you need to do is use $(O2)$ to show that $a > b$ if and only if
  $a + (-b) > 0$, or in more usual language $a - b > 0$. By the way, is it
  all right to write that?
TEACHER Yes. We can write $a + (-b) = a - b$ in the usual way, provided
  we remember what it means. But please continue.
ELEANOR Its all plain sailing from now on. $(O2)$ tells us that if $a - b > 0$
  and $b - c > 0$ then $a - c = (a - b) + (b - c) > 0$ and that does it.

Exercise 18.3.2 *Fill in the details of Eleanor's argument, making sure that everything can be deduced from the axioms.*

TEACHER So we have shown that the product of two negative numbers is positive and that if $a > b$ and $b > c$, then $a > c$. Do you want to prove anything else from these axioms?

STUART No, I think our thirst has been well and truly slaked.

TEACHER And have we succeeded in characterising the integers?

ELEANOR No, because the reals and the rationals obey the same rules.

TEACHER Did the introduction of the rules of order do any good at all in that direction?

ELEANOR Well, they certainly excluded $\mathbb{Z}_2$.

TEACHER Why?

ELEANOR If $1 > 0$ then, by $(O2)$, $0 = 1 + 1 > 0 + 0 = 0$ which contradicts $(O1)$. If $0 > 1$ then, by $(O2)$, $0 = 0 + 0 > 1 + 1 = 0$ which contradicts $(O1)$. By $(O1)$ that only leaves the possibility $1 = 0$ which isn't allowed since 0 and 1 are distinct.

STUART And they excluded the complex numbers.

TEACHER Why?

STUART Because, by using $(O3)$ and the fact that the product of two negative numbers is positive, we know that the square of any number (or at least any number in a system governed by our axioms) is positive. But $i^2 = -1$ and $1^2 = 1$ and, by $(O1)$, 1 and -1 cannot both be positive, so we have a contradiction. How many more axioms are we going to need?

TEACHER Only one, which because of its importance I will call 'The Fundamental Axiom for the Integers'.

STUART We can hear the capital letters. Tell us more.

TEACHER You know what a minimum is?

ELEANOR Yes, it's the smallest element of a set.

TEACHER *(Writes.)* Formally if A is a set (of integers, or indeed of any system satisfying the rules above) we say that b is a minimum (or a least member) of A if b belongs to A and $b \leq a$ whenever a is in A. *(Stops writing.)* Eleanor, does the set A of real numbers a with $a > 0$ have a minimum?

ELEANOR No, because if b is in A then so is $b/2$ and $b > b/2$.

TEACHER And the same goes for the set of rationals a with $a > 0$. Stuart, does the set of integers have a minimum?

STUART Obviously not. If b is an integer then so is $b - 1$ and $b - 1 < b$.

TEACHER But, if a set A does have a minimum, it is unique. Eleanor, would you like to prove that if b_1 and b_2 are minimum elements of A then $b_1 = b_2$?

ELEANOR Well, since b_1 is a least member of A, it is a member of A, and so, since b_2 is a least member of A, $b_2 \leq b_1$. By the same argument, $b_1 \leq b_2$ so, by Rule $(O1)$, $b_1 = b_2$.

TEACHER Thus exhibiting two useful techniques. *(Writes.)*

1 To show that something with a certain property is unique consider two objects with the same property and show that they are equal.

2 Sometimes a good way of showing that $a = b$ is to show that $a \geq b$ and $b \geq a$.

Now let me state our fundamental axiom

(FA) Every non-empty set of positive integers has a least element (minimum).

STUART What about empty sets?

TEACHER Do you really want to discuss what the smallest element of a set with no members should be?

ELEANOR I'm sure he does, but I don't. I'm worried about something more important. I can see that the fundamental axiom says something new (if only because it excludes the reals and the rationals) but I can't see how to use it. All the previous axioms are really just instructions about how to manipulate expressions — if you have $a + b$ then you can replace it by $b + a$, if $a > b$ then $a + c > b + c$ and so on. To prove things, you just do what you would normally do, but make sure that the rules permit it. The fundamental axiom is different.

TEACHER Yes it is, but it comes equipped with its own proof technique — 'form a set and examine its least member'. For example, when we discussed algorithms *(in Chapter 10.3 see in particular the remark following Algorithm 10.3.4)* we said that it was impossible to have 'an endless sequence of positive integers, each one strictly smaller than the one before'. See if you can prove it by looking for the least member of some set.

ELEANOR Well, there's only one set that I can see. Consider the set of all the integers in the sequence. Since it's a non-empty set of positive integers it has a least member. *(A smile lights up her face as the penny drops.)* But, whichever member of the sequence it is, the following one must be strictly smaller. Which contradicts the statement that it is the least — so there ain't no such animal and we are done.

TEACHER And that's not all. But first I need another coffee. Will you join me?

STUART No thanks, I'll sit this one out.

ELEANOR None for me, thank you.

18.4 A modern mathematics lesson III

TEACHER So, after some of us have enjoyed a refreshing coffee, we return to the fundamental axiom. Let's start with something simple. Can we show that, given any integer n and any integer $m > 0$, we can find d and r such that $n = md + r$ and $m > r \geq 0$? *(This result was assumed without proof in the statement of Lemma 10.3.2.)*

STUART According to you, we need a set of positive integers to take the least element of — but I can't see one.

TEACHER Why not try the set E of positive integers of the form $n - mD$? Start by showing that it is non-empty.

STUART If $n \geq 0$ set $D = 0$; if $n < 0$ then set $D = -n + 1$.

TEACHER Since it is a non-empty set of positive integers, it has a least element — call it $r = n - md$.

STUART And, following that broad hint, we examine r more closely. If $r \geq d$ then $r - d = n - (m+1)d$ is a strictly smaller member of E which contradicts the definition of r as the least element of E. Thus (since r is a member of E and so positive) we have $d > r \geq 0$ and (by our own definition of r) $n = md + r$.

I would say that was a very clever proof of something I didn't want to prove. I can't believe that things are that complicated.

TEACHER A little while ago you were all for rigour! If you can find a simpler proof using only our axioms please come and tell me. However, in deference to your feelings, let's move on to something a little less obvious. You remember that we used Euclid's algorithm to show that given any two integers a and b we can find integers R and S such that $Ra + Sb$ divides both a and b *(see Exercise 10.3.8)*. Eleanor, would you like to give a proof along the lines of the proof Stuart gave just now? Use pencil and paper.

ELEANOR Here goes. *(She writes.)*

Let E be the set of positive integers of the form $ra + sb$.

TEACHER Wait! What's the least element of your set?

ELEANOR Zero. That's not much use. But I can get round that. *(She crosses out what she has written and starts again.)*

Let E be the set of positive integers of the form $n = ra + sb$ with $n > 0$. It is non-empty because, if we look at the case $r = a$, $s = b$, we see that $n = aa + bb = a^2 + b^2 > 0$.

STUART What about $a = b = 0$?

ELEANOR What indeed? *(She writes at the top of the page.)*

(If $a = 0$ then, taking $R = S = 1$, we see that $b = Ra + Sb$ divides both a and b. Similar considerations apply if $b = 0$, so we need only now prove the result when both *a* and *b* are non-zero.)

(She continues from where she left off previously.)

Since E is a non-empty set of positive integers, it must have a least member $m = Ra + Sb$, say. Now $m > 0$, so setting $a = n$ in the theorem that Stuart didn't want to prove, we can find d and r such that $a = md + r$ and $m > r \geq 0$. But this means that $r = a - md = (1 - mR)a + (-S)b$, so that either r belongs to E (which is impossible since $m > r$ and m is the least member of E) or r fails to belong to E because it is not strictly positive. Thus r is not strictly positive and $r \leq 0$. Since, also, $r \geq 0$ it follows that $r = 0$ so that $a = md$ and m divides a. Similarly m divides b so we are done. *(She stops writing.)*

I must say that I just followed Stuart's proof without seeing where I was going.

TEACHER That's the point. The fundamental axiom (like some other particularly well chosen axioms) carries its own 'proof method' with it. To some extent, it thinks for you.

STUART But isn't mathematics about thinking for oneself? My school teachers said that we shouldn't think with our pens.

TEACHER Judging by the results they were excellent teachers, but, on this point, I disagree with them. Whitehead wrote that

> civilisation advances by extending the number of important operations which we can perform without thinking about them. Operations of thought are like cavalry charges in battle — they are strictly limited in number, they require fresh horses, and must only be made at decisive moments.

Mathematicians should only use thought as a weapon of last resort. It is too valuable to waste on what does not need it.

Stuart, would you like to show that there is no integer between 0 and 1?

STUART You mean that if $0 \leq n \leq 1$ then $n = 0$ or $n = 1$?

TEACHER Precisely.

STUART Well, to use the fundamental axiom we need a set of positive integers. The set we're interested in is the set, call it E, of integers n with $0 < n < 1$ which is certainly a set of positive integers. If it is empty, then we are done. If not, then E has a least member, call it m. *(He pauses.)* Now what?

TEACHER How about looking at m^2?

STUART Yes, I see. Since m is in E we have $0 < m < 1$ so, multiplying by m, we have $0 < m^2 < m$. Thus m^2 is in E and is smaller than m. This contradicts the definition of m as the least member of E. The only way out of this contradiction is to conclude that E is empty and so there is no integer between 0 and 1.

TEACHER Good. Do you think that the arguments involving the fundamental axiom are more or less interesting than those involving the earlier axioms?

ELEANOR More interesting, though that may be because the kind of argument is new.

TEACHER I would say not simply different to, but actually deeper than, those you are used to. But we seem to have become so involved in our axioms that we have forgotten what we wanted to use them for. What was our object?

STUART We wanted to prove that the square root of 2 was irrational.

TEACHER And will our axioms do this?

STUART I certainly hope so. You have led us on an interesting journey. It would be a pity if it was simply up the garden path.

TEACHER But our axioms concern the integers — the word irrational is nowhere mentioned.

STUART If that means more axioms, I shall go on strike.

TEACHER Is there any way forward other than industrial action?

ELEANOR Couldn't we rewrite it in terms of integers as follows? *(Writes.)*
If n and m are non-zero integers, then $2m^2 \neq n^2$.

TEACHER That seems fine to me. What do you think Stuart?

STUART Anything rather than more axioms. I agree that, if we have to argue in terms of integers, then this is the result we should aim for.

TEACHER As I remember it, Stuart's proof ran as follows: *(Writes.)*

Suppose that we can find non-zero integers n and m such that $2m^2 = n^2$.

(1) We can find non-zero integers p and q with no common factors which satisfy the equation $qn = pm$.

(2) Since $qn = pm$ and $2m^2 = n^2$ it follows that $2q^2 = p^2$.

(3) Thus p^2 is even, and, since the square of an odd number is odd, p must be even and so $p = 2r$ for some positive integer r.

(4) It follows that $q^2 = 2r^2$.

(5) Repeating our previous arguments we see that q is even.

(6) If both p and q are even, they have common factor 2. The assumption that the equation $2m^2 = n^2$ has a solution with n and m non-zero has thus led to a contradiction, so we conclude that there is no such solution.

(She stops writing.)

Do you agree that this was Stuart's proof?

STUART Essentially, yes.

TEACHER Can we now justify each part of the argument by appeal to our axioms? Eleanor, what about part (1)?

ELEANOR I think this needs the fundamental axiom.*(Writes.)*

Let E be the set of positive integers r with $r > 0$ such that there exists another integer s with $sn = rm$. The set E is non-empty (if $n > 0$ set $r = n$, $s = m$; if $n < 0$ set $r = -n$, $s = -m$) and so has a least element p, say. Since p is in E we can find an integer q with $qn = pm$. Suppose now that d is an integer with $d > 0$ which divides both p and q, in other words that there are integers P and Q such that $p = dP$ and $q = dQ$. Then $(dQ)n = (dP)m$ so, using the associative law of multiplication $(M2)$, $d(Qn) = d(Pm)$, using trichotomy $(O1)$, $d \neq 0$ and so, by the cancellation law $(M4)$, $Qn = Pm$. Next we show that $P > 0$. Observe that $d > 0$ and so, if $P \leq 0$, we have $dP \leq 0$ by much the same argument as we showed that the product of two negative integers is positive. Do you want the details?

TEACHER No, I think we can take them as read.

Exercise 18.4.1   *Supply the details.*

ELEANOR *(Continues writing.)* Thus, since $dP = p > 0$ we have $P > 0$ so that P is in E. Now look at d. Since, as we showed, there is no integer between 0 and 1, it follows that either $d = 1$ or $d > 1$. But if $d > 1$ then axioms $(O3)$, $(M1)$ and $(M3)$ show that $p = dP > 1P = P1 = P$ which means that p is not the least element of E. Since this contradicts the definition of p, we must have $d = 1$. In other words, p and q have no common factors.

TEACHER Very good. Stuart, what about (2)?

STUART I think that's easier. Since $qn = pm$, it follows, using axioms $(M1)$ and $(M2)$ repeatedly, that $q^2n^2 = p^2m^2$. But $n^2 = 2m^2$ so $q^2(2m^2) = p^2m^2$. Thus using, $(M2)$ and $(M1)$ again, we obtain $(2q^2)m^2 = p^2m^2$. Using the associative $(M2)$ this can be rewritten $((2q^2)m)m = (p^2m)m$. But $m \neq 0$, so applying the cancellation law $(M3)$ twice, we have $(2q^2)m = p^2m$ and $2q^2 = p^2$ as required.

TEACHER And (3)?

STUART That's easy too.

TEACHER Are you sure? How do you define an odd number?

STUART An integer is odd if it has the form $2r + 1$.

ELEANOR An integer is odd if it isn't even.

TEACHER So you have two possible definitions. Which one are you going to choose?

STUART But surely they both give the same answer?

TEACHER That's for you to prove.

STUART Unfortunately! Well, after your last coffee we proved that, given any integer n and any integer $m > 0$, we can find d and r such that $n = md + r$ and $m > r \geq 0$. In this case, taking $m = 2$, this means that, given any integer n, we can write it in the form $n = 2m + r$, with $2 > r \geq 0$. Now, either $2 > r \geq 1$ or $1 > r \geq 0$, so either (by subtracting 1 from both sides) $1 > (r - 1) \geq 0$ or $1 > r \geq 0$. But we showed just now that there is no integer between 0 and 1, so either $r - 1 = 0$ or $r = 0$ and either $n = 2m + 1$ or $n = 2m$. In other words, if an integer doesn't have the form $2m$ (that is it isn't even), it must have the form $2m + 1$ (that is it is odd in my sense).

Exercise 18.4.2 *Use the axioms for the integers to justify all the statements made by Stuart in the sentence which runs 'Now, either $2 > r \geq 1$ or $1 > r \geq 0$, so either (by subtracting 1 from both sides) $1 > (r - 1) \geq 0$ or $1 > r \geq 0$.'*

STUART *(Continues.)* But there's something I don't like about this.

TEACHER What is it?

STUART We're not really proving this for ourselves. You knew that we were going to need the result that there is no integer between 0 and 1, so you made us prove it first without telling us how it was to be used.

TEACHER And if I hadn't done so, what would have happened?

STUART We might have got stuck, or we might have worked it out for ourselves.

TEACHER I agree you're a bright pair and you might well have got over the difficulty yourselves. But was that the only place I guided you?

ELEANOR No. For example, simply by placing the problem of showing that there is no integer between 0 and 1 in the place that you did, you gave a strong hint about how to solve it.

TEACHER And do you think you could have got over all the difficulties yourselves?

ELEANOR Speaking only for myself, I very much doubt it.

STUART But that's not really the point. Even if we hadn't got as far without help as we did with your help, we would have got wherever we did get by ourselves.

TEACHER Why would that have been a good thing?

STUART Because education is about discovering things for ourselves — not about rote learning.

TEACHER I've already admitted that you're bright; but are you as clever as Gauss?

STUART Obviously not.

TEACHER How long do you think it would take you to do what Gauss could do in a week?

STUART Months — no, that's a stupid thing to say — you can't quantify things like that. Much longer, if at all.

TEACHER So if you wish to understand what Gauss did over his whole lifetime, are you going to be able to do that by rediscovering his results for yourself?

STUART Point taken.

TEACHER Erdős complains, 'Everybody writes and nobody reads.' Mathematicians find it harder to try to understand what somebody else has done than to do something themselves. But unless we take the trouble to learn from our predecessors, mathematics might as well be written on water.

Returning to our argument, would you like to complete the proof of (3)?

STUART That is easy now, even if it wasn't before. If p is not even, then it is odd and so $p = 2k + 1$ for some integer k. Using the axioms with gay abandon, we obtain $p^2 = 2(2(k^2 + k)) + 1$ so p^2 is not even. Thus, since, in fact, by (2), p^2 is even, it follows that p is even, that is $p = 2r$ for some positive integer r.

Exercise 18.4.3 *(i) Supply the details of the proof that if $p = 2k + 1$, then $p^2 = 2(2(k^2 + k)) + 1$.*

(ii) Prove, more generally, that the product of two odd numbers is always odd.

(iii) State and prove similar results about the product of two even numbers and about the product of an even and an odd number.

TEACHER What about parts (4) to (6) of Stuart's argument?

ELEANOR They seem quite straightforward to me.

TEACHER I agree. I don't see any need to discuss them in any further detail. What do you think, Stuart?

STUART I suppose so — but what would you have done if I hadn't agreed?

TEACHER We would have gone through them in detail.

STUART But, although we agree, perhaps someone else, who isn't here, might not.

TEACHER But I'm not teaching people who are not here. I'm teaching you and it would be a waste of your time (and, more importantly, mine) to go through something we agree is clear.

ELEANOR I know we're not supposed to talk about examinations, but how much of this will we need for the end-of-year examinations?

TEACHER Did you find it hard to prove things using the axioms?

ELEANOR It was hard to start with, and I'm not sure I've really understood how to use the fundamental axiom.

TEACHER But, leaving aside the fundamental axiom, do you think that proving things using the other axioms is basically hard?

STUART No. It's like shooting fish in a barrel.

TEACHER The difficulty of shooting fish in a barrel depends crucially on the size of the gun, the fish and the barrel. However, I agree with you — and so do the examiners. And, because it's so easy, they don't bother to test it. You are allowed to assume all the axioms for the integers (apart from the fundamental axiom) and all their consequences. You may have to do questions in which you are not allowed to assume the fundamental axiom or its consequences. In the same way and for much the same reasons your lecturers will assume all the axioms for the integers (apart from the fundamental axiom) and all their consequences, but will often go into details when they use the fundamental axiom.

Now let me produce a natural break by setting next week's work and then we'll go back to discussing what we've just done.

18.5 A modern mathematics lesson IV

(A few minutes later.)

TEACHER So, now you've had a little time to reflect, what did you think of our proof from first principles that the equation $2m^2 = n^2$ has no non-trivial integer solution?

STUART I was already convinced by the first version, so I don't see how I could be more convinced by dressing it all up in axiomatics.

TEACHER So you don't think you learnt anything from going back to first principles?

STUART I wouldn't say that. I just didn't learn anything new about the irrationality of the square root of 2.

TEACHER What did you learn if it wasn't about the square root of 2?

ELEANOR I think we learned about the integers — about what makes them tick.

TEACHER And what does make them tick?

ELEANOR I would say the fundamental axiom. But in school we seemed to manage quite well without. Was that all done with mirrors?

TEACHER Partly. But you did use a proof technique we haven't mentioned.

ELEANOR Induction?

STUART I was wondering where induction came in.

TEACHER Well, induction follows from the fundamental axiom.

STUART How?

TEACHER Tell me what you mean by induction.

STUART My mathematics teacher said we should always write it like this.
(*He writes.*)

The principle of induction states that if we have a mathematical statement $P(n)$ such that

1   If $P(n)$ is true then $P(n+1)$ is true, and

2    $P(0)$ is true,

then $P(n)$ is true for all positive integers n.

(*He stops writing.*) But I don't see what that has to do with the fundamental axiom.

TEACHER How do you apply the fundamental axiom?

ELEANOR Aha! Always look for the smallest counterexample.

(*She writes as she speaks.*)

If the set E of positive integers n for which $P(n)$ is false is non-empty, then by the fundamental axiom it has a least element N, say. Since $P(0)$ is true, $N \neq 0$ and $N - 1$ is a positive integer. Since $N - 1$ does not belong to E, $P(N - 1)$ is true, so, by condition 2, $P(N)$ is true which contradicts the statement that N is in E. Thus E must be empty and $P(n)$ must be true for all positive integers.

TEACHER Splendid!

ELEANOR So the fundamental axiom is more powerful than the principle of induction.

TEACHER No, because you can also deduce the fundamental axiom from the principle of induction.

STUART I don't see how. We need a $P(n)$ and the fundamental axiom doesn't contain any n.

TEACHER I won't go into the details, but what you do is prove, by induction on n, that any set of positive integers containing an integer $r \leq n$ has a least element. Since any non-empty set of integers contains an integer n this does it.

ELEANOR But to apply something like the fundamental axiom, you don't really need all the structure of the integers. All you need is an order to give least elements.

TEACHER That is a very acute remark. Computer scientists like to use 'divide and conquer' by reducing a problem to a number of simpler ones†.

STUART Like first doing the sky and then doing the grass in a jigsaw.

†Many excellent examples will be found in *Concrete Mathematics* [74].

TEACHER Exactly. We divide the jigsaw into smaller sub-jigsaws which we tackle one at a time.

Exercise 18.5.1 *How would you measure the difficulty of a jigsaw? How do you think the difficulty of an all white jigsaw would increase with the number of the pieces? Compare the difficulty of an all white jigsaw consisting of 1000 pieces with that of a jigsaw composed of 10 patches of distinct colours each containing 100 pieces.*

TEACHER *(Continues.)* In all the usual examples, you can produce an integer n to measure the complexity of the problem and apply induction that way; but it is more natural to look at the smallest counter-example — that is the simplest problem with which you cannot cope — and show that it does not exist. For this and other similar reasons, modern treatments of the integers tend to emphasise the fundamental axiom much more than the older ones which rely mainly on induction. A wise mathematician will learn to use both ideas.

STUART I have a question.

TEACHER Yes?

STUART When you gave the axioms for the integers, you started with the rules for addition and multiplication.

TEACHER Yes.

STUART But they weren't enough to characterise the integers because $\mathbb{Z}_2$ obeyed the same rules.

TEACHER Yes.

STUART So you added the laws of order; but that wasn't enough because the real numbers and the rationals also obey the same laws. You then added the fundamental axiom which is obeyed by the integers but not by the reals or by the rationals. But how do you know that you can stop there? Might there not be many different systems obeying the same axioms?

TEACHER That's a good question to which I answer yes and no.

STUART Fair enough. Why yes?

TEACHER Just take two exact copies of the integers and you have two different systems.

ELEANOR But they're not really different; two systems are only different if they have different properties.

TEACHER So I just paint one copy red and one copy blue. Then they will have different properties because one will have the property of being red and one of being blue.

STUART But that's not fair. We're not interested in the colour of our integers.

TEACHER Then what properties are we interested in?

STUART *(Hesitates.)* Addition, multiplication and order I suppose.

TEACHER If those are the only properties we are interested in, then the integers are essentially unique.

STUART What do you mean by 'essentially'?

TEACHER Let me state the precise theorem.

(She writes it down).

Suppose we have a system A with addition $+_A$, order $>_A$, and multiplication $\times_A$ and another system B with addition $+_B$, order $>_B$, and multiplication $\times_B$ both of which obey all our axioms for the integers. Then there exist functions $f : A \to B$ and $g : B \to A$ such that, for all n_A, m_A in A and all n_B, m_B in B,

(i) $g(f(n_A)) = n_A$ and $f(g(n_B)) = n_B$.

(ii) $f(n_A +_A m_A) = f(n_A) +_B f(m_A)$ and $g(n_B +_B m_B) = g(n_B) +_A g(m_B)$.

(iii) $f(n_A \times_A m_A) = f(n_A) \times_B f(m_A)$ and $g(n_B \times_B m_B) = g(n_B) \times_A g(m_B)$.

(iv) If $n_A >_A m_A$ then $f(n_A) >_B f(m_A)$. If $n_B >_B m_B$ then $g(n_A) >_A g(m_A)$.

STUART That seems as clear as mud to me.

ELEANOR Don't you see? The functions f and g lock the systems in step. If A does something then f forces B to do exactly the same.

TEACHER The Mullah Nasrudin was asked to preach in a small village. On the first day he asked the villagers, 'O People! Do you know what I am going to tell you?' 'No, Mullah, we do not.' 'Then I'm certainly not going to waste my time preaching to ignorant peasants,' said the Mullah and departed.

The next week he came again and again asked, 'O People! Do you know what I am going to tell you?' 'Yes, Mullah, we do .' 'Then my instruction is not needed,' said the Mullah and departed.

The third week came. 'O People! Do you know what I am going to tell you?' 'O, Mullah, some of us know and some of us do not.' 'Then let those who know tell those who do not.'

I'll leave it to you to convince Stuart. In any case it will all become clearer when you've met the notion of isomorphism in lectures.

ELEANOR I can see what it means, but I can't see how to prove it.

TEACHER It's not too hard, but, again, it will become easier once you have more experience of abstract algebra. The proof is written out in Chapter II of *Birkhoff and MacLane* [16]. I suppose *Birkhoff and MacLane* is considered a little old-fashioned now but Chapters I and VI still seem to me an excellent introduction to abstract algebra.

STUART There is another thing. You sing the praises of axiomatics but the lecturer in mathematical physics said he didn't need all this — I quote — pure mathematical rubbish.

TEACHER Nor does he. If a physical theory fails to predict the correct outcome of an experiment, then it is wrong, however rigorous the mathematics. On the other hand, if it does predict the correct outcome, then it doesn't matter how hair-raising the mathematics is, it must be, in some sense, correct. Pure mathematicians make statements (for example that 2 is not the square of a rational number) which cannot be checked by experiment and which must therefore be proved.

STUART But if all rigour does is check what we already know, then it really doesn't add much.

TEACHER When people climb a high mountain, they establish a base camp; from the base camp they carry supplies to a higher camp; from that to a higher camp and so on. From any point you can make a dash for the summit carrying as little as possible but, if that dash fails, the failure may well be final. At any stage in mathematics we know (or at least guess) much more than we can prove rigorously. If we abandon rigour and make a dash for the summit we may make it or not, but in either case we will have shot our bolt. The argument: ' Most experts think that A is true and A probably implies B so we expect B' is not a bad one, but a chain of such arguments as in '$A(1)$ is almost certainly true, and any gentleman would expect $A(2)$ to follow from $A(1)$, and if $A(2)$ is true, nature could not be so perverse as to allow $A(3)$ to be false, ... and if $A(n-1)$ is true then, for goodness sake, $A(n)$ must be true' ceases to carry any conviction when n becomes large.

ELEANOR So the higher the mountain, the more rigour we require.

TEACHER Precisely. And, since we are training you for the highest peaks, the course will contain a fair amount of rigour. But I must dash now. I have a meeting in three minutes. See you next week.

18.6 Epilogue

(The same time next week. The teacher has finished going through the students' exercises.)

TEACHER Not bad at all. Any questions?

STUART I've been thinking about what you said last week and I'm not happy.

ELEANOR Are you ever?

STUART You went on and on about rigour but you weren't really rigorous yourself.

TEACHER In what way?

STUART You gave all those axioms but you didn't give the laws of reasoning you used.

TEACHER Quite true. For example, I smuggled in the assumption that if $A = B$ and $B = C$ then $A = C$. To be perfectly correct, I ought to give not only my axioms and my laws of reasoning but also my laws of grammar.

ELEANOR What do you mean by grammar?

TEACHER If I were to tell you that 'Green hops like a happy'. Would that be true or false?

ELEANOR It's not true or false — it's just nonsense.

TEACHER Precisely. We need some method to exclude nonsense statements, so we give rules which define which statements we're prepared to consider.

ELEANOR So it's just like programming a computer. If the brackets don't match, it won't consider your program.

TEACHER The analogy is exact. The grammar tells us what statements we are allowed to make, the rules of reasoning tell us how we can manipulate them and the axioms are the statements we start with.

ELEANOR Then why can't computers do mathematics?

TEACHER I'm not sure I agree with the question. Computers can already do things that the man in the street calls mathematics, and I don't see how one can limit what they will be able to do in the future.

STUART But they don't do it now?

TEACHER The dictionary lists four kinds of intelligence — human, animal, military and artificial, and, for the moment, I agree with the dictionary.

On the other hand, although mathematically-talented machines may be a long way off, I don't see why mathematics-checking machines shouldn't be developed. Landau wrote a beautiful little book called *Foundations of Analysis* in which he develops the properties of the integers, the rationals, the real numbers and the complex numbers, starting from a set of axioms for the positive integers, and that has, apparently, been checked by machine.

STUART But you haven't answered my first point — that you preach rigour but don't practise it. And the same goes for our pure mathematics lecturers.

TEACHER Shall we leave my personal deficiencies aside for the moment and concentrate on the more interesting subject of the deficiencies of my colleagues? You say they are not rigorous. Do you mean that they have stated as theorems results which are false?

STUART Not as far as I can see.

TEACHER Or can you point out specific gaps in their reasoning?

STUART I can't point out specific gaps, but I know that there must be gaps.

TEACHER If you can't point out specific gaps, how do you know they are there?

STUART If I watch a pea and thimble game I may not be able to see the cheat but I know it's there.

TEACHER That is a good, but not a helpful answer. Shall I try to show you why you're uneasy by taking the proof that $2 + 2 = 4$, which we saw last time *(see Section 18.3)*, and making it 'rigorous'?

STUART OK.

TEACHER Here goes. *(She writes.)*

Recall that, by definition, $2 = (1 + 1)$ and so

$$2 + 2 = 2 + (1 + 1). \tag{18.1}$$

The associative law for addition tells us that $a + (b + c) = (a + b) + c$, so setting $a = 2$, $b = 1$ and $c = 1$, we have

$$2 + (1 + 1) = (2 + 1) + 1 \tag{18.2}$$

But if $a = b$ and $b = c$ then $a = c$, so setting $a = 2 + 2$, $b = 2 + (1 + 1)$ and $c = (2 + 1) + 1$, and using Equations (18.1) and (18.2), we obtain

$$2 + 2 = (2 + 1) + 1. \tag{18.3}$$

Next we recall that, by definition, $2 + 1 = 3$ and so

$$(2 + 1) + 1 = 3 + 1. \tag{18.4}$$

But if $a = b$ and $b = c$ then $a = c$, so setting $a = 2 + 2$, $b = (2 + 1) + 1$ and $c = 3 + 1$, and using Equations (18.3) and (18.4), we obtain

$$2 + 2 = 3 + 1. \tag{18.5}$$

Since, by definition, $3 + 1 = 4$, all that remains is to observe, for the last time that if $a = b$ and $b = c$ then $a = c$. Setting $a = 2 + 2$, $b = 3 + 1$ and $c = 4$, and using Equation (18.4) we obtain

$$2 + 2 = 4. \tag{18.6}$$

(Finishes writing.)

Does that satisfy you?

STUART I suppose so. But that's only my opinion. Someone cleverer than me might be able to show that it wasn't rigorous.

TEACHER And if that someone cleverer than you pointed out a gap, do you think that the gap could be filled?

STUART Of course.

TEACHER And how would the proof differ from the one I've just given?

ELEANOR It would just be even longer and more tedious.

TEACHER Do you agree, Stuart?

STUART Yes. I don't see how it would be basically different.

TEACHER In fact, my original proof is what I would call 'sufficiently rigorous'. At the end of the nineteenth century Euclid's axiomatic development of geometry was gone through with a fine-tooth comb by various mathematicians and was found to have certain subtle gaps.

ELEANOR Such as?

TEACHER For example, Euclid assumes implicitly that a straight line cannot pass through three sides of a triangle†.

†There is a very illuminating discussion of this point in Chapters II to IV of E. A. Maxwell's charming *Fallacies in Mathematics*.

STUART But it obviously can't.

TEACHER I agree, but Euclid should have stated it as an axiom. However, in spite of these flaws, none of his theorems turned out to be incorrect and, after appropriate modifications, all his proofs stand. So, although Euclid was not (as previous generations thought) perfectly rigorous, he was sufficiently rigorous.

STUART But why not aim for perfect rigour?

ELEANOR Because that would be like writing programs in machine code.

TEACHER Again, the computing analogy is exact. Now, do programs written in machine code usually work first time?

STUART No, nor the second, nor the third.

TEACHER Why? Are they hard to write?

STUART It's not so much that they are hard to write, as that they're damn nearly impossible to check.

TEACHER And why are they hard to check?

STUART Because they're meant for machines and not people.

TEACHER Precisely. Perfect rigour would produce mathematics for machines to check — not for humans to understand. We have to do mathematics in a way that is sufficiently rigorous to avoid error, but sufficiently relaxed for human beings to understand and communicate.

STUART I'm convinced.

ELEANOR For once.

STUART But ...

ELEANOR The Stuart we know and love continued ...

STUART ... that must mean that mistakes get made.

TEACHER When people measure the height of a mountain, they may get it wrong but that tells us only that humans make mistakes — it does not alter the nature of the mountain. People make mistakes in mathematics, but that does not alter the nature of mathematical truth.

STUART Also, it seems to me that if we are told to be absolutely rigorous, then we know what's wanted (even if we can't achieve it), or if we're told that we don't need any rigour, then we also know what's wanted (and we can achieve it). But how can we learn what constitutes 'sufficient rigour'?

TEACHER You learn it by watching other mathematicians at work (that's one of the main things lectures allow you to do) and from experience. When I look through your work, I point out the places where, in my opinion, you have not reasoned carefully enough and, more rarely, where you have been overcautious.

STUART So we learn it indirectly like the condemned men in Kafka's *The Penal Colony*.

TEACHER It's a good thing not all my students are so well read. I was going to add that, as a matter of observation, few of my students experience real difficulty in finding the right level. Perhaps, in theory, it ought to be hard but, in practice it isn't.

STUART But rigour is still a socially determined thing. You tell us what is rigorous and what is not.

TEACHER No, in mathematics the onus is always on the prover (however important a mathematician he or she may be) to prove the result to the satisfaction of the audience. In lectures, when someone asks a question the lecturer does not reply, 'You miserable worm, how dare you question your elders and betters,' but explains the point again†.

†Several mathematicians have commented that this is a somewhat idealised view. Oddly enough, they all gave the same distinguished professor as a counterexample.

There are university subjects in which everything is a matter of opinion. If you are a student of English literature who thinks that Lawrence is rubbish, then you have as much right to your opinion as your professor who thinks Lawrence is a god. There are others in which authority rules. Students of economics must follow one guru in Chicago and another in Harvard — and unless they are obviously brilliant, they would be well advised to stick to the party line.

Mathematics is different. You do not know as much as your lecturers and you are not as skilled in using what you do know as they are. Nonetheless it is their duty to prove things to your satisfaction. If they make a mistake (and everybody makes mistakes), you should point it out and they must correct it. If there is a gap in their argument which you point out, they must fill it in. There is authority in mathematics (Gauss was a great mathematician, I am not) but there is no appeal to authority — you are responsible for your own proofs and no one else.

STUART I surrender to superior rhetoric.

ELEANOR Can I ask a non-mathematical question?

TEACHER Ask away.

ELEANOR In the dialogue you started us with, Socrates leads the slave boy through an argument, but he doesn't seem to see anything wrong with having people as slaves. Surely someone as clever as he is supposed to have been should have seen that slavery is wrong.

TEACHER Since Socrates was executed, in effect, for holding and teaching unpopular opinions, we can be sure that, had he opposed slavery, he would have said so. Most of Plato's dialogues are literary devices for conveying his own views but, had his teacher Socrates been against slavery, he would have reported it. So neither Plato nor Socrates found slavery objectionable.

STUART But you don't mean to argue that slavery is wrong now but right then?

TEACHER No, I think it was wrong then, but that the Greeks accepted it as a natural part of society. Criticising Plato for not realising that slavery is a moral problem is like criticising him for not inventing the positional notation for arithmetic — he could have done so, but not even the cleverest of men can do everything. Furthermore, unless you ask as Socrates and Plato did, what a good society should look like, you are never going to ask whether slavery is good or bad. In classical Greece, as in the societies that followed it, women were assigned a different and usually inferior position to men. But in Plato's ideal society, women were to have the same place, duties and education as men. When Mill wrote *On The Subjection of Women*, he was consciously following Plato in this, and, still more importantly, in his view that everything is open to question and that positive good may come from rational discussion.

STUART And that is what university is all about.

TEACHER Not really.

STUART But that is what university ought to be all about.

TEACHER So you think the taxpayer is parting with large sums of money so that young ladies and gentlemen can sit around discussing life, the universe and everything. You are here to learn mathematics and more mathematics — not to row, play bridge, act or even to find yourselves — and that is what I am going to teach you.

STUART But, even if that is what the taxpayers want, is it what they ought to get? A university which just trains technicians is not a university; it is a technical college.

TEACHER Better a good technical college than a corrupt university. What ought you to learn at university besides mathematics?

STUART Students should learn to question received opinions.

TEACHER So, after I have made you write out 100 times: 'I must not accept authority,' what do we do next?

ELEANOR That's simple. You make us write out: 'I *really, really* must not accept authority.'

TEACHER Besides which, asking questions is the easy bit. It's finding good answers which is hard. A university is at least as much a repository for the accumulation of human experience and an instrument for passing on that experience as it is a device for adding to it.

STUART But just teaching mathematics is not enough. A lot of us will go on to be engineers and managers and will have to take moral decisions. So why don't you teach us ethics?

TEACHER But would you actually go to lectures on ethics?

STUART If the lecturer was good, yes.

TEACHER But anybody would go to hear Sir Isaiah Berlin lecturing on how to watch paint dry. The question is, would you go to listen to your ordinary lecturers talking about ethics?

ELEANOR Not unless it was for examination.

STUART So why not examine it?

TEACHER What would the examination questions look like? 'Is it wrong to steal from widows and orphans? Answer yes or no giving brief reasons.'

STUART There are lots of difficult and interesting moral problems.

TEACHER Yes, but the problems of the human race are not those of finding the answer to moral problems in hard cases but of acting on the answer in simple ones. American law schools now include courses on ethics, but the only observable result is that the defence in cases of fraud now begins 'My client's behaviour has throughout been not merely legal but ethical.'

The passage from Plato that I gave you is an extract from a longer piece which starts with Meno asking if virtue can be taught. Socrates holds that it cannot on the empirical grounds that good and wise fathers, who should surely wish their sons to be good and wise, so rarely have good and wise offspring.

If wisdom were teachable it would surely be our duty to teach it. Since it is not, we simply try to teach mathematics.

CHAPTER 19

Last thoughts

19.1 A mathematical career

[When I first started research, I and my fellow students were much in awe of Dusa MacDuff, a research student a couple of years older than us. We knew that she had just made an important breakthrough in a major problem left open by von Neumann concerning the existence of II_1 factors. (Here and in the passage that follows the reader will need to accept technical terms on trust.) To us, as I expect to most of my readers, this seemed the most splendid thing possible, and the climax of a mathematical career. As we shall see below, to someone who wanted to do mathematics at the very highest level, it could only represent a beginning.

The speech that follows was given by Dusa MacDuff on the occasion of her acceptance of the first Satter prize. Since this prize was created to 'recognise an outstanding contribution to mathematical research by a woman in the previous five years' she, quite properly, dwells on the problems of women in what is still a mainly male profession. However, many of the problems she considers — the difficulty of changing mathematical fields, mathematical isolation and failures of inspiration — are common to both sexes. Some readers may be surprised by the way professional and personal considerations are mingled, but at the highest level of endeavour (and often at lower levels), mathematical and private life must overlap and sometimes conflict. Mathematics is not a nine-to-five job.]

I am very honoured to be the first recipient of this prize and want to thank Joan Birman on behalf of the whole mathematical community for instituting it. I am particularly happy to get this prize because it is for my research. I grew up in a house in which creativity was much valued, but despite the achievements of the women in the family, males were seen to be more truly creative than females and it has taken me a long time to find my own creative voice. My life as a young mathematician was much harder than it needed to be because I was so isolated. I had no role models, and my first attempts at inventing a life style were not very successful. One important way of combating such isolation is to make both the achievements of women mathematicians and the different ways in which we live more visible. I hope that this will be one of the effects of the Satter Prize. I'll try to do my part by telling you something of my life.

I grew up in Edinburgh, Scotland, though my family was English. My father was a professor of genetics who travelled all over the world and wrote books on philosophy and art as well as on developmental biology and the uses of technology. My mother was an architect, who was also very talented, but had to make do with a civil service job since that was the best position which she could find in Edinburgh. Her having a career was very unusual: none of the other families I knew had mothers with professional jobs of any kind. There were other women on my mother's side of the family who led interesting and productive lives. I identified most with my maternal grandmother since I had her name: Dusa was a nickname given to her by H. G. Wells. She was most notable for creating a great scandal in the London of her time by running away with H. G. (this was before she married my grandfather) but she later wrote books, on Confucianism for example, and was active in left-wing politics. Her mother (my great-grandmother) was also distinguished: in 1911 she wrote a book about the working-class poor in London which I was pleased to find being used in [the University of] Stony Brook as a text-book. In discussing the women in my family I should mention my sister, who was the first Western anthropologist allowed to go on a field trip to Soviet Central Asia, and is now a Fellow of King's College, Cambridge, with a lectureship at the university.

I went to a girls' school and, although it was inferior to the corresponding boys' school it fortunately had a wonderful maths teacher. I always wanted to be a mathematician (apart from a time when I was eleven when I wanted to be a farmer's wife) and assumed that I would have a career, but I had no idea how to go about it: I didn't realise that the choices which one made about education were important and I had no idea that I might experience real difficulties and conflicts in reconciling the demands of a career with life as a woman.

When, as a teenager, I became more aware of my femininity, I rebelled *into* domesticity. I gladly started cooking for my boy-friend; I stayed in Edinburgh as an undergraduate to be with him instead of taking up my scholarship to Cambridge; and when I married I took his name. (My mother had kept her maiden name for professional purposes.) I did eventually go to Cambridge as a graduate student, this time followed by my husband. There I studied functional analysis with G. A. Reid and managed to solve a well known problem about von Neumann algebras, constructing infinitely many different II_1 factors. This was published in the *Annals of Mathematics* and for a long time was my best work.

After this, I went to Moscow for six months since my husband had to visit the archives there. In Moscow, I had the great fortune to study with I. M. Gel'fand. This was not planned: it happened that his was the only name which came to mind when I had to fill out a form in the Inotdel office. The first thing that Gel'fand told me was that he was much more interested in the fact that my husband was studying the Russian Symbolist poet Innokenty Annensky than that I had found infinitely many II_1 factors, but then he proceeded to open my eyes to the world of mathematics. It was a wonderful education, in which reading Pushkin's 'Mozart and Salieri' played as important a role as learning about Lie groups or reading Cartan and Eilenberg. Gel'fand amazed me by talking about mathematics as though it were poetry. He once said about a long paper bristling with formulas that it contained the vague beginnings of an idea which he could only hint at and which he had never managed to bring out more clearly. I had always thought of mathematics as being much more straightforward: a formula is a formula, and an algebra is an algebra, but Gel'fand found hedgehogs lurking in the rows of spectral sequences.

When I came back to Cambridge, I went to Frank Adam's topology lectures, read the classics of algebraic topology and had a baby. At the time, almost all the colleges in Cambridge were for men only†, and there was no provision at all for married students. I was very isolated with no one to talk to, and found that after so much reading I had no idea how to begin to do research again. After my postdoc, I got a job at York University. I was the family breadwinner and housekeeper and diaper changer (my husband said that diapers were too geometric for him to manage). At about this time I started working with Graeme Segal, and essentially wrote a second PhD with him. As this was nearing completion, I received an invitation to spend a year at MIT to fill a visiting slot which they had reserved for a woman. This was a turning point. While there I realised how far away I was from being the mathematician I felt that I could be, but also realised that I could do something about it. For the first time, I met some other women whom I could relate to and who also were trying to become mathematicians. I became less passive: I applied to the

†All Cambridge colleges and all Cambridge jobs are now open to women. (T. W. K.)

Institute for Advanced Study and got in and even had a mathematical idea again, which grew into a joint paper with Segal on the group-completion theorem. When back home I separated from my husband and, a little later, obtained a lectureship at Warwick. After two years at Warwick, I took an (untenured) assistant professorship at Stony Brook, so that I could live closer to Jack Milnor in Princeton. I went to Stony Brook sight unseen. I knew no one there and have always thought myself extremely lucky to have landed in such a fine department, although very foolhardy to have given up a tenured job for an untenured one.

After that, I had to do the work that everyone has to do to become an independent mathematician, building up on what one knows and following one's ideas. I spent a long time working on the relation between groups of diffeomorphisms and the classifying space for foliations: this grew out of my study of Gel'fand-Fuchs cohomology in Moscow and my work with Segal on classifying spaces of categories. I still worked very much in isolation and there are only a few people who are interested in what I did, but it was a necessary apprenticeship. I had some ideas and gained confidence in my technical abilities. Of course, I was influenced by the clarity of Jack Milnor's ideas and approach to mathematics, and was helped by his encouragement. I kept my job in Stony Brook, even though it meant a long commute to Princeton and a weekend relationship, since it was very important to me not to compromise on my job as my mother had done. After several years I married Jack and had a second child.

For the past eight years or so, I have worked in symplectic topology. Here again I have been very lucky. Just after I started getting interested in the subject, it was revitalised with new ideas from several sources. Most important to me was Gromov's work on elliptic methods. I took advantage of a sabbatical to spend the spring of 1985 at the IHES in Paris so that I could learn about Gromov's techniques, and the work I did then has been the foundation of all my recent research. At that time, our child was a few months old. So I worked rather short days, but found it easy to cope. Eventually he brought the family together. We didn't want to make him commute, and Jack did not like being left with him for the best part of each week. So Jack took a job at Stony Brook, where we are now enjoying life in one house.

In conclusion, I think that there is quite an element of luck in the fact that I have survived as a mathematician. I also got real help from the feminist movement, both emotionally and practically. I think things are somewhat easier now: there is at least a little more institutional support of the needs of women and families, and there are more women in mathematics so that one need not be so isolated. But I don't think that all the problems are solved.

[At the time of writing Dusa MacDuff is still at Stony Brook. She has just been elected to the Royal Society.]

19.2 The pleasures of counting

Writing a book is a long task and, as I come to the final chapter, it is natural to reflect on how far this book resembles the one I set out to write. On the one hand, it is substantially longer and, in places, contains harder mathematics than I had intended. On the other hand, I set out to write a book about mathematics and I have, I think, written a book of a type which could not be written about literary criticism, business, sport, music, politics or most of the useful and not so useful pursuits which mankind has found for itself. To that extent, I feel happy with the result.

It is true that mathematics is useful, that without mathematics much of modern civilisation could not exist and much of the rest would run only with great difficulty. Our compact disc players, our jumbo jets, our telephone services, our weather forecasts — all depend on subtle mathematics. But mathematicians would pursue mathematics even if it were not useful. Like poetry, like philosophy, like music, it gives point and zest to living.

I had various motives in writing this book but my chief aim has been to convey the pleasures of counting. If I have succeeded, or even if I have only partly succeeded, I should be happy to echo William Morris's Prologue to *The Earthly Paradise*.

> Of Heaven or Hell I have no power to sing,
> I cannot ease the burdens of your fears,
> Or make quick-coming death a little thing,
> Or bring again the pleasures of past years,
> Nor for my words shall ye forget your tears,
> Or hope again for aught that I can say,
> The idle singer of an empty day.

> But rather, when aweary of your mirth,
> From full hearts still unsatisfied ye sigh,
> And, feeling kindly unto all the earth,
> Grudge every minute as it passes by,
> Made the more mindful that the sweet days die –
> – Remember me a little then I pray,
> The idle singer of an empty day.

The heavy trouble, the bewildering care
That weighs us down who live and earn our bread,
These idle verses have no power to bear;
So let me sing of names remembered,
Because they, living not, can ne'er be dead,
Or long time take their memory quite away
From us poor singers of an empty day.

Dreamer of dreams, born out of my due time,
Why should I strive to set the crooked straight?
Let it suffice me that my murmuring rhyme
Beats with light wing against the ivory gate,
Telling a tale not too importunate
To those who in the sleepy region stay,
Lulled by the singer of an empty day.

Folk say, a wizard to a northern king
At Christmas-tide such wondrous things did show,
That through one window men beheld the spring,
And through another saw the summer glow,
And through a third the fruited vines a-row
While still unheard, but in its wonted way,
Piped the drear wind of that December day.

So with this Earthly Paradise it is,
If ye will read aright, and pardon me,
Who strive to build a shadowy isle of bliss
Midmost the beating of the steely sea,
Where tossed about all hearts of men must be;
Whose ravening monsters mighty men must slay,
Not the poor singer of an empty day.

APPENDIX I

Further reading

A1.1 Some interesting books

In this appendix I list a few books which the reader may find interesting.
I have not made a special search, so the choice is restricted to books
that I remember with pleasure.

There is, however, one problem.

At a literary dinner [during the Napoleonic wars, the poet] Campbell asked leave to
propose a toast, and gave the health of Napoleon Bonaparte. The war was at its
height, and the very mention of Napoleon's name, except in conjunction with some
uncomplimentary epithet, was in most circles regarded as an outrage. A storm of groans
broke out, and Campbell with difficulty could get a few sentences heard. 'Gentlemen,'
he said, 'you must not mistake me. I admit that the French Emperor is a tyrant. I
admit that he is a monster. I admit that he is the sworn foe of our nation, and, if you
will, of the whole human race. But, gentlemen, we must be just to our great enemy. We
must not forget that he once shot a bookseller.'

Without going quite so far, mathematicians may be permitted to mutter
a mild expletive when they consider the way that classic mathematical
texts are allowed to go out of print†. I expect that most of the titles
given will be accessible, I do not expect that all of them will be. Also,
because of the way such texts drift in and out of print many of them
have complicated publishing histories which I have not attempted to
trace.

Anthologies

The following are outstanding:

- *Mathematics: People, Problems, Results*, D. M. Campbell and
  J. C. Higgins. This contains several fine things including Pólya's
  'Some mathematicians I have known' from which I cannot resist
  quoting one anecdote. 'In working with Hardy, I once had an idea
  of which he approved. But afterwards I did not work sufficiently
  hard to carry out that idea, and Hardy disapproved. He did not

†Publishers have mixed
feelings about
mathematicians as well.
*The Chicago Manual of
Style* [77] says that
'Mathematics has long
been known in the printing
trade as *difficult,* or *penalty,*
copy because it is slower,
more difficult and more
expensive to set in type
than any other kind of
copy ... .' In future
mathematicians will
probably set their own
work using TEX but, no
doubt, both sides will be
able to find other things
to complain about.

tell me so of course, yet it came out when he visited a zoological garden in Sweden with Marcel Riesz. In a cage there was a bear. The cage had a gate, and on the gate there was a lock. The bear sniffed at the lock, hit it with his paw, then he growled a little, turned around and walked away. "He is like Pólya," said Hardy, "He has excellent ideas, but does not carry them out."'

- *The World of Mathematics*, J. R. Newman. If I had to choose one work (it is actually four volumes) from those mentioned in this chapter, this is the one I would recommend. Read Bernard Shaw on 'The vice of gambling and the virtue of insurance' and Poincaré's famous talk on 'Mathematical creation'† to start with, and then just browse. The essay by Whitehead on 'Mathematics as an element in the history of thought' contains the following splendid passage.

> I will not go so far as to say that to construct a history of thought without profound study of the mathematical ideas of successive epochs is like omitting Hamlet from the play which is named after him. That would be claiming too much. But it is certainly analogous to cutting out Ophelia. This simile is singularly exact. For Ophelia is quite essential to the play, she is very charming — and a little mad.

†The essay comes from his *Science and Method*. Hadamard wrote an interesting short book *The Psychology of Invention in the Mathematical Field* which is in part a commentary on Poincaré's views.

The mathematical life

The eminent French mathematician, Yves Meyer, once told me that, as a young man, he believed that any one who could do mathematics could do physics, that anyone who could do physics could do chemistry, anyone who could do chemistry could do biology and so on. This view of a descending chain of skills with mathematics at the top lasted until, as part of his military service, he had to sweep a floor. Less modest mathematicians retain the belief that mathematical ability implies philosophical and literary ability. This is not the case (as some of them have demonstrated all too well). However, the first four mathematicians cited unite mathematical and literary talents.

- *A Mathematician's Apology*, G. H. Hardy. Graham Greene knew 'of no writing — except perhaps Henry James's introductory essays — which conveys so clearly and with such an absence of fuss the excitement of the creative artist.' The second edition contains a foreword by C. P. Snow which should *not* be read at the same time as Hardy's essay. First look at the artist's work — then look at the artist's life (if you must).

- *A Mathematician's Miscellany*, J. E. Littlewood. Read in particular the essay 'A mathematical education'. The second edition contains, besides some material of interest mainly to those who believe that the fountain in Trinity Great Court is the centre of the universe, several worthwhile additions including a splendid lecture on 'The mathematician's art of work'. (During their prime, it was said that English mathematics was dominated by three great mathematicians — Hardy, Littlewood and Hardy-Littlewood.)

- *I Want to be a Mathematician*, P. R. Halmos. 'This book is about the career of a professional mathematician from the 1930s to the 1980s. ... It expresses prejudices, it tells anecdotes, it gossips about people, and it preaches sermons.' It contains sections on how to study, how to do research, how to be chairman of a mathematics department, and much else. ' ... the book is from the me of today to the me of yore, revealing some of the secrets that I desperately wanted to know then.' Remember, however, that Halmos is an outstanding mathematician and that you can be a useful member of the mathematical community without being quite as able and conscientious as his advice requires. There is a particularly fine index.

- *A Russian Childhood*, S. Kovalevskaya (translated by B. Stillman). In 1883, Stockholm University offered Sofya Kovalevskaya the post of 'Privatdozent' (the lowest, unpaid, rung of the academic ladder) and she became the first woman in modern times to hold a post at a European university. In 1889, in spite of the fact that she was a woman (with an unconventional private life), a foreigner, a socialist (or worse), and a practitioner of the new 'Weierstrassian analysis', she was appointed professor for life. Her memories of childhood are non-mathematical but form a fascinating human document. Beatrice Stillman includes an account of Kovalevskya's life, an autobiographical sketch and an article on her mathematics, all of which help place her person and her mathematics in context. For a detailed biography of someone whose whole life reads like a Russian novel, I refer the reader to *A Convergence of Lives* by A. H. Koblitz.

  There is a famous passage in her autobiographical sketch which, though long, is so apropos that I cannot resist quoting it. Sofya had an uncle who amused her first by telling fairy-tales and then by talking about what he had read — including mathematics. Then, when her parents' house was being redecorated, the new wallpaper ran out before the nursery had been done. The nursery was therefore papered with the nearest paper to hand, which proved to be old lithographed lecture notes on differential and integral calculus.

> I was then about eleven years old. As I looked at the nursery walls one day, I noticed that certain things were shown on them which I had already heard mentioned by Uncle. Since I was in any case quite electrified by the things he told me, I began scrutinising the walls very attentively. It amused me to examine these sheets, yellowed by time, all speckled over with some kind of hieroglyphics whose meaning escaped me completely, but which I felt must be wise and interesting. And I would stand by the wall for hours on end, reading and re-reading what was written there.
>
> I have to admit that I could make no sense of any of it at all then, and yet something seemed to lure me on toward this occupation. As a result of my sustained scrutiny I learned many of the writings by heart, and some of the formulas (in their purely external form) stayed in my memory and left a deep trace there. I remember particularly that on the sheet of paper which happened to be on the most prominent place on the wall, there was an explanation of the concepts of infinitely small numbers and of limits. The depth of that impression was evidenced several years later, when I was taking lessons from Professor A. N. Strannolyubsky in Petersburg. In explaining those very concepts he was astounded at the speed with which I assimilated them, and he said, 'You have understood them as though you knew them in advance.' And, in fact, much of the material had long been familiar to me from a formal standpoint.

- *Mathematical People*, edited by D. J. Albers and G. L. Anderson. This book consists mainly of extensive interviews with fifteen or so of the most creative and interesting mathematicians around during the 1980s. A treat.

- *The Mathematical Experience*, P. J. Davies and R. Hersh. A loving, sociological, look at the world of mathematics by two reflective mathematicians. Faces up to several problems which dog the modern mathematical endeavour but which optimistic books like the present one prefer to ignore.

Essays on mathematical topics

- *Mathematical Snapshots*, H. Steinhaus. 'One fine summer day,' the author begins, 'it happened that I was asked this question: "You claim to be a mathematician; well, what does one do all day when one is a mathematician?" We were seated in a park, my questioner and I, and I tried to explain to him a few geometric problems, solved and unsolved, using a stick to draw on the gravel pathway a

Jordan curve, or a Peano curve. ... That was how I conceived this book, in which the sketches, diagrams, and photographs provide a direct language and allow proofs to be avoided or at least reduced to a minimum.' Steinhaus succeeds admirably.

- *Mathematical Recreations and Essays*, W. W. Rouse Ball. Fashions change in mathematical recreations as in everything else — often for as little reason. Rouse Ball's book is an excellent collection of older topics like magic squares and Bachet's problem. Later editions have been tastefully modernised by H. S. M. Coxeter (though, as with all tasteful modernisations, there is loss as well as gain).

- *Game Set and Math*, I. Stewart. Ian Stewart is one of the most gifted expositors now writing, both at a technical and at a popular level. Everything he writes is worth reading and sometimes, as in his book on Galois theory [228], joy guides his pen. In this collection too, perhaps because it was initially written for translation into French, the high spirits shine through.

- *Penrose Tiles to Trapdoor Ciphers*, M. Gardner. This is the thirteenth collection of Martin Gardner's admirable *Scientific American* columns. It is extraordinary that someone could write so much and produce so little dross; the other twelve collections can be recommended just as confidently. Among his non-mathematical writings I particularly recommend *Fads and Fallacies in the Name of Science* but even his strange theological novel *The Flight of Peter Fromm* is worth reading.

- *Prisoner's Dilemma*, W. Poundstone. Consider the following 'Dollar Auction' in which a dollar note is auctioned under the following rules.

(1) Just as in an ordinary auction players may bid in any order, the only condition being that each new bid must be higher than the last. The auction ends when no one is prepared to make any further bids.

(2) Unlike an ordinary auction both the highest *and the second-highest* bidder must pay the auctioner the amounts of their last bids. The highest bidder gets the dollar in return but the second-highest bidder gets nothing.

What will happen in such an auction? What ought to happen?

Poundstone's book is in part a discussion of such problems, in part a biography of the brilliant mathematician von Neumann† and in part a study of the idea of preventive war. It is a meditation on a theme of von Neumann: 'The world ... is riding a very fine tiger. Magnificent beast, superb claws, etc. But do we know how to dismount?'

†Von Neumann was held in awed respect by his colleagues who claimed that 'While he was indeed a demi-god, he had made a detailed study of humans and could imitate them perfectly.'

- *What is Mathematics?*, R. Courant and H. Robbins. This is very tough going but very rewarding. I was helped through it by one of my mathematics teachers. At the time I took such help for granted, but now I am very grateful. (The discussion of the lever in the railroad car on page 312 contains a subtle flaw explained in Chapter 5 (question 7) of Stewart's book of essays [229] given above. The presence of such an error in such a book reminds us of how hard mathematics is, even for the best mathematicians.)

History of mathematics

- *Men of Mathematics*, E. T. Bell. If you ask a historian of mathematics their opinion of this book you will be lucky to get away with less than a half-hour's diatribe. But its many faults are irrelevant; generations of mathematicians have been inspired by this book, and generations will be inspired. Join them. It is not just a history of mathematics, it is now part of that history.
- *History of Mathematics: A Reader*, edited by J. Fauvel and J. Gray. A cheerful and instructive collection of documents from the history of mathematics which really does bring the past to life.
- *Mathematical Thought from Ancient to Modern Times*, M. Kline. The best general history in English, and likely to remain so for a long time to come. It is, however, intended for a mathematically prepared audience (indeed, this is one of its virtues) and readers of the present book may find it hard going. If you listen carefully, you will hear the faint sound of some of Kline's favourite axes being ground, but the noise is never obtrusive.

'How to' books

- *How to Solve It*, G. Pólya. A major mathematician sets out the standard strategies that mathematicians use to attack problems. He gives extra flesh to the bare bones in the longer work *Mathematics and Plausible Reasoning* and, in my view, less successfully in *Mathematical Discovery*.
- *How to Write Mathematics*, I. N. Steenrod, P. R. Halmos, and others. The copy in our departmental library is very battered from use. Since few of my readers will have any immediate intention of writing a mathematics book they may wonder what they can gain from reading such a manual. I hope they will gain insight into the way the texts that they read and the lecture courses they attend are organised. At the very least they may come a little closer

to realising that, even at its highest level, mathematics involves communication between two human beings.

- *How to Teach Mathematics*, S. G. Krantz. Another glimpse over the wall. If you think about how to teach, it may teach you how to learn. There are many books (mainly bad) about teaching in general but even the good ones are irrelevant to the special problems of teaching mathematics. This slim volume contains practically all the good advice I know for new university mathematics lecturers.

- *A Handbook for Scholars*, M.-C. Van Leunen. My generation was educated in blissful ignorance of projects and investigations. A less fortunate generation needs this kind of guide to the nuts and bolts of academic writing. Here is a sample†:

  > Most of us who went to American grade schools can remember long hours of copying articles out of encyclopaedias. 'The abode of the penguin is a hard and difficult one.' It was called doing research. Then in college we found that it was also called plagiarism.

- *How to Lie with Statistics*, D. Huff. A little masterpiece which can be read at a sitting.

- *The Visual Display of Quantitative Information*, E. R. Tufte. A marvellous Christmas present‡ for mathematicians, statisticians and artists alike. Contains Minard's map of Napoleon's retreat from Moscow which 'may well be the best statistical graphic ever drawn,' as well as (on page 118) what 'may well be the worst graphic ever to find its way into print' and (on page 95) a figure which 'achieves a graphical absolute zero.' The companion volume *Envisioning Information*, is hardly less fascinating, showing (on page 105) how to design a railway timetable and (on pages 56–7) the grimmest commentary on 'medical progress' that I know.

- *Mathematical Models*, H. M. Cundy and A. P. Rollett. A minority, but by no means an insignificant minority, find their way into mathematics by constructing models. This is the classic text for such students.

There are many other books that I could have mentioned: Rademacher and Toeplitz's *The Enjoyment of Mathematics*, Pedoe's *The Gentle Art of Mathematics*, the books of W. W. Sawyer (a little simpler than most of those mentioned above), Moritz's *On Mathematics and Mathematicians*

†I could also have chosen her views on the misuse of the footnote: 'There are scholars who footnote compulsively, six to a page, writing what amounts to two books at once. There are scholars whose frigid texts need some of the warmth and jollity they reserve for their footnotes and other scholars who write stale, dull footnotes like the stories brought inevitably to the minds of after-dinner speakers. There are scholars who write weasel footnotes, footnotes that alter the assertions in their texts. There are scholars who write feckless, irrelevant footnotes that leave their readers dumbstruck with confusion.'

‡I got my copy from my wife in this way.

(a collection of mathematical quotations and anecdotes) — the list, if not endless, still has some way to run. However, I draw the line well above a recent book on mathematics and humour which seems to have been conscientiously swept clean of jokes†.

If the reader is interested in the mathematical typesetting systems mentioned at the end of Section 12.1 the following remarks may be useful.

(1) TEX and its relatives are going to dominate mathematical text processing for some time to come. It is, in my belief, a waste of time to learn anything but a TEX-based system‡.

(2) There are, at the moment, three main dialects of TEX. They are TEX itself (for which the standard manual is Knuth's *The TEXbook*), LATEX (for which the standard manual is Lamport's *LATEX, A Document Preparation System*) and AMS-TEX (for which the standard manual is Spivak's *The Joy of TEX*). The newest version of LATEX allows you to use an extra package `amstex` which combines almost all the virtues of AMS-TEX and LATEX (but you should start by learning one or the other, for details see *The LATEX Companion* by Goosens, Mittelbach and Samarin). If you like doing things precisely right, then I suspect TEX is for you. If you are uninterested in the finer points of typography then LATEX is simpler and has the more comprehensible manual. If you have a friend who uses one system, follow her because all TEX systems are easier to learn if you have someone to help when things go wrong.

(3) It takes about five days of banging your head against your console to become fluent in a TEX dialect (the best way is just to try typing out a few pages of mathematics). After that, it is so easy that you wonder why you ever found it hard.

(4) Knuth put TEX in the public domain. Because of this there are good free implementations of TEX systems for most computers. Ask around.

A1.2 Some hard but interesting books

The books recommended in the previous section were all intended for the general (and persistent and intelligent) reader. But most of the mathematicians whom I know well admit to having tried to read much harder books during their school-days. Those whom I know really well also admit to having failed to understand the full import of what they read.

> Ah, but a man's reach should exceed his grasp
> Or what's a heaven for?

---

†When the reader has had a couple of years of university mathematics she will be ready for Linderholm's *Mathematics made Difficult* but until then she will have to take her jokes where she can find them.

‡I pronounce TEX in the same way as the first syllable of 'technology'. According to its author, TEX is an uppercase form of τεχ. 'Insiders pronounce the χ of TEX as a Greek chi, not an "x", so that TEX rhymes with the word "blecchh". ... When you say it correctly to your computer, the terminal may become slightly moist.' I pronounce LATEX as 'lay-TEX'.

Moreover, it is not necessary to understand everything in a book to derive great benefit from it. If a book shows us new patterns of thought and allows us to glimpse new ideas, however fleetingly, it lays down a foundation on which a later, fuller understanding can be built. The long quotation from Kovalevskaya which I gave in the previous section can, I am sure, be echoed by many other mathematicians. Moreover, one answer to the question, 'What do mathematicians do?' is, 'They try to understand things which are hard to understand.' If you sit by yourself for an afternoon trying to understand half a page of mathematics, then you are doing mathematics. If you listen to a lesson taught so as to be entirely within your grasp and then do homework carefully graded so that it will take between half-an-hour and an hour to do, then you are learning mathematics (and you must learn a great deal of mathematics to become a mathematician) but not doing mathematics.

I have given a fairly large selection, partly because not all will be readily available, but mainly because what appeals to one person may not appeal to another. If Hardy and Wright leave you unmoved, then perhaps Feynman will reach out from the page and drag you in. If one of these books catches your interest sufficiently to make you read 50 pages with attention, I shall have achieved my purpose. If this does not happen, then you may find your path into mathematics at a later time along a different road.

Number theory

Since it requires little previous knowledge, elementary number theory is a good way of seeing what mathematics is like. There are many excellent books including the following.

- *The Higher Arithmetic*, H. Davenport. Clear exposition aimed at beginners, but covering real mathematics.

- *An Introduction to the Theory of Numbers*, G. H. Hardy and E. M. Wright. A pleasure to read (the pleasure increased by the excellence of the mathematical printing). 'Our first aim,' say the authors, 'has been to write an interesting book, and one unlike other books. We may have succeeded at the price of too much eccentricity, or we may have failed; but we can hardly have failed completely, the subject matter being so attractive that only extravagant incompetence could make it dull.'

- *Number Theory in Science and Communication*, M. R. Schroeder. Although it varies wildly in difficulty and coherence, this is a fine book, full of ideas expressed with enthusiasm. Incidentally the author made a small fortune by applying number theory to hi-fi (see his Section 13.8 'Primitive roots and concert hall acoustics').

Analysis

Any really ambitious student ought to have a go at analysis before she meets it at university. The chances that she will really understand it first time round are slight. (The foundations of rigorous analysis are subtle and require time, discussion and effort to master.) However, if there is to be a second time round, there has to be a first. Several generations have cut their teeth on Hardy's *A Course of Pure Mathematics* but, splendid book though it is (Littlewood described the tone as that of 'a missionary talking to cannibals'), it no longer fits in well with the school preparation or the university courses of present-day students. Instead, I suggest one of the two following books.

- *Calculus*, M. Spivak. A careful and enthusiastic exposition which seems to me ideal for self-study.
- *A First Course in Mathematical Analysis*, J. C. Burkill. Though I always recommend Spivak, it must be admitted that my Trinity Hall pupils almost always prefer Burkill's concise and direct style.

Geometry

I believe that mathematicians have various sources of intuition — physical, probabilistic and geometric†. Unfortunately, the place of geometry in mathematical education has been downgraded to the point where few students can acquire geometric intuition. (There has been a great improvement in the teaching of probability, so not everything is getting worse.) I therefore strongly recommend the next book, not because it covers material met with at university, but because it covers material which is not.

†Of course, there are other sources of intuition. Conway and Knuth seem to have an 'intuition for pattern' which is rare and so, like all rare but useful things, extremely valuable. There is also an 'intuition for process,' a feeling for how the pattern of a complicated proof or algorithm 'ought to run', which, I suspect, is common but not universal.

- *Introduction to Geometry*, H. S. M. Coxeter. A classic work designed to be 'user-friendly' and permit easy browsing. Full of beautiful things.

Probability

There are now several excellent introductions to probability, but my favourite is still the grand-daddy of them all.

- *An Introduction to Probability Theory and Its Applications (Vol I)*, W. Feller. (There is enough in the first volume to occupy even the most probabilistically talented student. Much of the second volume requires advanced mathematical tools.) Feller deliberately varies the mathematical level within each chapter so the beginning student has to pick her way past thickets of advanced mathematics. However, the richness of Feller's probabilistic insight and the interest of his examples amply make up for this inconvenience. The third edition contains (in Chapter III) a discussion of coin tossing of which he was particularly, and rightly, proud. In his preface to that edition, he hopes that his book 'will continue to find readers who read it merely for enjoyment and enlightenment.' It will.

Physics

There are many popular books on 'physics without formulas' — the intellectual equivalent of watching sport on television — but few good introductions to physics. Fortunately there is one outstanding example.

- *The Feynman Lectures on Physics*, R. P. Feynman. Feynman was a great physicist, a great teacher and a great raconteur. In these lectures he shows how a physicist thinks about physics. The first volume, dealing with elementary physics, contains enough for many weeks (indeed, several months) of reading and thought, and the remaining two volumes are more or less accessible to the really determined student. His best stories (no doubt polished and repolished over a lifetime) are collected in *Surely You're Joking, Mr Feynman!*. Read in particular the essays 'The dignified professor' and 'Cargo cult science'.

Modern algebra

For my generation, the switch to modern abstract algebra was one of the most striking changes in the move from school to university. The school syllabus now contains some watered down† modern algebra, so preliminary study is less exciting and perhaps less urgent.

†When soup is watered down, it loses much of its taste and nutritional value.

- *A Survey of Modern Algebra*, G. Birkhoff and S. MacLane. This is the book that brought modern algebra into the Anglo-Saxon undergraduate syllabus. In my view, Chapters I, VI, VII and VIII remain as good a first introduction for the able student working by herself as can be found. It must be added that the word 'modern' is used here to mean 'modern in 1930,' and that more advanced university algebra involves a further layer of abstraction.

Foundations of mathematics

There is a good case for advising students not to worry about foundations until they know a fair amount about the building they intend to place on those foundations. However, students (particularly good ones) never listen to advice, so here are two suggestions, the first dealing with technical and the second with philosophical matters.

- *Naive Set Theory*, P. R. Halmos. A gem of a book which can be read at a couple of sittings and which will tell you all that a gentleman needs to know about set theory. (To grasp this useful Cambridge locution, you must understand that, whilst a gentleman knows roughly how a car works and could, in principle, drive one, in practice he leaves this to his chauffeur.)

- *The Philosophy of Mathematics*, S. Körner. In philosophy, as in mathematics, we stand on the shoulders of giants, and, if we cannot see so far, it is because philosophy is (at the highest level) harder than mathematics. My father's useful text summarises the three main philosophies of mathematics — logicism, formalism and intuitionism — and discusses the relation between pure and applied mathematics.

Algorithms and related topics

- *Numerical Methods That Work*, F. S. Acton. (If you look carefully at the cover of several of its editions, you will notice that the very faint word 'Usually' has been inserted after 'That'.) At its highest level, numerical analysis is a mixture of science, art and bar-room brawl. The most that a general mathematical course can hope to do for students in this direction is to make them 'educated consumers', able to make sensible use of packaged programs and able, if the need arises, to cobble together something that will work as it is supposed to (even if rather more slowly than an expert's version). Acton's book is an excellent, relaxed and good-humoured introduction to the problems and tools of the numerical analyst.

- *Concrete Mathematics*, R. L. Graham, D. E. Knuth and O. Patashnik. The authors claim that the material 'may seem at first to be a disparate bag of tricks, but practice makes it into a disciplined set of tools.' Even if the reader does not grasp the claimed unity of the whole, she will acquire many useful tricks and several essential ideas. As might be expected from these authors, there is hardly a page without a beautiful formula, an unexpected insight or a challenging problem. If you have access to a library with Knuth's epoch-making three volumes *The Art of Computer Programming* take them down and be enchanted.

- *Winning Ways*, E. R. Berlekamp, J. H. Conway and R. K. Guy. An advanced research text on how to play games like 'dots and boxes' (the authors are unable to give an unbeatable strategy). Only a small proportion of the text will be accessible to my readers — but that small proportion is worth the price of the whole work. Play 'sprouts' (page 564) with your friends, take time to understand the addictive game of 'philosopher's football' (page 688), think about how long 'sylver coinage' will last (page 576) and browse through Chapters 23, 24 and 25. Good hunting!

Any book that you can learn from is a good book. None the less, I believe that the books mentioned above have a special quality which sets them apart from more standard texts written to help the average student pass the average exam. At the end of the nineteenth century Jordan wrote one of the most influential text-books since Euclid (the second edition of his *Cours d'Analyse*). In it, he drew together the strands of rigorous analysis as it had developed in the last 100 years and wove it into what has become the standard presentation of the subject. Here is part of a review of a later edition by the great analyst Lebesgue.

When I was a rather disrespectful student at the École Normale we used to say that 'If Professor Jordan has four quantities which play exactly the same role in an argument he writes them as u, A'', λ and e_3'. Our criticism went a little too far but, nonetheless, we felt clearly how little Professor Jordan cared for the commonplace pedagogical precautions which we could not do without, spoiled as we were by our secondary schools.

The school teachers there, like some university teachers, deploy a great deal of ingenuity and often much talent to make things easy for their students. They wish to lead them effortlessly through to even the hardest results. It is wonderful how often they appear to succeed in this, but my opinion is that their success is always illusory.

To be properly understood these proofs, ingenious, clear and short as they are, must be scrutinised word by word. One word indicates, or rather obscures, a precaution to be taken or a particular fact to be observed without which the argument collapses and which must be taken to account. Another word recalls, or rather glosses over, a long intermediate argument which must be mentally substituted for it. In this way, the simpler the argument appears to be, the greater the mental gymnastics required to read it properly. I much prefer to read Professor Jordan's arguments which attack the main difficulty directly without trying to conceal or evade it and so make you understand what this difficulty consists in; when the difficulty is overcome we know why and how it has been overcome.

This type of proof has the additional advantage that, at the same time as it demonstrates those particular mathematical truths which we choose to state in the form of theorems, it also reveals their relation to the mathematical facts which surround them. Whatever some people may believe, mathematics is not a collection of theorems. If we only know a few theorems, however important in themselves, we have a view of mathematics as mistaken as those children who think of the Alps as four or five sugar cones because they know the names of four or five peaks.

To obtain a slick simple proof of a major theorem, we must remove everything that is not absolutely essential, everything that merely helps understand the result. We act like a man who tries to reach the summit of an unknown region whilst refusing to look around until he arrives at his goal. If someone guides him there, then, perhaps, he may see that he stands high above many things, but he will be unsure precisely what. We must also remember that usually you can see nothing from the very high peaks; mountaineers only climb them for the pleasure of the effort involved.

Professor Jordan's only object is to make us understand the facts of mathematics and their interrelations. If he can do this by simplifying the standard proofs, he does so; but he does not hesitate to use complex arguments if he thinks they are preferable. He multiplies examples and applications and, more generally, he omits nothing which might aid our understanding. But he never goes out of his way to reduce the reader's trouble or compensate for the reader's lack of attention.

Some notations

David Tranah has suggested that some comments on notation might be useful. Table A2.1 gives the Greek alphabet. The Greek letters enclosed in square brackets are identical in form with ordinary (Roman) letters and are rarely used.

The reader who feels inclined to complain should remember that users

Table A2.1. *Greek letters.*

Lower-case	Upper-case	Name	Corresponds to	Note
α	[A]	alpha	a	
β	[B]	beta	b	
γ	Γ	gamma	c	
δ	Δ	delta	d	
ϵ, ε	[E]	epsilon	e	Do not confuse with $\in$ (belongs to).
ζ	[Z]	zeta	—	
η	[H]	eta	—	
θ	Θ	theta	—	
ι	[I]	iota	i	Often reserved for 'identity' objects.
κ	[K]	kappa	k	
λ	Λ	lambda	l	
μ	[M]	mu	m	
ν	[N]	nu	n	
ξ	Ξ	xi	—	
[o]	[O]	omicron	o	Not used.
π	Π	pi	p	
ρ	[P]	rho	r	
σ, ς	Σ	sigma	s	
τ	[T]	tau	t	
υ	Υ	upsilon	u	Rarely used.
ϕ, φ	Φ	phi	—	
χ	[X]	chi	—	
ψ	Ψ	psi	—	
ω	Ω	omega	—	'I am the alpha and omega, the beginning and the ending.'

of other alphabets (for example, Russians and Japanese) have to use our Roman alphabet as their main mathematical alphabet. 𝔖𝔥𝔢 𝔰𝔥𝔬𝔲𝔩𝔡 𝔞𝔩𝔰𝔬 𝔱𝔞𝔨𝔢 𝔠𝔬𝔪𝔣𝔬𝔯𝔱 𝔣𝔯𝔬𝔪 𝔱𝔥𝔢 𝔣𝔞𝔠𝔱 𝔱𝔥𝔞𝔱 𝔪𝔞𝔱𝔥𝔢𝔪𝔞𝔱𝔦𝔠𝔦𝔞𝔫𝔰 𝔥𝔞𝔳𝔢, 𝔪𝔬𝔯𝔢 𝔬𝔯 𝔩𝔢𝔰𝔰, 𝔰𝔱𝔬𝔭𝔭𝔢𝔡 𝔲𝔰𝔦𝔫𝔤 𝔱𝔥𝔢 𝔉𝔯𝔞𝔨𝔱𝔲𝔯 (𝔬𝔯 𝔊𝔢𝔯𝔪𝔞𝔫, 𝔬𝔯 𝔅𝔩𝔞𝔠𝔨 𝔏𝔢𝔱𝔱𝔢𝔯) 𝔣𝔬𝔫𝔱. 𝔥𝔢𝔯𝔢 𝔦𝔰 𝔱𝔥𝔢 𝔣𝔲𝔩𝔩 𝔠𝔬𝔩𝔩𝔢𝔠𝔱𝔦𝔬𝔫 𝔬𝔣 𝔲𝔭𝔭𝔢𝔯𝔠𝔞𝔰𝔢 𝔞𝔫𝔡 𝔩𝔬𝔴𝔢𝔯 𝔠𝔞𝔰𝔢 𝔩𝔢𝔱𝔱𝔢𝔯𝔰 𝔱𝔬 𝔰𝔥𝔬𝔴 𝔴𝔥𝔞𝔱 𝔰𝔥𝔢 𝔥𝔞𝔰 𝔪𝔦𝔰𝔰𝔢𝔡.

$$\mathfrak{A}, \mathfrak{B}, \mathfrak{C}, \mathfrak{D}, \mathfrak{E}, \mathfrak{F}, \mathfrak{G}, \mathfrak{H}, \mathfrak{I}, \mathfrak{J}, \mathfrak{K}, \mathfrak{L}, \mathfrak{M}, \mathfrak{N}, \mathfrak{O}, \mathfrak{P}, \mathfrak{Q}, \mathfrak{R},$$
$$\mathfrak{S}, \mathfrak{T}, \mathfrak{U}, \mathfrak{V}, \mathfrak{W}, \mathfrak{X}, \mathfrak{Y}, \mathfrak{Z}.$$
$$\mathfrak{a}, \mathfrak{b}, \mathfrak{c}, \mathfrak{d}, \mathfrak{e}, \mathfrak{f}, \mathfrak{g}, \mathfrak{h}, \mathfrak{i}, \mathfrak{j}, \mathfrak{k}, \mathfrak{l}, \mathfrak{m}, \mathfrak{n}, \mathfrak{o}, \mathfrak{p}, \mathfrak{q}, \mathfrak{r}, \mathfrak{s}, \mathfrak{t}, \mathfrak{u}, \mathfrak{v}, \mathfrak{w}, \mathfrak{x}, \mathfrak{y}, \mathfrak{z}.$$

On the whole, the attempts of mathematicians to extend their repertoire of symbols by using other alphabets have not been particularly successful. (No doubt the natural human impulse to read any new symbol as 'splodge' has something to do with this.) The small band of those willing to read the revolutionary papers of Cantor in which he founded set theory was further reduced by his use of the Hebrew letters. The symbol ℵ (aleph, the first letter of the Hebrew alphabet) which he introduced is still used but, until quite recently, many printers set it upside down as ℵ.

The following remarks on notation are intended to help you if you are held up because you do not recognise a symbol.

- The symbol $\approx$ means 'approximately equal to'.
- $\sum_{r=1}^{n} a_r$ is a way of writing $a_1 + a_2 + \cdots + a_n$. To understand formulae containing expressions like this it is a good idea to replace n by some specific number like 4. Thus, for example,

$$\sum_{r=1}^{4} a_r = a_1 + a_2 + a_3 + a_4$$

- In the same way the product of n numbers $a_1, a_2, \ldots, a_n$ is written as $\prod_{r=1}^{n} a_r$.
- Thus, for example, we write

$$n! = 1 \times 2 \times 3 \times \cdots \times n = \prod_{r=1}^{n} r.$$

(In the civilized world $n!$ is called n factorial but in English schools it is called n bang or n shriek — a joke which never palls.)
- The binomial coefficient is given by

$$\binom{n}{r} = C_n^r = \frac{n!}{(n-r)!r!}.$$

It gives the number of ways of selecting r things from n (no attention being paid to order) and appears in the binomial expansion

$$(x+y)^n = \sum_{r=0}^{n} \binom{n}{r} x^r y^{n-r}.$$

- If x is a function of t the following are common expressions for the derivative of x at t:

$$\frac{dx}{dt} = \frac{d}{dt}(x(t)) = \dot{x}(t) = x'(t).$$

APPENDIX 3

Sources

The following tale is told of the Mullah Nasrudin.

> Jalal, an old friend of Nasrudin called one day. The Mullah said, 'I am delighted to see you after such a long time. I am just about to start on a round of visits, however. Come, walk with me, and we can talk.'
>
> 'Lend me a decent robe', said Jalal, 'because, as you see, I am not dressed for visiting.' Nasrudin lent him a very fine robe.
>
> At the first house Nasrudin presented his friend. 'This is my old companion, Jalal: but the robe he is wearing, that is mine!'
>
> On the way to the next village, Jalal said: 'What a stupid thing to say! "The robe is mine" indeed! Don't do it again.' Nasrudin promised.
>
> When they were comfortably seated at the next house, Nasrudin said: 'This is Jalal, an old friend come to visit me. But the robe: the robe is *his*!'
>
> As they left Jalal was as annoyed as before. 'Why did you say that? Are you crazy?'
>
> 'I only wanted to make amends. Now we are quits.'
>
> 'If you do not mind,' said Jalal, slowly and carefully, 'we shall say no more about the robe.' Nasrudin promised.
>
> At the third and final place of call, Nasrudin said: 'May I present Jalal, my friend. And the robe, the robe he is wearing ... but we mustn't say anything about the robe, must we?'

This book is not a work of scholarship. However, a few of my readers may wish to check some of my statements or simply to read further. This list of sources is intended to help them.

- Page ii *To judge in this [utilitarian] way ...* Quoted in [22].
- Page ii *We agreed ...* From *The Wrench* (Tiresias, page 53 of [143]).
- Page ii *To be placed on the title-page ...* Quoted in [194].
- Page ii *Mathematicians are ...* In [3].
- Page ix *some may assert ...* In *The Essays*, Book III, Essay 12.

- Page x *the habit of fitting* ...   [76], Chapter 11.
- Page x *I think you will like* ...   Sheridan, *The Rivals*, Act 1, Scene 1.
- Page 3 *The eyes surrounded by a dark circle* ...   Quoted in [164].
- Page 4 *in the last few years of his life* ...   Biographical introduction to [222].
- Page 4 *and the increase of the* ... *cholera poison* ...   All unattributed quotations in this section are from [222].
- Page 7 *which has passed through the kidneys* ...   Quoted in [164].
- Page 9 *were removed wholesale* ...   [97].
- Page 13 *The extraordinary eruption* ...   Quoted in [97].
- Page 14 *This practice was stopped and the epidemic ceased.* Account in [164].
- Page 14 *The New York epidemic* ...   Account in [203].
- Page 16 *graphic demonstration of the power of a German professor* ...   [54].
- Page 18 *from the ISIS-2 report* [100].
- Page 20 *A more dramatic example is given by* ...   [28].
- Page 19 *Figure 1.6* From [101].
- Page 21 *It is expressing it mildly* ...   [152], Volume 4, page 148.
- Page 22 *A decision must be reached* ...   [209], pages 258-62.
- Page 23 *The situation is very bad* ...   [237], page 47.
- Page 24 *The system of* ... *convoy is not recommended* ...   [237], pages 52–3.
- Page 25 *We have been upside down here* ...   [152], Volume 4, page 279.
- Page 25 *[The Admiralty] had not at any time* ...   [152], Volume 4, pages 132–3.
- Page 28 *The fears adumbrated* ...   [152], Volume 4, page 128.
- Page 29 *The oceans at once became bare* ...   [49], page 4.
- Page 30 *If a dredger sailed from Yarmouth to Felixstowe* ...   [248], Chapter 1.
- Page 30 *that branch of the art of lying* ...   [39], Preface.
- Page 30 *It was, in fact* ...   [152], Volume 4, page 151.
- Page 30 *Congestion at the top* ...   Vice-Admiral Dewar, quoted in [152], Volume 4, page 135.
- Page 31 *like a company managing-director* ...   Quoted in [152], Volume 4, page 288.
- Page 32 *Truth-loving Persians* ...   Graves, *The Persian Version*.
- Page 32 *Dönitz waved his arm to encompass all the ships* ...   [172], page 90, based on Dönitz's own accounts.
- Page 32 *Naval officers in previous centuries* ...   [258], page 115.
- Page 33 *formed part of a Technical History series* ...   [258], page 122.

- Page 33 *Another 'lost work'* ... Referred to in [237], pages 731–2.
- Page 34 *one of the most tragic situations in naval history* [49], page 45.
- Page 34 *A U-boat attacking a convoy* ... [49], page 4.
- Page 35 *His position, newly created,* ... [172] page 496.
- Page 36 *The price of petrol* ... This cartoon appeared in the *Daily Mirror* and is reproduced in [43], page 83.
- Page 37 *The only thing that ever really frightened me* ... [33], Volume II, page 529.
- Page 38 *the spot where the [battle-cruiser]* Queen Mary *had disappeared* ... [17],pages 27–8 and 73–4.
- Page 39 *Guns, big guns, long ranges* ... [152], Volume 1, page 414.
- Page 40 *If I am killed* ... Quoted in [152], Volume 3, page 215.
- Page 40 *The system of Proof* ... Quoted in [152], Volume 3, page 172.
- Page 41 *Those who used the many instances* ... [152] Volume 5, page 330.
- Page 42 *In the course of the discussion* ... [257], page 383.
- Page 42 *reported to his Captain* ... [152], Volume 5, page 325.
- Page 43 *He could not understand* ... [149].
- Page 43 *obituary notice by Sir Bernard Lovell* [149].
- Page 43 *Even without the far-sighted action* ... [149].
- Page 43 *In 1919 Sir Ernest Rutherford* ... [170].
- Page 44 *It seems to me that in some way* ... [174], page 363.
- Page 45 *Throughout the whole of its existence* ... [35], page 117.
- Page 45 *One idea which appeared to have a special fascination* ... [20], page 4.
- Page 46 *should not gravely interfere with the grouse shooting.* [20], page 27.
- Page 46 *Although the Air Defence Committee* ... [17], pages 106–7.
- Page 48 *An aircraft over the North Sea* ... [205], page 23.
- Page 49 *going home one day* ... [35], page 129.
- Page 49 *... his reasoning was crystal clear* ... [35], page 158.
- Page 55 *Including the two halfpennies* ... [102], Chapter 2.
- Page 56 *Some embarrassment was caused* ... See [98], page 3.
- Page 56 *According to what I heard at the time* ... [105], page 251.
- Page 57 *I expected them to be advanced* ... [190], Chapter 6.
- Page 57 *Apparently a start was made in 1942* ... See Chapter 15 of [26] and the accounts given in [191].
- Page 57 *R. V. Jones gives a rather chilling account* ... [104], Chapter 10.
- Page 57 *My immediate assignment* ... [17], page 206.

- Page 57 *Immense scientific and technical brilliance* ... [17], page 208.
- Page 60 *total budget expenditure* ... from [243], page 65. The values of tan come from [1].
- Page 61 *was to help work out in a week or two* ... [17], page 208.
- Page 61 *On looking into the rounds per bird* ... [17], page 211.
- Page 63 *New weapons for old* ... [17], pages 175–6.
- Page 64 *On a large wall map* ... [17], page 216.
- Page 65 *It appeared that in nearly* 40% *of the cases* ... [251], page 151.
- Page 65 *How then could one raise the chance* ... [17], pages 216–17.
- Page 66 *The Americans* ... *with their genius for nomenclature* ... [251].
- Page 66 *The number of escaping U-boats* ... [251], page 158.
- Page 67 *Congressman Jackson May* ... See [249], page 337.
- Page 67 *On the assumption that a U-boat would* ... [17], page 215.
- Page 68 *Captured German U-boat crews thought* ... [17], page 215.
- Page 69 *The distribution of the mid-points of the sticks* ... [251], page 188.
- Page 71 *[We] had always hoped* ... [251], page 255.
- Page 73 *could be likened to a golfer* ... [189].
- Page 73 *by merit raised* ... Milton, *Paradise Lost*, Book II.
- Page 74 *If I could, I would* ... Montaigne, *The Essays*, Book III, Essay 1. (See Revelation 12:7)
- Page 74 *the practical difficulty of changing* ... [17], page 229.
- Page 74 *Navy had to adopt some thirty* ... See [189], page 150.
- Page 74 *His recommendation* ... *horrified us* ... Quoted in [258], page 166.
- Page 74 *Lindemann also changed the method of classifying shipping losses* ... See [252].
- Page 75 *While it takes approximately 7000 hours* ... Quoted in [88], page 179.
- Page 75 *It is difficult to compare quantitatively* ... Quoted in [88], page 179.
- Page 76 *A long-range aircraft* ... [17].
- Page 76 *After some intense argument* ... [103], page 79.
- Page 79 *Such* a priori *investigations* ... [17], page 198.
- Page 80 *Under the exigencies of a very critical situation* ... [17], pages 230–2.
- Page 81 *sank on average one ship* ... See [252].
- Page 82 *before we advised the Navy* ... [17], page 115.
- Page 83 *Waters has written a fascinating study* ... In [252].

- Page 84 *that the US Navy did not enjoy ...* Quoted in [258], page 230.
- Page 84 *as many as possible of the rules ...* [17], page 210.
- Page 84 *a major military campaign ...* [17], page 122.
- Page 85 *The story goes ...* [17], page 110.
- Page 85 *When our brother Fire ...* MacNeice, *Brother Fire (1943)*.
- Page 85 *I confess to a haunting sense of personal failure ...* [17], page 126.
- Page 86 *In retrospect, it seems remarkable* [149].
- Page 86 *We persist in regarding ourselves ...* [73], page 229.
- Page 87 *I cannot praise a fugitive and cloistered virtue ...* Milton, *Areopagitica*.
- Page 87 *Ease and leisure is given thee ...* Milton, *The Reason of Church Government*.
- Page 87 *The long-range German aircraft ...* [17], pages 218-19.
- Page 88 *Officers who had been tempted into speaking ...* [251], page 207.
- Page 90 *An attempt was made to set a life ...* [251], page 70.
- Page 92 *Suppose the effort per aircraft were doubled ...* [251], page 42.
- Page 93 *to the great task ...* [17], page 199.
- Page 96 *'I weep for you' the Walrus said ...* Lewis Carroll, *The Walrus and the Carpenter*.
- Page 101 *I greatly doubt ...* [61], First Day.
- Page 104 *historical research reveals ...* See, for example, [50].
- Page 104 *among the safe ways to pursue truth ...* [50], page 409.
- Page 105 *From what has already been demonstrated ...* [61], Second Day.
- Page 106 *Giant Pope and Giant Pagan ...* [79].
- Page 107 *Why do we have 13 and 17 year cicadas ...* [72], page 102.
- Page 108 *Gravity, a mere nuisance to Christian ...* [79].
- Page 109 *Many methods have been used ...* [211], pages 80–2.
- Page 111 *a plastic strip weakened by a slot ...* [134].
- Page 114 *A 30 gram mouse ...* [211], page 146.
- Page 113 *Experiments suggest that climbing vertically ...* See [211].
- Page 114 *even if they were true ...* [178].
- Page 117 *The stone measured ...* [65], Chapter 15.
- Page 127 *There is no empirical method ...* [62], Introduction.
- Page 129 *did not seem to doubt ...* [234], page 527.
- Page 130 *went over the field ...* [234], page 527.
- Page 130 *My instructor was a flight sergeant ...* [234], pages 527–8.

- Page 136 *Escher was a Dutch artist* ... For an illustrated catalogue of his work see [146].
- Page 136 *a book of Escher's pictures* ... *The Times*, page 12, 21 August, 1995.
- Page 137 *Shut yourself up with some friend* ... [62], Second Day.
- Page 139 *Our whole progress* ... [155], Article 102.
- Page 139 *To the question* ... [92], Part B of the Introduction.
- Page 140 *The experiment is to me historically interesting,* ... [161], page 157.
- Page 141 *This invariance of the velocity* ... [173], Section 7a.
- Page 148 *The young scholar imagined* ... [173], Chapter 7.
- Page 149 *The chief cause of my failure* ... [173], Chapter 8.
- Page 149 *The views on space and time* ... [53], *Space and Time*.
- Page 150 *When in 1907* ... [173], Chapter 9.
- Page 151 *I had excellent teachers* ... [210], *Autobiographical Notes*.
- Page 153 *After dinner, the weather being warm,* ... [230].
- Page 153 *To see a World in a Grain of Sand* ... Blake, *Auguries of Innocence*.
- Page 155 *If a vessel, hung by a long cord,* ... [169], *Definitions*.
- Page 158 *Possibly he thinks* ... [63], *The Assayer*.
- Page 160 *That evening we attached a high-pitched whistle* ... Quoted in [7], page 40.
- Page 173 *Japanese mathematician Seki Kōwa* ... See [162].
- Page 174 *that whoever could make two ears of corn* ... Swift, *Voyage To Brobdingnag* in *Gulliver's Travels*.
- Page 177 *For steel spheres* ... [202], Volume 1, page 454.
- Page 181 *Convectional motions are hindered* ... [199], page 66.
- Page 181 *what can almost be regarded as an obsession* ... [187], page 71.
- Page 181 *Leonardo has filled the atmosphere* ... [34], page 245.
- Page 187 *Two slender bamboo poles* ... [202], Paper 1929:1.
- Page 187 *G. I. Taylor wrote a nice summary* ... In [233].
- Page 188 *I deeply appreciate the honour* ... Quoted in [7], page 122.
- Page 188 *Since the curve here* ... [233].
- Page 188 $\frac{4}{3}$ *was chosen partly as a rough mean* ... [202], Volume 1, 1926:1.
- Page 188 *As far as the War Office was concerned* ... Sutton in a letter printed in [182].
- Page 189 *there came a time of heart-break* ... Quoted in [181].
- Page 190 *eleven experts were required* ... to write Kolmogorov's obituary [113].
- Page 190 *The present author* ... In [130].

- Page 190 *In a later paper Kolmogorov says that ...* [126], Volume 1, Paper 58.

- Page 190 *In his first paper ...* [126], Volume 1, Paper 55.

- Page 191 *Kolmogorov's two papers on isotropic turbulence ...* Quoted in [113] pages 37–8.

- Page 192 *When we were schoolboys ...* Quoted in [181], page 547.

- Page 192 *The Richardson 4/3 diffusion law ...* [202], page 27.

- Page 193 *He loved his college ...* Quoted in [7], page 235.

- Page 197 *before a situation can be controlled ...* [200], page 226.

- Page 197 *It seemed that out of battle I escaped ...* Wilfred Owen, *Strange Meeting.*

- Page 206 *I tremble for my country ...* Jefferson, *Notes on Virginia,* Query XIIIV.

- Page 206 *had learnt to speak Esperanto ...* [7], page 57.

- Page 208 *He wrote a long paper ...* [202], Volume 2, Paper 1951:2.

- Page 218 *When every one is somebodee ...* Gilbert and Sullivan, *The Gondoliers.*

- Page 218 *asked by ... J. L. Synge ...* In [231]

- Page 218 *discussed at length ...* In [118]

- Page 220 *told anybody who would listen ...* [31], page 5.

- Page 220 *Mandelbrot's background ...* including, for example, [108].

- Page 223 *Murray starts out ...* in [166].

- Page 225 *The mechanism for the corrective adjustment ...* [250], pages 92–3.

- Page 225 *If a man does not keep pace ...* Thoreau, *Walden,* Conclusion.

- Page 226 *Now is this not very fine? ...* Quoted in [254] Chapter 10.

- Page 233 *Here begins the algorithm ...* Translated in [159].

- Page 234 *Experience has taught me ...* Quoted in [65].

- Page 237 *According to Knuth, ...* in Section 4.1 of Volume II of [121].

- Page 237 *'The Sand Reckoner' ...* In [90].

- Page 241 *upon a question which was wholly within the province of the administration ...* [29], pages 238–40.

- Page 246 *in any book on 'physics for musicians'* e.g. in Chapter 11 of [259].

- Page 249 *adepts find in mathematics delights ...* Quoted in [163].

- Page 258 *mobs of armed men ... .* Unfortunately this is undocumented and denied (see page 126 of [150]) but is probably a good reflection of his views. (Recall the denial that Bismarck

ever likened Lord Salisbury to a 'lath of wood painted to look
like iron.')

- Page 258 *first with the most*, Reported saying of Nathan Bedford
Forrest, a Southern cavalry commander.
- Page 258 *'Build no more fortresses, build railways,'* ... [242], Chap-
ter 6.
- Page 268 *Braess produced the following example* ... For references
see [112].
- Page 292 *He who fights with monsters* ... Nietzsche, *Beyond Good
and Evil*, Chapter IV, 146.
- Page 292 *Both direct and indirect approaches were tried* ... [37].
- Page 293 *Mathematicians are like Frenchmen* ... Maxim 1279,
quoted in [163].
- Page 298 *accidents were not as frequent as might be expected.* [119].
- Page 300 *named after the accident* ... [171].
- Page 301 *estimated that it would take five years* ... [6].
- Page 301 *Studies have shown that for every six* ... [68].
- Page 304 *The author feels that* ... [122], Preface.
- Page 305 *Here a difficulty presents itself* ... [61], First Day.
- Page 311 *Hodge's splendid biography of Turing* [96],
- Page 325 *the major codes of the major powers* ... [110], page 49.
- Page 327 *followed these instructions to the letter* ... See [110],
page 119.
- Page 331 *We intend to begin unrestricted submarine warfare* ...
Text given in [241].
- Page 335 *We developed a very friendly feeling* ... [253], page 132.
- Page 340 *A retired banker named Burberry* ... [160], page 35.
- Page 341 *sensed something strange in an Italian intercept* ... [110],
page 139.
- Page 341 *a procedure known as "gardening"* ... [94], page 122.
- Page 342 *Difficulty arises in remunerating the patentees* ... Quo-
tations taken from [236], pages 40–1.
- Page 343 *Preventing A Code Book From Falling Into Enemy Hands*.
The cartoon comes, via page 485 of [109], from the British
Admiralty, *Merchant Ships Signal Book*, III.
- Page 343 *Did nothing in particular* ... Gilbert and Sullivan, *Iolan-
the*.
- Page 345 *We were amazed to be presented with twenty* ... [94],
page 130.
- Page 346 *the Swiss, Spanish, and Italian governments* ... According
to [132].
- Page 348 *a stunning achievement* ... [110], page 66.

- Page 353 *though the cryptological agencies ... grew in size ...* [109], page 455.
- Page 353 *As far as mathematical training goes ...* [132].
- Page 362 *after the meeting, Dilly returned ...* [94], page 127.
- Page 364 *Our decodes must have given early warning ...* [253], page 96.
- Page 366 *I discovered, however, that the intercept operators ...* [253], page 93.
- Page 366 *The old procedures ...* [165], page 139.
- Page 366 *You know, the Germans don't mean you to read their stuff ...* [94], page 236.
- Page 368 *In Britain , Cambridge students ...* [110], page 280.
- Page 368 *viewed with considerable awe ...* [160], pages 77–8.
- Page 368 *There was always a sense of immense power ...* and succeeding quotations [93], page 48–52.
- Page 370 *assembled out of standard parts for special purposes ...* [109], page 440.
- Page 378 *The conditions of the work were disclosed ...* and succeeding quotations [94] Chapter 17.
- Page 379 *Helen Rance, one of the hundreds ...* [102], pages 395–6.
- Page 380 *It was possible to choose a sequence ...* [253], page 144.
- Page 380 *the British Navy reacted strongly and unfavourably ...* See [11], page 208.
- Page 381 *The oldest hath borne most ...* Shakespeare, *King Lear.*
- Page 381 At a rough estimate ... Based, for example, on the figures given on pages 216 and 217 of [110].
- Page 381 *the sense of shock ...* [94], Chapter 3.
- Page 383 *By April 1943 the average kill ...* Quoted in [248], page 444.
- Page 384 *The Admiralty ... recorded that ...* [204], Volume 2, pages 367–8.
- Page 384 *It seems not unlikely that a quarter ...* [12].
- Page 384 *poverty line for a family of five ...* Based on the discussion in [21].
- Page 384 *These, in the day when heaven was falling, ...* Housman, *Epitaph on an Army of Mercenaries.*
- Page 385 *Not only did returning U-boats always signal ...* [95], Volume 2, pages 549–50.
- Page 388 *seemed to be lined with spy novels ...* Quoted in [86], page 639.
- Page 388 *Some U-boat commanders returning from patrol ...* [258], page 250.

- Page 388 *I received a buff-coloured envelope ...* Quoted in [189], pages 123–4.
- Page 391 *The nations of Western Europe ...* See [198].
- Page 391 *It was lucky that the security people didn't know ...* [160], page 34.
- Page 392 *supposing there were a machine ...* Section 17 of the Monadology quoted in [129].
- Page 393 *It is clear, I think, that we may account ...* [238], page 271.
- Page 393 *Laplace once went in form ...* [47], Volume 2, pages 1–2.
- Page 398 *Shannon thought they should play music to it ...* [96], page 251.
- Page 399 *into single letters using tables ...* [110], Appendix.
- Page 399 *The redundancy of language ...* [216].
- Page 400 *Normal infants begin to acquire real speech ...* [139], Chapter 23.
- Page 407 *they swept the contents of his desk into sacks ...* An exageration, but not too far from the truth, see [239].
- Page 410 *[a]t quarter past midnight ...* [239], page 123.
- Page 413 *Generations pass while some trees stand ...* Sir Thomas Browne, *Urn Burial (Hydriotaphia)*, Chapter V.
- Page 413 *The decay of the families of men ...* [64] reprinted in [219].
- Page 417 *The Microbe is so very small ...* Belloc, *The Microbe* in *Cautionary Verses.*
- Page 430 *Table 17.1 ...* taken from a paper of Radovich ([69], pages 110–11),
- Page 433 *Between 1964 and 1968 Norwegians invested ...* [69], page 53.
- Page 435 *rapidly changed to dark wings.* Details are given in [116].
- Page 435 *[t]he identity of many plants ...* Darwin [44], Chapter XI.
- Page 445 *The Indians die so easily ...* [157], page 211.
- Page 445 *Two Union soldiers ...* See, for example [158], Chapter 15.
- Page 445 *The worst case was presented ...* [36], page 104.
- Page 446 *Our business was done at the river's brink ...* Browning, *The Pied Piper of Hamelin.*
- Page 449 *In scarcely any house did only one die ...* Quoted in [218], page 48.
- Page 450 *Perhaps one third of the population ...* See [89] for a discussion of the English figures.

- Page 450 *not a disease of human beings at all* ... [25], page 225.
- Page 450 *It is possible that we have acquired measles* ... See [36].
- Page 450 *The most likely forecast* ... [25], page 263.
- Page 450 *During the last two decades* ... [48], Introduction.
- Page 451 *He knew the tale he had to tell* ... Translated by Stuart Gilbert in the Penguin Modern Classics series.
- Page 460 *I have a cat called Socrates.* Ionesco, *Rhinoceros*, Act 1.
- Page 465 *Ah! Why, ye Gods* ... Pope, *The Dunciad*, Book 2.
- Page 468 *The method of "postulating"* ... [207], Chapter VII.
- Page 472 *civilisation advances by extending* ... [255].
- Page 482 *checked by machine* ... [247].
- Page 488 *I am very honoured to be the first recipient* ... [156].
- Page 494 *At a literary dinner* ... From *The Life and Letters of Lord Macaulay* by G. O. Trevelyan, Volume II, Chap.XII.
- Page 495 *of no writing — except perhaps Henry James's* ... [75].
- Page 501 *Insiders pronounce the χ of T$_E$X* ... Chapter 1 of [122].
- Page 501 *man's reach* ... Browning, *Andrea del Sarto*.
- Page 506 *When I was a rather disrespectful student* ... [140], Volume 5, pages 310–11.
- Page 508 *I am Alpha and Omega*, Revelation 1:7
- Page 511 *Jalal, an old friend* ... Told as *The Robe* in [214].
- Page 517 *lath of wood* ... See the second edition of the *Oxford Dictionary of Quotations*.
- Page 521 *A man will turn over half a library to make one book.* Boswell, *Life of Johnson* April, 1775.

BIBLIOGRAPHY

[1] M. Abramowitz and I. A. Stegun. *Handbook of Mathematical Functions*. Dover, New York, 1965. The various Dover printings of this are reprints of various Government printings.

[2] F. S. Acton. *Numerical Methods That Work*. Harper and Row, New York, 1970.

[3] D. J. Albers and G. L. Anderson, editors. *Mathematical People*. Birkhäuser, Boston, 1985.

[4] R. M. Anderson and R. M. May. *Infectious Diseases of Humans*. OUP, Oxford, 1991.

[5] V. I. Arnol'd. *Catastrophe Theory*. Springer, Berlin, 1983. Translated from the Russian. Later editions contain interesting extra material.

[6] C. Arthur. Pressurised managers blamed for ambulance failure. *New Scientist*, page 5, 6 March 1993.

[7] O. M. Ashford. *Prophet or Professor? (The Life and Work of Lewis Fry Richardson)*. Adam Hilger, Bristol, 1985.

[8] N. T. J. Bailey. *The Mathematical Theory of Epidemics*. Charles Griffin, London, 1957.

[9] W. W. Rouse Ball and H. S. M. Coxeter. *Mathematical Recreations and Essays*. University of Toronto, Toronto, 12th edition, 1974.

[10] G. K. Batchelor. G. I. Taylor. *Obituary Notices of Fellows of the Royal Society*, 30:565–633, 1985.

[11] P. Beesly, editor. *Very Special Intelligence*. Sphere, London, revised edition, 1978.

[12] C. B. A. Behrens. *Merchant Shipping and the Demands of War*. HMSO, London, 1955.

[13] E. T. Bell. *Men of Mathematics*. Simon and Schuster, New York, 1937. 2 vols.

[14] E. R. Berlekamp, J. H. Conway, and R. K. Guy. *Winning Ways*. Academic Press, London, 1982. 2 vols.

[15] G. Birkhoff. *Hydrodynamics*. Princeton University Press, Princeton, N. J., 1950.

[16] G. Birkhoff and S. MacLane. *A Survey of Modern Algebra*. Macmillan, New York, revised edition, 1953.

[17] P. M. S. Blackett. *Studies of War*. Oliver and Boyd, Edinburgh, 1962.

[18] M. Born. *Einstein's Theory of Relativity*. Dover, New York, revised edition, 1962.

[19] A. Bourke. *The Visitation of God? The Potato and the Irish Famine*. Lilliput Press, Dublin, 1993.

[20] E. G. Bowen. *Radar Days*. Adam Hilger, Bristol, 1987.

[21] A. Briggs. *A Social History of England*. Weidenfeld and Nicholson, London, 1983.

[22] W. K. Bühler. *Gauss, A Biographical Study*. Springer, Berlin, 1981.

[23] J. P. Bunker, B. A. Barnes and F. Mosteller, editors. *Costs, Risks and Benefits of Surgery*. OUP, Oxford, 1977.

[24] J. C. Burkill. *A First Course in Mathematical Analysis*. CUP, Cambridge, 1962.

[25] M. Burnett and D. O. White. *Natural History of Infectious Disease*. CUP, Cambridge, 4th edition, 1971.

[26] R. Burns, editor. *Radar Development to 1945*. Peter Peregrinus (on behalf of IEE), London, 1988.

[27] D. M. Campbell and J. C. Higgins. *Mathematics: People, Problems, Results*. Wadsworth, Belmont, Calif, 1984. 3 vols.

[28] CAST. Preliminary Report: Effect of encaide and flecainide on mortality in a randomised trial of arrythmia suppression after myocardial infarction. *New England Journal of Medicine*, 312:406–12, 1989.

[29] Somerset De Chair, editor. *Napoleon's Memoirs*. Faber and Faber, London, 1958.

[30] A. Charlesworth. Infinite loops in computer programs. *Mathematics Magazine*, 52:284–91, 1979.

[31] G. Cherbit, editor. *Fractals*. Wiley, New York, 1991.

[32] L. Childs. *A Concrete Introduction To Higher Algebra*. Springer, Berlin, 1989.

[33] W. S. Churchill. *The Second World War*. Cassel, London, 1948-53. 6 vols.

[34] K. Clark. *Leonardo da Vinci*. Penguin, Harmondsworth, England, 1989. Revised Pelican edition of a work first published by CUP in 1939.

[35] R. W. Clark. *Tizard*. Methuen, London, 1965.

[36] A. Cliff, P. Hagett, and M. Smallman-Raynor. *Measles, An Historical Geography*. OUP, Oxford, 1988.

[37] W. H. Cockroft *et al*. *Mathematics Counts*. Her Majesty's Stationery Office, London, 1982. Report of a Committee of Enquiry into the Teaching of Mathematics in Schools.

[38] E. Colerus. *From Simple Numbers to the Calculus*. Heinemann, London, 1955. English translation from the German.

[39] F. M. Cornford. *Microcosmographia Academica*. Bowes and Bowes, Cambridge, 2nd edition, 1922.

[40] R. Courant and H. Robbins. *What is Mathematics?* OUP, Oxford, 1941.

[41] H. S. M. Coxeter. *Introduction to Geometry*. Wiley, New York, 1961.

[42] H. M. Cundy and A. P. Rollett. *Mathematical Models*. OUP, Oxford, 1951.

[43] J. Darracott. *A Cartoon War*. Leo Cooper, London, 1989.

[44] C. Darwin. *On the Origin of Species by Means of Natural Selection*. Murray, London, 1859.

[45] H. Davenport. *The Higher Arithmetic*. Hutchinson, London, 1952.

[46] P. J. Davies and R. Hersh. *The Mathematical Experience*. Birkhäuser, Boston, 1981.

[47] A. De Morgan. *A Budget of Paradoxes*. Books for Libraries Press, Freeport, New York, second, reprinted edition, 1969.

[48] R. S. Desowitz. *The Malaria Capers*. Norton, New York, 1991.

[49] K. Dönitz. *Memoirs*. Weidenfeld and Nicholson, London, 1954. English translation.

[50] S. Drake. *Galileo at Work*. University of Chicago Press, Chicago, 1978.

[51] C. V. Durell. *General Arithmetic for Schools*. G.Bell, London, 1936.

[52] A. Einstein and L. Infeld. *The Evolution of Physics*. CUP, Cambridge, 1938.

[53] A. Einstein, H. A. Lorentz, H. Weyl and H. Minkowski. *The Principle of Relativity*. Dover, 1952. A collection of papers first published in English by Methuen in 1923.

[54] R. J. Evans. *Death in Hamburg*. OUP, Oxford, 1987.

[55] J. Fauvel and J. Gray, editors. *History of Mathematics: A Reader*. Macmillan, Basingstoke, England, 1987.

[56] W. Feller. *An Introduction to Probability Theory and its Applications*, volume I. Wiley, New York, 3rd edition, 1968.

[57] R. P. Feynman. *The Character of Physical Law*. BBC Books, London, 1965.

[58] R. P. Feynman. *QED, The Strange Theory of Light and Matter*. Princeton University Press, Princeton, N. J., 1985.

[59] R. P. Feynman. *Surely You're Joking, Mr Feynman!* W. W. Norton, New York, 1985.

[60] R. P. Feynman *et al*. *The Feynman Lectures on Physics*. Addison-Wesley, Reading, Mass., 1963. 3 vols.

[61] Galileo. *Dialogues Concerning Two New Sciences*. Macmillan, London, 1914. Translated by H. Crew and A. De Salvio.

[62] Galileo. *Dialogues Concerning the Two Chief World Systems*. University of California Press, Berkeley, Calif, 1953. Translated by S. Drake.

[63] Galileo. *Discoveries and Opinions of Galileo*. Anchor Books (Doubleday), New York, 1957. Translated by S. Drake.

[64] F. Galton and H. W. Watson. On the probability of the extinction of families. *Journal of The Anthropological Institute*, 4:138–44, 1874.

[65] M. Gardner. *Fads and Fallacies in the Name of Science*. Dover, New York, 1957.

[66] M. Gardner. *The Flight of Peter Fromm*. William Kaufman, Los Altos, California, 1973.

[67] M. Gardner. *Penrose Tiles to Trapdoor Ciphers*. W. H. Freeman, New York, 1989.

[68] W. W. Gibbs. Software's chronic crisis. *Scientific American*, pages 72–81, September 1994.

[69] M. H. Glantz and J. D. Thompson, editors. *Resource Management and Environmental Uncertainty*. Wiley, New York, 1989.

[70] E. Gold. L. F. Richardson. *Obituary Notices of Fellows of the Royal Society*, 9:217–35, 1954.

[71] M. Goosens, F. Mittelbach, and A. Samarin. *The LATEX Companion*. Addison-Wesley, Reading, Mass., 1994.

[72] S. J. Gould. *Ever Since Darwin*. Penguin, Harmondsworth, England, 1980.

[73] M. Gowing. *Independence and Deterrence*, volume I. Macmillan, London, 1974.

[74] R. L. Graham, D. E. Knuth, and O. Patashnik. *Concrete Mathematics*. Addison-Wesley, Reading, Mass., 1979. The second edition contains interesting new material.

[75] G. Greene. The austere art. *The Spectator*, 165:682, 1940.

[76] J. Gross. *The Rise and Fall of the Man of Letters*. Weidenfeld and Nicholson, London, 1969.

[77] J. Grossman, editor. *The Chicago Manual of Style.* Chicago University Press, Chicago, 14th edition, 1993.

[78] J. Hadamard. *The Psychology of Invention in the Mathematical Field.* Princeton University Press, Princeton, N. J., 1945.

[79] J. B. S. Haldane. *On Being the Right Size.* OUP, Oxford, 1985.

[80] P. R. Halmos. *Naive Set Theory.* Van Nostrand, Princeton, N. J., 1960.

[81] P. R. Halmos. *I Want to be a Mathematician.* Springer, Berlin, 1985.

[82] J. M. Hammersley. On the enfeeblement of mathematical skills by 'modern mathematics' and by similar soft intellectual trash in schools and universities. *Bulletin of the Institute of Mathematics and its Applications,* 4:68–85, 1968.

[83] G. H. Hardy. *A Course of Pure Mathematics.* CUP, Cambridge, 1914.

[84] G. H. Hardy. *A Mathematician's Apology.* CUP, Cambridge, 1940.

[85] G. H. Hardy and E. M. Wright. *An Introduction to the Theory of Numbers.* OUP, Oxford, 1938.

[86] A. D. Harvey. *Collision of Empires.* Phoenix, London, paperback edition, 1994.

[87] J. Hašek. *The Good Soldier Švejk.* Penguin, Harmondsworth, England, 1973. English translation by C. Parrott.

[88] M. Hastings. *Bomber Command.* Michael Joseph, London, 1979.

[89] J. Hatcher. *Plague, Population and the British Economy 1348–1530.* MacMillan, London, 1977.

[90] T. L. Heath. *The Works of Archimedes.* Dover, New York, reprint edition, 1953.

[91] T. L. Heath. *The Thirteen Books of Euclid's 'Elements'.* Dover, New York, reprint edition, 1956. 3 vols.

[92] H. Hertz. *Electric Waves.* Dover, New York, reprint edition, 1963. English translation by D. E. Jones first published in 1893.

[93] P. J. Hilton. Algebra and logic, Lecture notes in mathematics 450. In J. N. Crossley, editor, *Algebra and Logic,* Berlin, 1975. Springer.

[94] F. H. Hinsley and A. Stripp, editors. *Codebreakers.* OUP, Oxford, paperback edition, 1994.

[95] F. H. Hinsley, E. E. Thomas, C. F. G. Ransome and R. C. Knight. *British Intelligence in the Second World War.* HMSO, London, 1979-88. 5 volumes.

[96] A. Hodges. *Alan Turing, The Enigma of Intelligence.* Hutchinson, London, 1983.

[97] M. Howell and P. Ford. *The Ghost Disease.* Penguin, Harmondsworth, England, 1986.

[98] D. Howse. *Radar At Sea.* Macmillan, London, 1993.

[99] D. Huff. *How to Lie with Statistics.* W. W. Norton, New York, 1954.

[100] ISIS-2 Collaborative Group. Randomised trial of intravenous streptokinase, oral aspirin, both, or neither among 17 187 cases of suspected acute myocardial infarction. *The Lancet,* pages 349–60, August 1988.

[101] ISIS-3 Collaborative Group. A randomised comparison of streptokinase versus tissue plasminogen activator versus anistreptase and of aspirin plus heparin versus aspirin alone among 41 299 cases of suspected acute myocardial infarction. *The Lancet,* 339:753–70, March 1992.

[102] B. Johnson. *The Secret War.* BBC Publications, London, 1978.

[103] R. V. Jones. Winston Churchill. *Obituary Notices of Fellows of the Royal Society,* 12:35–106, 1966.

[104] R. V. Jones. *Most Secret War.* Hamish Hamilton, London, 1978.

[105] R. V. Jones. *Reflections On Intelligence.* Mandarin, London, paperback edition, 1990.

[106] C. M. Jordan. *Cours d'Analyse de l'Ecole Polytechnique.* Gauthier-Villars, Paris, 2nd edition, 1893–6. 3 vols.

[107] M. Kac. *Selected Papers.* MIT, Cambridge, Mass, 1979.

[108] J.-P. Kahane and R. Salem. *Ensembles parfaits et séries trigonométriques.* Herman, Paris, 1963.

[109] D. Kahn. *The Codebreakers.* Macmillan, 1967.

[110] D. Kahn. *Seizing the Enigma.* McGraw Hill, 1982.

[111] J. Keegan. *The Price of Admiralty.* Hutchinson, London, 1988.

[112] F. P. Kelly. Network routing. *Philosophical Transactions of the Royal Society A,* 337:343–67, 1991.

[113] D. Kendall *et al.* Obituary of A. N. Kolmogorov. *Bulletin of the London Mathematical Society,* 22:31–100, 1990.

[114] D. G. Kendall. Branching processes since 1873. *Journal of The London Mathematical Society,* 41:385–406, 1966.

[115] D. G. Kendall. The genealogy of genealogy . . . . *Bulletin of The London Mathematical Society,* X:225–53, 1975.

[116] B. Kettlewell. *The Evolution of Melanism.* OUP, Oxford, 1973.

[117] C. Kinealy. *This Great Calamity.* Gill and Macmillan, Dublin, 1994.

[118] J. F. C. Kingman. The thrown string. *Journal of the Royal Statistical Society, Series B,* 44(2):109–38, 1982. With discussion.

[119] G. M. Kitchenside and A. Williams. *British Railway Signalling*. Ian Allan, Shepperton, Surrey, UK, 3rd edition, 1979.

[120] M. Kline. *Mathematical Thought from Ancient to Modern Times*. OUP, Oxford, 1972.

[121] D. E. Knuth. *The Art of Computer Programming*. Addison-Wesley, Reading, Mass., 1968–73. 3 volumes. This was intended to be 7 volumes. As I go to press there are persistent rumours that Knuth will take up the task again.

[122] D. E. Knuth. *The TEXbook*. Addison-Wesley, Reading, Mass., 1984.

[123] D. E. Knuth. *Computers and Typesetting*. Addison-Wesley, Reading, Mass., 1986. 5 volumes. Volume A of the set is *The TEXbook*.

[124] A. H. Koblitz. *A Convergence of Lives*. Birkhäuser, Boston, 1983.

[125] N. Koblitz. *A Course in Number Theory and Cryptography*. Springer, Berlin, 1987.

[126] A. N. Kolmogorov. *Selected Works*. Kluwer Academic Publishers, Dordrecht, The Netherlands, 1991. 3 vols. Annotated translation of the Russian.

[127] A. G. Konheim. *Cryptography*. Wiley, New York, 1981.

[128] S. Körner. *The Philosophy of Mathematics*. Hutchinson, London, 1960.

[129] S. Körner. *Fundamental Questions of Philosophy*. Penguin, Harmondsworth, England, 1969.

[130] T. W. Körner. Uniqueness for trigonometric series. *Annals of Mathematics*, 126:1–34, 1987.

[131] S. Kovalevskaya. *A Russian Childhood*. Springer, Berlin, 1978. Translated from the Russian by B. Stillman.

[132] W. Kozaczuk. *Enigma, How the German Cypher Machine was Broken*. Arms and Armour, London, 1984. Translated from the Polish by C. Kasparek.

[133] S. G. Krantz. *How to Teach Mathematics: A Personal Perspective*. AMS, Providence, Rhode Island, 1993.

[134] E. Laithwaite. *Invitation to Engineering*. Blackwell, Oxford, 1984.

[135] L. Lamport. *LATEX, A Documement Preparation System*. Addison-Wesley, Reading, Mass., 2nd edition, 1994.

[136] F. W. Lanchester. *Aircraft in Warfare*. Constable, London, 1916.

[137] E. Landau. *Foundations of Analysis*. Chelsea, New York, 1951. Translated from the German by F. Steinhardt.

[138] H. Lauwrier. *Fractals*. Princeton University Press, Princeton, N. J., 1991. Translated from the Dutch.

[139] P. Leach. *Babyhood*. Penguin, Harmondsworth, England, 2nd edition, 1983.

[140] H. Lebesgue. *Oeuvres Scientifiques*. L'Enseignement Mathématique, Geneva, 1972–3. 5 vols.

[141] I. Lengyel, S. Kádár and I. R. Epstein. Transient Turing structures in a gradient-free closed system. *Science*, 259:493–5, 1993.

[142] M.-C. Van Leunen. *A Handbook for Scholars*. Alfred A. Knopf, Inc., New York, 1978.

[143] P. Levi. *The Wrench*. Abacus (part of the Penguin group), London, 1988. English translation by W. Weaver.

[144] C. E. Linderholm. *Mathematics made Difficult*. Wolfe Publishing Ltd, London, 1971.

[145] J. E. Littlewood. *A Mathematician's Miscellany*. CUP, Cambridge, 2nd edition, 1985. Editor B.Bollobás.

[146] J. L. Locher *et al. Escher*. Thames and Hudson, London, 1982. Translated from the Dutch. Contains a complete illustrated catalogue of his graphic works.

[147] J. D. Logan. *Applied Mathematics*. Wiley, New York, 1987.

[148] E. N. Lorenz. *The Essence of Chaos*. University of Washington Press, 1994.

[149] B. Lovell. P. M. S. Blackett. *Obituary Notices of Fellows of the Royal Society*, 21:1–115, 1975.

[150] J. Luvaas. *The Military Legacy of the Civil War*. Kansas University Press, Kansas, 1988.

[151] C. H. Macgillavry. *Symmetry Aspects of M. C. Escher's Periodic Drawings*. A. Oosthoek's Uitgeversmaatschappij NV for the International Union of Crystallography, Utrecht, 1965.

[152] A. J. Marder. *From the Dreadnought to Scapa Flow*. OUP, Oxford, 1961–70. 5 vols.

[153] E. Marshall, editor. *Longman Crossword Key*. Longman, Harlow, 1982.

[154] E. A. Maxwell. *Fallacies in Mathematics*. CUP, Cambridge, 1959.

[155] J. C. Maxwell. *Matter and Motion*. Dover, New York, 1991. Reprint of the 1920 edition, the first edition was published in 1877.

[156] D. McDuff. Satter prize acceptance speech. *Notices of the AMS*, 38(3):185–7, March 1991.

[157] W. H. McNeill. *Plagues and Peoples*. Blackwell, Oxford, 1976.

[158] J. M. McPherson. *Battle Cry of Freedom*. Blackwell, Oxford, 1993.

[159] K. Menninger. *Number Words and Number Symbols*. MIT Press, Cambridge, Mass, 1969.

[160] N. Metropolis, J. Howlett, and Gian-Carlo Rota, editors. *A History of Computing in the Twentieth*

Century. Academic Press, London, 1980.

[161] A. A. Michelson. *Light Waves and Their Uses.* Chicago University Press, Chicago, 1903.

[162] Y. Mikami. *The Development of Mathematics in China and Japan.* Open Court, Chicago, 1914. There is a Chelsea reprint.

[163] R. E. Moritz. *On Mathematics and Mathematicians.* Dover, New York, 1958. Reprint of the original 1914 edition.

[164] R. J. Morris. *Cholera 1832.* Croom Helm, London, 1976.

[165] M. Muggeridge. *Chronicles of Wasted Time*, volume 2. Collins, London, 1973.

[166] C. D. Murray. The physiological principle of minimum work (Part I). *Proceedings of the National Acadamy of Sciences of the USA*, 12:207–14, 1926.

[167] J. D. Murray. *Mathematical Biology.* Springer, Berlin, 1989.

[168] J. R. Newman, editor. *The World of Mathematics.* Simon and Schuster, New York, 1956. 4 vols. Reprinted by Tempus Books, Washington in 1988.

[169] I. Newton. *Principia.* University of California Press, Berkeley, Calif, 1936. Motte's translation revised by Cajori.

[170] Nobel Foundation. *Nobel Lectures in Physics 1942–62*, Amsterdam, 1964. Elsevier.

[171] O. S. Nock. *Historic Railway Disasters.* Ian Allan, Shepperton, Surrey, UK, 1966.

[172] P. Padfield. *Dönitz.* Gollancz, London, paperback edition, 1993.

[173] A. Pais. *Subtle Is the Lord ... .* OUP, Oxford, 1982.

[174] A. Pais. *Inward Bound.* OUP, Oxford, 1986.

[175] D. Pedoe. *The Gentle Art of Mathematics.* English Universities Press, London, 1958.

[176] H.-O. Peitgen and P. H. Richter. *The Beauty of Fractals.* Springer, Berlin, 1986.

[177] R. Penrose. *The Emperor's New Mind.* OUP, Oxford, 1989.

[178] Petrarch. On his own ignorance and that of many others. In *The Renaissance Philosophy of Man.* Chicago University Press, Chicago, 1948.

[179] S. Pinker. *The Language Instinct.* Penguin, Harmondsworth, England, 1994.

[180] Plato. *Meno.* Penguin, Harmondsworth, England, 1956. Translated by W. K. Guthrie.

[181] G. W. Platzman. A retrospective view of Richardson's book on weather prediction. *Bulletin of the American Meteorological Society*, 48:514–50, 1967.

[182] G. W. Platzman. Richardson's weather prediction. *Bulletin of the American Meteorological Society*, 49:496–500, 1968.

[183] H. Poincaré. *Science and Method.* Dover, New York, reprint edition, 1952. Translated by F.Maitland.

[184] G. Pólya. *Mathematics and Plausible Reasoning.* Princeton University Press, Princeton, N. J., 1954. 2 vols.

[185] G. Pólya. *How to Solve It.* Princeton University Press, Princeton, N. J., 2nd edition, 1957.

[186] G. Pólya. *Mathematical Discovery.* Wiley, New York, 1962. 2 vols.

[187] A. E. Popham. *The Drawings of Leonardo da Vinci.* Cape, London, 1946.

[188] W. Poundstone. *Prisoner's Dilemma.* Doubleday, New York, 1992.

[189] A. Price. *Aircraft versus Submarine.* William Kimber, London, 1973.

[190] A. Price. *Instruments of Darkness.* Granada, London, 1979.

[191] S. Pritchard. *The Radar War.* Patrick Stephens, Wellingborough, Northamtonshire, England, 1989.

[192] H. Rademacher and O. Toeplitz. *The Enjoyment of Mathematics.* Princeton University Press, Princeton, N. J., 1957.

[193] A. S. Ramsey. *Dynamics (Part 1).* CUP, Cambridge, 1929.

[194] J. H. Randall. *The Making of the Modern Mind.* The Riverside Press, Cambridge, Massachusetts, 1940.

[195] E. Raymond. *The New Hacker's Dictionary.* MIT Press, Cambridge, Mass, 1991.

[196] J. Reader. *Missing Links.* Penguin, Harmondsworth, England, 2nd edition, 1988.

[197] C. Reid. *Hilbert.* Springer, Berlin, 1970.

[198] D. J. Revelle and L. Lumpe. Third World submarines. *Scientific American*, pages 16–21, August 1994.

[199] L. F. Richardson. *Weather Prediction by Numerical Process.* CUP, Cambridge, 1922.

[200] L. F. Richardson. *Arms and Insecurity.* Boxwood, Pittsburg, 1960.

[201] L. F. Richardson. *Statistics of Deadly Quarrels.* Boxwood, Pittsburg, 1960.

[202] L. F. Richardson. *Collected Works.* CUP, Cambridge, 1993. 2 vols.

[203] C. E. Rosenberg. *The Cholera Years.* University of Chicago Press, Chicago, 1962.

[204] S. W. Roskill. *The War At Sea, 1939–1945.* HMSO, London, 1954–61. 3 volumes in 4 parts.

[205] A. P. Rowe. *One Story of Radar.* CUP, Cambridge, 1948.

[206] M. J. S. Rudwick. *The Great Devonian Controversy.* University of Chicago Press, Chicago, 1985.

[207] B. Russell. *Introduction to Mathematical Philosophy*. George Allen and Unwin, London, 1919.

[208] T. Sandler and K. Hartley. *The Economics of Defence*. CUP, Cambridge, 1995.

[209] Admiral R. Scheer. *Germany's High Sea Fleet in the World War*. Cassell, London, 1920.

[210] P. A. Schillp, editor. *Albert Einstein: Philosopher-Scientist*. Library of Living Philosophers, La Salle, Ill, 1949.

[211] K. Schmidt-Nielsen. *Scaling*. CUP, Cambridge, 1984.

[212] W. A. Schocken. *The Calculated Confusion of Calendars*. Vantage Press, New York, 1976.

[213] M. R. Schroeder. *Number Theory in Science and Communication*. Springer, Berlin, 1984.

[214] I. Shah. *The Exploits of the Incomparable Mulla Nasrudin*. Jonathan Cape, London, 1966.

[215] C. E. Shannon. *Collected Papers*. IEE Press, 445 Hoes Lane, PO Box 1331, Piscataway, NJ, 1993.

[216] C. E. Shannon and W. Weaver. *The Mathematical Theory of Communication*. University of Illinois Press, Urbana, Ill, 1949.

[217] R. Sheckley. *Dimension of miracles*. Dell, New York, 1968.

[218] J. F. D. Shrewsbury. *A History of the Bubonic Plague in the British Isles*. CUP, Cambridge, 1971.

[219] D. Smith and N. Keyfitz. *Mathematical Demography*. Springer, Berlin, 1977.

[220] J. Maynard Smith. *Mathematical Ideas in Biology*. CUP, Cambridge, 1968.

[221] C. P. Smyth. *Our Inheritance in the Great Pyramid*. Alexander Strahan and Co., London, 1864.

[222] J. Snow. *Snow on Cholera*. The Commonwealth Fund, New York, 1936. Facsimile Reprint.

[223] South West Thames Regional Health Authority, 40 Eastbourne Terrace, London W2 3QR. *Report of the Inquiry into the London Ambulance Service*, 1993.

[224] M. Spivak. *Calculus*. Bejamin, New York, 1967.

[225] M. D. Spivak. *The Joy of TEX*. AMS, Providence, Rhode Island, 2nd edition, 1990.

[226] I. N. Steenrod, P. R. Halmos, *et al. How to Write Mathematics*. AMS, Providence, Rhode Island, 1973.

[227] H. Steinhaus. *Mathematical Snapshots*. OUP, New York, 3rd edition, 1969.

[228] I. N. Stewart. *Galois Theory*. Chapman and Hall, London, 1973.

[229] I. N. Stewart. *Game, Set and Math*. OUP, Oxford, 1989.

[230] W. Stukeley. *Memoirs of Sir Isaac Newton's Life*. Taylor and Francis, Red Lion Court, Fleet Street, London, 1936.

[231] J. L. Synge. Letter to the editor. *The Mathematical Gazette*, LII:165, February 1968.

[232] G. I. Taylor. *Scientific Papers of Sir Geoffrey Ingram Taylor*. CUP, Cambridge, 1958.

[233] G. I. Taylor. The present position in the theory of turbulent diffusion. *Advances In Geophysics*, 6:101–11, 1959.

[234] G. I. Taylor. Aeronautics before 1919. *Nature*, 233:527–9, 1971.

[235] G. I. Taylor. The history of an invention. *Eureka*, 34:3–6, 1971. c/o Business Manager, Eureka, The Arts School, Bene't Street, Cambridge, England.

[236] J. Terraine. *The Right of the Line*. Hodder and Stoughton, London, 1985.

[237] J. Terraine. *Business in Great Waters*. Leo Cooper Ltd, London, 1989.

[238] D. W. Thompson. *On Growth and Form*. CUP, Cambridge, 1966. This is an abridged edition edited by J. T. Bonner.

[239] T. M. Thompson. *From Error-Correcting Codes through Sphere Packings to Simple Groups*. Mathematical Association of America, Washington, 1983.

[240] J. Thurber. *The Beast in Me*. Hamish Hamilton, London, 1949.

[241] B. W. Tuchman. *The Zimmermann Telegram*. Constable, London, 1959.

[242] B. W. Tuchman. *August 1914*. Constable, London, 1962.

[243] E. R. Tufte. *The Visual Display of Quantitative Information*. Graphics Press, PO Box 430, Cheshire, Connecticut 06410, 1982.

[244] E. R. Tufte. *Envisioning Information*. Graphics Press, PO Box 430, Cheshire, Connecticut 06410, 1990.

[245] A. M. Turing. On computable numbers with an application to the Entscheidungsproblem. *Proceedings of the London Mathematical Society (2)*, 42:230–65, 1937. There are some minor corrections noted in the next volume of the *Proceedings*.

[246] A. M. Turing. *Collected Works*. North Holland, Amsterdam, 1992. 3 vols.

[247] L. S. van B. Jutting. *Checking Landau's "Grundlagen" in the AUTOMATH System*. Mathematisch Centrum, PO Box 4079, 1009 AB Amsterdam, The Netherlands, 1979.

[248] D. Van der Vat. *The Atlantic Campaign*. Hodder and Stoughton, London, 1988.

[249] D. Van der Vat. *The Pacific Campaign*. Hodder and Stoughton, London, 1991.

[250] S. Vogel. *Vital Circuits*. OUP, Oxford, 1992.

[251] C. H. Waddington. *OR in World War 2*. Elek Science, London, 1973.

[252] D. W. Waters. The science of admiralty. *The Naval Review*, LI:395–410, 1963. Continued in Volume LII, 1964, on pages 15–26, 179–94, 291–309 and 423–37.

[253] G. Welchman. *The Hut Six Story*. Allen Lane, London, 1982.

[254] R. S. Westfall. *Never At Rest*. CUP, Cambridge, 1980.

[255] A. N. Whitehead. *An Introduction to Mathematics*. Williams and Norgate, London, 1911.

[256] C. M. Will. *Was Einstein Right?* OUP, Oxford, 1988.

[257] D. Wilson. *Rutherford*. Hodder and Stoughton, London, 1983.

[258] J. Winton. *Convoy*. Michael Joseph, London, 1983.

[259] A. Wood. *The Physics of Music*. Methuen, London, 1944.

INDEX

al-Khowârizmî, gives two words to
European languages, 232
algorithm
binary chop, 286
Euclid's
and Bezout's theorem, 246–248,
396, 471
competitor, 248
exposition of, 242–245
remark, 470
speed of, 245–246, 254–257
finding largest
bubble max, 277
knock-out, 277–281
for checking all algorithms, *see*
Turing theorem
Ford–Fulkerson, 260–267
general requirements for, 275
sorting
bubble sort, 282
comparative speed, 284
in general, 276
knock-out, 283–284, 288
maximum speed, 286–289
Stein's, 248–249
alphabet soup, 508–509
American Civil War, 200, 206, 258, 445
Appleyard, R., lost work of, 32, 82
arms races, 194–197
arteries
as Richardson–Kolmogorov cascade,
221
Murray's law, 221–225
ASDIC, 33
axioms
for integers
addition, 465
are they sufficient?, 479–480
fundamental, *see* fundamental
axiom

multiplication, 465
order, 467
needed, 463
no cause for alarm, 477

bamboos and primes, 107
bearing from range, 50, 53
Bezout's theorem, 246–248, 396, 471
Bismarck, battleship, 58, 365
Bismarck, O. von, statesman, 517
Blackett, P. M. S.
AA adviser, 57–61, 63
and bomber offensive, 84–86
and convoy, 73–84
and nuclear weapons, 86
at Cambridge, 43
at Coastal Command, 63–73, 87
circus, 57, 92
in navy, 38–39
joins Tizard committee, 45
long range aircraft to close 'Gap', 76
Nobel Prize, 43
on operational research, 63
radio direction finding, 385
research, 43–44, 87
social conscience, 43, 44, 87, 93
sources of information, 381
Bletchley, *see also* Turing bombe
and Churchill, 366
breaks SHARK, 387
conditions of work, 378–381
exploits procedural failures, 364
German Naval Codes, 366, 380
Polish gift, 362, 363, 366
recruitment, 363, 366
secret kept, 340, 391
Short Weather Cipher, 380
text setting problem, 377–378
U-boat blackout, 381–382

value of results, 365–366, 381, 387,
389
worst failure, 388
Zygalski sheets
exploited, 363
rendered useless, 363
blimp, 74, 88
bombe, *see* Turing bombe
bomber offensive, 84–86
Braess's paradox, 268–275
British Army, indifference to elements,
129
British Navy
and radar, 49
battles against British Air Force, 33,
75
code problems, 35, 343, 389
shell problems, 39–41
strengths, 24, 41, 43, 84
traditions of, 319, 379
weaknesses, 25, 30, 32, 33, 38–42

calendar, reform of, 295–297
can machines think?, 391–392, 398, 482
Cantor, G.
aleph, ℵ, 509
diagonal argument, *see* diagonal
argument
Carroll, Lewis, favourite puzzle of, 251,
462
Cassini's identity, 251–252
catastrophe theory, 227
causes of World War I, 195, 205,
258–259
cavity magnetron of Boot and Randall,
55
centre of universe
misplaced by 250 metres, 496
non-existent, 157
chaos, 227
Cherwell, Lord, *see* Lindemann, F. A.

Chesterfield, Lord, worldly wisdom, 295–296

cholera, 3–14, 16–17

Churchill, W. S., 36, 76, 366, 398

cipher, 233, *see also* code

clot busters, 14–20

coast of Britain, 210

code, *see also* Shannon theorem
  and cipher, 319
  Caesar, 319–320
  combined rotation and substitution, 325–330
  cycle representation, 349–352
  decipherment
    by frequency count, 321–322, 324, 325, 330, 345
    by probable word, 327, 335, 341–342, 367, 377
  Enigma, *see under that heading*
  error correcting, 407, 408
  error detecting, 408
  Hamming, 407, 408
  ISBN, 407
  mechanisation, 370
  Rivest, Shamir and Adleman, 395–398
  rotation, 324
  Short Weather Cipher, 380, 387, 399
  simple substitution, 320–324

compound interest, 96, 238

computation, plunging cost of, 173, 283

conversation, art of, 226

convoy
  human cost, 36, 384
  introduced, 26, 28, 31
  losses reduced by Bletchley, 381
  system in crisis, 382–387
  theory of, 23, 25–29, 71–73, 79–84
  time scales, 338
  victory, 387–390
  Waters' study of, 83

Conway, J. H.
  changes life, 410
  exercises self-denial, 136
  knows day of the week, 297
  rare intuition, 503

crossword puzzles, why possible, 399

Darwin C., *Origin of Species*, 293, 435

death ray, 45

diagonal argument

applied to reals, 307–309
applied to suite of programs, 310
in Turing's theorem, 314–315
new mode of argument, 309

differential geometry, founded by Gauss and Riemann, 151

dimensional analysis
  atomic explosion, 133
  bifilar pendulum, 122
  flow in pipe, 121, 223
  four-thirds rule, 191
  helicopter, 120
  rationale for, 116–118
  ship designs, 124
  simple pendulum, 120, 126, 174
  water waves, 121
  wind tunnels, 122, 124

diseases, *see also* epidemics
  origins, 448–450

Dönitz, K., 28, 32, 34, 37, 72, 73, 75, 381, 382
  admits defeat, 387
  anti-convoy tactics, 34–35, 385
  possible weakness, 35, 388
  strengths, 35

drug resistance, 435, 451

drunkard's walk, 183–186

Einstein, A.
  genesis of
    general relativity, 150–152
    special relativity, 139, 141
  mass energy equivalence, 44, 148
  on relative motion, 139–141
  photon theory, 128
  splendid biography of, 153
  unsatisfactory student, 149

Enigma, *see also* Bletchley *and* plugboard
  advantages, 331, 334
  and the Poles
    bomby, 337
    females, 354–358
    gift to Britain, 362
    mathematical attack, 319
    recover daily setting, 354–358
    recover wiring, 348, 362
    resources outstripped, 362
    Zygalski sheets, 358, 362
  and Typex, 342

Commercial, 340, 342
development
  fourth rotor, 381
  interchangeable rotors, 342
  invented by Scherbius, 331–332
  plugboard, 344–345
  reflector added, 339
initial British failure, 333, 348, 362
intelligence used, 346
Italian, broken, 341, 346
Military, 344
mock
  in bombe, 370
  use of, 335–337
non-self-encipherment, weakness, 341–342, 344
operator weaknesses, 341, 345
represented algebraically, 338–340, 344, 348–352, 355–356
secrets partially revealed, 348
self-inverse, weakness, 341, 344, 371
slow rotors, fundamental weakness, 334–337, 341, 370
text setting
  as separate problem, 377–378
  German Army procedure, 347
  Polish attack on, 354–358
  problem, 346
three-rotor Primitive, 332, 333

epidemics
  and Spanish conquest of New World, 444–445
  Black Death, 449
  cholera, 3, 14
  measles, 444, 448
  polio, 445
  progress of, 424, 437–443
  smallpox, 446–448
  start, 423

Euclid, *Elements*, 242, 452

evolution, Darwin's theory of, 433

ex falso quod libet, 372

Falklands, naval battle, 38, 156

falling bodies
  Einstein's view, 150
  Galileo's view, 101–105, 149
  Newton's view, 153

Farr, W., 5, 7, 14

FDA (Food and Drug Administration), 14

females and Enigma, 354–358

Fermat–Euler little theorem, 397

Feynman, R. P., great expositor, 129, 154, 504

Fibonacci numbers
and Euclid's algorithm, 254–257
definition, 251
origin, 250
properties, 251–254

fighter interception, problems of, 47–49

fishing and over fishing, 426–433

flood defences, 65

Foucault's pendulum, 156

four-thirds rule, 181–192

fractals, 221, 227

fundamental axiom
and 'divide and conquer', 478
and induction, 477–478
associated proof technique, 470–473
discussion, 469–470
used, 470–474

Galileo, G.
on falling bodies, 101–105, 149
on infinity, 305–307
on relative motion, 137–138
on size of structures, 105–106
pendulum, 126
truth not a matter of convention, 158

Galton, F., surname problem, 413

Gap, Atlantic, 75–77, 388

Gauss, C. F., 152, 281

general good not good of each individual, 94, 241, 273, 432–433

genetic change, 434–437

greatest common divisor, gcd, 242

Greek alphabet, 508

Haldane, J. B. S., on scaling in biology, 106, 108

Halmos, P. R.
gem of book, 505
index of which this is a pale shadow, 496
sage advice, 496
tombstone ■, 244

Hamming code, 407–410

Harris, A., 'Bomber', 75, 76

Hausdorff dimension, 218

heart
as pump, 106
beats per lifetime, 114

disease, 14–20, 122
size governed by cost of blood, 224

Heisenberg's inequality, 52

highest common factor, *see* greatest common divisor

historians, killjoy, routed, 104, 153

how to
deal with cranks, 234
do lots of things, 496, 499–500
get things done (sometimes), 292–296
keep fraudulent accounts, 31
make conversation, 226
pronounce TEX, 501

Huff-duff, 385, 388

human desires, conflicting and not well defined, 94

ice age, 436

immunisation, 95, 437, 445–448

inclusion–exclusion formulae, 359

induction and fundamental axiom, 477–478

interest rates, 95–97, 431–432

intuition, various kinds of, 503

ISBN, International Standard Book Number, 407

ISIS trials, 17–20

isolation, and origin of species, 435–437

Jack the Giant Killer, Haldane unimpressed by, 106

Jones, R. V., 56, 57

Jutland, naval battle, 38–40, 49

Kac, M., 176

Knuth, D. E.
Art of Computer Programming, 242, 276, 303, 505
TEX, 304, 501

Koch, R., 14, 16

Kolmogorov, A. N., and the four-thirds rule, 190–191

Kovalevskaya, S., 496–497

Lanchester's *N*-square Rule, 77

language
American, vivid, 47, 66, 389
highly redundant, 398–400
not just for communication, 399

LATEX, *see* TEX

law of large numbers, 401

Lebesgue, H., on mathematical teaching, 506

length of curve, existence of
empirical studies, 208–214
von Koch curve, 215–218

Leonardo da Vinci
branching of blood vessels, 222
fascinated by motion of water, 181

light
both wave and photon, 129
composed of photons, 128
electromagnetic wave, 128, 139
speed a constant, 139–141

Lindemann, F. A., 46, 57, 74, 75, 84, 130

linearisation, 74, 164, 178

London Ambulance Service disaster, 300–301

Lorentz transformation
addition of velocities, 146
consequences, 145–148
derivation, 141–145
time dilation, 148
twin paradox, 157–158

Lorenz, E. N., and chaos, 227, 433

lungs, as surfaces of nearly infinite area, 218

Macduff, D., mathematical career, 488–492

maintenance problems, 89–92

MakeIndex, 289

Mandelbrot, B., 220

marriage lemma of Philip Hall, 266

matrix inversion, perils of, 252

max flow, min cut theorem, 261

Maxwell, J. C.
electromagnetism and light, 128, 139
equations, 139, 140, 148
on relative motion, 139

menagerie, mathematical
angel fish, 394
bullfrogs, 361
butterflies, 137, 433
dodo, 96
elephants (and mice), 109–112
fish, 125, 426–429, 431–432
fleas, 113
great auk, 28
gulls, 66
lemmings, 97
mathematicians, 159
Microbe, 417
microbes, *see* epidemics
moths, 435

oysters, 96, 173
rabbits, 250
SHARK, 382, 387
sharks, 434
snakes, 109
squirrels, 113
whales, 113
Wrens, 378–380
merchant seamen, casualties, 384
Michelson A. A., and speed of light, 140
militaristic powers, categorisation, 201
military spending to reduce casualties, 202
morphogenesis, 392–394
mouse-to-elephant curve, 109–112, 221
Murray's law, 221–225, 393

Nasrudin, Mullah, 480, 511
networks
    Braess's paradox, 268–275
    Ford Fulkerson algorithm, 260–268
Newton, I.
    apple, 153
    bucket, 155
    complaint, 226
nowhere differentiable function, 178
number system
    alternative, 235–237
    choice of base, 234–235
    flip-flap-flop, 237
    Indian, 232–234
    Roman, 231–232

obiter dicta
    Besicovitch, A. S., 260
    Bohr, N., 177
    Dieudonné, J., 463
    Einstein, A., 127
    Erdős, P., 476
    Knuth D. E., 304
    Poincaré, H., 249
    Russell, B., 234, 468
    Swinnerton-Dyer, H. P., 244
    Whitehead, A. N., 472, 495
one-to-one correspondence, 307–308
operational research, 56
    aircraft maintenance, 89
    aircraft utilisation, 87, 92
    balance of ship types, 73
    bombing accuracy, 68–70

convoy size, 79–82
convoy speed, 74
depth charge setting, 67–68
in peacetime, 89, 93
preliminary questions, 66
submarine spotting, 64–66
outbreaks of war, absence of pattern, 203–206
Oxford, keeps barbarians at bay, 56

P–NP problem, 394
Pacific sardine, cautionary tale of, 430–431
parsnips, float better than golf balls, 190
Pearson, K., 205
persuasion, art of, 294–296
physics and mathematics, differences between, 176, 481
placebo, 15, 17
Plato
    dialogue, 452–459
    opinions, 486
plugboard
    advantages, 344
    beaten, 370–376
    described, 344
    perhaps not invulnerable, 349–352
Poincaré, H., 210, 249, 495
population growth, 418, 426
predators
    large, 107
    small, 448
prediction problems, 47, 57–61, 65, 90, 433
professors, German, power of, 16
publishers, viii, 494
pyramid inch, 117
Pythagoras's theorem, 460

queues, behaviour of, 422

radar
    airborne, 49, 54, 365
    British and German, 55
    British development of, 45–55
    multiple invention of, 55, 160
    naval, 49, 52
    needed, 44
    proposed by Watson-Watt, 45
    success helps hide other successes, 388

wavelength, 49–55
radio direction finding
    ship based, 387, 388
    shore based, 35, 385
railways
    in war, 258–260
    safety of, 298–299
redundancy, good and bad, 347, 398
Rejewski, M.
    versus the plugboard, 348–352
    works out Enigma wiring, 348, 352
relative motion
    Galileo, 137–138
    Maxwell, 139
relativity
    as symbol, 154
    books on, 153
    general, 149–152
    special
        initial discussion, 139–141
        mathematics of, *see* Lorentz transformation
Reynolds number, 125, 135
Richardson, L. F.
    deferred approach to the limit, 164–174
    doubts existence of wind velocity, 177–180
    four-thirds rule, 181–189
    lengths of coasts and frontiers, 208–215
    life, 159–163
    looks back on life, 192
    mathematical theory of arms races, 194–197
    measures wind velocity, 177
    number for curves, 217
    numerical weather forecasting, 160–164, 433
    scientific work conflicts with conscience, 188
    Statistics of Deadly Quarrels, 192, 198–208
    studies under Pearson, 205
Richardson–Kolmogorov cascades, 181, 190, 220, 221
rigour
    students' misleading idea of, 419
    sufficient, 482–485
    why needed, 268, 481
Roosevelt, F. D., 76, 398
rotation, reality of, 154–157
rumours, progress of, 424–426
Rutherford, E., 33, 42–44

scaling
  climbing, 113
  diving, 113
  falling bodies, 104
  Galileo on, 105–106
  Haldane on, 106, 108
  heartbeats per lifetime, 114
  in biology, 105–114, 221–225
  jumping, 113
  metabolic rates, 109–112
  size of structures, 105–106
self-steckered, 344
sex, why on earth?, 448
Shannon C. E.
  and Turing, 368, 398
  theorem
    discussion, 400–402, 406–410
    precise statement, 405
    proof, 402–406
    strong version stated, 406
SHARK, 382, 387
Short Weather Cipher, 380, 387, 399
smallpox, conquest of, 446–448
Snow, J., 4–14
Socrates
  dialogue, 452–459
  opinions, 486
SONAR, *see* ASDIC
square root of two, irrational, 464
staircase to extinction, 421
steckel, *see* plugboard
Stirling type formulae, 288, 290
submarine, *see also* convoy
  between the wars, 33–35
  in World War I, 21–33
  in World War II, 36
  Japan fails to master, 73
  performance of, 21, 73, 125
  post World War II, 73, 125, 391
  versus aircraft, 33, 35, 62–74
surname problem, *see also* Watson
  and bombe, 373
  stated, 413
Swordfish torpedo-bomber, 50, 58

taxpayer US, money of, not wasted, 158
Taylor, G. I.
  at Farnborough, 129–131
  atomic explosion, 133
  CQR anchor, 131–133
  on four-thirds rule, 191
  on Richardson, 187
  photon experiment, 127–129
  swimming micro-organisms, 135
TEX, 304–305, 501
Thompson, D'Arcy W., on
  morphogenesis, 393
time, as healer, 206
Tizard, H.
  angle, 48
  at Farnborough, 130
  committee, 45, 47, 57
  disagreements with Lindemann, 46,
    57, 84
  farseeing, 49, 86
  mission to U. S., 55
  operational research, 56
trapezium rule
  discussed, 166–171
  stated, 167
travelling salesman problem, 395
tree, as Richardson–Kolmogorov
  cascade, 220
turbulence, 181
  four-thirds rule, 181–192
Turing, A. M.
  and Shannon, 368, 398
  bombe, 367
    and Polish bomby, 337, 367, 370
    and surname problem, 373, 413
    as physical object, 372, 378–380
    four-rotor, 382
    in short supply, 382
    menus, 372–375
    problem of the fourth rotor, 382,
      387
    why possible, 370–376
    work of many people, 370
  can machines think?, 391, 398
  character, 368–370
  homosexuality, 391
  on morphogenesis, 392–394
  splendid biography of, 391
  theorem
    background, 303, 310–312
    discussion, 394
    proof, 312–316

twin paradox, 157–158
two and two, make four, 465

U-boat, *see* submarine

velocity, meaning of, 177–180
viscosity, 121, 135, 190, 223
von Koch snowflake, 215–218
von Neumann, J.
  demi-god, 498
  numerical meteorology, 163
  problem solved by MacDuff, 488

Waddington, C. H., 64, 394, 489
Watson, H. W.
  attacks surname problem, 413–417
  paradox
    African Eve, 418
    for surnames, 418
    possible engine for evolution,
      435–437
    queues, 423
    stated, 417
  staircases, 420–422
  theorem, 416
    epidemics, 423, 437
    gap in proof, 419–422
    genetic change, 434–437
    queues, 422
Watson-Watt, R. A., 45, 56
weights and measures, British
  divine origin, 117
  in all their glory, 237–241
Welchman, G., 335, 358, 364, 366, 370
Wilkins, A. F., 45
wind tunnels, 122, 124
wisdom, not teachable, 487
Wrens at Bletchley, 378–380
wrong envelope problem, 358

x, Joy of, viii

Yahoody the Impossible, 66

zero, great invention, 232, 233
Zimmermann Telegram, 23, 331, 333
Zygalski sheets, 358, 362, 363

Acknowledgements

Figure 1.5 reproduced with permission from *The Lancet*, August 1988, 349–360.

Figure 1.6 reproduced with permission from *The Lancet*, March 1992, 753–770.

Figure 2.1 reproduced from *The Daily Mirror* 1941.

Figures 4.1, 4.2, 4.3, 4.4, 4.5, 4.6, 4.7 are Crown copyright and reproduced with permission of the Controller of HMSO.

Figure 6.2 is © 1995 M. C. Escher/ Art–Baarn–Holland, reproduced with permission.

Figures 8.6, 8.7, 9.8 from the Royal Collection © Her Majesty Queen Elizabeth II and are reproduced with permission.

Figure 9.1 is reproduced from *Arms and Insecurity*, by L. F. Richardson, Boxwood Press, Pittsburg, 1960.

Figures 9.2 and 9.3 are reproduced from *Statistics of Deadly Quarrels*, by L. F. Richardson, Boxwood Press, Pittsburg, 1960.

Figure 12.1 is from the *Report of the Inquiry into the London Ambulance Service*, and is reproduced with the permission of the South West Thames Regional Health Authority.

Figure 13.4 is reproduced with permission from the British Admiralty's *Merchant Ships Signal Book III*.

Figure 15.3 is taken from *The Price of Admiralty* by J. Keegan, and is reproduced with permission of Hutchinson.

Figure 15.4 is taken from *The War at Sea* by S. W. Roskill, is Crown copyright and is reproduced with permission of the Controller of HMSO.

Permission to quote *The Persian Version* by Robert Graves, taken from his *Collected Poems* has been granted by Carcanet Press Limited.

Permission to quote *Brother Fire* by L. MacNiece, taken from *The Collected Poems of Louis MacNiece* has been granted by Faber and Faber Ltd.

Permission to quote *The Microbe* taken from *The Complete Verse* of Hilaire Belloc has been granted by Peters, Fraser and Dunlop.

Permission to quote from *Meno*, translated by W. K. Guthrie, has been granted by Penguin Books Ltd.

Permission to quote from the *Satter Prize Acceptance Speech* has been granted by D. McDuff and the American Mathematical Society.